MW00760181

Singular and Chiral Nanoplasmonics

Singular and Chiral Nanoplasmonics

edited by
Svetlana V. Boriskina
Nikolay I. Zheludev

Pan Stanford Publishing

Published by

Pan Stanford Publishing Pte. Ltd.
Penthouse Level, Suntec Tower 3
8 Temasek Boulevard
Singapore 038988

Email: editorial@panstanford.com
Web: www.panstanford.com

British Library Cataloguing-in-Publication Data
A catalogue record for this book is available from the British Library.

Singular and Chiral Nanoplasmonics

ISBN 978-981-4613-17-0 (Hardcover)
ISBN 978-981-4613-18-7 (eBook)

Printed in the USA

Contents

Preface xv

1 Chiral Nanostructures with Plasmon and Exciton Resonances **1**
Zhiyuan Fan, Hui Zhang, Robert Schreiber, Tim Liedl, Gil Markovich, Valérie A. Gérard, Yurii K. Gun'ko, and Alexander O. Govorov

1.1 Introduction 2
1.2 Interactions between Metal Nanoparticles and Chiral Molecules 7
1.2.1 Theory of Circular Dichroism in Complex Nanomaterials 7
1.2.2 Molecular Resonance Overlapping the Plasmonic Resonance 10
1.2.3 Molecular Resonance Far from the Plasmonic Resonance 11
1.2.4 Anisotropy in Circular Dichroism Measurements 12
1.2.5 Optical Chirality and Circular Dichroism Enhancement 13
1.3 Plasmon–Plasmon Interactions 16
1.3.1 Formalism: Quasi-static Model of Scattering by Metal Spheres 17
1.3.2 3D Chiral Nanoparticle Assemblies 18
1.3.3 Defects and Robustness of CD in Nanocrystal Assemblies 20
1.4 Electrodynamic Effects in Chiral Molecular-Plasmonic Structures 22

1.4.1 Model: Maxwell's Equations Incorporating a Chiral Parameter 22
1.4.2 Circular Dichroism in the Quasi-Static Limit 25
1.4.3 Electrodynamic Effects in Large Structures 26
1.5 Plasmonic Nanocrystals with Chiral Shapes 28
1.5.1 Formalism 29
1.5.2 Origins of Plasmonic CD: Mode Mixing 30
1.6 Experimental Measurements of Plasmonic Circular Dichroism 32
1.7 Chiral Semiconductor Quantum Dots with Exciton Resonances 41
1.7.1 Short Theory Introduction 41
1.7.2 Experiments with Chiral Quantum Dots 42
1.8 Conclusion and Perspectives 46

2 Reciprocity and Optical Chirality **57**
Aurélien Drezet and Cyriaque Genet
2.1 Introduction 57
2.2 The Reciprocity Theorem and the Principle of Path Reversal 61
2.2.1 The Lorentz Reciprocity Relation 61
2.2.2 Rotation of the Optical Medium and Reciprocity: Conserving the Handedness of the Reference Frame 63
2.2.3 Time-Reversal versus Reciprocity 66
2.3 Optical Chirality 67
2.3.1 Chiral Jones Matrix 67
2.3.2 Optical Activity 70
2.3.3 Planar Chirality 74
2.3.4 Generalization 77
2.3.5 Eigenstates and Chirality: Time Reversal versus Reciprocity 78
2.4 Discussion and Examples 81
2.4.1 Surface Plasmons and Archimedean Spirals: Planar Chirality Gives a Twist to Light 83
2.4.2 Chiral Surface Plasmons and Singular Optics: Tailoring Optical Vortices 86
2.5 Outlook 90

3 Chiral Effects in Plasmonic Metasurfaces and Twisted Metamaterials **97**
Yang Zhao, Amir Nader Askarpour, and Andrea Alù
3.1 Introduction 97
3.2 Wave Interaction with Planar Metasurfaces: Theoretical Basis 100
3.2.1 Transverse-Magnetic Excitation 103
3.2.2 Transverse Electric Plane Wave Incidence 105
3.2.3 Circularly Polarized Plane Wave Incidence 107
3.2.4 Surface Impedance Model 109
3.3 Twisted Metamaterials 111
3.3.1 Transmission through a Stack of Two Rotated Metasurfaces 111
3.3.2 Optimized Pair of Rotated Metasurfaces for Maximal Chiral Response 115
3.3.3 Chiral Effects in Twisted Metamaterials 116
3.4 Conclusion 119

4 Engineering of Radiation of Optically Active Molecules with Chiral Nano-Meta Particles **127**
Vasily Klimov and Dmitry Guzatov
4.1 Introduction 127
4.1.1 Chiral (Optically Active) Molecules and Their Importance 128
4.1.2 Chiral Metamaterials 131
4.2 Radiation of Chiral Molecule Near Chiral Half-Space 136
4.2.1 Spontaneous Radiation of a Chiral Molecule Placed near a Chiral Half-Space 136
4.2.2 Graphical Illustrations and Discussion of the Results Obtained 143
4.3 Radiation of a Chiral Molecule Near a Chiral Spherical Particle 147
4.3.1 Spontaneous Emission of a Chiral Molecule Placed Near a Chiral Spherical Particle 148
4.3.2 Analysis of the Results Obtained and Graphical Illustrations 152
4.4 Radiation of a Chiral Molecule Near a Cluster of Two Chiral Spherical Particles 158

4.4.1 Spontaneous Radiation of a Chiral Molecule Located Near a Cluster of Two Chiral Spherical Particles 159
4.4.2 The Analysis of the Results Obtained and Graphical Illustrations 163
4.5 Applications 167
4.6 Conclusion 170
4.A Appendix: Chiral Molecule Near a Chiral Half-Space 171
4.A.1 Vector Cylindrical Harmonics 171
4.A.2 Electromagnetic Field of a Chiral Molecule in Cylindrical Coordinates 172
4.B Appendix: A Chiral Molecule Near Chiral Microspheres 173
4.B.1 Vector Spherical Harmonics 173
4.B.2 Electromagnetic Field of a Chiral Molecule in Spherical Coordinates 174
4.B.3 Elements of the Translational Addition Theorem for Vector Spherical Harmonics 176

5 Unusual Optical Properties of Helical Metallic Photonic Crystals and Chiral Channels in Dielectric Photonic Crystals 185
Hongqiang Li, Jian Wen Dong, and Che Ting Chan
5.1 Introduction 186
5.2 Metallic Helix Array 188
5.2.1 Band Theory of Metallic Helix Array 188
5.2.2 MST Approach to Solve for the Band Structure of Helix Crystals 189
5.2.3 Band Structure of Metallic Helix Array 195
5.2.4 Robustness of the MST Approach for Band Structure 196
5.2.5 Wide Polarization Gap and Negative Refraction at Low Frequencies 196
5.2.6 Transverse Propagation in Helix Array: A Functionality of Broadband Wave Plate 199
5.3 Dielectric Photonic Crystals 203
5.3.1 Robust Transport of Light in Chiral Channels in a Dielectric Photonic Crystal 203
5.3.2 Sample Construction 205

5.3.3 Calculation Method and Experimental Setup 206
5.3.4 Band Structure, Transmissions, and Polarization 208
5.3.5 Robust Transport in the Chiral Channel 210
5.4 Summary 214

6 Chiral Surface Plasmon Polaritons on One-Dimension Nanowires 221
Shunping Zhang and Hongxing Xu
6.1 Introduction 221
6.2 Surface Plasmon Polaritons on Metallic Nanowires 223
6.3 Generation of Chiral Surface Plasmon Polaritons 225
6.4 Influence of External or Structual Parameters 228
6.5 Applications of Chiral Surface Plasmon Polaritons 232
6.6 Summary and Perspective 234

7 Manipulation of Surface Plasmon Patterns with Chirality of Metallic Structure 239
Il-Min Lee, Seung-Yeol Lee, Yohan Lee, and Byoungho Lee
7.1 Introduction: Optical Chirality and Plasmonic Chiral Patterns 240
7.2 Plasmonic Apertures with Chirality: A Simple Analogical Approach 241
7.3 Dependence of Chirality on the Bent Angle of a Gammadion Metal Slit 245
7.4 Plasmonic Chirality in Complex Chiral Patterns: Vogel's Spiral Apertures 251
7.5 Summary 255

8 Local Field Topology behind Light Localization and Metamaterial Topological Transitions 259
Jonathan Tong, Alvin Mercedes, Gang Chen, and Svetlana V. Boriskina
8.1 Introduction 260
8.2 Back to Basics: Surface Plasmon Polariton 265
8.3 SPP Coupling via Shared Circulating Powerflow 270
8.4 Hyperbolic Metamaterials: Global Field Topology Defined by Local Topological Features 272
8.5 Conclusion and Outlook 277

9 Nano-Fano Resonances and Topological Optics **285**
Boris Luk'Yanchuk, Zengbo Wang, Andrey E. Miroshnichenko, Yuri S. Kivshar, Arseniy I. Kuznetsov, Dongliang Gao, Lei Gao, and Cheng-Wei Qiu
9.1 Introduction 286
9.2 Fano Resonance and Its Mechanical Analogue 287
9.3 Scaling of the Fano Resonances within the Mie Theory 291
9.4 Formation of Vortices within the Mie Theory 294
9.5 Scaling of Fano Resonances in Plasmonic Nanostructures 298
9.6 Vortices Near Fano Resonances in Plasmonic Nanowires 303
9.7 Concluding Remarks 306

10 Chiral Nanostructures Fabricated by Twisted Light with Spin **311**
Takashige Omatsu and Ryuji Morita
10.1 Introduction 312
10.2 Radiation Force of the Light 313
10.2.1 Angular Momentum Density 314
10.2.2 How Can One Produce Optical Vortices? 317
10.3 Optical Vortex Laser Ablation 320
10.3.1 Nanoneedle Fabrication 321
10.3.2 Chiral Nanoneedle Fabrication 325
10.4 Conclusion 329

11 Engineering the Orbital Angular Momentum of Light with Plasmonic Vogel Spiral Arrays **335**
Luca Dal Negro, Nate Lawrence, and Jacob Trevino
11.1 Introduction to Aperiodic Optical Structures 336
11.2 Rotational Symmetry: From Tilings to Vogel Spirals 338
11.2.1 Structural Properties of Vogel Spiral Arrays 341
11.3 Engineering Orbital Angular Momentum of Light 352
11.3.1 Diffracted Beam Propagation from Aperiodic Chiral Spirals 356
11.3.2 Experimental Demonstration of OAM Generation and Control 361
11.4 Outlook and Conclusion 367

12 Probing Magnetic Plasmons with Vortex Electron Beams **375**
Reuven Gordon
12.1 Introduction 375
12.2 Nanoplasmonics on a Cylindrical Wire 376
12.2.1 Surface Waves 376
12.2.2 Localized Surface Plasmons 378
12.2.3 Magnetic Plasmons 380
12.2.4 Summary 381
12.3 Electron Energy Loss Spectroscopy 381
12.3.1 Theory 381
12.3.2 EELS for Nanoplasmonics 382
12.4 Vortex-EELS and the Magnetic Plasmon 384
12.4.1 Vortex Electron Beams 384
12.4.2 Effective Magnetic Charge for a Vortex Beam 385
12.4.3 Vortex-EELS Scattering Loss Probability 386
12.4.4 Probing the Magnetic Plasmon 387
12.4.5 Summary 388
12.5 Outlook 388

13 Electromagnetic Optical Vortices of Plasmonic Taiji Marks **393**
Wei Ting Chen, Pin Chieh Wu, and Din Ping Tsai
13.1 Introduction: An Explanation of the Process and Approach 393
13.2 Electromagnetic Energy Vortex without Optical Singularity 395
13.2.1 Fabrication Method 395
13.2.2 Taiji Mark Design 396
13.2.3 Measurement and Simulation Results 397
13.2.4 Vortex-Like Poynting Vector Profile 399
13.2.5 Taiji Mark for Opto-Mechanical Oscillator 403
13.3 Summary 405

14 Passive and Active Nano-Antenna Systems **409**
Samel Arslanagić, Sawyer D. Campbell, and Richard W. Ziolkowski
14.1 Introduction 410
14.2 Coated Nanoparticles Excited by a Plane Wave 411

14.3 Coated Nanoparticles Excited by Electric Hertizan Dipoles 427
14.3.1 Problem Formulation 427
14.3.2 Theoretical Considerations 428
14.3.3 CNP Materials and Gain Model 431
14.3.4 Resonance Effects of Active CNPs for Single EHD Excitation 432
14.3.4.1 NRR results 433
14.3.4.2 Near-field distributions and directivity 434
14.3.4.3 Influence of EHD orientation 436
14.3.5 Jamming Effects of Active CNPs for Single and Multiple EHD Excitations 438
14.3.5.1 Configuration 439
14.3.5.2 Far-field results 440
14.3.5.3 Power-flow density distribution 443
14.4 Conclusion 444

15 Plasmonic Nanostructures for Nanoscale Energy Delivery and Biosensing: Design Fabrication and Characterization 451
Remo Proietti Zaccaria, Alessandro Alabastri, Andrea Toma, Gobind Das, Andrea Giugni, Salvatore Tuccio, Simone Panaro, Manohar Chirumamilla, Anisha Gopalakrishnan, Anwer Saeed, Hongbo Li, Roman Krahne, and Enzo Di Fabrizio
15.1 Introduction 452
15.2 A Plasmonic Device for Multidisciplinary Investigation 453
15.2.1 Historical Perspective 454
15.2.2 Description of the Device 456
15.2.3 Design 457
15.2.3.1 Optical singularity in adiabatic compression regime 460
15.2.4 Fabrication 462
15.2.5 Characterization 463
15.2.6 Conclusions 464

15.3 Coupling Colloidal Nanocrystal Emission to Plasmons Propagating in Metallic Nanowire Structures 465
15.3.1 Optical Properties of Rod-Shaped Colloidal Core–Shell Nanocrystals 466
15.3.2 Device Fabrication and Experimental Setup 467
15.3.3 Theoretical Modeling of Plasmon Propagation in Au Nanowire 468
15.3.4 Experimental Results 469
15.3.5 Conclusion and Outlook 472
15.4 Plasmonics SERS Devices 472
15.4.1 Nanocuboids 473
15.4.2 Nanostars 478
15.5 Nanoantenna-Based Devices for Highly Efficient Hot Spot Generation 480
15.5.1 Introduction 480
15.5.2 Stacked Optical Antenna 483
15.5.3 L-Shaped Antenna 486
15.5.4 Resonant Terahertz Dipole Nanoantenna 489

Index 503

Preface

Plasmonics—the branch of science studying light-driven collective oscillations of charge carriers in materials—has matured significantly during the last few decades. Applications enabled by plasmonics span many areas of science and technology, including cancer research, super-resolution imaging, classical and quantum communications, ultra-sensitive bio(chemical) sensing and spectroscopy, renewable energy generation, and nanoscale heat management, including heat-assisted magnetic recording (HAMR). Furthermore, engineering of electromagnetic properties of photonic metamatherials crucially depends on the exploitation of plasmonic resonances. The majority of the conventional applications of plasmonics make use of the well-understood phenomena of light localization and the high-density of optical states in metal nanostructures and materials. However, a new field has recently emerged where plasmonic effects are used to harness and manipulate photon angular momenta. With the field still in its infancy, understanding the unique role of plasmonics in this new area is just beginning to form.

This book focuses on two types of angular momentum–sensitive light-matter interactions enhanced by plasmonics—*chirooptical and vortical effects*. Chirooptical effects arise in the scattering of light carrying angular momentum (e.g., left- or right-handed circularly polarized light) by chiral objects, that is, objects that cannot be superimposed onto their mirror images through any rotation or translation. Molecular chiroptical effects are typically very small due to a large mismatch in length scales between the wavelength of light and the molecule size, but can be dramatically enhanced via plasmonic effects on chiral metal nanostructures. Vortical effects—molding and recirculating the optical powerflow around a landscape of phase singularities within the electromagnetic field—

can also be generated by light interaction with specially designed plasmonic nanostructures or materials. Plasmonic vortical effects include generation of twisted light—optical beams with orbital angular momentum—as well as trapping and guiding optical energy via a sequence of coupled nanoscale optical vortices "pinned" to plasmonic nanostructures.

Recent studies have proven that plasmonic nanostructures can both harness and generate angular momentum of light. The different optical response of chiral plasmonic nanostructures to light beams with different spin angular momenta shows great promise in molecular imaging. Plasmonic nanostructures, metasurfaces, and metamaterials also offer unique nanoscale solutions to the generation of beams with orbital angular momentum and to the development of new types of optical vortex-pinning light-trapping platforms. On the other hand, light carrying angular momentum can be used to fabricate new types of plasmonic and photonic nanostructures with novel complex geometries and unusual symmetry properties.

This book is a collective effort by several international research groups to push the frontiers of plasmonics research into the emerging areas of plasmonics-enabled chiral and vortical effects. The book is a collection of 15 chapters written by leading experts in the field of photonics and plasmonics. The chapters cover various aspects of chiral and singular nanoplasmonics including electromagnetic analysis and design of new plasmonic nanostructures and materials, discovery of novel physical effects, development of nanofabrication and characterization techniques, and expansion of plasmonics into new application domains. Chapters 1–7 focus on novel plasmonics-enabled chirooptical effects and Chapters 8–15 detail the physics and applications of vortex-trapping and vortex-generating plasmonic components.

The discussion on the plasmonic-enhanced chirooptical effects starts in Chapter 1, where Zhiyuan Fan and colleagues discuss design and applications of chiral nanostructures supporting plasmon and exciton resonances. The authors predict several mechanisms to transfer and induce circular dichroism in the visible wavelength range using plasmonic nanostructures.

Aurelien Drezet and Cyriaque Genet review fundamental optical properties associated with chiral nanoplasmonic systems via a rigorous algebraic approach in Chapter 2. The authors reveal the underlying relation between the concepts of chirality and reciprocity, which define global classes of chiral elements.

Chapter 3, by Yang Zhao and colleagues, focuses on modeling the physical mechanisms behind extrinsic chiral effects induced in plasmonic metasurfaces and metamaterials, and explores potential new applications of these effects.

In Chapter 4, Vasily Klimov and Dmitry Guzatov discuss the radiation of optically active molecules in various chiral meta-environments. The authors demonstrate a high level of control over their emission characteristics, paving the way for potential applications of chiral nanostructures in drug discovery and mass production.

H. Q. Li and colleagues demonstrate new types of helical photonic crystals with unique photonic dispersion characteristics in Chapter 5. The authors show robust transport of light via chiral-guided modes, which is immune to scattering by isotropic homogenous impurities and is phenomenologically similar to robust transport of electrons in topological insulators.

In Chapter 6, Shunping Zhang and Hongxing Xu introduce chiral collective electron motions associated with electromagnetic waves sustained on the metal–dielectric interface of a 1D nanowire. The authors discuss potential applications of such chiral surface plasmon polariton waves in enhancing chirooptical effects and in boosting the information transmission capacity of plasmonic networks.

Chapter 7, by Il-Min Lee and coworkers, discusses the surface plasmon–enhanced chirooptical effects in light scattering on metallic apertures with chiral shapes or distributions. The authors demonstrate conversion of the spin angular momentum of the incident light into the orbital angular momentum of light transmitted through a metallic plate with a chiral aperture pattern.

The second part of the book, focusing on the phenomena of optical singularities and vortical effects in nanoplasmonics, starts with Chapter 8 by Jonathan Tong and colleagues. The authors show that formation of optical vortices is a hidden mechanism behind many unique characteristics of surface plasmon polariton modes and reveal that global topological transitions in hyperbolic metamaterials are governed by the rearrangement of local topological phase singularities.

In Chapter 9, Boris Luk'yanchuk and colleagues provide the link between two well-known interference phenomena associated with the scattering of light—Fano resonances and optical vortices—and demonstrate that Fano resonances are accompanied by the generation of optical vortices with the characteristic core size well beyond the diffraction limit.

Takashige Omatsu and Ryuji Morita propose a new approach to next-generation materials processing with vortex laser beams in Chapter 10. The authors demonstrate that optical vortices can twist material to fabricate chiral nanostructures for potential applications in nanoscale imaging for selective identification of the chirality and optical activity of molecules and chemical composites.

In Chapter 11, Luca Dal Negro and colleagues present their work on the generation of twisted light beams by using aperiodic arrays of plasmonic nanoparticles arranged into spiral geometries. The authors demonstrate the successful encoding of specific numerical sequences, determined by the aperiodic geometry, with the azimuthal angular momentum values of diffracted optical beams.

Chapter 12, by Reuven Gordon, discusses the use of vortex electron beams with orbital angular momentum to experimentally probe magnetic plasmons at the nanometer scale. This is in analogy to the local probing of plasmonic resonances by monitoring how fast electrons lose their energy to metal nanostructures.

Wei Ting Chen and colleagues demonstrate formation of an optical vortex in the near-field region of a low-symmetry plasmonic nanostructure with a Taiji pattern in Chapter 13. The authors study the vortex-induced optical torque for potential applications in the opto-mechanical oscillators.

In Chapter 14, Samel Arslanagi and colleagues review interesting effects in light manipulation with passive and active core–shell nano-particles. The authors provide design rules for engineering nano-scatterers with cross sections significantly larger than their geometrical size, and nano-antennas that can either amplify or jam the emission of localized dipole sources.

Chapter 15, by Remo Proietti Zaccaria and co-authors, closes the volume with a broad discussion of applications that exploit surface plasmon generation, propagation, and concentration. The authors reveal the role of optical singularities in the process of adiabatic compression, which allows the concentration of energy and the induction of high intensity electric and magnetic fields in nanoscale regions.

We thank all the authors for contributing their high-profile work to this volume, many other colleagues for helping to shape our views on the subject via stimulating discussions, and the editorial team of Pan Stanford Publishing for their hard work in bringing the book to the readers.

Svetlana V. Boriskina
Nikolay I. Zheludev
Summer 2014

Chapter 1

Chiral Nanostructures with Plasmon and Exciton Resonances

Zhiyuan Fan,[a] Hui Zhang,[a] Robert Schreiber,[b] Tim Liedl,[b] Gil Markovich,[c] Valérie A. Gérard,[d] Yurii K. Gun'ko,[d,e] and Alexander O. Govorov[a]

[a] *Department of Physics and Astronomy, Ohio University, Athens, Ohio 45701, USA*
[b] *Fakultät für Physik and Center for Nanoscience, Ludwig-Maximilians-Universität, Geschwister-Scholl-Platz 1, 80539 Munich, Germany*
[c] *School of Chemistry, Raymond and Beverly Sackler Faculty of Exact Sciences, Tel Aviv University, Tel Aviv 69978, Israel*
[d] *School of Chemistry and CRANN, University of Dublin, Trinity College, Dublin 2, Ireland*
[e] *St. Petersburg National Research University of Information Technologies, Mechanics and Optics, St. Petersburg 197101, Russia*
govorov@ohiou.edu

This chapter is focused on chiral nanostructures with plasmon and exciton resonances. Electromagnetic interactions between chiral and achiral building blocks in nanomaterials are modeled using classical or semiclassical theories. The theory predicts several mechanisms to transfer and induce circular dichroism (CD) in the visible wavelength region using nanostructures. In particular, enhanced molecular CD can be achieved in metal nanostructures using Coulomb interaction and plasmon enhancement. Additional amplification of molecular CD appears in plasmonic nanostructures with hot spots and anisotropic geometries. The chapter also reviews

Singular and Chiral Nanoplasmonics
Edited by Svetlana V. Boriskina and Nikolay I. Zheludev

ISBN 978-981-4613-17-0 (Hardcover), 978-981-4613-18-7 (eBook)
www.panstanford.com

several experimental papers showing successful fabrications of optically active chiral nanostructures and nano-assemblies with new plasmonic CD signals.

1.1 Introduction

Chirality can also be referred to as handedness. Chiral objects are not superimposable onto their mirror images. Looking around us, our hands, snail shells, screws, and so on, are all chiral. It often provides a smooth, efficient, and strong coupling between objects with the same chirality, such as the tightening between a bolt and a nut or a firm handshake using both right hands. Chirality also plays an important role in the origin of life. For example, all natural amino acid monomers have L-configuration, while sugars are D-isomers.

If two chiral molecules are mirror images of each other, we call them "enantiomers." The chemical properties may be dramatically different for each enantiomer of the same molecule. In the human body, for example, sometimes biological stereoselectivity only allows a particular type of enantiomers in a racemic drug to react with the targeted biomolecules [12]. Therefore, chirality is of great interest in the pharmaceutical industry. In order to not only improve the safety of drugs but also to reduce their production costs, efforts have been made into manufacturing drugs with only one kind of enantiomers, as the other kind of enantiomers may be less active, inactive, or even responsible for adverse effects in treatments, a distressing example being the sedative drug thalidomide.

Circular dichroism (CD) refers to the difference in absorption of left- and right-circularly polarized (LCP and RCP) light in chiral medium. It has been shown that CD spectroscopy is a powerful tool to study chirality of materials. Most of the homochiral materials had been found created in nature, until CD was demonstrated in the lab from a ligand-protected metal nanocluster system more than a decade ago [60, 61]. The interest in such artificial chirality grows exponentially. It was summarized [24, 53] that the CD from a nanocluster can be induced by intrinsically chiral metal core; chiral environment; or an intermediate mechanism called footprint effect. It was also shown that a mixture of these mechanisms is possible [59]. A different study demonstrated CD in the plasmon band of 10 nm silver nanoparticles (NPs) that were grown on a DNA template [65]. It was suggested that this NP system as well as its CD mechanism may be different from a nanocluster system [53, 65]. Since then, more NP-based nanostructures have been created using a variety of techniques and plasmonic CD was observed [1, 25, 26, 35, 36, 42, 48, 56, 66, 67, 72].

Nanomaterials involving metal NPs (MNPs) showed the potential of transferring a CD signal to the visible band and of modifying molecular CD signals that are naturally mostly found in the ultraviolet (UV) band. To explain the origin of these plasmonic CD phenomena, we look into the electromagnetic interactions inside NP-based nanostructures [18, 20, 31, 34]. Generally speaking, the plasmon-exciton, plasmon-plasmon interactions, and electrodynamic Mie effect are physically different from a proximity effect [27] or from an orbital hybridization effect of surface states [23]. Since localized surface plasmons (LSP) on NPs typically show wide continuous bands while nanoclusters have discrete atomic-like lines, the physics of chiral effects in nanocrystals (NCs) is quite different to the case of nano-clusters [57]. Regardless of these differences in the physical mechanisms, the electromagnetic theories provide relatively fast, flexible, and reliable simulations as well as assistance in design of novel nanostructures. Yet in experiments, MNPs featuring well-defined structures, easy preparation, and reusability are good candidates for applications such as sensors [37, 47, 83].

In this chapter, we are going to introduce the fundamental physical models of electromagnetic interactions, experimental realizations, and potential applications of chiral nanostructures with plasmon and exciton resonances. First of all, let us introduce the physical models and theories for plasmonic CD mechanisms [18, 20, 31, 34]. Generally, we describe CD in an engineered nanomaterial by three pieces using Eq. (1.1):

$$CD_{\text{total}} = CD_{\text{abs,molecule}} + CD_{\text{abs,NC}} + CD_{\text{sca}}, \tag{1.1}$$

where the sub-indices indicate CD contributions from the absorption by a molecular chiral media and NCs, and from scattering by the whole complex comprising molecules and NCs.

In the CD spectrum, both terms, $CD_{\text{abs,molecule}}$ and $CD_{\text{abs,NC}}$, contribute to the total CD spectra. The plasmon–exciton interaction will be studied within a semiclassical model in Section 1.2 [31, 32, 81]. If molecules absorb light in the UV band, their interaction with a plasmon on a gold NP is most likely to be off-resonance. Interestingly, CD signals show up both in the plasmon band and in the molecular band. This can be explained as follows. The dipole of a molecule influences the charge distribution on an MNP. Then, the motion of the charges and the corresponding heat dissipation influenced by the dipole of a chiral molecule will be different under LCP and RCP illumination. By maximizing the differential absorption at plasmon frequencies, we will be able to see an induced plasmonic CD signal in the visible band. The theory of off-resonance interaction can successfully address the origin of plasmon CD in some experiments [26, 48, 67]. Simultaneously, the possibility of enhancing molecular CD signal is of great interest to many researchers in the fields of physics, chemistry, and biology. A convenient parameter C to estimate optical chirality enhancement was adopted by Cohen et al. [70]. We will see its contribution to $CD_{\text{abs,molecule}}$ and its physical origin in a discussion in Section 1.2. When the plasmon band and molecular band are in resonance, the CD spectra will demonstrate interesting Fano-like characteristics. This is because the CD signals coming from molecules and NCs in Eq. (1.1) have strong interference in the overlapping bands. The total effect on CD can be an enhancement or a quenching, accompanied by a shape distortion compared with the original CD of free molecules.

In addition, $CD_{\text{abs,NC}}$ can also be a result of plasmon–plasmon interaction between NPs on a chiral assembly. In Section 1.3, we will show theoretically that a helical chain of gold NPs is able to generate significant plasmonic CD in the visible band [18, 19, 42]. Plasmon–plasmon interactions will be treated using classical interacting dipole theories [74, 82]. The interaction between MNPs in a chiral configuration creates chiral collective dipole excitations. A flipped geometry leads to an opposite handedness in such excitations. In a perspective [55], J. B. Pendry showed that metamaterials with a negative refractive index can potentially be created with chiral structures.

In NP-molecular complexes with sizes ranging from tens to hundreds of nanometers, scattering is able to compete with or even exceed absorption. The contribution from CD_{sca} becomes significant at electrodynamic resonances. In an isotropic chiral molecular medium, the constitutive relations are [40]:

$$\vec{D} = \varepsilon \vec{E} + i\xi \vec{B}, \tag{1.2}$$

$$\vec{H} = \frac{\vec{B}}{\mu} + i\xi \vec{E}. \tag{1.3}$$

where ξ is introduced as a chiral parameter. Solving the coupled Maxwell's equations with constitutive relations of a chiral medium,

$$\begin{aligned} \nabla \times \vec{E} &= -\frac{1}{c}\frac{\partial \vec{B}}{\partial t}, \\ \nabla \times \vec{H} &= \frac{1}{c}\frac{\partial \vec{D}}{\partial t}, \\ \nabla \cdot \vec{D} &= 0, \\ \nabla \times \vec{B} &= 0, \end{aligned} \tag{1.4}$$

we obtain

$$\nabla \times \begin{pmatrix} \vec{E} \\ \vec{H} \end{pmatrix} = i\frac{\omega}{c} \begin{pmatrix} -i\xi & 1 \\ -\varepsilon - \xi^2 & -i\xi \end{pmatrix} \begin{pmatrix} \vec{E} \\ \vec{H} \end{pmatrix}. \tag{1.5}$$

We will see that the degeneracy between RCP and LCP is broken by finding eigenvalues of the matrix in Eq. (1.5),

$$\begin{pmatrix} k_+ \\ k_- \end{pmatrix} = \frac{\omega}{c} \begin{pmatrix} \xi + \sqrt{\varepsilon + \xi^2} \\ -\xi + \sqrt{\varepsilon + \xi^2} \end{pmatrix}. \tag{1.6}$$

Therefore, those light beams will propagate with different propagation constants. The difference between the propagation constants, k_+ and k_-, results from the chiral parameter ξ of a chiral medium. Generally speaking, it will shift electromagnetic resonances and influence the strength of absorption and scattering. A CD signal with an electrodynamic origin can be identified from the spectra. This process is an interplay strongly dependent on the geometric parameters such as the size of the structure and the thickness of shells, and the dielectric functions of component materials. In Section 1.4, a Mie solution incorporating a chiral parameter will be used to solve the light scattering problem of multilayer core–shell structures [6]. Multipole vector harmonics are included to address this scattering problem [7].

The systems above can be created using well-defined building blocks such as metal or semiconductor NPs, bio linkers, various kinds of protein molecules, and DNA origami structures. In Section 1.5, we also find that plasmonic CD can be created by single metal NPs that have a chiral shape [20]. Such a shape can be created by top-down lithography or bottom-up wet chemistry methods. For NPs with chiral shapes, the plasmonic mechanism is essentially different from chiral NP assemblies wherein CD signals can be dominated by a dipolar plasmon–plasmon interaction. In chiral NCs in contrast, multipole electromagnetic fields are created and mixed by surface distortion and a CD signal is induced by mixing of multipole plasmons.

In Section 1.6, some of the recently performed experiments will be demonstrated. With the theories developed, we try to present a physical picture of these interactions in the experimental plasmonic systems.

Semiconductor quantum dots (QDs) demonstrated tunability of their optical properties through chemical control of size and shape. Other than light-emitting diodes (LEDs) or photovoltaic devices, more biosensing applications are anticipated for chiral QD complexes. Along with the development in research of plasmonic CD, a different type of chiral NP structure was made from semiconducting materials [29, 50]. CD was also measured from D- and L-penicillamine stabilized QDs, while a racemic mixture of the two only showed a weak CD signal. It was attributed to the chiral shell created by chiral molecular adsorbates [52], or an exciton–exciton interaction [31, 33]. In Section 1.7, a discussion on exciton–exciton interaction will be presented and the experimental realization of NCs stabilized by chiral molecules or with intrinsic chirality will be introduced.

Finally in Section 1.8, we conclude that the CD mechanisms have been well developed theoretically and that a couple of experiments provided solid supporting evidence. However, the experimental realization still remains challenging when incorporation of various materials and realization of interesting geometries are desired. In perspective, we show that anisotropy and plasmonic hot spots effects are of great advantages to enhance plasmonic and molecular

CD signals. In our opinion, they will have promising applications in the future.

1.2 Interactions between Metal Nanoparticles and Chiral Molecules

Several experimental papers reported new CD lines from nanostructures assembled from chiral molecules and achiral NCs [11, 36, 45, 61, 65]. Lots of work has been focused on the description of the underlying mechanisms, namely Coulomb interactions. Also, this section will mainly focus on the Coulomb interaction between a plasmon and an exciton [11, 31, 32].

The Coulomb interaction between an NP and a molecule can be dipolar [30] and multipolar [78]. In an MNP-molecule complex, the plasmon–exciton interaction will give rise to a CD signal in the visible optical band, while CD enhancement can be achieved in the molecular band. In the formalisms of solving such a light-matter interaction problem, the molecular component is treated using a density matrix [80], although the metals and the incident light are still treated classically. From the interaction, we will see that there are two major mechanisms of CD generated by a plasmon–exciton interaction, the non-resonant and the resonant interaction. Also, the influence of several geometrical configurations will be tested.

1.2.1 *Theory of Circular Dichroism in Complex Nanomaterials*

A density matrix describes a quantum system in a mixed state. The density matrix formalism is applied when a large number of quantum mechanical objects are involved. The Hamiltonian of an object interacting with an external field is

$$H = H_0 + H', \tag{1.7}$$

where H_0 is an operator for an exciton and H' describes the interaction between the exciton and the electric field [80]. The classical light-matter interaction term in Hamiltonian is

$$H' = V_\omega e^{-i\omega t} + V_\omega^\dagger e^{i\omega t}, \tag{1.8}$$

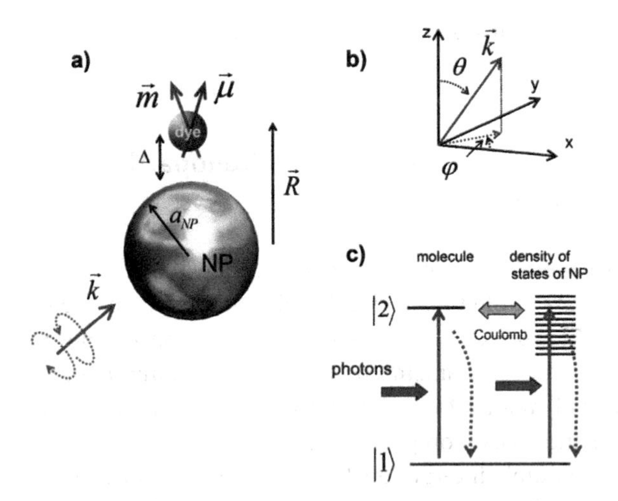

Figure 1.1 (a) Schematics of MNP-dye molecule interaction. (b) The wave vector $\vec{k}$ is given in the coordinate system associated with the MNP-dye hybrid. (c) Transition diagram of Coulomb interaction and relaxation processes in this hybrid. Reprinted with permission from Ref. [31]. Copyright 2010, American Chemical Society.

where $V_\omega = -\frac{e}{mc}\vec{A}_\omega \cdot \vec{p}$, $\vec{E}_\omega = \frac{i\omega}{c}\vec{A}_\omega$. We define ω as the frequency of the external field and ω_0 as the transition frequency of an exciton.

To include relaxation mechanisms, the equation of motion of density matrix elements ρ_{nm} is:

$$\frac{d}{dt}\rho_{nm} = -\frac{i}{\hbar}[H, \rho]_{nm} - \sum_{kl} \Gamma_{nm,kl}\,\rho_{kl}. \tag{1.9}$$

The terms $\Gamma_{nm,kl}\,\rho_{kl}$ are called collision terms that describe different channels of energy relaxation. The non-zero elements in our model are $\Gamma_{22,22} = \Gamma_{22}\Gamma_{11,22} = -\Gamma_{22}$ and $\Gamma_{12,12} = \Gamma_{21,21} = \Gamma_{12}$

When the detuning between the light and the exciton is small, that is, $|\omega - \omega_0| \ll \omega + \omega_0$, a rotating wave approximation will be applied. Then, the density matrix elements can be written as

$$\rho_{21} = \sigma_{21}(t)\,e^{-i\omega t}, \quad \rho_{11} = \sigma_{11}(t), \quad \rho_{22} = \sigma_{22}(t). \tag{1.10}$$

After an MNP is added into the complex, as is shown in Fig. 1.1a, the Hamiltonian H' in light-matter interaction Eqs. (1.7), (1.8), and (1.9) has all information of the electric fields induced by an incident beam, a plasmon, and a dipole image of the molecule.

Using a rotating wave approximation (1.10), we are able to solve for ρ_{22} and ρ_{21}. Correspondingly,

$$\sigma_{22} = \frac{2e^2}{\omega_0^2 m^2 \Gamma_{22}} \frac{\Gamma_{21}(\rho_{11}-\rho_{22}) \left|\langle 2| \left(\vec{E}_{\omega,laser} + \vec{E}_{\omega,NP}\right) \cdot \vec{p}\,|\rangle\right|^2}{|(\hbar\omega - \hbar\omega_0 + i\Gamma_{12}) - (\rho_{11}-\rho_{22})\,G_\omega|^2},$$

$$\sigma_{21} = -\frac{e}{i\omega_0 m} \frac{(\rho_{11}-\rho_{22}) \langle 2| \left(\vec{E}_{\omega,laser} + \vec{E}_{\omega,NP}\right) \cdot \vec{p}\,|\rangle}{(\hbar\omega - \hbar\omega_0 + i\Gamma_{12}) - (\rho_{11}-\rho_{22})\,G_\omega}, \tag{1.11}$$

where $\vec{p}$ is a linear momentum operator, $\vec{\Phi}_\omega$ is a vector potential created by the dipole moment of the molecule near the MNP, $\vec{p}_{12} = \langle 1|\,\vec{p}\,|2\rangle$ and finally

$$G_\omega = \frac{e^2}{\omega_0^2 m^2} \left(\nabla \cdot \left(\vec{p}_{12} \cdot \vec{\Phi}_\omega\right)\right) \cdot \vec{p}_{21}. \tag{1.12}$$

Each term has been well explained above. We see that this is a self-consistent equation. After solving it numerically, we can obtain the absorption from these equations:

$$Q_{\text{molecule}} = \omega_0 \sigma_{22} \Gamma_{22}, \tag{1.13}$$

and

$$Q_{\text{NP}} = \text{Im}\,(\varepsilon_m)\,\frac{\omega}{2\pi} \int dV\,\vec{E}_\omega^{in} \cdot \vec{E}_\omega^{in*}. \tag{1.14}$$

In colloidal systems, MNP-molecule complexes are floating in a solution with random orientations. Then, the CD signal is defined as:

$$CD = \langle Q_+ - Q_- \rangle_\Omega, \tag{1.15}$$

where the $+/-$ signs indicate that incident beam is either LCP or RCP. The average is needed when light is coming from all directions, which is equivalent to the situation when hybrid nanostructures are oriented randomly in a solution. Total absorption has two terms formally, the absorption in a molecule (1.13) and that in a metal NP (1.14). Then, the absorption and CD of an MNP-molecule complex system can be written as

$$Q = Q_{\text{metal}} + Q_{\text{molecule}}, \tag{1.16}$$

$$CD = CD_{\text{metal}} + CD_{\text{molecule}}. \tag{1.17}$$

To be more specified, the CD signal within a dipolar limit is formally contributed by the molecules and the NPs:

$$CD_{\text{molecule}} = E_0^2 \frac{8ck}{3} \frac{\Gamma_{12}}{\left|\hbar\omega - \hbar\omega_0 + i\Gamma_{12} - G(\omega)\right|^2} \text{Im}\left(\hat{P}\vec{\mu}_{12} \cdot \vec{m}_{21}\right), \tag{1.18}$$

$$CD_{\text{NP}} = \frac{8}{9} a_{\text{NP}}^3 \left|\frac{3\varepsilon_0}{\varepsilon_{\text{NP}} + 2\varepsilon_0}\right|^2 E_0^2 \text{Im}\left(\varepsilon_{\text{NP}}\right) \frac{ck}{R^3 \varepsilon_0} \text{Im} \times \left(\frac{\mu_{12x} m_{21x} + \mu_{12y} m_{21y} - 2\mu_{12z} m_{21z}}{\hbar\omega - \hbar\omega_0 + i\Gamma_{12} - G(\omega)}\right). \tag{1.19}$$

where the electric and magnetic dipole moments of a molecule are defined as

$$\vec{\mu}_{12} = -\left|e\right| \langle 2 \left|\vec{r}\right| 1 \rangle,$$
$$\vec{m}_{21} = -\frac{|e|}{2mc} \langle 2 \left|\left[\vec{r} \times \vec{p}\right]\right| 1 \rangle. \tag{1.20}$$

Now, these equations are ready for the following investigation of resonant and non-resonant plasmon–exciton interactions.

1.2.2 *Molecular Resonance Overlapping the Plasmonic Resonance*

The complex is composed of a silver NP (AgNP) and a chiral molecule. When molecular resonance matches the plasmonic resonance, it shows a Fano signature in the extinction in the inset of Fig. 1.2a and in the CD spectra (Fig. 1.2b). This indicates a strong interference between the molecule and the AgNP. Experimentally however, the extinction of AgNPs is overwhelmingly large. It is hard to distinguish the extinction from the molecule in an extinction spectrum by dividing the extinction into two terms such as those in Eq. (1.16). Fortunately, CD spectroscopy may help us to resolve the presence of a molecule near the AgNP. In Fig. 1.2b, the CD spectra were demonstrated for two different configurations when the electric dipole of the molecule is perpendicular or parallel to the AgNP surface. The shape of the CD spectra is dramatically different, indicating strong configurational dependence of the CD signal. We also found that the strength of the CD signal was amplified nearly three times. Note that this moderate enhancement is achieved within a dipole limit.

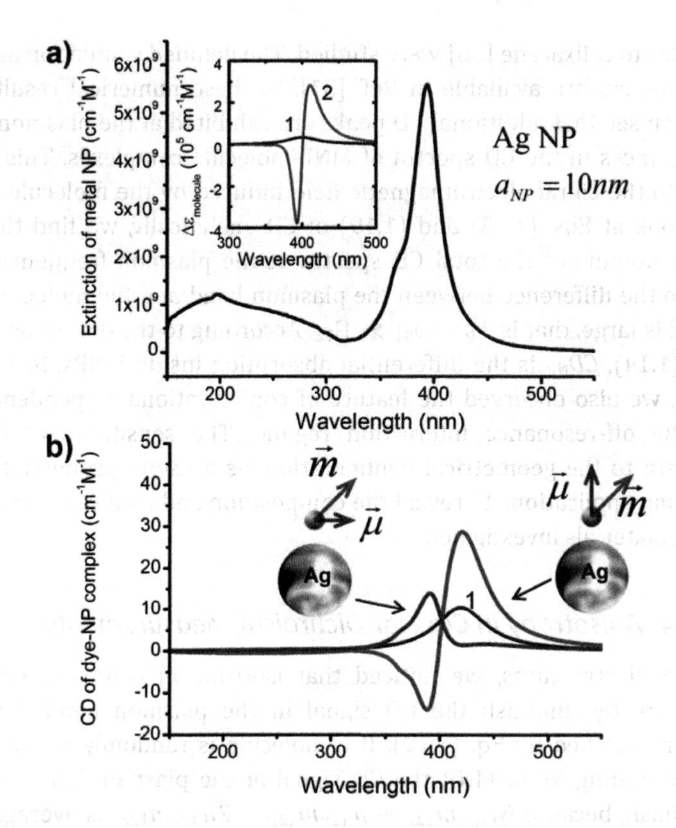

Figure 1.2 The examples shown here are complexes made from an AgNP and a molecule represented by a pair of an electric dipole and a magnetic dipole. The plasmon resonance is close to the resonance of the molecule around 390 nm. In other words, this is a case when exciton and plasmon bands overlap. (a) Extinction spectra of AgNP. The inset shows the contribution to the total extinction from the dye molecule when it is alone (1) and when the AgNP is present (2). (b) CD spectra are plotted for different configurations; (1) shows CD of the molecule without an AgNP. Reprinted with permission from Ref. [31]. Copyright 2010, American Chemical Society.

1.2.3 *Molecular Resonance far from the Plasmonic Resonance*

In applications, most molecules will have absorption bands in the UV, whereas the metal NPs show plasmon resonances in the visible band. As real examples, an alpha-helix [75] and a molecule similar to calixarene [36] were studied. The detailed simulation and parameters are available in Ref. [31]. In these numerical results, we can see that additional CD peaks are exhibited at the plasmonic resonances in the CD spectra of MNP-molecule complexes. This is due to the chiral electromagnetic field induced by the molecule. If we look at Eqs. (1.18) and (1.19) of CD analytically, we find that CD_{NP} dominates the total CD spectra at the plasmon frequencies when the difference between the plasmon band and the molecular band is large, that is, $|\omega - \omega_0| \gg \Gamma_{12}$. According to the definition of Eq. (1.14), CD_{NP} is the differential absorption inside MNPs. In Fig. 1.3d, we also observed the feature of configurational dependence in the off-resonance interaction regime. The sensitivity of CD spectra to the geometrical configurations is a useful property for sensing applications to reveal the composition and configuration of nanomaterials investigated.

1.2.4 *Anisotropy in Circular Dichroism Measurements*

In small structures, we noticed that isotropy in chiral medium will greatly diminish the CD signal in the plasmon band. This can be verified by Eq. (1.19). If a molecule is randomly oriented after binding to an MNP, the CD signal at the plasmon band will diminish, because $\langle \mu_{12x} m_{12x} + \mu_{12y} m_{12y} - 2\mu_{12z} m_{12z} \rangle$ is averaged to 0. Meanwhile, the enhancement factor in the molecular band will return to 1. This result shows consistency with a tiny plasmonic CD given by chiral Mie solutions [6, 34] for an MNP coated with a thin layer of optically chiral medium, as is shown in Fig. 1.4a. But both theories [6, 34] show that nanostructures with oriented molecular adsorption are good candidates for plasmonic CD measurements, while the enhancement in the molecular band depends on the orientation of a molecule.

It was also suggested [22] that in order to achieve significant enhancement in molecular CD, if randomly oriented, molecules should be adsorbed at the bottom and the top of a spherical MNP, while a beam of light propagates along the line connecting the poles of the sphere. Note that the averaging over the incident light direction or the averaging over the orientation of a molecular dipole is needed to describe experiments in solution. In other cases, the

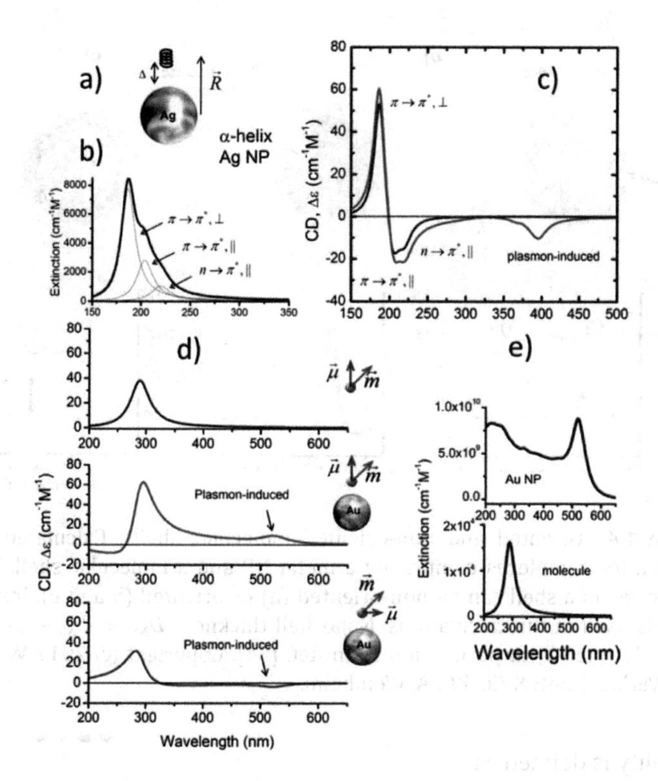

Figure 1.3 Off-resonance enhancement of CD. (a) Schematics of the α-helix-AgNP complex. (b) Theoretical fit of the extinction of an α-helix. (c) Calculated CD spectrum of helix-AgNP complexes, which shows an induced plasmonic CD band. (d) Schematics of the dye-AuNP hybrid, with different orientation of the molecular dipole moment with respect to the AuNP, and corresponding CD spectra. (e) Extinction spectra of dye molecules and AuNPs. Adapted with permission from Ref. [31]. Copyright 2010, American Chemical Society.

orientation of nanostructures is fixed with respect to the direction of light propagation. Such cases include, for example, planar NC arrays covered with chiral molecules [37, 47].

1.2.5 *Optical Chirality and Circular Dichroism Enhancement*

The quantity of optical chirality was previously adopted in Ref. [70] to estimate the enhancement of CD of biomolecular sensing. Optical

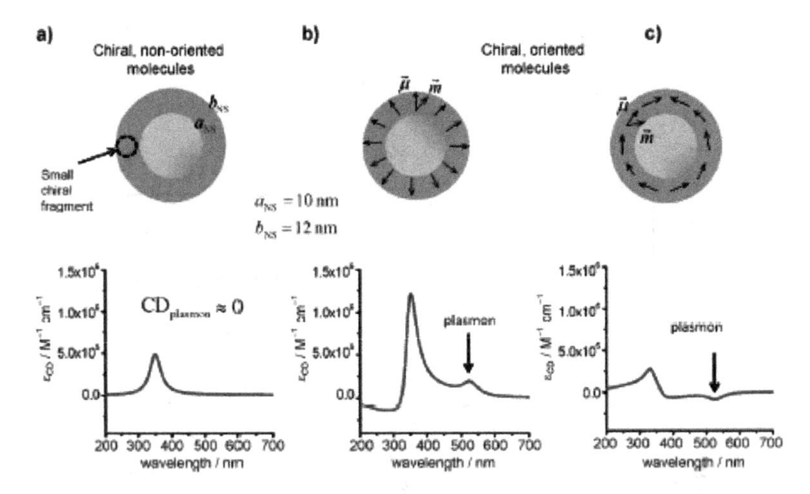

Figure 1.4 Oriented and non-oriented molecular shells. Calculated CD spectra for complexes comprising a metal NP and a molecular shell. The molecules in a shell can be non-oriented (a) or oriented (b and c). Insets: Models used in the calculations. Nanoshell thickness $D_{NS} = b_{NS} - a_{NS} = 2\,\text{nm}$. Adapted with permission from Ref. [34]. Copyright © 2012 Wiley-VCH Verlag GmbH & Co. KGaA, Weinheim.

chirality is defined as

$$C = \frac{\varepsilon_0}{2}\vec{E}\cdot\nabla\times\vec{E} + \frac{1}{2\mu_0}\vec{B}\cdot\nabla\times\vec{B} = -\frac{\omega\varepsilon_0}{2c}\text{Im}\left(\vec{E}^*_\omega\cdot\vec{B}_\omega\right), \quad (1.21)$$

where the electric field was represented as

$$\vec{E} = \frac{1}{2}\left(\vec{E}_\omega e^{-i\omega t} + \vec{E}^*_\omega e^{i\omega t}\right).$$

The parameter C is convenient in many cases and was employed in several recent papers [22, 37, 62]. Here, we would like to describe the physical origin of this quantity within a few paragraphs.

From Ref. [3], the absorption in a medium is proposed as

$$Q_{\text{abs}} = 2\omega\int_{V_0}\text{Im}\left(\vec{P}\cdot\vec{E}^*_\omega + \vec{M}\cdot\vec{B}^*_\omega\right)dV, \quad (1.22)$$

where $\vec{D} = \vec{E} + 4\pi\vec{P}$ and $\vec{H} = \vec{B} - 4\pi\vec{M}$. Using constitutive relations of a chiral medium (1.2) and (1.3), Eq. (1.22) can be written as a sum of two terms [34]:

$$
\begin{aligned}
Q_{\text{abs}} &= Q_{\text{abs},\varepsilon} + Q_{\text{abs},\xi} \\
&= \omega \int_{V_0} \text{Im}\,(\varepsilon)\, \frac{\vec{E}_\omega \cdot \vec{E}^*_\omega}{2\pi} dV \\
&\quad + \omega \int_{V_0} \text{Re}\left[\frac{\xi}{2\pi}\left(\vec{B}_\omega \cdot \vec{E}^*_\omega - \vec{E}_\omega \cdot \vec{B}^*_\omega\right)\right] dV \\
&= \frac{\omega}{2\pi} \int_{V_0} \text{Im}\,(\varepsilon)\, \vec{E}_\omega \cdot \vec{E}^*_\omega dV - \frac{\omega}{\pi} \int_{V_0} \text{Im}\,(\xi) \text{Im}\left(\vec{E}^*_\omega \cdot \vec{B}_\omega\right) dV
\end{aligned}
\tag{1.23}
$$

The physical meaning of this equation is that absorption occurs due to both properties of the system—the dissipative property ($Q_{\text{abs},\varepsilon}$) and the chiral property ($Q_{\text{abs},\xi}$). The first term of Eq. (1.23) comes from a dielectric function of an NC or a molecular medium and the second term results from the so-called optical chirality that interacts with a chiral medium. We find the second term is linear in the chiral parameter ξ_c. It is indeed proportional to the optical chirality parameter $C\,(\vec{r})$, if we compare it with Eq. (1.21) that has been derived from the Maxwell's equations (1.4).

The absorption CD signal will be the difference in Eq. (1.23) for the incident beams of LCP and RCP:

$$
CD_{\text{abs}} = CD_{\text{abs},\varepsilon} + CD_{\text{abs},\xi}. \tag{1.24}
$$

Then, the CD contribution from the chiral properties of a molecular medium in Eq. (1.24) can be expressed as:

$$
\begin{aligned}
CD_{\text{abs},\xi} &= -\frac{\omega}{\pi} \int_{V_0} \text{Im}\,(\xi_c) \text{Im}\left(\vec{E}^*_{\omega+} \cdot \vec{B}_{\omega+}\right) dV \\
&\quad + \frac{\omega}{\pi} \int_{V_0} \text{Im}\,(\xi_c) \text{Im}\left(\vec{E}^*_{\omega-} \cdot \vec{B}_{\omega-}\right) dV \\
&\propto \int_{V_0} \text{Im}\,(\xi_c)\left(C_+\,(\vec{r}) - C_-\,(\vec{r})\right) dV
\end{aligned}
\tag{1.25}
$$

where the signs $+/-$ indicate that the incident beam is LCP or RCP. For several specific core/shell metal-molecular structures, explicit analytic expressions and as well numerical results for both $CD_{\text{abs},\varepsilon}$ and $CD_{\text{abs},\xi}$ can be found in Ref. [34].

When we look closely at (1.23), we should note that the electromagnetic fields $\vec{E}$ and $\vec{B}$ are the self-consistent electromagnetic fields solved from Maxwell's equations and corresponding boundary conditions. The information carried by these electromagnetic fields comes from both the metal NC and the chiral medium. Hence, the optical chirality $C_{+/-}(\vec{r})$ in (1.25) should also include an optical response of the chiral medium. If the system is weakly interacting, the optical property of a chiral medium may play a much less active role in the optical chirality parameter. Then, it is very convenient to estimate a possible enhancement just by studying a free-standing NC. On the contrary, if the system is strongly interacting, such as the case when plasmon and exciton are in resonance, we have seen that the CD signal becomes not predictable, as the Fano effect will distort the CD signal. More discussion on electromagnetic effects in chiral nanostructures with large sizes (NC size $\sim$ wavelength of light) will be given in Section 1.4 of this chapter. The related properties can be understood by looking at Eq. (1.1). This tells us that the chiral parameter $C(\vec{r})$ determines only one of three contributions to the CD and all three contributions in a general case are of similar magnitude. Theoretical calculations supporting this picture were performed in Ref. [34] for several model nanostructures.

1.3 Plasmon–Plasmon Interactions

Several experiments with successfully realized chiral NP assemblies have been reported [10, 11, 35, 42, 49, 56, 66, 72, 79]; some of these experiments demonstrated the anticipated CD signals. Here, we provide a description for another mechanism of plasmonic CD, resulting from the dipolar plasmon–plasmon interaction between MNPs in a chiral configuration instead of a plasmon–exciton interaction.

We show theoretically that an MNP-based complex system with a chiral arrangement of MNPs will lead to non-zero CD signal near plasmonic resonances [18, 19]. A complex with dimensions around 40 nm consisting of small MNPs with radii around 5 nm will, in fact, exhibit relatively strong CD signals according to our calculations. Importantly, this has also been demonstrated experimentally [42, 66]. For densely packed complexes, it was observed [40] that the

CD signal in the visible optical range can be very strong, greatly overcoming the typical molecular CD in the UV range.

1.3.1 *Formalism: Quasi-Static Model of Scattering by Metal Spheres*

From electrostatics, the polarizability of a dielectric sphere is given by Ref. [43]:

$$\alpha = \frac{\varepsilon_{\mathrm{m}} - \varepsilon_0}{\varepsilon_{\mathrm{m}} + 2\varepsilon_0} r_a^3, \tag{1.26}$$

where ε_{m}, ε_0 are the dielectric constants of metal and matrix respectively, and r_a is the radius of a sphere. The electric dipole moment is given by

$$\vec{p} = \alpha \vec{E}_0. \tag{1.27}$$

Due to dissipation in metals, heat can be calculated from work done to the oscillating charges inside a particle by an electric field:

$$Q(t) = \int \vec{j}(t) \cdot \vec{E}(t)\, dV, \tag{1.28}$$

where $\vec{E}(t)$ is a real total electric field inside the particle, and $\vec{j}(t)$ is the induced current. Then, the time averaged heat dissipation over a period T is

$$Q = \langle Q(t) \rangle_t = \frac{\int Q(t)\, dt}{T}. \tag{1.29}$$

Other simplified forms of equations of heat dissipation and extinction are given here:

$$\begin{aligned} Q_{\mathrm{abs},i} &= \frac{1}{2}\omega\varepsilon_0 \mathrm{Im}\left(\frac{\vec{p}_i^{\,*}}{\alpha_i^*} \cdot \vec{p}_i\right), \\ Q_{\mathrm{ext},i} &= \frac{1}{2}\omega\varepsilon_0 \mathrm{Im}\left(\vec{E}_{\mathrm{ext\_}i}^{\,*} \cdot \vec{p}_i\right), \\ Q_{\mathrm{sca},i} &= Q_{\mathrm{ext},i} - Q_{\mathrm{abs},i}. \end{aligned} \tag{1.30}$$

These equations also work for multi-particle complexes. The sub-index i represents i-th particle inside that complex. $\vec{p}_i$'s are resultant electric dipole moments, and $\vec{E}_{\mathrm{ext\_}i}$'s are the incident fields at the i-th particle. The total extinction or absorption inside that system will require a summation over all particles.

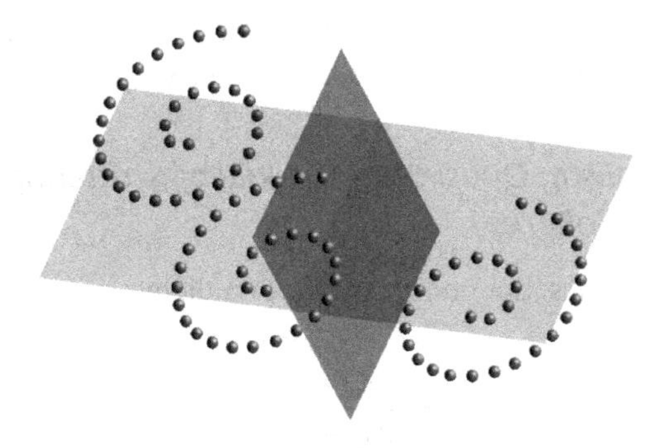

Figure 1.5 A 2D chiral object is not able to be matched with its mirror image in the same plane after any rotational and translational operation. When it is lifted off from the plane, then it will be moved onto its mirror image easily. Note that strictly speaking, the dimension of the NPs shown in the figure is not negligible; hence, in 3D space, there is a mirror plane cutting through the centers of the NPs.

However, when an NP is large or its geometry is not spherical, the plasmonic resonance will be quite different. It requires more sophisticated models [17, 77], because in this case both multipoles and scattering effects are important. Alternatively, an open source software DDSCAT [16] is very reliable in solving problems of light scattering by objects with arbitrary shapes.

1.3.2 *3D Chiral Nanoparticle Assemblies*

The 2D chiral object presented in Fig. 1.5 cannot be matched with its mirror image by any operation unless it is lifted off the plane. Consequently, we confirmed that an ideal 2D chiral object will demonstrate 0 extinction CD if the incident beam is normal to the plane and they are suspended in a uniform medium.

3D chiral complexes are "real" chiral complexes that exhibit non-zero CD signals from isotropic systems. We investigated several geometries of complexes, including helices with different lengths, pyramids, and tetramers with symmetric or asymmetric frames, with identical or differing particles. All CD signals show bisignate features. In this dipole regime, the radius dependence of the CD strength was found to be r_a^{12}. The CD signal also shows sensitive dependence on the overall geometry of a complex. Among all

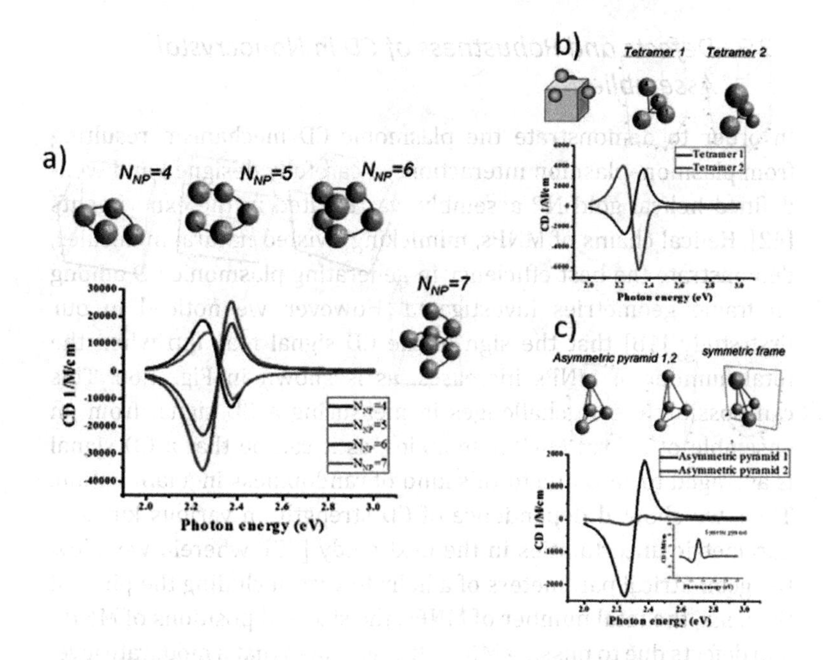

Figure 1.6 (a) Helical complexes of AuNP with identical sizes. CD is calculated for such complexes with different numbers of NPs. (b) Tetramers: two structures with overall similar parameters can show inverted CD signals. (c) In pyramidal structures, our study shows that the CD signal is extremely sensitive to the size of particles and complexes with a symmetric frame will show a very low CD signal in the dipole regime. Adapted with permission from Ref. [18]. Copyright 2010, American Chemical Society.

structures investigated, the helical structure is found to be the most efficient structure to create CD responses (Fig. 1.6). On the contrary, a significant CD signal is not found for a tetrahedral tetramer that has a symmetric frame but differing particles on its vertices. This is similar to a pyramid with a symmetric frame that exhibits a very small CD signal. In the dipole interaction regime, we conclude that a chiral MNP complex with a symmetric frame only exhibits weak CD signal. An interesting extension of the theory was made for another type of metal NCs, nanorods, in Ref. [2]. As nanorods typically have strong optical dipoles, chiral assemblies composed of nanorods may produce strong plasmonic CD signals.

1.3.3 *Defects and Robustness of CD in Nanocrystal Assemblies*

In order to demonstrate the plasmonic CD mechanism resulting from plasmon–plasmon interactions, a carefully designed and well-defined helical gold NP assembly was created in the experiments [42]. Helical chains of MNPs, mimicking twisted natural molecules, demonstrate the best efficiency in generating plasmonic CD among all frame geometries investigated. However, we noticed in our first study [18] that the sign of the CD signal may flip when the total number of MNPs increases, as is shown in Fig. 1.6a. This can possibly create challenges in measuring a CD signal from an ensemble of helical MNP assemblies, as it can be that a CD signal is averaged to zero due to this kind of randomness in a fabrication. Then, we showed dependence of CD strength on various kinds of parametric uncertainties in the next study [19], wherein we allow the geometrical parameters of a helix to vary, including the pitch of the helix, the total number of MNPs, the size and positions of MNPs, and defects due to missing MNPs. It was shown that a moderate level of randomness in these parameters does not diminish or change the shape of the CD signals. In Fig. 1.7, two cases are presented that demonstrate the consequences of randomized particle sizes and randomized particle positions. We can see that the influence of such disorder is weak. In nature, partially disordered molecular complexes such as random-coil proteins still demonstrate significant CD signals. From this point of view, it is not surprising that a helical chain of MNPs has a stable CD response. After these careful verifications, we were confident that a plasmonic CD from helical MNP chains with carefully selected geometric parameters should be strong.

To explain the bisignate CD signal in the plasmon band, our study [42] showed that it originates from the splitting between a transverse (xy) mode and a longitudinal (z) mode. Such collective excitations result from a plasmon–plasmon interaction in the dipole model. When circularly polarized light propagates vertically, the transverse (xy) modes in a standing helix are excited, while longitudinal (z) modes will be excited in a lying helix. The strength of an excitation will be slightly stronger if its handedness matches

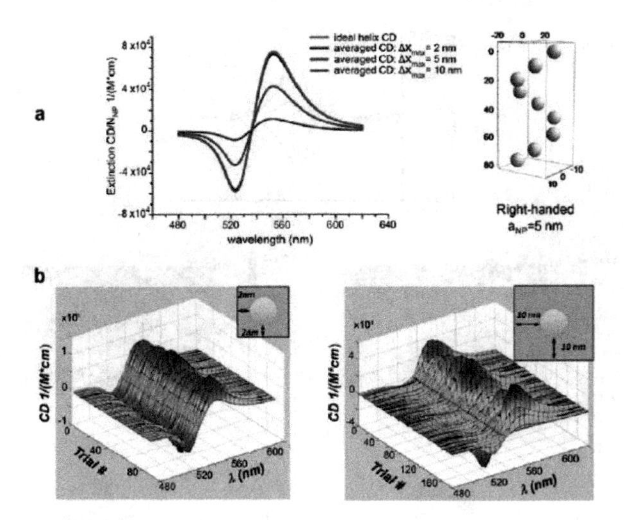

Figure 1.7 Effects of disorder on the CD spectra of nine-particle helices. (a) Calculated averaged CD spectra for the ideal helix and for three non-ideal helices. For the non-ideal helices, the CD spectra were averaged over a set of trials. Overall, ~80 trials with generating random positions of particles were made. Inset: The model to scale. (b) Calculated CD spectra for randomized positions of helically arranged NPs as a function of the trail number. Insets show how the NP position fluctuations were introduced. Reprinted with permission from SI of Ref. [42]. Copyright © 2012, Rights Managed by Nature Publishing Group.

the handedness of an incident light beam. Hence, it generates more heat dissipation. In Fig. 1.8a, we show that a total CD signal can be mathematically separated. The wavelength of a longitudinal (z) mode is slightly longer than a transverse (xz) mode. At 522 nm, we observe negative maxima at the poles of the sphere in Fig. 1.8c for a right-handed helix. At the poles, it indicates that incident beams propagate along the axis of a helix. Therefore, a right-handed xy-mode is slightly stronger under illumination of RCP light, contributing a negative CD signal to the total CD spectrum. At 552 nm, although there is no direct analogy between handedness of the dipole excitations and handedness of the helix, we observed that the CD signal flipped between a right-handed helix and a left-handed helix in the simulations reported in the SI of Ref. [42]. This indicates

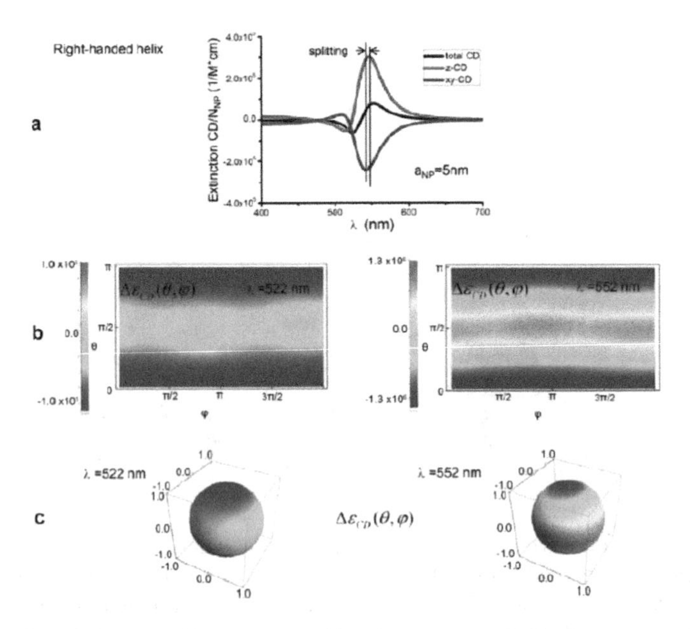

Figure 1.8 Calculated CD properties of the right-handed helix composed of 9 NPs. (a) Decomposition of the total CD (black spectrum) into parallel *z*-CD (red spectrum) and perpendicular *xy*-CD (blue spectrum) components. (b) and (c) Calculated directional CD spectra at two wavelengths that correspond to the extremes of the CD in graph (a). The similar simulation results are available for left-handed helices in SI of Ref. [42]. Reprinted with permission from Ref. [42]. Copyright © 2012, Rights Managed by Nature Publishing Group.

a direct correspondence between the CD signal and the handedness of a helix.

1.4 Electrodynamic Effects in Chiral Molecular-Plasmonic Structures

1.4.1 *Model: Maxwell's Equations Incorporating a Chiral Parameter*

Generally speaking, the light-scattering problem of core–shell multilayer structures can be solved exactly using the Mie solution. When a chiral medium is involved, its chiral property will break the symmetry of a system and the degeneracy between RCP and LCP electromagnetic waves. It is hence convenient to expand electromagnetic fields using circular polarized vector harmonics as basis. The coupled Maxwell's Eqs. (1.5) can be diagonalized:

$$\nabla \times \begin{pmatrix} \vec{P} \\ \vec{Q} \end{pmatrix} = \frac{\omega}{c} \begin{pmatrix} \xi + \sqrt{\varepsilon + \xi^2} & 0 \\ 0 & \xi - \sqrt{\varepsilon + \xi^2} \end{pmatrix} \begin{pmatrix} \vec{P} \\ \vec{Q} \end{pmatrix}, \tag{1.31}$$

where $\vec{P}$ and $\vec{Q}$ are RCP and LCP spherical waves, which are linear combinations of vector harmonics $\vec{M}$ and $\vec{N}$ [7]. The vector harmonics are defined as:

$$\begin{aligned}
\vec{M}_{emn} &= \frac{-m}{\sin\theta} \sin m\phi \, P_n^m(\cos\theta)\, z_n(\rho)\vec{e}_\theta - \cos m\phi \frac{dP_n^m(\cos\theta)}{d\theta} z_n(\rho)\vec{e}_\phi, \\
\vec{M}_{omn} &= \frac{m}{\sin\theta} \cos m\phi \, P_n^m(\cos\theta)\, z_n(\rho)\vec{e}_\theta - \sin m\phi \frac{dP_n^m(\cos\theta)}{d\theta} z_n(\rho)\vec{e}_\phi, \\
\vec{N}_{emn} &= \frac{z_n(\rho)}{\rho} n(n+1) \cos m\varphi \, P_n^m(\cos\theta)\, \vec{e}_r \\
&\quad + \cos m\varphi \frac{dP_n^m(\cos\theta)}{d\theta} \frac{1}{\rho}\frac{d}{d\rho}(\rho z_n(\rho))\, \vec{e}_\theta \\
&\quad - \sin m\varphi \frac{P_n^m(\cos\theta)}{\sin\theta} \frac{1}{\rho}\frac{d}{d\rho}(\rho z_n(\rho))\, \vec{e}_\varphi, \\
\vec{N}_{omn} &= \frac{z_n(\rho)}{\rho} n(n+1) \sin m\varphi \, P_n^m(\cos\theta)\, \vec{e}_r \\
&\quad + \sin m\varphi \frac{dP_n^m(\cos\theta)}{d\theta} \frac{1}{\rho}\frac{d}{d\rho}(\rho z_n(\rho))\, \vec{e}_\theta \\
&\quad + m \cos m\varphi \frac{P_n^m(\cos\theta)}{\sin\theta} \frac{1}{\rho}\frac{d}{d\rho}(\rho z_n(\rho))\, \vec{e}_\varphi.
\end{aligned} \tag{1.32}$$

Explicitly,

$$\begin{aligned}
\vec{P} &= \vec{M}_{o/e,mn}{}^{(3)/(1)}(k_+ r) + \vec{N}_{o/e,mn}{}^{(3)/(1)}(k_+ r), \\
\vec{Q} &= \vec{M}_{o/e,mn}{}^{(3)/(1)}(k_- r) - \vec{N}_{o/e,mn}{}^{(3)/(1)}(k_- r).
\end{aligned} \tag{1.33}$$

$\vec{P}$ and $\vec{Q}$ waves have propagation constants k_+ and k_-, respectively. The total electric or magnetic fields will be linear combinations of $\vec{P}$ and $\vec{Q}$. This new basis is called odd/even LCP/RCP vector harmonics.

In Fig. 1.9, several typical examples of the core–shell multilayer structures are shown. In each layer, the dielectric functions ε and chiral parameters ξ are specified. Eigen modes of each layer are

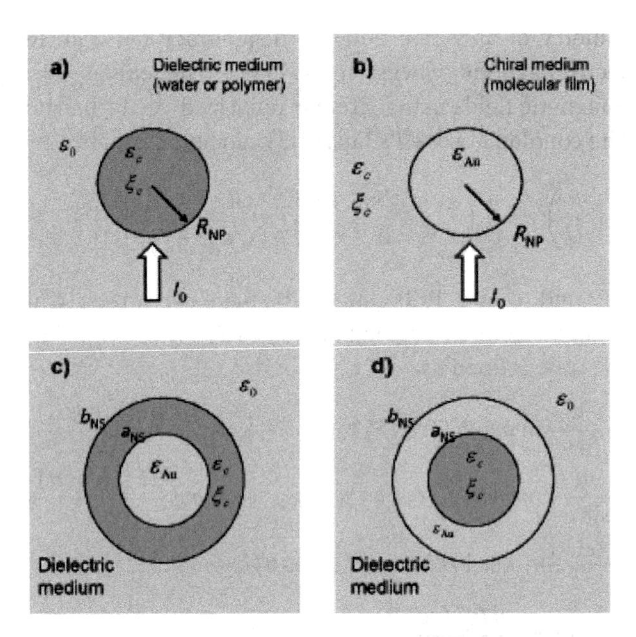

Figure 1.9 Models of NPs and core–shell NCs used in the calculations. Reprinted with permission from Ref. [34]. Copyright © 2012 Wiley-VCH Verlag GmbH & Co. KGaA, Weinheim.

conveniently identified using Eqs. (1.32) and (1.33). The total electric and magnetic fields are given as:

$$\vec{E}_i = \sum_{n=1,2,3,\ldots} A^{(3)/(1)}_{o/e,1n}\left(\vec{M}_{o/e,1n}{}^{(3)/(1)}(k_+) + \vec{N}_{o/e,1n}{}^{(3)/(1)}(k_+)\right)$$
$$+B^{(3)/(1)}_{o/e,1n}\left(\vec{M}_{o/e,1n}{}^{(3)/(1)}(k_-) - \vec{N}_{o/e,1n}{}^{(3)/(1)}(k_-)\right),$$
$$\vec{H}_i = -i\sqrt{\frac{\varepsilon_0 + \mu\xi^2}{\mu}} \sum_{n=1,2,3,\ldots}\left(A^{(3)/(1)}_{o/e,1n}\left(\vec{M}_{o/e,1n}{}^{(3)/(1)}(k_+)+\vec{N}_{o/e,1n}{}^{(3)/(1)}(k_+)\right)\right.$$
$$\left.-B^{(3)/(1)}_{o/e,1n}\left(\vec{M}_{o/e,1n}{}^{(3)/(1)}(k_-) - \vec{N}_{o/e,1n}{}^{(3)/(1)}(k_-)\right)\right). \tag{1.34}$$

where the sub-index i represents the i-th layer, (3) and (1) indicate that the function $z_n(\rho)$ in Eq. (1.32) is the spherical Hankel function of the first type [$h_n^{(1)}(\rho)$] or the spherical Bessel function [$j_n(\rho)$]. On the boundary of each layer, the transverse component of electric and magnetic fields is continuous. The coefficients $A^{(3)/(1)}_{o/e,1n}$ and $B^{(3)/(1)}_{o/e,1n}$ of those modes can be found by solving the matrix equations given in

Ref. [44]. Using a set of convenient notations for the coefficients, the scattered field can be written in the following form:

$$\vec{E}_{\text{sca}} = \sum_{n=1,2,3,\ldots} a_{o/e,1n}\left(\vec{M}_{o/e,1n}{}^{(3)}(k_+) + \vec{N}_{o/e,1n}{}^{(3)}(k_+)\right)$$
$$+b_{o/e,1n}\left(\vec{M}_{o/e,1n}{}^{(3)}(k_-) - \vec{N}_{o/e,1n}{}^{(3)}(k_-)\right),$$

$$\vec{H}_{\text{sca}} = -i\sqrt{\frac{\varepsilon_0 + \mu\xi^2}{\mu}}$$
$$\times \sum_{n=1,2,3,\ldots}\left(a_{o/e,1n}\left(\vec{M}_{o/e,1n}{}^{(3)}(k_+) + \vec{N}_{o/e,1n}{}^{(3)}(k_+)\right)\right.$$
$$\left.-b_{o/e,1n}\left(\vec{M}_{o/e,1n}{}^{(3)}(k_-) - \vec{N}_{o/e,1n}{}^{(3)}(k_-)\right)\right).$$

The coefficients of scattered fields carry the information of extinction, absorption, and scattering of a nanostructure [7, 34]. The extinction and scattering cross-sections are given in terms of scattering coefficients:

$$C_{\text{ext},\pm} = \frac{2\pi}{k^2}\text{Re}\left\{\sum_n \frac{-a_{o,1n} \mp i a_{e,1n}}{E_n/\sqrt{2}}(2n+1)\right\}, \tag{1.35}$$

$$C_{\text{sca}} = \frac{2\pi}{k^2}\text{Re}\left\{\sum_n (2n+1)\left(\left|\frac{a_{o,1n}}{E_n/\sqrt{2}}\right|^2 + \left|\frac{a_{e,1n}}{E_n/\sqrt{2}}\right|^2\right.\right.$$
$$\left.\left.+\left|\frac{b_{o,1n}}{E_n/\sqrt{2}}\right|^2 + \left|\frac{b_{e,1n}}{E_n/\sqrt{2}}\right|^2\right)\right\}, \tag{1.36}$$

where $E_n = (-i)^n E_0 \frac{2n+1}{n(n+1)}$.

1.4.2 *Circular Dichroism in the Quasi-Static Limit*

Alternatively, convenient for small nanostructures in particular, the absorption is defined for the whole system using these symbolic equations:

$$Q_{\text{abs}} = 2\int_{V_0} \text{Re}\left(\vec{j}_{\text{eff}} \cdot \vec{E}^*_{\text{eff}}\right) dV;$$

$$\vec{j}_{\text{eff}} = \sigma\vec{E}_\omega - \frac{c_0 i}{4\pi}\nabla \times \left(\xi_c \vec{E}_\omega\right) + \frac{\omega\xi_c}{4\pi}\vec{B}_\omega. \tag{1.37}$$

or [3]

$$Q_{\text{abs}} = 2\omega \int_{V_0} \text{Im}\left(\vec{P} \cdot \vec{E}^*_{\omega} + \vec{M} \cdot \vec{B}^*_{\omega}\right) dV, \qquad (1.38)$$

where $\vec{D} = \vec{E} + 4\pi\vec{P}$ and $\vec{H} = \vec{B} - 4\pi\vec{M}$. These two equations are essentially equivalent. Using Eqs. (1.2)–(1.3), Eq. (1.22) can be written as

$$Q_{\text{abs}} = \omega \int_{V_0} \text{Im}(\varepsilon) \frac{\vec{E}_{\omega} \cdot \vec{E}^*_{\omega}}{2\pi} dV + \omega \int_{V_0} \text{Re}\left[\frac{\xi_c}{2\pi}\left(\vec{B}_{\omega} \cdot \vec{E}^*_{\omega} - \vec{E}_{\omega} \cdot \vec{B}^*_{\omega}\right)\right] dV. \qquad (1.39)$$

Equations (1.22) and (1.23) have been discussed briefly in Section 1.2. In formula (1.23), the integration is taken over the whole system. It can be formally distributed into two terms if we integrate it respectively inside a molecular medium and a metal NC. As usual, the CD is defined using Eq. (1.17), where averaging over the incident angles is necessary for colloidal systems.

For the nanostructure that is optically small, the consistency between a semiclassical quasi-static solution and the Mie solution was shown in Ref. [34]. In Section 1.2.4 devoted to the plasmon–exciton interaction, we have already seen that both methods give a vanishing plasmon CD signal when the layer of a chiral medium is thin and isotropic. This is because chiral molecules in an isotropic layer have different orientations and, therefore, give zero resultant CD in the plasmon band (1.19). Analytically, the Mie solution shows that the plasmonic CD will be:

$$CD_{\text{metal}} = \text{Im}(\varepsilon_{\text{metal}})\left(k_0 \text{Re}\left(\xi_c\left(\varepsilon_c - \varepsilon_0\right)\right) \text{f}_1\left(\varepsilon_c, \varepsilon_{\text{metal}}\right)\right) + k_0^3 \text{Re}\left(\xi_c\right) \text{f}_2\left(\varepsilon_c, \varepsilon_{\text{metal}}\right). \qquad (1.40)$$

The first term is due to a near-field interaction, and the second term is a retardation effect, which is small for small complexes. As the presence of chiral molecule does not change much the dielectric constant of the chiral shell ε_c, the plasmonic CD vanishes when the size of a nanostructure is small.

1.4.3 *Electrodynamic Effects in Large Structures*

When the structure is large, for example, $k_0 b \sim 1$, or even greater, the full Mie solution has to be used to simulate the extinction,

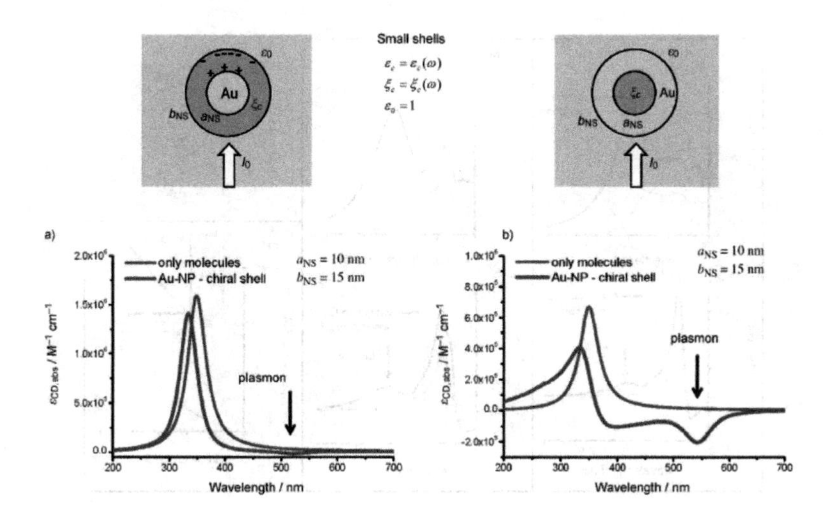

Figure 1.10 (a) Calculated CD spectra for a chiral nanoshell–metal NP structure; we also show the signal from the molecular nanoshell in the absence of a metal NP. (b) The inverted case: A metal nanoshell–chiral particle structure; for comparison, we again show the CD signal from the molecules alone. Insets: Nanostructure models. Reprinted with permission from Ref. [34]. Copyright © 2012 Wiley-VCH Verlag GmbH & Co. KGaA, Weinheim.

scattering, and CD spectra. Some examples of such nanostructures are given in Fig. 1.11. From the extinction spectra, the resonances mostly come from scattering at the plasmon band around 600 nm. It is worth noting that this is an electrodynamic effect originating from resonances in the Mie coefficients. When we compute its extinction spectrum with Eq. (1.35), the extinction is directly related to the real part of amplitudes of scattered waves. As the nanostructure is large, we are not able to predict resonances from an expansion of spherical Bessel functions with arguments $k_0 a\sqrt{\varepsilon_{\text{metal}}}$ or $k_0 b\sqrt{\varepsilon_{\text{c}}}$. However, we know that these resonances in the extinction spectra are results of interplay between the dielectric functions of component materials and all other geometric parameters. In the particular case displayed in Fig. 1.11, the resonances come from the a coefficients, which corresponds to the TM modes or the dynamic electric multipoles in the language of Ref. [7]. For convenience and consistency, we name the associated CD signals as plasmonic CD,

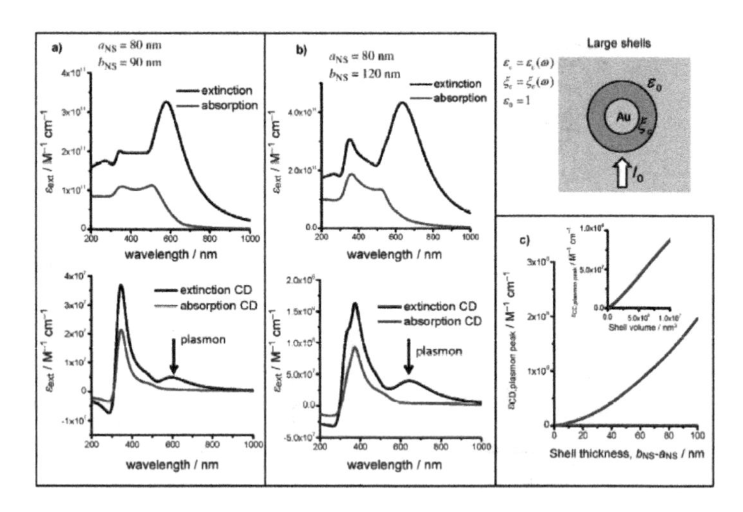

Figure 1.11 (a) and (b) Calculated CD spectra for large core–shell structures. (c) The plasmon peak CD as a function of the shell thickness, $b_{NS}-a_{NS}$. Inset shows the plasmon peak CD as a function of the volume of the chiral shell. General inset: Model of the chiral plasmonic core–shell structure. Reprinted with permission from Ref. [34]. Copyright © 2012 Wiley-VCH Verlag GmbH & Co. KGaA, Weinheim.

as they appear at the plasmon wavelength. In comparison, when the core is dielectric, the scattering and interference effects become more interesting. Such a case was considered recently in Ref. [2]. For example, in silicon particles, it supports magnetic resonances (coming from the *b* coefficients) due to its large dielectric constant ($\varepsilon_{Si} = 11.7$). The electromagnetic waves are able to oscillate for at least one cycle inside the nanostructures, creating interferences that maximize scattered electromagnetic fields. As a consequence, a few resonances will be found in the extinction spectrum. The corresponding CD bands may come from differential absorption and differential scattering as well. Also, they are associated with either TM or TE modes. The CD spectra exhibit interesting shapes.

1.5 Plasmonic Nanocrystals with Chiral Shapes

Anisotropy in metal NCs leads to interesting optical properties, such as tunability of the plasmon band or significant enhancements in plasmonic fields. Chiral metal NCs constitute a new type of nanoscale systems. Here, we present a study of chiral metal NPs, using a multipole expansion method. The calculations will show a new plasmonic CD mechanism coming from a mixing of multiple plasmon modes.

1.5.1 *Formalism*

The formalism to describe chiral MNPs follows a quasi-static approximation. The solution for the electric fields induced by a chiral NP can be found by solving the Poisson equation:

$$\nabla \cdot \varepsilon(r) \nabla \delta\varphi_{\text{ind}} = f(r). \tag{1.41}$$

On the boundary,

$$\varphi_{\text{ind}}|_- = \varphi_{\text{ind}}|_+,$$
$$\varepsilon_{\text{Au}} \vec{n} \cdot (\vec{E}_0 + \delta\vec{E})\Big|_- = \varepsilon_0 \vec{n} \cdot (\vec{E}_0 + \delta\vec{E})\Big|_+, \tag{1.42}$$

where $\vec{n}$ is a vector normal to the NC's surface, $\vec{E}_0$ is an incident field, $\delta\vec{E}$ is an induced electric field related to φ_{ind} and

$$\varphi_{\text{ind}}^{\text{in}} = \sum_{\substack{l\geq 0 \\ l\geq m\geq -l}} A_{l,m} \left(\frac{r}{a_{NC}}\right)^l Y_{l,m}(\theta,\varphi), \quad (r \leq R_{\text{NC}})$$
$$\varphi_{\text{ind}}^{\text{out}} = \sum_{\substack{l\geq 0 \\ l\geq m\geq -l}} B_{l,m} \left(\frac{a_{NC}}{r}\right)^{l+1} Y_{l,m}(\theta,\varphi). \quad (r > R_{\text{NC}}) \tag{1.43}$$

Therefore, we end up solving these linear equations:

$$\sum_{l\geq 0,\ -l\geq m\geq l} A_{l,m} a_{l',m';l,m} + B_{l,m} b_{l',m';l,m} = 0,$$
$$\sum_{l\geq 0,\ -l\geq m\geq l} A_{l',m'} \cdot \tilde{a}_{l'm',lm} + \sum_{l\geq 1,m} B_{l,m} \cdot \tilde{b}_{l'm',lm} = F_{\text{ext},l'm'}, \tag{1.44}$$

where the matrix elements a, b, $\tilde{a}$, $\tilde{b}$, and F_{ext} are defined in the supplementary information of Ref. [20]. Once the electric fields are explicitly expressed through these coefficients, the heat absorption that we are seeking is defined as:

$$\text{Im}\varepsilon_{\text{NC}} \frac{\omega}{2\pi} \int_{NC} E_\omega \cdot E_\omega^* dV, \tag{1.45}$$

where E_ω is the field inside NC.

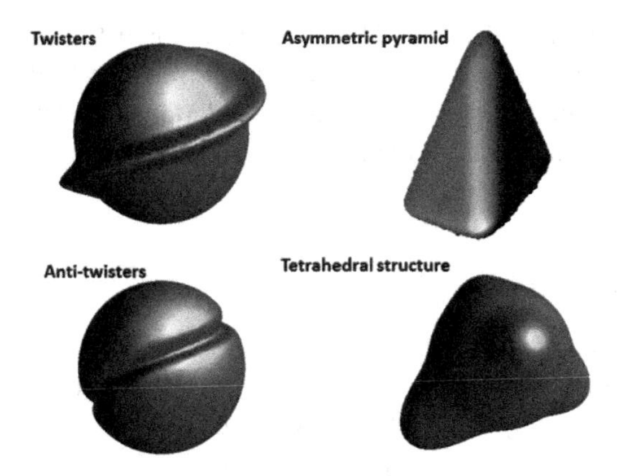

Figure 1.12 Schematics of chiral objects generated by various equations $r = R_{\mathrm{NC}}(\theta, \varphi)$. Adapted with permission from Ref. [20]. Copyright 2012, American Chemical Society.

In a small NC, the CD signal can be expanded as a power series of k, using a long wave approximation [20]:

$$CD = ak^2 + bk^4 + \ldots, \tag{1.46}$$

where a and b are coefficients related to the dielectric functions and the geometric parameters. Leading by k^2, CD is much smaller than absorption, as absorption is of the power of k. On the contrary, CD is much stronger than scattering of k^4.

The chiral shape of an NC is defined using equations $r = R_{\mathrm{NC}}(\theta, \varphi)$. It will be more straightforward to just list these examples at the beginning of the section. The plots of these objects are shown in Fig. 1.12, while the functions describing these objects are available in the SI of Ref. [20]

1.5.2 *Origins of Plasmonic CD: Mode Mixing*

In order to learn more about the physical process underlying the interactions between plasmon modes of a chiral NC and incident light, we now use a modal analysis and an artificial dielectric constant from the Drude model. The Drude model parameters are selected to provide a very narrow absorption peak so that we can

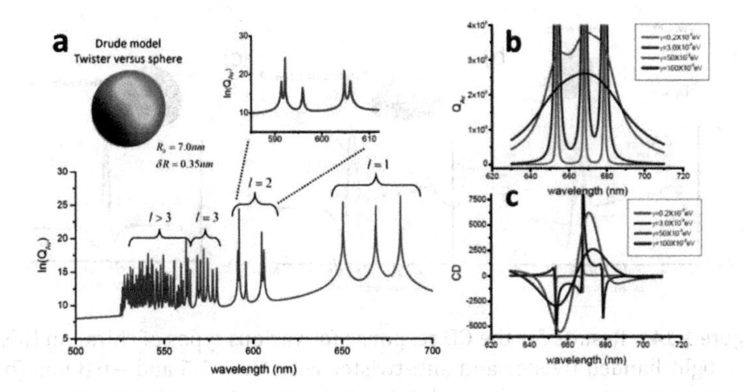

Figure 1.13 Modal analysis of plasmon resonances of a chiral NC using the Drude model. (a) The spectrum of absorption for a small plasmon broadening. Inset: Model of the chiral NC and its comparison with an ideal sphere. (b) The lowest triplet modes ($l = 1$) for various plasmon broadenings. (c) CD spectra of the triplet modes, plotted again for different broadenings. Adapted with permission from Ref. [20]. Copyright 2012, American Chemical Society.

see the splitting of multipole plasmons clearly

$$\varepsilon_{\text{metal}} = 1 - \frac{\omega_{\text{p}}^2}{\omega\left(\omega + i\gamma_{\text{p}}\right)}, \tag{1.47}$$

in which we choose $\omega_{\text{p}} = 4eV$. Figure 1.13 shows the results for a twister with a very small surface variation $\delta R/R_0 = 0.05$ and a small plasmon broadening ($\gamma_{\text{p}} = 10^{-5}eV$). The l-mode will split into $2l+1$ lines, as it is shown in Fig. 1.13. The l- multipole is centered at

$$\omega_{\text{p}}\sqrt{1 + \frac{l+1}{l}\varepsilon_0}.$$

By gradually increasing the broadening parameter γ_{p}, both the absorption line and CD lines are broadened and then merged, as is demonstrated in Fig. 1.13b and Fig. 1.13c. These spectra are focused on the dipolar plasmon band ($l = 1$). We see clearly how a bisignate CD signal is formed. It should be noted that in our computation, a CD signal will arise in this band only when we include more harmonics than just $l = 1$. In other words, the mixing between multipoles is critical for the formation of CD signals in chiral NCs.

Using a real dielectric function taken from Ref. [39], we have tested various functions $r = R_{\text{NC}}\left(\theta, \varphi\right)$ that generate chiral surface

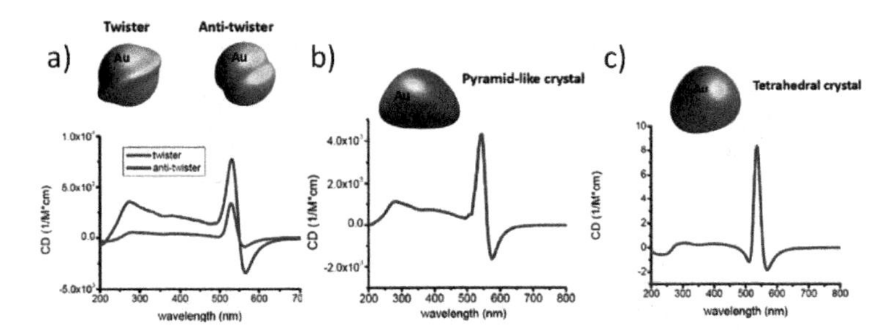

Figure 1.14 Results for the CD response for various types of chiral Au NCs. (a) Right-handed twister and anti-twister with δR $= 1.5$ and -0.8 nm. (b) Asymmetric pyramid with $\alpha = 1$. (c) Chiral tetrahedral structure with $a0 =$ 2 nm. In all cases, $R0 = 7$ nm. Adapted with permission from Ref. [20]. Copyright 2012, American Chemical Society.

distortion. The function describes a perturbation in geometry created to a sphere. Perturbation can be modulated by a parameter δR, seen in the SI of Ref. [20]. The formalism can be applied to generally all kinds of distortions described by such functions. In our study, we are interested in helical twisters, pyramid-like NPs, and tetrahedron-like NPs, on which four bumps of different sizes are created at the vertices to mimic a tetrahedral framework. Coincidently, the helical structure again demonstrated the strongest potential in plasmonic CD generation, though we have discussed in the last paragraph that the mechanism is essentially different from helical AuNP chains. This suggested an interesting connection between chiral plasmon excitations and chiral geometries.

1.6 Experimental Measurements of Plasmonic Circular Dichroism

Chiral plasmonic NC assemblies so far involved spherical NPs, nanorods, nanofibers, chiral mesoporous films, chiral inorganic molecules, biomolecules, and so on [1, 25, 26, 35, 36, 42, 48, 56, 66, 67, 72]. In this section, we will give a few examples of experimental works on chiral nanoscale systems.

In the field of plasmonic CD research of NPs, a pioneer study was carried out on chiral silver NPs assembly by Shemer et al. [65]. In

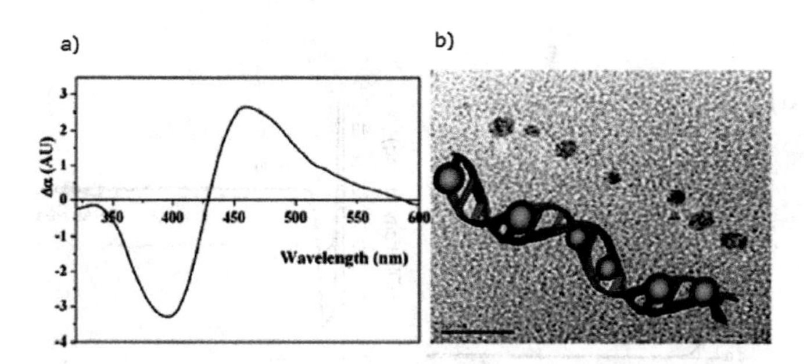

Figure 1.15 The plasmonic CD band of Ag NPs assembled on DNA (a) and the geometry of the system including a TEM image of the Ag NP chain (b). Reprinted with permission from Ref. [65]. Copyright 2006, American Chemical Society.

these experiments, silver ions were complexed with nucleic acids and reduced from the solution. DNA served as a template for the growth of silver NPs. The resulting complex is shown in the TEM image in Fig. 1.15b. Interestingly, the CD signal of DNA at 260 nm was inversed, while a new CD band was created in the plasmon band as shown in Fig. 1.15a. In a control experiment, it was shown that no plasmonic CD could be provoked by merely mixing DNA molecules and premade silver NPs. It was concluded that DNA had directed the asymmetric growth of silver NPs [65]. Now we know that the asymmetry can be in the overall geometry of the assembly or stem from surface distortions on the NPs. It has to be noted that the plasmonic CD can also come from the other mechanisms such as the orbital hybridization or the exciton–plasmon interaction.

Previously, it was proposed that both plasmon–plasmon and plasmon–exciton interactions could induce CD signals at plasmon bands. In order to see the physical pictures more clearly, alternative designs of nanostructures were realized in the following experiments.

On colloidal AuNPs coated with chiral peptides [67], a moderately strong plasmonic CD signal can be identified from the spectra of E5-AuNP in Fig. 1.16 and of FlgA3-AuNP that can be found in Ref. [67]. This induced CD signal was attributed to the plasmon–exciton interaction. An α-helical peptide (E5) has a stronger dipole moment than an unstructured peptide (FlgA3). Therefore, the α-helical

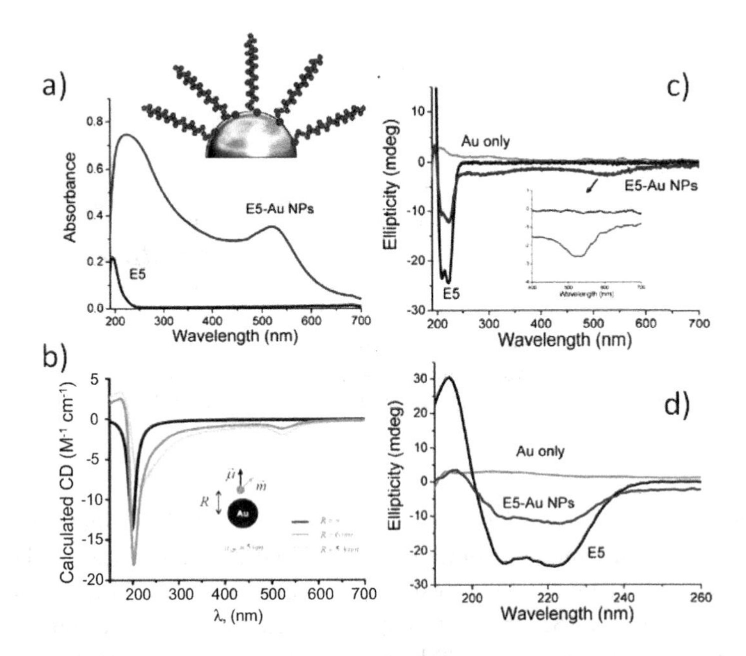

Figure 1.16 Optical characterization of gold NPs functionalized with the E5 peptide (5.0 μM) with 1.14×10^{13} particles/ml. (a) UV vis spectra of E5 and E5 gold NPs. (Inset) Illustration of gold NP surface covalent linked to the E5 peptide via the thiol linkage (red circle). (b) The calculated CD spectra for a dipole of chiral molecule (black curve for $R = \infty$) and for a dipole-NP complex with two separations (R = 5.3 and 6 nm); the absorption wavelength of the dipole $\lambda_{\text{dipole}} = 200$ nm. Inset shows the model used. (c) CD spectra and (d) CD spectra of UV region of E5 peptide gold NPs with or without E5 peptide. Adapted with permission from Ref. [67]. Copyright 2012, American Chemical Society.

peptide's interaction with the AuNP is expected to be stronger than an unstructured peptide. However, it is interesting to note that in this experiment [67], the CD signals measured from both peptides are comparable. It was assumed that the adsorption of unstructured peptides to AuNPs is in a multidentate fashion, while an α-helical E5 coil protein bonds with AuNPs through a single cysteine residue at its end. In this picture, an effective distance between the chromophores of an α-helical peptide and the NP surface is longer and therefore transfer of chirality for this peptide becomes reduced.

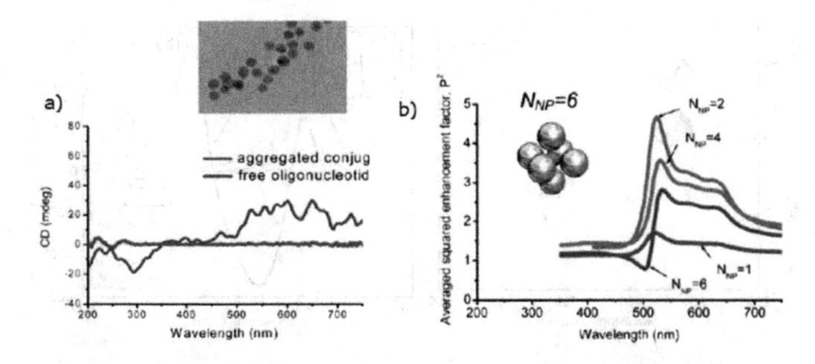

Figure 1.17 (a) CD spectra measured by Gerard and co-workers for oligonucleotides alone and for aggregates of 10 nm Au NPs conjugated with oligonucleotides. Inset: TEM image of oligonucleotide-conjugated gold NPs [26]. Taken and adapted with permission from Ref. [26]. Copyright 2011, The Royal Society of Chemistry. (b) Calculated averaged plasmonic enhancement factors ($\bar{P}_2$) for the plasmonic hot spot in the center of an assembly of N_{Au} NPs. Inset: Geometry of one particular complex with N_{NP} = 6; the plasmonic hot spot is located in the center of the system indicated as a blue sphere. Adapted with permission from Ref. [30]. Copyright 2006, American Chemical Society.

Plasmon–exciton interaction induced CD was also identified in another experiment [26], in which Gerard et al. demonstrated that the plasmonic CD can be strongly enhanced. To start with a most simple system, spherical gold NPs were conjugated to oligonucleotides via their 3′ end T6 phosphorothioate groups. The UV-visible absorption spectra of bare and oligonucleotide-conjugated particles showed a 7 nm red shift of the plasmon band, from 519 to 526 nm. This may be held as a first evidence of conjugation [69], as the red shift is induced by a modification of the dielectric constant of the environment and electronic interactions of the oligonucleotide molecules with the metal core [51]. The presence of oligonucleotides on the particle surface also resulted in an increase in the hydrodynamic radius, which was assessed by dynamic light scattering (DLS) measurements. Bare gold NPs had an average hydrodynamic diameter of 20 nm, whereas for conjugated particles, it was 70 nm. A strong CD appeared only after centrifugation and washing of the conjugates. The process caused partial aggregation of the particles as revealed by a further 5 nm red-shift in the plasmon band and TEM images. This also

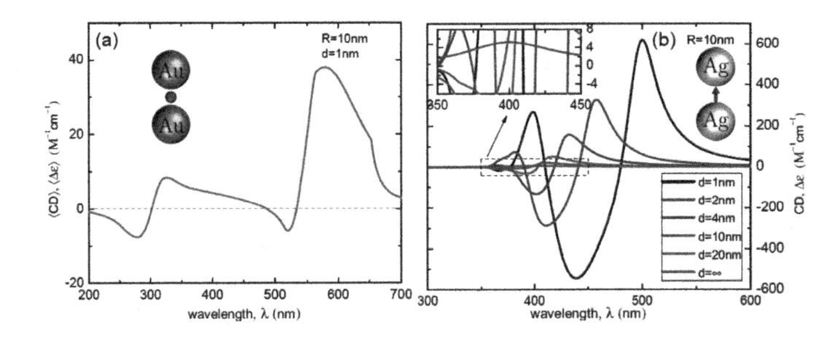

Figure 1.18 (a) CD signal for the molecule-Au dimer complex averaged over the molecular dipole orientation at molecular resonance $\lambda_0 = 300$ nm; (b): CD spectrum of molecule-Ag dimer with molecular orientation $\mu \parallel z$ at $\lambda_0 = 400$ nm. Adapted from Ref. [81]. Copyright (2013) by the American Physical Society.

explained the change in CD response, as it was reported that larger particles and aggregates induced a stronger CD signal at the plasmon wavelength [11]. Similar effects were measured in Ref. [26] also for oligonucleotides conjugated to gold nanorings and gold-silver nanoboxes. The process of aggregation and related CD generation was found to be reversible.

We stress that, in a solution of gold NPs decorated with oligonucleotides, plasmonic CD was observed only when these gold NPs were partially aggregated [26]. This observation suggests that the plasmonic hot spots may be the physical reason for the observation of strong CD signals at the plasmon wavelength in Ref. [26]. Recent calculations [81] have shown that, in NP aggregates, the electromagnetic enhancement ($|E/E_0|^2$) in the hot spot of MNP dimer can be very large for small inter-NP separations that can lead to giant CD signals [81]. For instance, the field enhancement in a hot spot can reach up to 10^4 for an Au-NP dimer with a NP radius $R = 10$ nm and a separation $d = 1$ nm. Therefore, by employing the NP dimer, one can expect the appearance of giant CD signals at the plasmonic wavelength with enhancement factors nearly 10^2 [81].

Over the last two decades, nanotechnology [64] has taken a rapid development from being a niche playground for fundamental researchers to now providing a robust technology platform for the assembly of nanostructures of arbitrary shapes [14, 58], or for the spatial organization of nanoscale objects such as fluorophores

[68], NPs [13], and biologically relevant molecules [15, 63], with nanometer precision. It was possible to harness this assembly power for the construction of gold NP assemblies of perfectly controlled geometry and chirality [42]. Using the DNA origami method, in which a virus-derived DNA "scaffold" single strand of nearly 8000 bases is folded into shape with the help of nearly 200 short synthetic "staple" oligonucleotides, a rigid bundle of 24 tightly connected parallel double helices was built that together adopted the shape of a cylinder with a length of 100 nm and a diameter of 16 nm. On the surface of this DNA origami object, either nine positions forming a left-handed or nine positions forming a right-handed "spiral staircase" around the cylinder were chosen. At each of these positions, the ends of the three nearest staple strands were synthesized with 15 extra bases to form a "handle" for the gold NPs. The 10 nm gold NPs in these experiments were each covered with multiple (>20) thiol-modified DNA sequences that were complementary to the handle sequence. When those particles were mixed with one of the two DNA origami cylinder versions (with left or right-handed staircase attachment positions), the handle sequences were occupied by the NPs resulting in the formation of a helix of plasmonic particles of defined chirality (Fig. 1.19a). Importantly, attachment yields greater than 97% were achieved (Fig. 1.19b).

The assembled gold particle helices exhibited exactly the theoretically predicted optical activity. In Fig. 1.20a, the calculated CD signals for a plasmonic helix with the geometrical parameters of the DNA-templated helices are shown. The characteristic bisignate shape—dip-peak for a right-handed helix or peak-dip for a left-handed helix—at the specific plasmon frequency of gold NPs (525 nm) is reproduced by the experiments (Fig. 1.20b). To demonstrate the dependence of the signal strength from the NP diameter, this parameter was changed by depositing additional gold from solution onto the pre-assembled particle helices in a process called electroless deposition. A dramatic increase of the CD signal (~500-fold) and a shift of the plasmon frequency to a longer wavelength was found if the particle diameter was changed from 10 to 16 nm (Fig. 1.20c). In further experiments, silver was deposited on the gold particles and a shift of the plasmon resonance

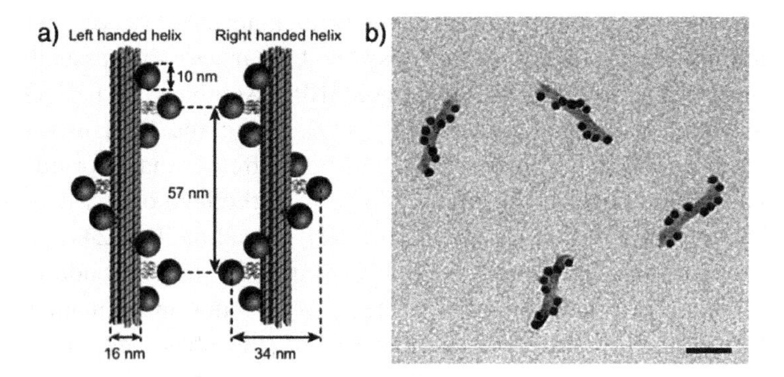

Figure 1.19 Gold particle helices templated with DNA origami. (a) 24 parallel DNA double strands form the scaffold for nine gold particles that are arranged either in a left-handed or a right-handed helix. (b) Transmission electron micrograph of the DNA origami constructs carrying the gold particles. Scale bar: 50 nm. Adapted with permission from Ref. [42]. Copyright © 2012, Rights managed by Nature Publishing Group.

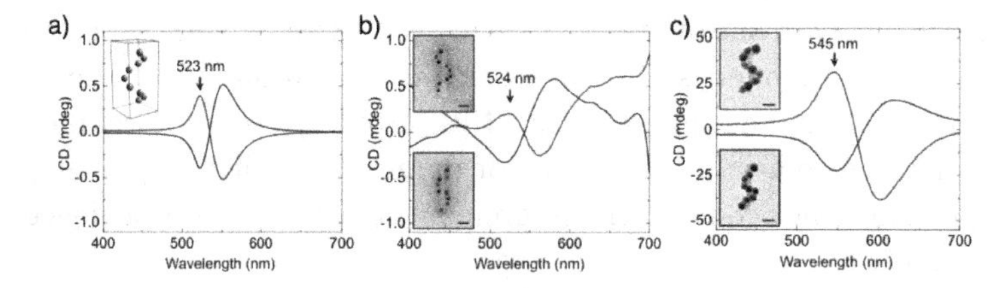

Figure 1.20 Calculated and experimental CD signals of DNA-origami templated gold particle helices (left-handed helix: red lines; right-handed helix: blue lines). (a) Calculated CD signal of nine gold particles (10 nm diameter) arranged in a helix with a diameter of 34 nm and a helical pitch of 57 nm. (b) Experimental CD for solutions containing the 10 nm gold particle helices presented in Fig. 1.19. (c) The signal strength increases dramatically with increasing particle diameter (here: 16 nm). Scale bars: 20 nm. Adapted with permission from Ref. [42]. Copyright © 2012, Rights managed by Nature Publishing Group.

to shorter wavelengths was observed. The collective plasmon–plasmon interactions were also strong enough to be detected as frequency-dependent optical rotatory dispersion when droplets of the samples were placed between two crossed linear polarizers [42].

Another possibility to create chiral plasmonic resonances is to use a large chiral molecule as a template for growth of plasmonic NCs. In Ref. [48], silver NPs were prepared in aqueous solutions of

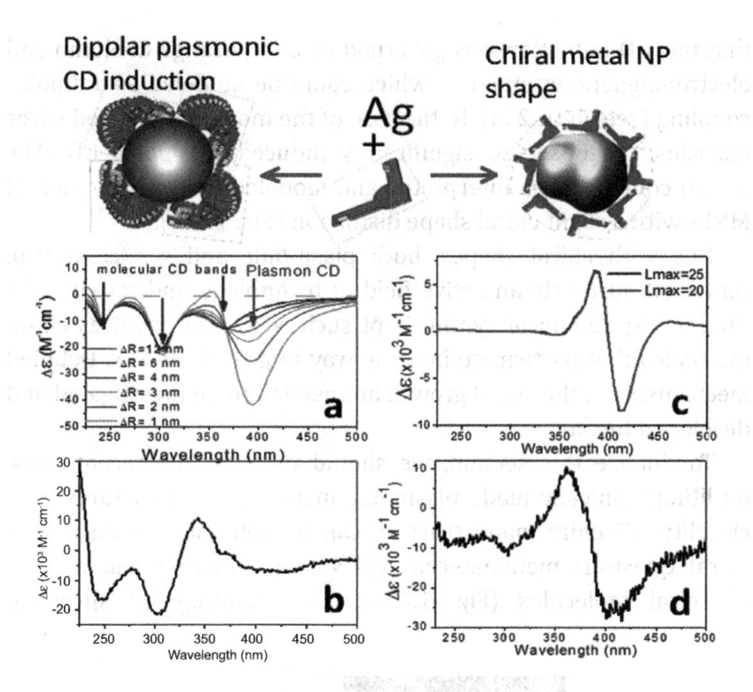

Figure 1.21 Top: Schematic of the two possible extreme cases of chiral molecule coated Ag NPs: Left: Monolayer-coated particles for Ag:molecule ratio 4:1; Right: molecular stack-coated particles for Ag:molecule 32:1. (a) Calculation of the CD spectrum for a dipolar particle-stack interaction at different particle-molecule average separation distances. (b) The experimental CD spectrum for an Ag:molecule 4:1 sample, showing the broad negative plasmonic CD around 400 nm. (c) A calculation of the plasmonic CD for a particle with a chiral surface distortion taking the first 20 and 25 spherical harmonics into account. (d) The experimental-induced plasmonic CD spectrum for the Ag:molecule 32:1 sample. Adapted with permission from Ref. [48]. Copyright 2012, American Chemical Society.

chiral supramolecular structures made of chiral molecular building blocks. By changing the metal-atom:molecule ratio, it was possible to obtain either a molecular monolayer coating of the NPs (see the top right scheme of Fig. 1.21) or chiral stacks of molecules coating the NPs (see the top left scheme of Fig. 1.21). In the latter case, the chiral stack-plasmon coupling induced a CD signal at the surface plasmon resonance absorption band of the silver NPs (see Fig. 1.21b). The sensitivity of this induction effect to temperature, which affected the stacking of the molecules (i.e., the level of chirality), indicated that this CD induction was governed by a long-range Coulomb and electromagnetic interaction, which could be modeled as a dipolar coupling (see Fig. 1.21a). In the case of the monolayer-coated silver particles, the observed significantly induced plasmonic CD (Fig. 1.21d) could only be interpreted and modeled by the formation of MNPs with a slight chiral shape distortion (Fig. 1.21c).

NCs with chiral shapes, both plasmonic and semiconductor, represent currently an active field of technology and research [9, 46, 56, 76]. Chemical synthesis of such NCs often involves chiral molecules that participate in some way in growth of NCs. Detailed mechanisms of the chiral growth are needed to be investigated and developed further.

To finalize this section, we should mention the recent work on lithographically made plasmonic metamaterial structures with chirality. 3D chiral nanostructure can be fabricated using a non-chiral quasi-2D metamaterials (plasmonic crosses) and a layer of chiral molecules (Fig. 1.22a). The resulting CD spectrum

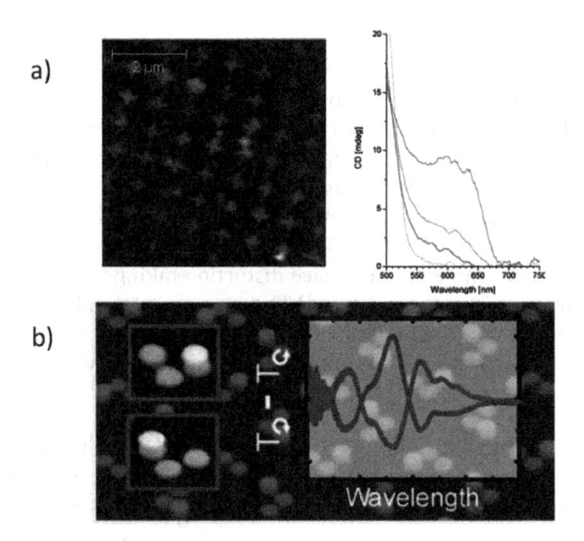

Figure 1.22 (a) Plasmonic Au crosses covered with a layer of chiral molecules exhibit a plasmon peak in the CD spectrum. Adapted with permission from Ref. [1]. Copyright 2012, American Chemical Society. (b) Lithographical plasmonic oligomer structures exhibiting strong directional CD. Adapted with permission from Ref. [38]. Copyright 2012, American Chemical Society.

of this structure contained a plasmon peak due to long-range electromagnetic interactions between the plasmonic and molecular parts [1]. Another possibility to create a chiral nanostructure is to build a purely plasmonic system with a 3D chiral architecture. One example is lithographically made two-layer structure with plasmonic oligomers [38] exhibiting strong directional CD. Another example is plasmonic helices realized in Ref. [21]. Chirality can also be created using 2D chiral structures and the configuration of normal incidence of light [41, 54, 71].

1.7 Chiral Semiconductor Quantum Dots with Exciton Resonances

1.7.1 *Short Theory Introduction*

It was previously mentioned that chirality and the corresponding CD come from several types of dissymmetry, the chiral core or distorted surface effect and induction from a chiral environment. The dissymmetry may induce CD signal on nanocluster and NP systems, though the underlying physical mechanisms of the interaction are different. In the case of semiconductor NPs, these mechanisms also exist.

First of all, an orbital coupling was proposed as a possible chirality-transfer mechanism. From quantum computations, it was suggested that the electronic states of a QD could be hybridized with the orbital wave functions of chiral molecules. Chirality of the molecules is transferred through such hybridization onto the QD. Although, similar to a plasmon–exciton interaction, the exciton–exciton interaction can also induce a new CD signal onto QDs in a QD-chiral molecular system [31, 33]. Excitons in a QD can be coupled to a molecular exciton through the Coulomb interactions. The CD strength is given as [33]:

$$CD_{\mathrm{QD}}(\omega) \sim \frac{\mathrm{Im}\left[\left(\hat{F}\cdot\vec{\mu}_{12}\right)\cdot\vec{m}_{12}\right]}{R^3}\frac{\omega-\omega_0}{(\omega-\omega_0)^2+\Gamma^2}\cdot\mathrm{Im}\left[\alpha_{\mathrm{QD}}(\omega)\right], \tag{1.48}$$

where ω and ω_0 are frequencies of incident photon and molecular transition, respectively. $\hat{F}$ is a dipole orientation matrix. $\alpha_{\mathrm{QD}}(\omega)$ is the polarizability of QD. As $\mathrm{Im}\alpha_{\mathrm{QD}}(\omega)$ is proportional to the

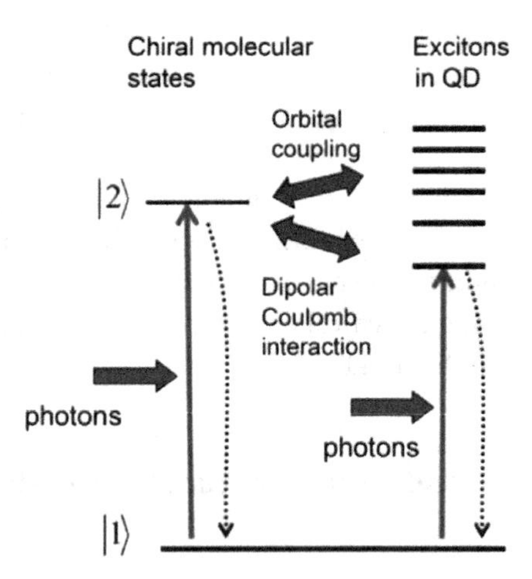

Figure 1.23 Interactions and transitions in the chiral molecule-QD system. The solid vertical arrows represent light-induced transitions, and the horizontal blue arrows depict the orbital and Coulomb couplings. The dotted vertical arrows show relaxation processes. Reprinted with permission from Ref. [33]. Copyright © 2011, Royal Society of Chemistry.

absorption cross-section of QD, it is expected that the CD signal will have a peak in the exciton band of QD. In the molecular band, the CD is given as:

$$CD_{\text{mol}}(\omega) \sim \text{Im}\left[\left(\hat{P}\cdot\vec{\mu}_{12}\right)\cdot\vec{m}_{12}\right]. \qquad (1.49)$$

The QD, on the contrary, will modify the CD signal in the molecular band through its high refractive index effect on $\hat{P}$. Figure 1.23 illustrates both mechanisms of interaction mentioned above.

1.7.2 *Experiments with Chiral Quantum Dots*

Chirality can be induced on excitons in semiconductor QDs using bio-conjugation with chiral ligands. Figure 1.24 shows the absorption spectra of such QDs involving enantiomers (D-Pen and L-Pen). Methods of preparation of such chiral conjugates can be found in Refs. [29] and [50]. CD studies of the particles gave particularly striking results [50]. D- and L-penicillamine stabilized particles produced corresponding mirror image CD scans (Fig. 1.25), while the particles prepared with a racemic mixture showed only a weak

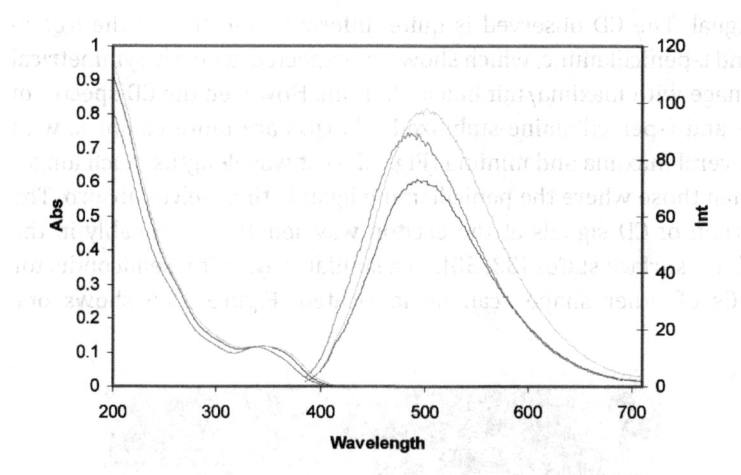

Figure 1.24 Optical spectra (UV-vis absorption—left and emission—right) of CdS NCs stabilized with D-Pen (Blue), L-Pen (Green), and *Rac*-Pen (Red). Excitation wavelength for all emission spectra is 365 nm [50]. Reprinted with permission from Ref. [50]. Copyright © 2010, Royal Society of Chemistry.

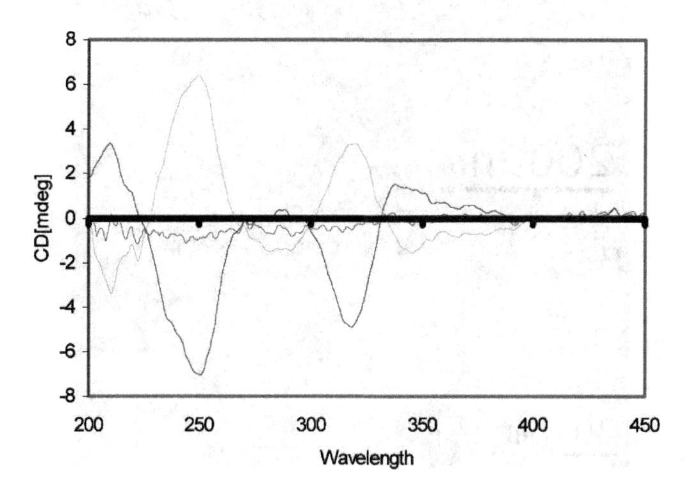

Figure 1.25 CD scans of the D-Pen (Blue), L-Pen (Green), and *Rac*-Pen (Red) stabilized CdS QDs. Reprinted with permission from Ref. [50]. Copyright © 2010, Royal Society of Chemistry.

signal. The CD observed is quite different from that of the free D- and L-penicillamine, which shows, as expected, a nearly symmetrical image with maxima/minima at 234 nm. However, the CD spectra of D- and L-penicillamine-stabilized CdS QDs are more complex, with several maxima and minima (Fig. 1.25) at wavelengths much longer than those where the penicillamine ligands themselves absorb. The origin of CD signals at the exciton wavelengths is probably in the chiral surface states [33, 50]. In a similar way, chiral semiconductor NCs of other shapes can be fabricated. Figure 1.26 shows one

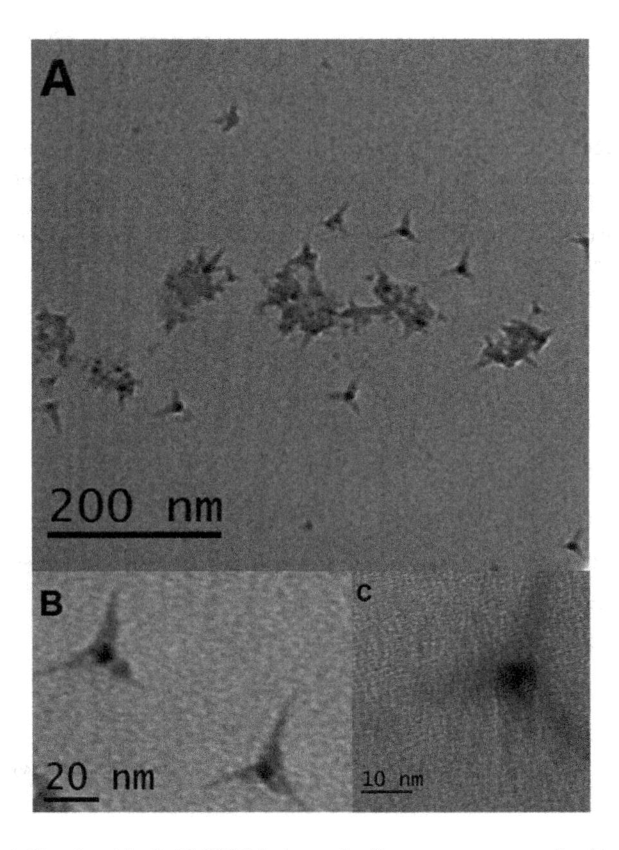

Figure 1.26 A wide-field TEM image of D-Pen nano-tetrapods showing a number of tetrapods (A), a closer look of two tetrapods (B), and a high-resolution image of a single tetrapod (C). Reprinted with permission from Ref. [29]. Copyright © 2010, Royal Society of Chemistry.

interesting example—chiral tetrapods. These NCs also exhibited CD signals at the excitonic transitions [29]. Detailed studies on a size-dependence of CD signals of chiral bio-conjugated QDs can be found in Ref. [5].

Here, it is important to mention another very recent study on the QDs with an intrinsically chiral crystal structure [4]. Although the various plasmonic or excitonic NPs interactions with chiral molecules lead to small CD induction effects, as mentioned above, a different concept that may lead to large chiroptical effects for single NPs would be to use certain types of inorganic materials that form chiral crystals, that is, their lattices correspond to chiral symmetry

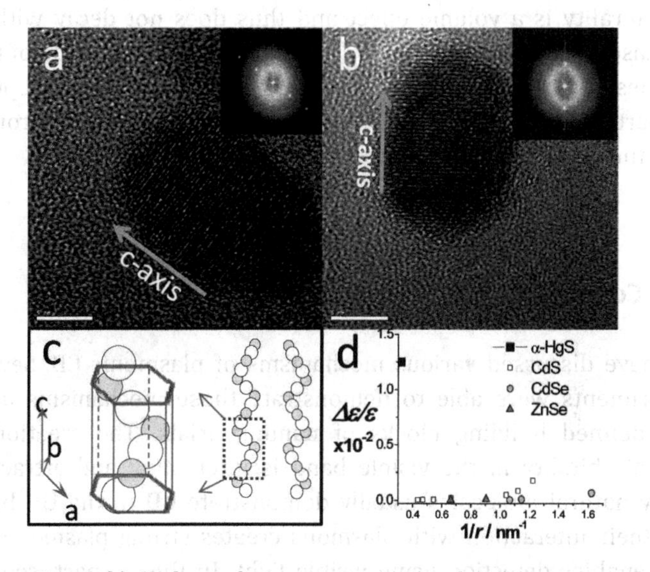

Figure 1.27 (a,b) High-resolution transmission electron microscopy images showing the single crystal nature of the α-HgS NCs. The long axis is always identified as the c-axis, which is the axis of the helical Hg-S spirals, as sketched in (c), which schematically shows the unit cell structure of these crystals. (d) A comparison of the dissymmetry factor at the first exciton transition for NCs of four different semiconductor NCs at various sizes as a function of surface/volume ratio ($1/r$), where r is the NC radius. In achiral semiconductors, the surface-sensitive CD induction strongly decays with r, contrary to the intrinsically chiral α-HgS, wherein relatively large particles exhibit large dissymmetry. Adapted with permission from Ref. [4]. Copyright © 2012 Wiley-VCH Verlag GmbH & Co. KGaA, Weinheim.

groups. The first demonstration of such NCs was recently done with α-HgS (cinnabar) NCs. In that work, the asymmetric synthesis of *L*- or *D*-cinnabar NCs was driven using a simple chiral capping molecule, penicillamine [4]. The cinnabar NCs exhibited giant CD at their absorption threshold in the visible range. Figure 1.27 displays TEM images of such NCs, a scheme of the crystal structure of these crystals, consisting of helical chains of mercury and sulfur atoms held together by van der Waals forces, and a comparison of the dissymmetry factor obtained for these NCs relative to other semiconductor NCs with achiral crystal structures. The dissymmetry of regular cubic or hexagonal NCs decays with size, as the surface-to-volume ratio is decreased, unlike the case of the α-HgS NCs, wherein the chirality is a volume effect and thus does not decay with an increase in size. This discovery [4] may lead to a multitude of new syntheses of size and shape-controlled chiral NCs with chiroptical properties at selected wavelengths, which would be much stronger than those attainable in molecular systems.

1.8 Conclusion and Perspectives

We have discussed various mechanisms of plasmonic CD. Several experiments were able to demonstrate those mechanisms using well-defined building blocks of nanomaterials. The creation of optical chirality in the visible band is interesting and attractive. Many natural molecules usually demonstrate CD in the UV band. But their interaction with plasmons creates strong plasmonic CD that enables detection using visible light. In this respect, sensing of biomolecules may become more convenient and inexpensive. On the contrary, plasmonic CD can also be a very good method to identify materials other than biomolecules, such as the ions of a heavy element Hg^{2+} [83]. It enables an easy but useful application with high sensitivity, by using Hg^{2+}-mediated aggregation of Au nanorods. Efforts are also being made in finding metamaterials with negative refractive index [55], with attempted application to build novel optical devices. In the future, induced plasmonic CD may have many more applications that are currently not foreseeable.

In experiments on plasmonic structures, we have observed enhanced CD in the molecular band [8]. In addition, it was suggested that an enhanced vibrational molecular CD in the infrared band is able to show more details of molecular structures [22]. Our general conclusion is that the amplification of the molecular CD requires special efforts in the design and fabrication. For example, anisotropy of a chiral nanostructure can be crucial for creation of strong plasmonic CD. In particular, we have shown theoretically that enhanced molecular CD appears for oriented molecules attached to a metal NP [31, 32]. Also, a metal NP dimer, which is a highly anisotropic nanostructure, is a good candidate to enhance molecular CD signals due to a strong field enhancement effect in a plasmonic hot spot [32]. A hot spot effect in such a dimer structure can more effectively enhance the molecular CD when the NPs approach each other closely [81]. However, a hot spot effect from random NP aggregates generally needs to be verified carefully, as complicated features may appear in the CD spectra. Stronger anisotropy can be generated by introducing 3D chiral NCs or assemblies. The consideration of the optical chirality factor in this chapter led us to the general conclusion that the near-field enhancement and twisting effects in light beams should be combined in sensing of biomolecules [62].

Despite several interesting recent developments and potential of chiral NCs for future applications, the fabrication of artificial chiral nanostructures is still in its infancy and remains challenging. In particular, more efforts are needed to achieve the structural yield and homochirality of NCs in an ensemble. Nowadays, the technologies available to construct chiral nanomaterials may be divided into two categories. In the first category, the nanostructures are assembled on an atomic level. This will include metal/semiconductor nanoclusters either with an intrinsic chiral structure or a surface chiral molecular adsorption. In the second category, on which this chapter is mostly focused, the fabrication incorporates nanomaterials that interact through electromagnetic interactions. The building blocks are better defined and the techniques of assembling the nanostructures are more generic and easier to adapt to new applications. Informative reviews are available in Refs. [28, 73], which introduced the self-assembly technologies to create NP structures through mechanisms of functionalized building blocks, driving external fields or a directionality by templates. In addition, focused ion beam lithography or electron beam lithography is available to build larger-scale structures for optical chirality studies, such as chiral or achiral metal particles arranged in a 2D array.

In conclusion, the field of fabrication of chiral nanoscale structures with NPs and molecules is rapidly developing and has attracted attention of researchers of diverse fields such as biology, chemistry, and physics. Chiral NP systems have demonstrated interesting optical properties, and these days, we are in the exciting situation to see new achievements being frequently reported in theory, novel designs, experimental realizations, and applications. A better understanding will surely help us to find better design strategies and to manipulate the optical properties further, although the ongoing development of nanotechnology is another important factor that drives the creation of new and interesting chiral nanostructures.

Acknowledgments

This work was supported by the Science Foundation Ireland, the Volkswagen Foundation (Germany), the NSF (USA), and the ISF (Israel).

References

1. Abdulrahman, N. A., Fan, Z., Tonooka, T., Kelly, S. M., Gadegaard, N., Hendry, E., Govorov, A. O. and Kadodwala, M. (2012). Induced chirality through electromagnetic coupling between chiral molecular layers and plasmonic nanostructures, *Nano Letters*, **12**, pp. 977–983.
2. Auguié, B., Alonso-Gómez, J. L., Guerrero-Martínez, A. s. and Liz-Marzán, L. M. (2011). Fingers crossed: optical activity of a chiral dimer of plasmonic nanorods, *The Journal of Physical Chemistry Letters*, **2**, pp. 846–851.
3. Barron, L. D. (2004). *Molecular Light Scattering and Optical Activity*, 2nd Ed., Cambridge University Press, Cambridge.

4. Ben-Moshe, A., Govorov, A. O. and Markovich, G. (2012). Enantioselective synthesis of intrinsically chiral mercury sulfide nanocrystals, *Angewandte Chemie International Edition*, **52**, pp. 1275–1279.
5. Ben Moshe, A., Szwarcman, D. and Markovich, G. (2011). Size dependence of chiroptical activity in colloidal quantum dots, *ACS Nano*, **5**, pp. 9034–9043.
6. Bohren, C. F. (1975). Scattering of electromagnetic waves by an optically active spherical shell, *The Journal of Chemical Physics*, **62**, p. 1566.
7. Bohren, C. F. and Huffman, D. R. (1983) *Absorption and Scattering of Light by Small Particles*, Wiley, New York.
8. Carmeli, I., Lieberman, I., Kraversky, L., Fan, Z., Govorov, A. O., Markovich, G. and Richter, S. (2010). Broad band enhancement of light absorption in photosystem I by metal nanoparticle antennas, *Nano Letters*, **10**, pp. 2069–2074.
9. Cathcart, N. and Kitaev, V. (2011). Monodisperse hexagonal silver nanoprisms: synthesis via thiolate-protected cluster precursors and chiral, ligand-imprinted self-assembly, *ACS Nano*, **5**, pp. 7411–7425.
10. Chen, C.-L., Zhang, P. and Rosi, N. L. (2008). A new peptide-based method for the design and synthesis of nanoparticle superstructures: construction of highly ordered gold nanoparticle double helices, *Journal of the American Chemical Society*, **130**, pp. 13555–13557.
11. Chen, W., Bian, A., Agarwal, A., Liu, L., Shen, H., Wang, L., Xu, C. and Kotov, N. A. (2009). Nanoparticle superstructures made by polymerase chain reaction: collective interactions of nanoparticles and a new principle for chiral materials, *Nano Letters*, **9**, pp. 2153–2159.
12. Davies, N. M., Teng, X. W. (2003). Importance of chirality in drug therapy and pharmacy practice: implications for psychiatry, *Advances in Pharmacy*, **1**, pp. 242–252.
13. Ding, B., Deng, Z., Yan, H., Cabrini, S., Zuckermann, R. N. and Bokor, J. (2010). Gold nanoparticle self-similar chain structure organized by DNA origami, *Journal of the American Chemical Society*, **132**, pp. 3248–3249.
14. Douglas, S. M., Dietz, H., Liedl, T., Hogberg, B., Graf, F. and Shih, W. M. (2009). Self-assembly of DNA into nanoscale three-dimensional shapes, *Nature*, **459**, pp. 414–418.
15. Douglas, S. M., Bachelet, I. and Church, G. M. (2012). A logic-gated nanorobot for targeted transport of molecular payloads, *Science*, **335**, pp. 831–834.
16. Draine, B. T. and Flatau, P. J. (2012). The Discrete Dipole Approximation for Scattering and Absorption of Light by Irregular Particles, The

DDA code was taken from the open source at http://www.astro.princeton.edu/~draine/DDSCAT.html (accessed April 2013).

17. Eremin, J. A., Orlov, N. V. and Rozenberg, V. I. (1994). Scattering by non-spherical particles, *Computer Physics Communications*, **79**, pp. 201–214.
18. Fan, Z. and Govorov, A. O. (2010). Plasmonic circular dichroism of chiral metal nanoparticle assemblies, *Nano Letters*, **10**, pp. 2580–2587.
19. Fan, Z. and Govorov, A. O. (2011). Helical metal nanoparticle assemblies with defects: plasmonic chirality and circular dichroism, *The Journal of Physical Chemistry C*, **115**, pp. 13254–13261.
20. Fan, Z. Y. and Govorov, A. O. (2012). Chiral nanocrystals: plasmonic spectra and circular dichroism, *Nano Letters*, **12**, pp. 3283–3289.
21. Gansel, J. K., Thiel, M., Rill, M. S., Decker, M., Bade, K., Saile, V., von Freymann, G., Linden, S. and Wegener, M. (2009). Gold helix photonic metamaterial as broadband circular polarizer, *Science*, **325**, pp. 1513–1515.
22. García-Etxarri, A. and Dionne, J. A. (2012). Surface-enhanced circular dichroism spectroscopy mediated by nonchiral nanoantennas, *Physical Review B*, **87**, p. 235409.
23. Gautier, C. and Bürgi, T. (2006). Chiral *N*-isobutyryl-cysteine protected gold nanoparticles:? preparation, size selection, and optical activity in the UV—vis and infrared, *Journal of the American Chemical Society*, **128**, pp. 11079–11087.
24. Gautier, C. and Bürgi, T. (2009). Chiral gold nanoparticles, *Chem Phys Chem*, **10**, pp. 483–492.
25. George, J. and Thomas, K. G. (2010). Surface plasmon coupled circular dichroism of au nanoparticles on peptide nanotubes, *Journal of the American Chemical Society*, **132**, pp. 2502–2503.
26. Gerard, V. A., Gun'ko, Y. K., Defrancq, E. and Govorov, A. O. (2011). Plasmon-induced CD response of oligonucleotide-conjugated metal nanoparticles, *Chemical Communications*, **47**, pp. 7383–7385.
27. Goldsmith, M. R., George, C. B., Zuber, G., Naaman, R., Waldeck, D. H., Wipf, P. and Beratan, D. N. (2006). The chiroptical signature of achiral metal clusters induced by dissymmetric adsorbates, *Physical Chemistry Chemical Physics: PCCP*, **8**, pp. 63–67.
28. Gong, J., Li, G. and Tang, Z. Self-assembly of noble metal nanocrystals: fabrication, optical property, and application, *Nano Today*, **7**, pp. 564–585.

29. Govan, J. E., Jan, E., Querejeta, A., Kotov, N. A. and Gun'ko, Y. K. (2010). Chiral luminescent CdS nano-tetrapods, *Chemical Communications*, **46**, pp. 6072–6074.
30. Govorov, A. O., Bryant, G. W., Zhang, W., Skeini, T., Lee, J., Kotov, N. A., Slocik, J. M. and Naik, R. R. (2006). Exciton—plasmon interaction and hybrid excitons in semiconductor—metal nanoparticle assemblies, *Nano Letters*, **6**, pp. 984–994.
31. Govorov, A. O., Fan, Z., Hernandez, P., Slocik, J. M. and Naik, R. R. (2010). Theory of circular dichroism of nanomaterials comprising chiral molecules and nanocrystals: plasmon enhancement, dipole interactions, and dielectric effects, *Nano Letters*, **10**, pp. 1374–1382.
32. Govorov, A. O. (2011). Plasmon-induced circular dichroism of a chiral molecule in the vicinity of metal nanocrystals. Application to various geometries, *The Journal of Physical Chemistry C*, **115**, pp. 7914–7923.
33. Govorov, A. O., Gun'ko, Y. K., Slocik, J. M., Gerard, V. A., Fan, Z. and Naik, R. R. (2011). Chiral nanoparticle assemblies: circular dichroism, plasmonic interactions, and exciton effects, *Journal of Materials Chemistry*, **21**, pp. 16806–16818.
34. Govorov, A. O. and Fan, Z. (2012). Theory of chiral plasmonic nanostructures comprising metal nanocrystals and chiral molecular media, *ChemPhysChem*, **13**, pp. 2551–2560.
35. Guerrero-Martínez, A., Auguié, B., Alonso-Gómez, J. L., Džolić, Z., Gómez-Graña, S., Žinić, M., Cid, M. M. and Liz-Marzán, L. M. (2011). Intense optical activity from three-dimensional chiral ordering of plasmonic nanoantennas, *Angewandte Chemie International Edition*, **50**, pp. 5499–5503.
36. Ha, J.-M., Solovyov, A. and Katz, A. (2008). Postsynthetic modification of gold nanoparticles with calix[4]arene enantiomers: origin of chiral surface plasmon resonance, *Langmuir*, **25**, pp. 153–158.
37. Hendry, E., Carpy, T., Johnston, J., Popland, M., Mikhaylovskiy, R. V., Lapthorn, A. J., Kelly, S. M., Barron, L. D., Gadegaard, N. and Kadodwala, M. (2010). Ultrasensitive detection and characterization of biomolecules using superchiral fields, *Nature Nanotechnology*, **5**, pp. 783–787.
38. Hentschel, M., Schäferling, M., Weiss, T., Liu, N. and Giessen, H. (2012). Three-dimensional chiral plasmonic oligomers, *Nano Letters*, **12**, pp. 2542–2547.
39. Johnson, P. B. and Christy, R. W. (1972). Optical constants of the noble metals, *Physical Review B*, **6**, pp. 4370–4379.
40. Kong, J. A. (1975). *Theory of Electromagnetic Waves*, Wiley, New York.

41. Kuwata-Gonokami, M., Saito, N., Ino, Y., Kauranen, M., Jefimovs, K., Vallius, T., Turunen, J. and Svirko, Y. (2005). Giant optical activity in quasi-two-dimensional planar nanostructures, *Physical Review Letters*, **95**, p. 227401.
42. Kuzyk, A., Schreiber, R., Fan, Z., Pardatscher, G., Roller, E.-M., Hogele, A., Simmel, F. C., Govorov, A. O. and Liedl, T. (2012). DNA-based self-assembly of chiral plasmonic nanostructures with tailored optical response, *Nature*, **483**, pp. 311–314.
43. Landau, L. D., Pitaevskii, L. P. and Lifshitz, E. M. *Electrodynamics of Continuous Media*, 2nd Ed., Butterworth–Heinemann, Oxford.
44. Li, L. W., Dan, Y., Leong, M. S. and Kong, J. A. (1999). Electromagnetic scattering by an inhomogeneous chiral sphere of varying permittivity: a discrete analysis using multilayered model, *Progress In Electromagnetics Research*, **23**, pp. 239–263.
45. Lieberman, I., Shemer, G., Fried, T., Kosower, E. M. and Markovich, G. (2008). Plasmon-resonance-enhanced absorption and circular dichroism, *Angewandte Chemie International Edition*, **47**, pp. 4855–4857.
46. Liu, S., Han, L., Duan, Y., Asahina, S., Terasaki, O., Cao, Y., Liu, B., Ma, L., Zhang, J. and Che, S. (2012). Synthesis of chiral TiO2 nanofibre with electron transition-based optical activity, *Nature Communications*, **3**, p. 1215.
47. Maoz, B. M., Chaikin, Y., Tesler, A. B., Elli, O. B., Fan, Z., Govorov, A. O., Vaskevich, A., Rubinstein, I. and Markovich, G. (2012). Amplification of chiroptical activity of chiral molecules by its induction on surface plasmons, *Nano Letters*, **13**, pp. 1203–1209.
48. Maoz, B. M., van der Weegen, R., Fan, Z., Govorov, A. O., Ellestad, G., Berova, N., Meijer, E. W. and Markovich, G. (2012). Plasmonic chiroptical response of silver nanoparticles interacting with chiral supramolecular assemblies, *Journal of the American Chemical Society*, **134**, pp. 17807–17813.
49. Mastroianni, A. J., Claridge, S. A. and Alivisatos, A. P. (2009). Pyramidal and chiral groupings of gold nanocrystals assembled using DNA scaffolds, *Journal of the American Chemical Society*, **131**, pp. 8455–8459.
50. Moloney, M. P., Gun'ko, Y. K. and Kelly, J. M. (2007). Chiral highly luminescent CdS quantum dots, *Chemical Communications*, pp. 3900–3902.
51. Moores, A. and Goettmann, F. (2006). The plasmon band in noble metal nanoparticles: an introduction to theory and applications, *New Journal of Chemistry*, **30**, pp. 1121–1132.

52. Nakashima, T., Kobayashi, Y. and Kawai, T. (2009). Optical activity and chiral memory of thiol-capped CdTe nanocrystals, *Journal of the American Chemical Society*, **131**, pp. 10342–10343.
53. Noguez, C. and Garzon, I. L. (2009). Optically active metal nanoparticles, *Chemical Society Reviews*, **38**, pp. 757–771.
54. Papakostas, A., Potts, A., Bagnall, D. M., Prosvirnin, S. L., Coles, H. J. and Zheludev, N. I. (2003). Optical manifestations of planar chirality, *Physical Review Letters*, **90**, p. 107404.
55. Pendry, J. B. (2004). A chiral route to negative refraction, *Science*, **306**, pp. 1353–1355.
56. Qi, H., Shopsowitz, K. E., Hamad, W. Y. and MacLachlan, M. J. (2011). Chiral nematic assemblies of silver nanoparticles in mesoporous silica thin films, *Journal of the American Chemical Society*, **133**, pp. 3728–3731.
57. Román-Velázquez, C. E., Noguez, C. and Garzón, I. L. (2003). Circular dichroism simulated spectra of chiral gold nanoclusters:? a dipole approximation, *The Journal of Physical Chemistry B*, **107**, pp. 12035–12038.
58. Rothemund, P. W. K. (2006). Folding DNA to create nanoscale shapes and patterns, *Nature*, **440**, pp. 297–302.
59. Sánchez-Castillo, A., Noguez, C. and Garzón, I. L. (2010). On the origin of the optical activity displayed by chiral-ligand-protected metallic nanoclusters, *Journal of the American Chemical Society*, **132**, pp. 1504–1505.
60. Schaaff, T. G., Knight, G., Shafigullin, M. N., Borkman, R. F. and Whetten, R. L. (1998). Isolation and selected properties of a 10.4 kDa gold:glutathione cluster compound, *The Journal of Physical Chemistry B*, **102**, pp. 10643–10646.
61. Schaaff, T. G. and Whetten, R. L. (2000). Giant gold—glutathione cluster compounds? Intense optical activity in metal-based transitions, *The Journal of Physical Chemistry B*, **104**, pp. 2630–2641.
62. Schäferling, M., Dregely, D., Hentschel, M. and Giessen, H. (2012). Tailoring enhanced optical chirality: design principles for chiral plasmonic nanostructures, *Physical Review X*, **2**, p. 031010.
63. Schüller, V. J., Heidegger, S., Sandholzer, N., Nickels, P. C., Suhartha, N. A., Endres, S., Bourquin, C. and Liedl, T. (2011). Cellular immunostimulation by CpG-sequence-coated DNA origami structures, *ACS Nano*, **5**, pp. 9696–9702.

64. Seeman, N. C. (2010). Nanomaterials based on DNA, *Annual Review of Biochemistry*, **79**, pp. 65–87.

65. Shemer, G., Krichevski, O., Markovich, G., Molotsky, T., Lubitz, I. and Kotlyar, A. B. (2006). Chirality of silver nanoparticles synthesized on DNA, *Journal of the American Chemical Society*, **128**, pp. 11006–11007.

66. Shen, X., Song, C., Wang, J., Shi, D., Wang, Z., Liu, N. and Ding, B. (2011). Rolling up gold nanoparticle-dressed DNA origami into three-dimensional plasmonic chiral nanostructures, *Journal of the American Chemical Society*, **134**, pp. 146–149.

67. Slocik, J. M., Govorov, A. O. and Naik, R. R. (2011). Plasmonic circular dichroism of peptide-functionalized gold nanoparticles, *Nano Letters*, **11**, pp. 701–705.

68. Steinhauer, C., Jungmann, R., Sobey, T. L., Simmel, F. C. and Tinnefeld, P. (2009). DNA Origami as a nanoscopic ruler for super-resolution microscopy, *Angewandte Chemie International Edition*, **48**, pp. 8870–8873.

69. Storhoff, J. J., Elghanian, R., Mucic, R. C., Mirkin, C. A. and Letsinger, R. L. (1998). One-pot colorimetric differentiation of polynucleotides with single base imperfections using gold nanoparticle probes, *Journal of the American Chemical Society*, **120**, pp. 1959–1964.

70. Tang, Y. and Cohen, A. E. (2010). Optical chirality and its interaction with matter, *Physical Review Letters*, **104**, p. 163901.

71. Valev, V. K., Baumberg, J. J., Sibilia, C. and Verbiest, T. (2013). Plasmonic nanostructures: chirality and chiroptical effects in plasmonic nanostructures: fundamentals, recent progress, and outlook (Adv. Mater. 18/2013), *Advanced Materials*, **25**, pp. 2509–2509.

72. Wang, R.-Y., Wang, H., Wu, X., Ji, Y., Wang, P., Qu, Y. and Chung, T.-S. (2011). Chiral assembly of gold nanorods with collective plasmonic circular dichroism response, *Soft Matter*, **7**, pp. 8370–8375.

73. Wang, Y., Xu, J., Wang, Y. and Chen, H. (2013). Emerging chirality in nanoscience, *Chemical Society Reviews*, **42**, pp. 2930–2962.

74. Weber, W. H. and Ford, G. W. (2004). Propagation of optical excitations by dipolar interactions in metal nanoparticle chains, *Physical Review B*, **70**, p. 125429.

75. Woody, W. W. (1996) *Circular Dichroism and the Conformational Analysis of Biomolecules*, Plenum, New York.

76. Xie, J., Duan, Y. and Che, S. (2012). Chirality of metal nanoparticles in chiral mesoporous silica, *Advanced Functional Materials*, **22**, pp. 3784–3792.

77. Xu, Y.-l. (1995). Electromagnetic scattering by an aggregate of spheres, *Appl. Opt.*, **34**, pp. 4573–4588.

78. Yan, J.-Y., Zhang, W., Duan, S., Zhao, X.-G. and Govorov, A. O. (2008). Optical properties of coupled metal-semiconductor and metal-molecule nanocrystal complexes: role of multipole effects, *Physical Review B*, **77**, p. 165301.

79. Yan, W., Xu, L., Xu, C., Ma, W., Kuang, H., Wang, L. and Kotov, N. A. (2012). Self-assembly of chiral nanoparticle pyramids with strong R/S optical activity, *Journal of the American Chemical Society*, **134**, pp. 15114–15121.

80. Yariv, A. (1975) *Quantum Electronics*, 2nd Ed., Quantum Electronics, New York.

81. Zhang, H. and Govorov, A. O. (2013). Giant circular dichroism of a molecule in a region of strong plasmon resonances between two neighboring gold nanocrystals, *Physical Review B*, **87**, p. 075410.

82. Zhen, Y.-R., Fung, K. H. and Chan, C. T. (2008). Collective plasmonic modes in two-dimensional periodic arrays of metal nanoparticles, *Physical Review B*, **78**, p. 035419.

83. Zhu, Y., Xu, L., Ma, W., Xu, Z., Kuang, H., Wang, L. and Xu, C. (2012). A one-step homogeneous plasmonic circular dichroism detection of aqueous mercury ions using nucleic acid functionalized gold nanorods, *Chemical Communications*, **48**, pp. 11889–11891.

Chapter 2

Reciprocity and Optical Chirality

Aurélien Drezet[a] and Cyriaque Genet[b]

[a]*Institut Néel, CNRS and Université Joseph Fourier (UPR 2940), 25 Rue des Martyrs, 38000 Grenoble, France*

[b]*ISIS, CNRS and Université de Strasbourg (UMR 7006), 8 Allée Gaspard Monge, 67000 Strasbourg, France*

drezet@grenoble.cnrs.fr, genet@unistra.fr

2.1 Introduction

Chirality (or handedness as the word stems from the greek $\chi\epsilon\acute{\iota}\rho$ meaning hand), refers to the lack or absence of mirror symmetry of many systems [1–3]. It is a fascinating property having important consequences in every areas of science. For example, it is connected to several fundamental problems such as the apparition of life, the origin of homochirality (that is of single handedness) of many biomolecules [4], and also to the asymmetry between left and right-handed fermions with respect to electroweak interaction [5]. Historically, I. Kant was one of the first eminent scholar to point out the philosophical significance of mirror operation. Already in 1783 in his celebrated "Prolegomena to any future metaphysics," he wrote

Singular and Chiral Nanoplasmonics
Edited by Svetlana V. Boriskina and Nikolay I. Zheludev

ISBN 978-981-4613-17-0 (Hardcover), 978-981-4613-18-7 (eBook)
www.panstanford.com

"Hence the difference between similar and equal things, which are yet not congruent (for instance, two symmetric helices), cannot be made intelligible by any concept, but only by the relation to the right and the left hands which immediately refers to intuition" [6].

W. Thomson (Lord Kelvin), who was one of most important figure in physics at the end of the XIXth century, defined chirality more precisely in the following way:

"I call any geometrical figure or group of points, chiral and say it has chirality, if its image in a plane mirror, ideally realized cannot be brought to coincide with itself" [7].

The interest of Kelvin for chirality is not surprising. He played himself a critical role in the foundation of electromagnetism and thermodynamics, and he was well aware of the work presented in 1896 by M. Faraday on what is now known as the Faraday effect, which is intimately related to optical activity and chirality [8]. In this context, it is interesting to observe that like Pasteur after him, Faraday also had fruitless attempts to establish some relations between electricity, chirality, and light; it was a letter from Thomson in 1845 that actually led Faraday to repeat his experiments with a magnetic field and to discover nonreciprocal gyrotropy (i.e., magnetic optical rotation).

Remarkably, this description of Kelvin provides an operational definition of chirality particularly suited to optics, as we illustrate below. In optics indeed, since the pioneer work of Arago [9] in 1811 and Biot in 1812 [10], chirality is associated with optical activity (natural gyrotropy), which is the rotation of the plane of polarization of light upon going through a 3D chiral medium such as a quartz crystal or an aqueous solution of sugar. The first mathematical description of optical activity has arisen from the work of Fresnel in 1825 [11] who interpreted phenomenologically the effect in terms of circular birefringence, that is, as a difference in optical index for left and right-handed circularly polarized light (written LCP and RCP, respectively) passing through the medium. However, the intimate relationship between optical activity and chirality became more evident after the work of Pasteur [12, 13] in 1848 concerning the change in sign of the optical rotatory power for enantiomorphic solution of left and right-handed chiral molecules of tartaric acid. In 1874, Le Bel and van't Hoff [14, 15] related rotatory power to the unsymmetrical arrangements of substituents at a saturated atom, thus identifying the very foundation for stereochemistry. Since then, optical activity, including circular birefringence and dichroism, the so-called Cotton effect [16], that is the difference in absorbtion for LCP and RCP, have become very powerful probes of structural chirality in a variety of media and environments.

With the recent advent of metamarials, which are artificially structured photonic media, a resurgence of interest concerning optical activity is observed. Current inspirations can be traced back to the pioneer work of Bose [17] who, as early as 1898, reported on the observation of rotatory power for electromagnetic microwaves propagating through a chiral artificial medium (actually left and right-handed twisted jute elements). In the context of metamaterials, Lindman in 1920 [18] (see also Tinoco and Freeman in 1957 [19]) reported a similar rotatory dispersion effect through a system of copper helices in the giga-Hertz (GHz) range. Very recently, and largely due to progress in micro and nanofabrication techniques, researchers have been able to tailor compact and organized optically active metamerials in the GHz and visible ranges. It was for instance shown that planar chiral structures made of *gammadions*, that is, equilateral crosses made of four bented arms, in metal or dielectric, can generate optical activity [20–36] with giant gyrotropic factors [24, 28–30]. Important applications in opto-electronics and also for refractive devices with a negative optical index for RCP and LCP have been suggested in this context [20, 35, 37].

These studies have raised an important debate on the genuine meaning of planar chirality [21–23]. Indeed, intuitively, a two-dimensional (2D) chiral structure, which is by definition a system that cannot be put into congruence with itself until left from the plane, is not expected to display any chiral optical characteristics due to the fact that simply turning the object around leads to the opposite handedness. More precisely, it was shown that optical activity being a reciprocal property (i.e., obeying to the principle of reciprocity of

Lorentz, see below), it necessarily implies that reversing the light path through the medium must recreate back the initial polarization state. However, as the sense of twist of a 2D chiral structure changes when looking from the second side, this polarization reversal is impossible. It would otherwise lead to the paradoxical conclusion that a left and right-handed structure generates the same optical activity in contradiction with the definition, finally meaning that optical activity must vanish in strictly planar chiral systems. This behavior strongly contrasts with what is actually observed for 3D chiral objects having a helicoidal structure (like a quartz crystal [9, 12, 13], a twisted jute element [17], or a metal helix [18, 19]), in full agreement with the principle of reciprocity of Lorentz [1, 38–41], as the sense of twist of an helix is clearly conserved when we reverse back the illumination direction. The experimental observation of optical activity in gammadion arrays forces one to conclude that such systems must present a form of hidden 3D chirality, which turns fully responsible for the presence of optical rotation that rules over the dominant 2D geometrical chiral character possessed by the system.

Things could have stopped here, but the understanding of 3D chirality was recently challenged in a pioneering study in which it was shown that chirality has a distinct signature from optical activity when electromagnetic waves interact with a genuine 2D chiral structure and that the handedness can be recognized [42]. Although the experimental demonstration was achieved in the GHz (mm) range for extended 2D structures (the so-called fish-scale structures[42]), the question remained whether this could be achieved in the optical range, as the optical properties of materials are not simply scalable when downsizing to the nanometer level. Theoretical suggestions were provided to overcome this difficulty by using localized plasmon modes excited at the level of the nanostructures [43]. Surface plasmons (SPs) [44–46] are indeed hybrid photon/electron excitations, which are naturally confined in the vicinity of a metal structure. As evanescent waves, SPs are very sensitive to local variations of the metal and dielectric environments [44]. This property was thus used to tune some 2D chiral metal structure to optical waves. Two series of experiences made in the near infrared [47] with fishscale structure and in the visible with Archimidian spirals [48, 49] confirmed the peculiarity of genuine 2D chirality at the nanoscale.

In this chapter, we will review some fundamental optical properties associated, in full generality, with chiral systems. An algebraic approach will allow us to reveal the underlying connexions between the concepts of chirality and reciprocity in a simple way from which global classes of chiral elements will be drawn. These classes will be described with the framework of Jones matrices, enabling a clear discussion on their respective optical properties. Finally, a few examples taken from our recent work will be discussed as illustrative examples of the relevant of planar chirality in the context of nanophotonics.

2.2 The Reciprocity Theorem and the Principle of Path Reversal

2.2.1 *The Lorentz Reciprocity Relation*

In order to understand the physical meaning and implications of the different chiral matrices we will discuss, we need to introduce the principle of reciprocity of Lorentz [1, 38, 39]. First, we remind that, under the validity conditions of the paraxial approximation, the properties of light going through an optical medium are fully characterized by the knowledge of the 2×2 Jones matrix

$$J := \begin{pmatrix} J_{\mathrm{xx}} & J_{\mathrm{xy}} \\ J_{\mathrm{yx}} & J_{\mathrm{yy}} \end{pmatrix}, \tag{2.1}$$

which ties the incident electric field $\mathbf{E}^{(\mathrm{in})} = E_x^{(\mathrm{in})}\hat{\mathbf{x}} + E_y^{(\mathrm{in})}\hat{\mathbf{y}}$ to the transmitted electric field $\mathbf{E}^{(\mathrm{out})} = E_x^{(\mathrm{out})}\hat{\mathbf{x}} + E_y^{(\mathrm{out})}\hat{\mathbf{y}}$ (defined in the cartesian basis x, y of the transverse plane). This corresponds to the transformation $\mathbf{E}^{(\mathrm{out})} = \hat{J}\,\mathbf{E}^{(\mathrm{in})}$ where $\hat{J}$ is the operator associated with Eq. (2.1).

From the point of view of Jones matrices it is then possible to give a simple formulation of the reciprocity principle. Consider then a localized system illuminated by a plane wave $\mathbf{E}_{\rightarrow}^{(\mathrm{in})}$ propagating along the $+z$ direction (see Fig. 2.1). The arrow indicates the direction of propagation. The transmitted field (in the paraxial approximation) is given by $\mathbf{E}_{\rightarrow}^{(\mathrm{out})} = \hat{J}\,\mathbf{E}_{\rightarrow}^{(\mathrm{in})}$. Same, we also define an incoming

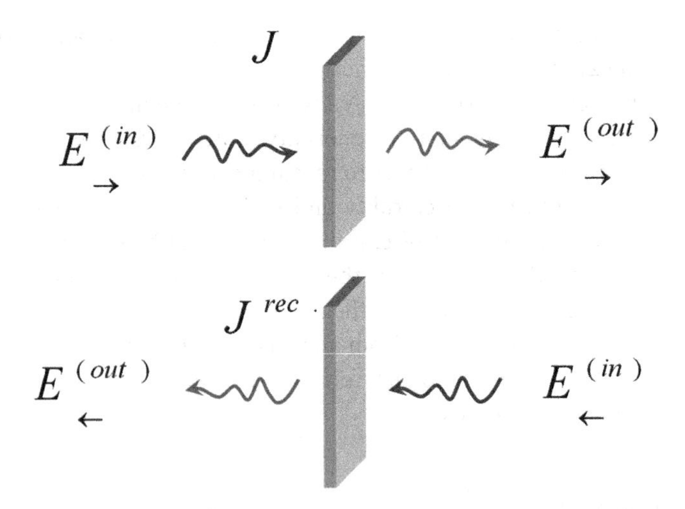

Figure 2.1 The two different reciprocal histories for a light beam through a medium. The initial path going from left to right is represented by the Jones matrix, J while the reversed path that goes from right to left corresponds to the matrix J^{rec}. (See Eqs. (2.5) and (2.10)).

plane wave propagating along the $-z$ direction $\mathbf{E}^{(\text{in})}_{\leftarrow}$ impinging from the other side of the system and which after transmission gives the output state $\mathbf{E}^{(\text{out})}_{\leftarrow} = \hat{J}^{\text{rec}}\mathbf{E}^{(\text{in})}_{\leftarrow}$. Here, by definition, $\hat{J}^{\text{rec}}$ is the reciprocal Jones operator associated with $\hat{J}$. Using Maxwell equations, one can easily show that $\hat{J}^{\text{rec}} = \hat{J}^{\text{T}}$, where T denotes the transposition in the cartesian basis $\hat{\mathbf{x}}$, $\hat{\mathbf{y}}$. The proof reads as follow: first, from Maxwell equations, one deduces [1] the reciprocity theorem of Lorentz, which states that if in a passive and linear environment, we consider two space points A and B, then the vector field $\mathbf{E}(B)$ in B produced by a (harmonic) point-like dipole source $\mathbf{P}(A)$ located in A is linked to the field $\mathbf{E}'(A)$ in A produced by a second point-like dipole $\mathbf{P}'(B)$ located in B through the formula

$$\mathbf{P}(A) \cdot \mathbf{E}'(A) = \mathbf{P}'(B) \cdot \mathbf{E}(B) \tag{2.2}$$

(the time harmonic dependency $e^{-i\omega t}$ has been dropped everywhere).

Now, the electric field produced in M by a point-like dipole located in M' is written $\mathbf{E}(M) = \underline{\mathbf{G}}(M, M') \cdot \mathbf{P}(M')$ where $\underline{\mathbf{G}}(M, M')$ denotes the dyadic Green's function for this environment [46].

Equation (2.2) thus reads

$$\mathbf{P}(A) \cdot (\underline{\mathbf{G}}(A, B) \cdot \mathbf{P}'(B)) = \mathbf{P}'(B) \cdot (\underline{\mathbf{G}}(B, A) \cdot \mathbf{P}(A)) \quad (2.3)$$

or equivalently in tensorial notation $\sum \sum_{i,j} (G_{ij}(A, B) - G_{ji}(B, A)) P_i(A) P_j'(B) = 0$ $(i, j = 1, 2, 3)$. This relation is valid for every point dipoles in A and B and implies consequently

$$G_{ij}(A, B) = G_{ji}(B, A). \quad (2.4)$$

Actually, this relation constitutes a Maxwellian formulation of the principle of light path reversal used in optical geometry. In the next step of the proof, we consider A and B located in $z = \pm\infty$; the fields can be then considered asymptotically as plane waves and in the paraxial approximation $G_{ij}(-\infty, +\infty)$ identifies with the Jones matrix J_{ij}. We immediately see that the matrix $G_{ij}(+\infty, -\infty)$ identifies with the reciprocal matrix J_{ij}^{rec}. In other words, from the point of view of Jones formalism, the principle of reciprocity states

$$J^{\text{rec}} := J^{\text{T}} = \begin{pmatrix} J_{\text{xx}} & J_{\text{yx}} \\ J_{\text{xy}} & J_{\text{yy}} \end{pmatrix}. \quad (2.5)$$

In this context, it is relevant to point out the similarity between the reasoning given here for establishing the reciprocity theorem and the one used in textbooks and articles [1, 2, 39] for establishing the symmetry of the permittivity tensor $\epsilon_{i,j}$ $(i, j = 1, 2, 3)$ in solids. In particular, by taking into account spatial non-locality, it is possible to obtain a version of the reciprocity theorem that reads

$$\epsilon_{i,j}(\omega, -\mathbf{k}) = \epsilon_{j,i}(\omega, \mathbf{k}) \quad (2.6)$$

where $\mathbf{k}$ is the wavevector of the monochromatic plane wave. The analogy with Eq. (2.5) is complete if we choose the wave vector along the z axis and if i, j corresponds to either x or y. Because of these similarities, many reasoning done for the Jones matrix through this chapter could be easily restated for the electric permittivity ϵ or magnetic permeability μ tensors.

2.2.2 *Rotation of the Optical Medium and Reciprocity: Conserving the Handedness of the Reference Frame*

Using the previous formalism, the reciprocity principle gives us a univocal way to calculate the transmitted light beam propagating in the $-z$ direction through a structure if we know the transmission Jones matrix for propagation in the $+z$ direction. However, we must remark that this formulation is not always the most convenient, as we compare the Jones matrix from a situation in which the triplet of unit vectors built by $\hat{\mathbf{x}}$, $\hat{\mathbf{y}}$, and the wavevector $\mathbf{k}$ of the light wave (along z in the paraxial regime) constitute a right-handed trihedra to a situation in which the same trihedra of vectors is left handed (as the sign of $\mathbf{k}$ is opposite in the two situations). This in particular means that for an observer watching from one side B of the medium, a light beam coming from the other side A, the definition for LCP and RCP light as $\hat{\mathbf{L}}, \hat{\mathbf{R}} = (\hat{\mathbf{x}} \pm i\hat{\mathbf{y}})/\sqrt{2}$ is different from the one obtained by a reciprocal observer watching from the side B, a light incident from the A side, that is, $\hat{\mathbf{R}}, \hat{\mathbf{L}} = (\hat{\mathbf{x}} \pm i\hat{\mathbf{y}})/\sqrt{2}$. In order to remove this ambiguity, the principle of reciprocity can be reformulated by considering a reference frame transformation, which is a global flip of the optical medium. More precisely, directly comparing J^{T} with J necessitates a rotation R_x of the plane $y - z$ by an angle π around x. Mathematically, this three-dimensional transformation reads

$$R_x = \begin{pmatrix} 1 & 0 & 0 \\ 0 & -1 & 0 \\ 0 & 0 & -1 \end{pmatrix}. \tag{2.7}$$

The handedness of the three-axis coordinate system is kept unchanged after the application of the reference frame transformation R_x. It also implies that the full structure of Maxwell equation is also conserved, that is, the optical effect is rigorously equivalent to a π-rotation of the system around x. In particular, the wavevector

$$\mathbf{k} = \begin{pmatrix} 0 \\ 0 \\ -|k| \end{pmatrix} \tag{2.8}$$

of the light going through the system transforms through R_x into

$$\mathbf{k}' = \begin{pmatrix} 0 \\ 0 \\ +|k| \end{pmatrix}. \tag{2.9}$$

From the point of view of the 2×2 Jones matrix R_x simply reduces to $\Pi_x = \begin{pmatrix} 1 & 0 \\ 0 & -1 \end{pmatrix}$, that is, to a planar mirror symmetry with reflection axis parallel to x. Through R_x, the field vectors transform as $\mathbf{E}_{\Pi_x} = \hat{\Pi}_x \mathbf{E}$, so that the Jones matrix becomes $\Pi_x \cdot J^{\mathrm{T}} \cdot {\Pi_x}^{-1}$.

Fundamentally, it is interesting to note that a symmetry operation, which is defined geometrically (and independently from any set of physical laws), manifests itself specifically according to a particular physical environment or context. From the point of view of Maxwell's equations, this symmetry operation is indeed implemented at the level of the susceptibility of the medium interaction with light, that is, at the level of the Jones matrices.

Actually, the reciprocal transformation defined above constitutes a rigorous and operational optical definition of the medium flipping

$$J^{\,\mathrm{flip}} := \Pi_x \cdot J^{\,\mathrm{rec}} \cdot {\Pi_x}^{-1} \text{ with } J^{\,\mathrm{rec}} = J^{\mathrm{T}}. \tag{2.10}$$

It will be also convenient in the following to express the electric field in the left (L) and right (R) circularly polarized light basis defined by $\hat{\mathbf{L}}, \hat{\mathbf{R}} = (\hat{\mathbf{x}} \pm i\hat{\mathbf{y}})/\sqrt{2}$. We write $\overline{J}$ the Jones matrix in such basis, and we have

$$\overline{J} = \begin{pmatrix} J_{ll} & J_{lr} \\ J_{rl} & J_{rr} \end{pmatrix} = U\,J\,U^{-1}, \tag{2.11}$$

where $U = \frac{1}{\sqrt{2}} \begin{pmatrix} 1 & -i \\ 1 & i \end{pmatrix}$ defines the unitary matrix associated with this vector basis transformation.

With these definitions, Eq. (2.10) reads in the RCP and LCP basis

$$\overline{J}^{\mathrm{flip}} = \overline{J}^{\mathrm{T}} = \begin{pmatrix} J_{ll} & J_{rl} \\ J_{lr} & J_{rr} \end{pmatrix} \tag{2.12}$$

(remark that $\overline{J^{\mathrm{T}}} = \begin{pmatrix} J_{rr} & J_{lr} \\ J_{rl} & J_{ll} \end{pmatrix} \neq \overline{J}^{\mathrm{T}}$ as the transposition T and the transformation $\hat{U}$ are not commutative operations).

To summarize, we showed that $J^{\,\mathrm{flip}}$ and $\overline{J}^{\mathrm{flip}}$ define the genuine representation of reciprocity and path reversal in a coordinate system having the same handedness as the original one (i.e., as seen from the other side of the object). This corresponds to the axes transformation $x' = x$, $y' = -y$, and $z' = -z$ and it implies an exchange in the role of LCP and RCP.

2.2.3 *Time-Reversal versus Reciprocity*

We point out that the reciprocity relations (2.10) and (2.12) should not be confused with the time-reversal transform. Time-reversal is a fundamental symmetry that dictates the invariance of physical laws between exchange of past and future. In classical mechanics, this symmetry corresponds to a system described by its position and momentum $(\mathbf{q}, \mathbf{p})$, which equations of motion are invariant through the transformation $(\mathbf{q}, \mathbf{p}, t) \to (\mathbf{q}, -\mathbf{p}, -t)$. Although an isolated mechanical system is time-reversal, a real system is obviously always coupled with its environment, resulting in friction that immediately breaks the time-reversal symmetry.

In the presence of electromagnetic fields $\mathbf{E}(\mathbf{x}, t)$, $\mathbf{B}(\mathbf{x}, t)$, this time-reversal invariance is preserved if in addition to the variables $(\mathbf{q}, \mathbf{p})$, one also transforms the field into $\mathbf{E}'(\mathbf{x}, t) = \mathbf{E}(\mathbf{x}, -t)$, $\mathbf{B}'(\mathbf{x}, t) = -\mathbf{B}'(\mathbf{x}, -t)$. In order that Maxwell's equation be automatically satisfied by the new solutions $\mathbf{E}'$, $\mathbf{B}'$, the electric current and charge distributions are changed accordingly into $\mathbf{J}'(\mathbf{x}, t) = -\mathbf{J}(\mathbf{x}, -t)$, $\rho'(\mathbf{x}, t) = \rho(\mathbf{x}, -t)$. In the monochromatic regime where fields, electric currents, and charges are conveniently described by their time-Fourier transforms at the positive pulsation ω, the time-reversal operation reads:

$$\begin{aligned} \mathbf{E}'_\omega(\mathbf{x}) = \mathbf{E}^*_\omega(\mathbf{x}),\ \mathbf{B}'_\omega(\mathbf{x}) = -\mathbf{B}^*_\omega(\mathbf{x}) \\ \mathbf{J}'_\omega(\mathbf{x}) = -\mathbf{J}^*_\omega(\mathbf{x}),\ \rho'_\omega(\mathbf{x}) = \rho^*_\omega(\mathbf{x}). \end{aligned} \tag{2.13}$$

The time dependence is restored after integration over the pulsation Spectrum, for example, $\mathbf{E}(\mathbf{x}, t) = \int_0^{+\infty} d\omega \mathbf{E}_\omega(\mathbf{x}) e^{-i\omega t} + \int_0^{+\infty} d\omega \mathbf{E}^*_\omega(\mathbf{x}) e^{+i\omega t}$ etc.

For optical situations in which a modal expansion into plane waves of vector $\mathbf{k} = k_x\hat{\mathbf{x}} + k_y\hat{\mathbf{y}}$, $k_z = \sqrt{(\frac{\omega}{c})^2\epsilon_\omega - \mathbf{k}^2}$ is considered (ϵ_ω being the complex-valued dielectric permittivity of the medium), time-reversal implies new modal components such as

$$\epsilon'_\omega = \epsilon^*_\omega,\ k'_z = k^*_z,\ \mathbf{E}'_{\omega,\pm}(\mathbf{k}) = \mathbf{E}^*_{\omega,\mp}(-\mathbf{k}),\ \mathbf{B}'_{\omega,\pm}(\mathbf{k}) = -\mathbf{B}^*_{\omega,\mp}(-\mathbf{k}), \tag{2.14}$$

where $\mathbf{E}_\omega(\mathbf{x}) = \int d^2\mathbf{k} e^{i\mathbf{kx}}[\mathbf{E}_{\omega,+}(\mathbf{k})e^{ik_z z} + \mathbf{E}_{\omega,-}(\mathbf{k})e^{-ik_z z}]$ etc. In particular, if the medium is lossless, that is, $\text{Imag}[\epsilon_\omega] = 0$, the time-reversal operation dictates (i.e., for $|\mathbf{k}| \leq \omega\sqrt{\epsilon_\omega}/c$) a change in the sign of the wave vectors in the propagative sector corresponding to a reversal of propagation direction for every plane-wave of the modal expansion. Going back to the Jones matrix formalism, we can transform the relation $\mathbf{E}^{(\text{out})}_{\rightarrow} = \hat{J}\,\mathbf{E}^{(\text{in})}_{\rightarrow}$ into $\mathbf{E}^{(\text{in}),*}_{\rightarrow} = \hat{J}^{\,-1,*}\mathbf{E}^{(\text{out}),*}_{\rightarrow}$.

Now, from the previous discussion concerning time reversibility, the complex conjugated input field $\mathbf{E}^{(\text{in}),*}_{\rightarrow}$ at $z = -\infty$ corresponds to the time-reversed output $\mathbf{E}'^{(\text{out})}_{\leftarrow}$ computed at $z = -\infty$, whereas the complex conjugated output field $\mathbf{E}^{(\text{out}),*}_{\rightarrow}$ at $z = +\infty$ corresponds to the time-reversed output $\mathbf{E}'^{(\text{in})}_{\leftarrow}$ computed at $z = +\infty$. The Jones matrix associated with time-reversal is therefore

$$J^{\,\text{inv}} := J^{\,-1,*} = \frac{1}{J^*_{\text{xx}}J^*_{\text{yy}} - J^*_{\text{xy}}J^*_{\text{yx}}}\begin{pmatrix} J^*_{\text{yy}} & -J^*_{\text{xy}} \\ -J^*_{\text{yx}} & J^*_{\text{xx}} \end{pmatrix}. \tag{2.15}$$

As it is clear from its definition, $J^{\,\text{inv}}$ is in general different from $J^{\,\text{rec}}$, exemplifying the importance of losses and dissipation in the relation between time reversibility and reciprocity in optics. The two operators are indeed identical if, and only if, J is unitary, that is, $J^{\,-1} = J^{\,\dagger}$, meaning that an optical system through which energy is conserved and which is simultaneously reciprocal will be the only optical system to be time-reversal invariant. This reveals the non-equivalence between time reversibility and reciprocity. The latter is more general: reciprocity can hold for systems in which irreversible processes take place, as a fundamental consequence of Onsager's principle of microscopic reversibility [50]. In the context of planar chirality, this subtle link plays a fundamental role, as it will be discussed in Section 2.3.3.

2.3 Optical Chirality

2.3.1 *Chiral Jones Matrix*

Following the operational definition of Lord Kelvin, the study of chirality demands to characterize the optical behavior of the considered system through a planar mirror symmetry Π_ϑ. By definition, an in-plane symmetry axis making an angle $\vartheta/2$ with respect to the x-direction is associated with transformation matrices

$$\Pi_\vartheta = \begin{pmatrix} \cos\vartheta & \sin\vartheta \\ \sin\vartheta & -\cos\vartheta \end{pmatrix}, \ \overline{\Pi_\vartheta} = \begin{pmatrix} 0 & e^{-i\vartheta} \\ e^{+i\vartheta} & 0 \end{pmatrix}, \tag{2.16}$$

respectively, written in cartesian and circular bases with $\Pi_{\vartheta=0} = \Pi_x$. Through such Π_ϑ symmetry operation, the Jones matrix transforms as $\hat{J}_\Pi = \hat{\Pi}_\vartheta \hat{J}\, \hat{\Pi}_\vartheta^{-1}$ in the cartesian basis and as

$$\overline{J}_\Pi = \overline{\Pi_\vartheta} \cdot \overline{J} \cdot \overline{\Pi_\vartheta}^{-1} = \begin{pmatrix} J_{rr} & J_{rl}e^{-i2\vartheta} \\ J_{lr}e^{+i2\vartheta} & J_{ll} \end{pmatrix}. \tag{2.17}$$

in the circular basis.

With Kelvin's definition, a system will be optically non-chiral if, and only if, it is invariant under Π_ϑ, meaning that $J = J_\Pi$ or equivalently that the operators, respectively, associated with the Jones matrix and the mirror-symmetry matrix commute as $[\hat{J}, \hat{\Pi}_\vartheta] = \hat{J}\,\hat{\Pi}_\vartheta - \hat{\Pi}_\vartheta \hat{J} = 0$. The invariance condition $\overline{J} = \overline{J}_{\text{mirror}}$ enforces two constraints on the Jones matrix coefficients, namely that

$$J_{ll} = J_{rr} \text{ and } J_{rl} = J_{lr}e^{2i\vartheta}. \tag{2.18}$$

This implies that the Jones matrix associated with a non-chiral optical system has the following general form

$$J_{\text{mirror}} = \begin{pmatrix} A + B\cos\vartheta & B\sin\vartheta \\ B\sin\vartheta & A - B\cos\vartheta \end{pmatrix} \tag{2.19}$$

By contrapositive of conditions (2.18), we see that

Theorem:

Optical chirality is possible if, and only if,

$J_{ll} \neq J_{rr}$ OR $|J_{lr}| \neq |J_{rl}|$,

OR being the logical disjunction.

This constitutes a theorem equivalent to Kelvin's statement that an optically chiral system has no mirror symmetry, with $J \neq J_\Pi$ or, equivalently, with non-commuting operators, respectively, associated with the Jones matrix and the mirror-symmetry matrix as

$$[\hat{J}, \hat{\Pi}_\vartheta] = \hat{J}\,\hat{\Pi}_\vartheta - \hat{\Pi}_\vartheta \hat{J} \neq 0 \text{ for any } \vartheta. \tag{2.20}$$

Such a Jones matrix can be written in the following form:

$$\overline{J} = \begin{pmatrix} (J_{ll} + J_{rr})/2 & J_{lr} \\ J_{rl} & (J_{ll} + J_{rr})/2 \end{pmatrix} + \frac{(J_{ll} - J_{rr})}{2} \begin{pmatrix} 1 & 0 \\ 0 & -1 \end{pmatrix}. \tag{2.21}$$

In addition for an optically chiral system, the application of a mirror symmetry Π_ϑ provides new chiral optical structures called enantiomers, which are characterized by their Jones matrices $\overline{J}^{\text{enant}}(\vartheta)$. By definition, we have

$$\overline{J}^{\text{enant}}(\vartheta) = \overline{\Pi_\vartheta} \cdot \overline{J} \cdot \overline{\Pi_\vartheta}^{-1} \neq \overline{J}. \tag{2.22}$$

In general, the lack of rotational invariance of $\overline{J}$ implies that these enantiomorphic matrices depend specifically on the mirror reflection $\overline{\Pi_\vartheta}$ chosen in the definition given by Eq. (2.21). Thus, there are actually infinite numbers of such enantiomers.

Three important classes of chiral systems can be derived from the truth table associated with the theorem:

i) A first class satisfying $J_{ll} \neq J_{rr}$ but with $|J_{lr}| = |J_{rl}|$. For reasons presented below, this class will be named the "optical activity class" or $E_{\text{o.a.}}$. This class is, until recently, the class essentially discussed in the literature.
ii) A second class corresponding to $J_{ll} = J_{rr}$ but satisfying the constraint $|J_{lr}| \neq |J_{rl}|$. This class will be named in the following the "(genuine) planar chirality class" $E_{2\text{D}}$, for reasons also to be given further down.
iii) A third class associated with $J_{ll} \neq J_{rr}$ and $|J_{lr}| \neq |J_{rl}|$. This is the most general class of optically chiral system coined as the "optical chirality class" and which will be described below in details.

Our point is that these chiral classes correspond to specific spatial relations of chiral systems with respect to 3D space. Such relations are fundamental to the characterization of chiral objects, which depends on the shape of the objects and the dimension of the space within which the objects are probed [51]. In optics, these relations can be unveiled through reciprocity: as light propagates in 3D space, the effect of optical path reversal through any chiral object will reveal its relation with surrounding space. Analyzing the behavior of the objects concerning reciprocity allow characterizing the relation of any chiral objects with respect to 3D space, from which a classification of the chirality type can be drawn.

2.3.2 *Optical Activity*

One of the most illustrating example of geometrical chirality in nature is the helix. The helix is intimately linked to the most known form of optical chirality, namely optical activity or natural gyrotropy. For example, several natural systems such as sugar molecules and quartz crystal possess a helicoidal structure and indeed show optical activity properties such as rotatory power, that is, circular birefringence, or circular dichroism (i.e., a differential absorption for RCP and LCP light).

For the present Purpose, one of the most relevant property of helices concerns their sense of "twist." It is indeed a basic fact that the twist orientation of a helix with its axis along z is invariant through the rotation R_x: such a helix looks actually quite the same when watched by an observer in the $+z$ or $-z$ direction. This is mathematically rooted in the fact that the helix is a 3D object observed in a 3D space.

A particular application of the geometrical analogy is the case of an isotropic and homogenous distribution of helices, which is indeed an extreme limit in which the system cannot be physically (in particular) distinguished when watched from the front or the back side. This is the case of the sugar molecules solution considered by Arago and Pasteur in their pioneer works on optical activity [9, 12]. However, it would be an oversimplification to limit optical activity to such totally invariant system, as in general, even a helix with its axis oriented along z is not completely invariant through R_x (although the sense of twist obviously is). Indeed, due to its finite length, one will have to rotate the helix by a given supplementary angle ϑ after application of R_x in general around z in order to return to the original helix (as seen from the front side). The analogy with the helix will give us a simple way to generalize our discussion and to define criteria for optical activity.

More precisely, in the limit of the paraxial approximation considered here, the question we should ask to ourself is: What must be the precise structure of the Jones matrix J if we impose that J^{flip} (see Eq. (2.10)) is, up to a rotation R_ϑ by an angle ϑ around the Z axis, identical to J?

This last condition reads actually in the cartesian basis

$$J^{\text{flip}} = \Pi_x \cdot J^{\text{T}} \cdot \Pi_x^{-1} = R_\vartheta \cdot J \cdot R_\vartheta^{-1}. \tag{2.23}$$

It is preferable to use the **L**, **R** basis and the previous condition becomes

$$\overline{J}^{\text{T}} = \overline{R_\vartheta} \cdot \overline{J} \cdot \overline{R_\vartheta}^{-1}. \tag{2.24}$$

The rotation matrix is defined by

$$R_\vartheta = \begin{pmatrix} \cos\vartheta & \sin\vartheta \\ -\sin\vartheta & \cos\vartheta \end{pmatrix}, \ \overline{R_\vartheta} = \begin{pmatrix} e^{+i\vartheta} & 0 \\ 0 & e^{-i\vartheta} \end{pmatrix}. \tag{2.25}$$

From equations (2.10, 2.17, 2.23), we deduce directly that the previous condition imposes $J_{lr}e^{i2\vartheta} = J_{rl}$ (i.e., $|J_{lr}| = |J_{rl}|$). Therefore, the Jones matrix takes the following form

$$J = \begin{pmatrix} J_{ll} & J_{lr} \\ J_{lr}e^{+i2\vartheta} & J_{rr} \end{pmatrix}. \tag{2.26}$$

By comparing with the condition for the absence of mirror symmetry (see Eq. (2.20)) and our theorem, we see that this J matrix is chiral if and only if $J_{ll} \neq J_{rr}$. The class of all the matrices

$$\overline{J}_{\text{o.a.}} = \begin{pmatrix} J_{ll} & J_{lr} \\ J_{lr}e^{+i2\vartheta} & J_{rr} \end{pmatrix}, \tag{2.27}$$

fulfilling these conditions is physically associated with the phenomenon of optical activity. This justifies the name given to $E_{\text{o.a.}}$. Equivalently stated, this result means that Eq. (2.27) defines the most general Jones matrices, which are chiral and such that the optical signature of chirality is, up to a rotation R_ϑ, invariant after reversal of the direction of propagation through the system. Clearly, reciprocity here dictates the rules. Importantly, Eq. (2.27) can also be written

$$\overline{J}_{\text{o.a.}} = \begin{pmatrix} (J_{ll}+J_{rr})/2 & J_{lr} \\ J_{lr}e^{+i2\vartheta} & (J_{ll}+J_{rr})/2 \end{pmatrix} + \frac{J_{ll}-J_{rr}}{2}\begin{pmatrix} 1 & 0 \\ 0 & -1 \end{pmatrix}, \tag{2.28}$$

that is as the sum of a matrix $\overline{J}_{\text{mirror}}$ obeying Eqs. (2.18, 2.19) (i.e., having an in-plane mirror symmetry axis) and of a matrix $\begin{pmatrix} \delta & 0 \\ 0 & -\delta \end{pmatrix}$

(with $\delta \neq 0$,) which actually induces the chiral behavior. In the cartesian basis, we can equivalently write

$$J_{\text{o.a.}} = \begin{pmatrix} A + B\cos\vartheta & B\sin\vartheta \\ B\sin\vartheta & A - B\cos\vartheta \end{pmatrix} + \begin{pmatrix} 0 & i\gamma \\ -i\gamma & 0 \end{pmatrix}, \quad (2.29)$$

with $A = (J_{ll} + J_{rr})/2$, $B = J_{lr}e^{i\vartheta}$ and $\gamma = (J_{rr} - J_{ll})/2$. The presence of the antisymmetrical part is the signature of chirality and the coefficient $\gamma \neq 0$ is called (natural) gyromagnetic factor.

An important particular case concerns Jones matrices, which are invariant through a rotation by an angle ϑ around the Z axis. Such a rotation is defined by the matrix R_ϑ with

$$R_\vartheta = \begin{pmatrix} \cos\vartheta & \sin\vartheta \\ -\sin\vartheta & \cos\vartheta \end{pmatrix}, \ \overline{R_\vartheta} = \begin{pmatrix} e^{+i\vartheta} & 0 \\ 0 & e^{-i\vartheta} \end{pmatrix}. \quad (2.30)$$

The invariance by rotation implies that the matrix $\overline{J}_R = \overline{R_\vartheta} \cdot \overline{J} \cdot \overline{R_\vartheta}^{-1}$ equals $\overline{J}$, that is:

$$\overline{J}_{\text{rotation axis}} = \begin{pmatrix} J_{ll} & 0 \\ 0 & J_{rr} \end{pmatrix}. \quad (2.31)$$

If $J_{ll} \neq J_{rr}$ then Eq. (2.31) is clearly a particular case of Eq. (2.27) that actually describes optical activity in isotropic media, such as quartz crystals or molecular solutions, and corresponds to the circular birefringence (and dichroism) introduced by Fresnel in 1825. It is interesting to observe that this is also the matrix which is associated with the gammadions artificial structure considered in [21–24] and which have a four-fold rotational invariance around Z.

Following its definition, the Jones enantiomorphic matrix associated with $\overline{J}_{\text{o.a.}} = \begin{pmatrix} A & 0 \\ 0 & B \end{pmatrix}$ writes as $\overline{J}_{\text{o.a.}}^{\text{enant}}(\vartheta) = \begin{pmatrix} B & 0 \\ 0 & A \end{pmatrix}$. Because of rotational invariance, $\overline{J}_{\text{o.a.}}^{\text{enant}}(\vartheta)$ is independent of ϑ. These two enantiomorphic matrices are associated with opposite optical rotatory powers.

Consider for example the Jones matrix associated with optical activity in an isotropic medium, such as a random distribution of helices for example. From Eqs. (2.10, 2.22, 2.31) it is immediately seen that $\overline{J}^{\text{flip}} = \overline{J}$. This invariance means that an observer illuminating such a system cannot distinguish the two sides from one other. This well known property explains in particular why an optically active medium cannot be used as an optical isolator: reciprocity prohibits such a scenario. In this context, we point out that nothing here forbid unitarity to hold. In the particular case of a Jones matrix represented by a rotation (see Eq. (2.25)), we have indeed $J^{-1} = J^{\dagger}$. We will see in the next section that this is not the case for 2D chirality where losses are unavoidable.

It is finally useful to remark that the ensemble of all the matrices $\overline{J}_{\text{o.a.}}$, that is, $E_{\text{o.a.}}$, is not closed for the addition and the product of matrices (i.e., the sum or the product of chiral matrix belonging to $E_{\text{o.a.}}$ is not necessary contained in $E_{\text{o.a.}}$). For example, by combining two enantiomers characterized by the matrices $\begin{pmatrix} A & 0 \\ 0 & B \end{pmatrix}$ and $\begin{pmatrix} B & 0 \\ 0 & A \end{pmatrix}$, we get

$$\begin{pmatrix} A & 0 \\ 0 & B \end{pmatrix} + \begin{pmatrix} B & 0 \\ 0 & A \end{pmatrix} = (A+B)\begin{pmatrix} 1 & 0 \\ 0 & 1 \end{pmatrix},$$

$$\text{and,} \begin{pmatrix} A & 0 \\ 0 & B \end{pmatrix} \cdot \begin{pmatrix} B & 0 \\ 0 & A \end{pmatrix} = (A.B)\begin{pmatrix} 1 & 0 \\ 0 & 1 \end{pmatrix}, \tag{2.32}$$

which are obviously not chiral and correspond to what is called a racemic medium (i.e., a mixture or a juxtaposition of opposite enantiomers).

We point out that the rotation matrices considered in the present discussion involve only an axis of rotation oriented along the z direction. It could be interesting to consider more general 3D rotation with an axis arbitrarily oriented. However, the case of planar chirality to be considered in the next section would thus be problematic in our classification, since as stated in the introduction, a planar chiral structure can be brought into congruence with its mirror image if it is lifted from the plane. The classification used here that explicitly considers distinct the two ensembles as distinct E_{2D} and $E_{\text{o.a.}}$ will appear actually very convenient, as (as shown in Section 3.4) any chiral matrix can be split into a first matrix belonging to E_{2D} and a second matrix belonging to $E_{\text{o.a.}}$.

2.3.3 *Planar Chirality*

Despite its fundamental importance, the previous analysis of optical activity does not exhaust the problem of chirality in optics. As we wrote in the Introduction, 2D chirality characterizes, by definition, a system *which cannot be put into congruence with itself until left from the plane* (for a more mathematical discussion see [51, 52]). This corresponds to a different chirality class than optical activity, as it can be simply seen. If instead of a 3D helix one considers a 2D spiral contained within the (x, y) plane, one has obviously a system that has a dimension lower than the dimension of its surrounding space. A flip of the structure is now possible, when it was not for the helix. As we discussed above, this discussion corresponds optically to a change of twist orientation when the light path is reversed, and clearly motivates the experimental demonstration for this second class of chiral objects which are planar.

Two geometrical examples of such planar chiral object are the Archimedean spiral that does not have point symmetry and the gammadion that possesses rotational invariance. As optically gammadion is associated with gyrotropy, that is, essentially the 3D effect as discussed in the previous section, we can already think that such gammadions are not genuine 2D chiral object from the optical point of view (even though it is obviously the case from basic geometrical considerations). We will here consider this 2D chirality class of Jones matrices in more detail, that is, $E_{2\text{D}}$.

The planar chirality class $E_{2\text{D}}$ was until very recently completely ignored in the literature and concerns chiral systems characterized by the conditions $J_{ll} = J_{rr}$ and a Jones matrix of the form:

$$\overline{J}_{2\text{D}} = \begin{pmatrix} J_{ll} & J_{lr} \\ J_{rl} & J_{ll} \end{pmatrix} \text{ with } |J_{lr}| \neq |J_{rl}|. \tag{2.33}$$

The condition $|J_{lr}| \neq |J_{rl}|$ actually leads to chirality. As we will show below, the equality condition on diagonal elements corresponds to reciprocity.

It is also important to observe that, as Eq. (2.33) is different from Eqs. (2.18, 2.31), the matrix $\overline{J}_{2\text{D}}$ not only has no mirror symmetries, but it has additionally no rotational invariance. This means that $\overline{J}_{2\text{D}}$ can only be associated with chiral systems without any point symmetries, such as an Archimedean spiral or a fish-scale structure.

A gammadion structure with its four-fold rotational invariance cannot display such optical property. It should also be remarked that the fish-scale structure considered in Ref. [42] actually has a central point symmetry, that is, a two-fold rotation axis. However, from the point of view of Jones matrix, such transformation is equivalent to the identity, as it is immediately seen by writing $\vartheta = \pi$ in Eq. (2.30), and consequently, the structure of $\overline{J}$ is not constrained by such transformation.

Same as for $\overline{J}_{\text{o.a.}}$, one can define enantiomers structures by the relation

$$\overline{J}_{2D}^{\text{enant}} = \overline{\Pi_\vartheta} \cdot \overline{J}_{2D} \cdot \overline{\Pi_\vartheta}^{-1} = \begin{pmatrix} J_{ll} & J_{rl}e^{-2i\vartheta} \\ J_{lr}e^{2i\vartheta} & J_{ll} \end{pmatrix}. \tag{2.34}$$

$\overline{J}_{2D}^{\text{enant}}$ of course belongs to E_{2D}.

We will now go back to the reciprocity theorem and consider the properties of planar chiral system from the point of view of path reversal. Same as for optical activity, the geometrical analogy appears very convenient for characterizing the reciprocal properties of chiral planar systems. The archetype of planar chiral objects is, as we already mentioned, the Archimedean spiral with no point symmetry. Watching such a spiral from one side or the other changes obviously the sense of twist. This contrasts strongly with the case of the helix of gammadion discussed before. This suggests the following definition: the chiral system characterized by the Jones matrix J is plan chiral if and only if Eq. (2.20) is satisfied and if

$$J^{\text{flip}} = \Pi_x \cdot J^{\text{T}} \cdot \Pi_x = R_\vartheta \cdot (\Pi_x \cdot J \cdot \Pi_x) \cdot R_\vartheta^{-1}. \tag{2.35}$$

where $\Pi_x \cdot J \cdot \Pi_x$ corresponds to a particular enantiomorphic Jones matrix of J (see Eq. (2.34)) parameterized by the angle ϑ.

This condition is in the **L**, **R** basis equivalent to

$$\begin{pmatrix} J_{ll} & J_{rl} \\ J_{lr} & J_{rr} \end{pmatrix} = \begin{pmatrix} J_{rr} & J_{rl}e^{2i\vartheta} \\ J_{lr}e^{-2i\vartheta} & J_{ll} \end{pmatrix}, \tag{2.36}$$

which admits a solution if and only if $J_{ll} = J_{rr}$ and $\vartheta = 0$ or π. Since $R_0 = -R_\pi = I$ (i.e., the identity operator), it means that the rotation here plays no role in the definition and that we could have reduced our reasoning to the alternatives conditions

$$J^{\text{flip}} = \pm\Pi_x \cdot J \cdot \Pi_x \text{,i.e., } J^T = \pm J. \tag{2.37}$$

In addition, in order to satisfy Eq. (2.20), we must necessarily have $|J_{lr}| \neq |J_{rl}|$. This is rigorously equivalent to the definition of the class E_{2D} given above.

In the cartesian basis, this means

$$J_{2D} = \begin{pmatrix} \varepsilon_+ & \Gamma \\ \Gamma & \varepsilon_- \end{pmatrix} = J_{2D}^T \tag{2.38}$$

with $\varepsilon_\pm = J_{ll} \pm (J_{lr} + J_{rl})/2$, and $\Gamma = i(J_{lr} - J_{rl})/2$. The condition for chirality $|J_{lr}| \neq |J_{rl}|$ implies $\varepsilon_+ \neq \varepsilon_-$ and $\Gamma \neq 0$. This condition for chirality also implies a stronger restriction.

Indeed, writing the non-diagonal coefficients J_{lr},J_{rl} in the polar form $J_{lr} = ae^{i\phi}$ and $J_{rl} = be^{i\chi}$ (with a, b the amplitudes and ϕ, χ the phases), we can define the ratio

$$\eta = \frac{\Gamma}{(\varepsilon_+ - \varepsilon_-)} = i\frac{J_{lr} - J_{rl}}{J_{lr} + J_{rl}}$$

which thus becomes

$$\eta = i\frac{1 - \frac{b^2}{a^2}}{1 + \frac{b^2}{a^2}} - 2\frac{\frac{b}{a}}{1 + \frac{b^2}{a^2}} \sin(\phi - \chi). \tag{2.39}$$

The condition for chirality therefore implies:

$$\text{Imag}\left[\frac{\Gamma}{(\varepsilon_+ - \varepsilon_-)}\right] \neq 0. \tag{2.40}$$

In the context of planar chirality, it is also useful to check whether E_{2D} is closed with respect to the matrices addition and multiplication. Same as for $E_{\text{o.a.}}$, it is obvious that this is not the case, as by summing

$$\overline{J}_{2D} = \begin{pmatrix} \alpha & \beta \\ \gamma & \alpha \end{pmatrix} \tag{2.41}$$

and

$$\overline{J}_{2D}^{\text{enant}} = \begin{pmatrix} \alpha & \gamma \\ \beta & \alpha \end{pmatrix} \tag{2.42}$$

one gets

$$\begin{pmatrix} 2\alpha & \beta + \gamma \\ \gamma + \beta & 2\alpha \end{pmatrix}, \tag{2.43}$$

which does not belong to E_{2D} (indeed, we have $|J_{lr}| = |J_{rl}|$). Similarly, for the product of $\overline{J}_{2D}$ and $\overline{J}_{2D}^{\text{enant}}$ one obtain $\begin{pmatrix} \alpha^2 + \beta^2 & \alpha(\gamma + \beta) \\ \alpha(\gamma + \beta) & \alpha^2 + \gamma^2 \end{pmatrix}$ which for the same reasons does not also belong to E_{2D}.

Finally, we point out that all genuine 2D chiral systems share another important property, namely a breaking of time invariance or reversal [42, 43, 47, 48]. To understand this peculiarity, we go back to our discussion of time reversal and consider the condition that the Jones matrix $\overline{J}_{2D}$ defined by Eq. (2.33) should fulfill in order to be unitary, that is, $\overline{J}_{2D}^{-1} = \overline{J}_{2D}^{\dagger}$. From Eq. (2.41), this leads to the relation:

$$\frac{1}{\alpha^2 - \beta\gamma}\begin{pmatrix} \alpha & -\beta \\ -\gamma & \alpha \end{pmatrix} = \begin{pmatrix} \alpha^* & \gamma^* \\ \beta^* & \alpha^* \end{pmatrix}. \tag{2.44}$$

This implies $\frac{\alpha}{\alpha^2-\beta\gamma} = \alpha^*$, $\frac{\beta}{\alpha^2-\beta\gamma} = -\gamma^*$ and $\frac{\gamma}{\alpha^2-\beta\gamma} = -\beta^*$. By taking the norms of each terms we deduce $|\alpha^2 - \beta\gamma| = 1$ and $|\beta| = |\gamma|$. This last equality contradicts the definition of $\overline{J}_{2D}$ and consequently such a planar chiral Jones matrix cannot be unitary. This implies that $J_{2D}^{\mathrm{T}} \neq J_{2D}^{-1,*}$ and that therefore J_{2D}^{rec} is different from J_{2D}^{inv}.

In other words, a 2D chiral system provides a perfect illustration that time-reversal is necessarily different from reciprocity, that is, path reversal. As time-reversal is a key property of fundamental physical laws at the *microscopic* level, the only solution is to assume that this breaking of time-reversal at the level of 2D chiral objects is associated with *macroscopic* irreversibility. Indeed, the imaginary part of the permittivity, for example, is connected to losses and dissipation into the environment (seen as a thermal bath) and the condition for its positivity implies a strong irreversibility in the propagation. Similarly here, 2D optical chirality means that some sources of irreversibility must be present in order to prohibit unitarity of the Jones matrix. This is an interesting example where two fundamental aspects of nature, namely chirality and time irreversibility (intrinsically linked to the entropic time arrow,) are intimately connected.

2.3.4 *Generalization*

The most general Jones matrix J characterizing a chiral medium can be written:

$$\overline{J}_{\text{chiral}} = \begin{pmatrix} (J_{ll} + J_{rr})/2 & J_{lr} \\ J_{rl} & (J_{ll} + J_{rr})/2 \end{pmatrix} + \frac{(J_{ll} - J_{rr})}{2}\begin{pmatrix} 1 & 0 \\ 0 & -1 \end{pmatrix}, \tag{2.45}$$

where both conditions $J_{ll} \neq J_{rr}$, $|J_{lr}| \neq |J_{rl}|$ are satisfied. This defines the optical chirality class E_{chiral}.

An important property is that the sum of a matrix belonging to $E_{\text{o.a.}}$ with a matrix of E_{2D} belongs to E_{chiral}. To see that this is obviously the case, it is sufficient to remark that the sum of a matrix $\overline{J}_{\text{2D}}$ with a matrix $\overline{J}_{\text{o.a.}}$ can be written

$$\overline{J}_{\text{2D}} + \overline{J}_{\text{o.a.}} = \overline{J}_{\text{2D}} + \overline{J}_{\text{mirror}} + \begin{pmatrix} \alpha & 0 \\ 0 & -\alpha \end{pmatrix}. \tag{2.46}$$

However, we also have

$$\overline{J}_{\text{2D}} + \overline{J}_{\text{mirror}} = \begin{pmatrix} \alpha + \alpha' & \beta + \beta' \\ \gamma + \beta' e^{i\phi} & \alpha + \alpha' \end{pmatrix} \tag{2.47}$$

with $|\beta| \neq |\gamma|$. This is necessary of the form $\overline{J}_{\text{2D}}$, as otherwise we should have $|\beta + \beta'| = |\gamma + \beta' e^{i\phi}|$ in contradiction with the condition $|\beta| \neq |\gamma|$. Reciprocally, any matrices $\overline{J}'_{\text{2D}}$ can be written as a sum $\overline{J}_{\text{2D}} + \overline{J}_{\text{mirror}}$ since from the previous result for any $\overline{J}_{\text{mirror}}$ the difference $\overline{J}'_{\text{2D}} - \overline{J}_{\text{mirror}}$ belongs to E_{2D}. This means that a matrix belonging to E_{chiral} can always be written as the sum of a matrix belonging to the class $E_{\text{o.a}}$ with a matrix belonging to E_{2D}. Interestingly, the combination of a spiral and a helix leads to the geometrical shape of the screw. Finally, one can observe that E_{chiral} is not close with respect to the matrix addition and product, as it is already not the case for the subclasses $E_{\text{o.a.}}$ and E_{2D}.

2.3.5 *Eigenstates and Chirality: Time Reversal versus Reciprocity*

In the context of reciprocity, it is of practical importance to consider the backward propagation of light through the medium along the z direction after path reversal by a mirror located after it (see Fig. 2.2). This corresponds to the following succession of events: A) the initial state that we write here $|in\rangle = E_x^{(\text{in})}|x\rangle + E_y^{(\text{in})}|y\rangle = E_L^{(\text{in})}|L\rangle + E_R^{(\text{in})}|R\rangle$ (instead of $\mathbf{E}^{(\text{in})} = E_x^{(\text{in})}\hat{\mathbf{x}} + E_y^{(\text{in})}\hat{\mathbf{y}}$ used in Section 2.2.1), propagates through the chiral medium and we obtain afterward the new state $|2\rangle = \hat{J}\,|in\rangle$. B) The reflection by the mirror induces a change in the electric field sign and also reverses the path propagation direction. By rotating the coordinate axes by an angle π around the x axis, we preserve (as explained before) the handedness of such coordinate

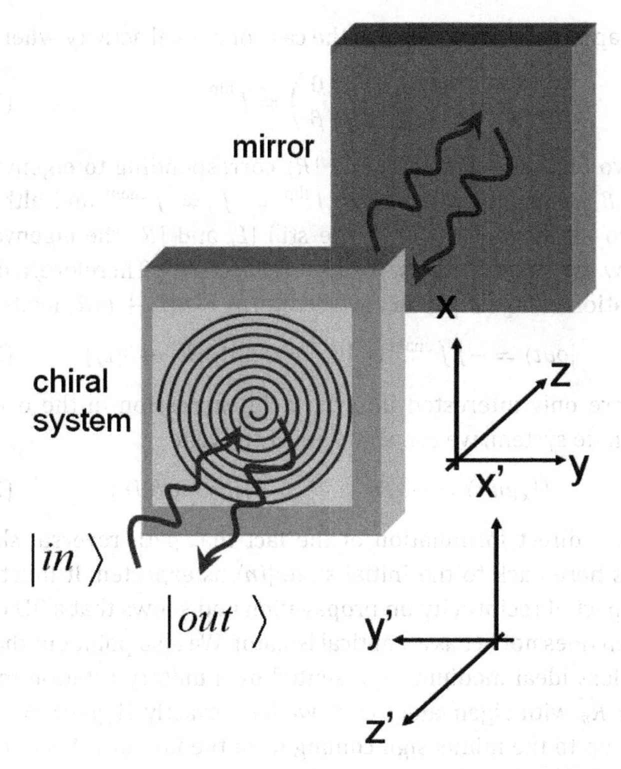

Figure 2.2 Principle of path reversal by a mirror through a chiral structure (e.g., a planar chiral system such as those considered in Ref. [60]) in more detail. The two coordinate systems x, y, z and $x' = x, y' = -y, z' = -z$ discussed in the text are represented.

system as well as the positive sign of the propagation direction. This means that in this new basis, the vector $|2\rangle$ evolves as $|3\rangle = -\hat{\Pi}_x|2\rangle$. C) The backward propagation through the medium is described by the flip operator $\hat{J}^{\text{flip}}$ and the final state (in the new coordinate system) reads $|out\rangle = \hat{J}^{\text{flip}}|3\rangle$. We have consequently

$$|out\rangle = -\hat{J}^{\text{flip}}\hat{\Pi}_x\hat{J}\,|in\rangle = -\hat{J}^{\text{flip}}\hat{J}^{\text{enant}}\hat{\Pi}_x|in\rangle \tag{2.48}$$

where by definition $\hat{J}^{\text{enant}} = \hat{\Pi}_x\hat{J}\,\hat{\Pi}_x$.

In order to analyze the effect of Eq. (2.48), we will first study the eigenstates and eingenvalues of the chiral Jones matrix discussed in this chapter. First, we consider the case of optical activity where

$$\overline{J}_{\text{o.a.}} = \begin{pmatrix} \alpha & 0 \\ 0 & \beta \end{pmatrix} = \overline{J}^{\text{flip}}. \tag{2.49}$$

The two eigenstates are $|L\rangle$ and $|R\rangle$ corresponding to eigenvalues α and β, respectively. However, $\hat{J}^{\text{flip}} = \hat{J} \neq \hat{J}^{\text{enant}}$ and although the two eigenstates of $\hat{J}^{\text{enant}}$ are still $|L\rangle$ and $|R\rangle$ the eigenvalues are now exchanged, that is, β and α, respectively. Therefore, a direct application of Eq. (2.48) on any vector $|in\rangle = a|L\rangle + b|R\rangle$ leads to

$$|out\rangle = -\hat{J}\hat{J}^{\text{enant}}\hat{\Pi}_x|in\rangle = -\alpha\beta(a|R\rangle + b|L\rangle). \tag{2.50}$$

If we are only interested into the field expression in the original coordinate system, we can alternatively rewrite

$$\hat{\Pi}_x|out\rangle = -\alpha\beta(a|L\rangle + b|R\rangle) = -\alpha\beta|in\rangle. \tag{2.51}$$

This is a direct formulation of the fact that path reversal should lead us here back to the initial state $|in\rangle$ as expected. It illustrates the impact of reciprocity on propagation and shows that a 3D chiral medium does not act as an optical isolator. We also point out that for a loss-less ideal medium represented by a unitary rotation matrix $\overline{J}_{\text{o.a.}} = \overline{R_\vartheta}$ with eigenvalues $e^{\pm i\vartheta}$, we have exactly $\hat{\Pi}_x|out\rangle = -|in\rangle$, which, up to the minus sign coming from the mirror reflection, is a perfect illustration of time reversal and symmetry for natural optical activity.

We now consider 2D planar chirality. The eigenstates and eingenvalues of the chiral Jones matrix

$$\overline{J}_{\text{2D}} = \begin{pmatrix} \alpha & \beta \\ \gamma & \alpha \end{pmatrix} \tag{2.52}$$

are by definition states $|\pm\rangle$ defined by $\hat{J}_{\text{2D}}|\pm\rangle = \lambda_\pm|\pm\rangle$. After straightforward calculations, we obtain

$$\lambda_\pm = \alpha \pm \sqrt{(\beta\gamma)}, \tag{2.53}$$

and

$$|\pm\rangle = \frac{\sqrt{\beta}|L\rangle \pm \sqrt{\gamma}|R\rangle}{|\beta| + |\gamma|}. \tag{2.54}$$

Using similar methods, we can easily find eigenstates and values of the reciprocal matrix $\overline{J}_{\text{2D}}^{\text{flip}} = \overline{J}_{\text{2D}}^{\text{enant}}$ such as $\hat{J}_{\text{2D}}^{\text{flip}}|\pm\rangle_{\text{enant}} = \lambda'_\pm|\pm\rangle_{\text{enant}}$.

The eigenvalues are the same as for $\overline{J}_{2D}$, that is, $\lambda'_{\pm} = \lambda_{\pm}$, but the eigenvectors are now

$$|\pm\rangle_{\text{enant}} = \frac{\sqrt{\gamma}|L\rangle \pm \sqrt{\beta}|R\rangle}{|\beta| + |\gamma|}. \tag{2.55}$$

Importantly her,e $\hat{J}^{\text{flip}} = \hat{J}^{\text{enant}}$ and therefore by applying Eq. (2.48) on the initial states $|in\rangle = |\pm\rangle$, we get

$$|out\rangle = -(\hat{J}^{\text{flip}})^2 \hat{\Pi}_x |in\rangle = \mp\lambda_{\pm}^2 |\pm\rangle_{\text{enant}} \tag{2.56}$$

where we used the relations $\hat{\Pi}_x|L\rangle = |R\rangle$, $\hat{\Pi}_x|R\rangle = |L\rangle$, and $\hat{\Pi}_x|\pm\rangle = \pm|\pm\rangle_{\text{enant}}$. Like we did for optical activity, we can go back to the initial coordinate system x, y, z and the final states read now

$$\hat{\Pi}_x|out\rangle = -\lambda_{\pm}^2|\pm\rangle \tag{2.57}$$

As before that again illustrates the effectiveness of the reciprocity principle. Furthermore, as the transformation is not unitary, we could not obtain such a result using J^{inv} as defined by Eq. (2.15).

2.4 Discussion and Examples

As we mentioned in the Introduction, it is very interesting to observe that the property concerning the change of twist for genuine 2D chiral systems when watched from two different sides stirred a considerable debate in the recent year in the context of metamaterials.

To understand this in more detail, we remind that partly boosted by practical motivations, such as the quest of negative refractive lenses [20] or the possibility to obtain giant optical activity for applications in optoelectronics, there is currently a renewed interest [20–30] in the optical activity in artificial photonic media with planar chiral structures. It was shown for instance that planar gammadionic structures, which have by definition no axis of reflection but a four-fold rotational invariance [21, 23], can generate optical activity with giant gyrotropic factors [24, 28–30]. Importantly, and in contrast to the usual three-dimensional (3D) chiral medium (like quartz and its helicoidal structure [3, 17]), planar chiral structures change their observed handedness when the direction of light is reversed through the system [21, 40]. This paradoxically challenged Lorentz principle of reciprocity [1] (which is known to hold for any linear non magneto-optical media) and stirred up considerable debate [21, 22, 24, 41] which came to the conclusion that optical activity cannot be a purely 2D effect and always requires a small dissymmetry between the two sides of the system [24, 28–30].

This point becomes more clear from the previous definitions and discussion concerning chirality and reciprocity. Indeed, a gammadion being rotationally invariant its optical properties must be characterized by a Jones matrix belonging to $E_{\text{o.a.}}$ that is,

$$\overline{J}_{\text{o.a.}} = \begin{pmatrix} A & 0 \\ 0 & B \end{pmatrix}, \tag{2.58}$$

with $A \neq B$ However, since geometrically the 2D gammadion change its sense of twist when watched from the other side, the discussion concerning reciprocity and change of twist developed between Eqs. (2.35, 2.40) imposes that the Jones matrix should also belong to $E_{2\text{D}}$, that is,

$$\overline{J}_{2\text{D}} = \begin{pmatrix} \alpha & \beta \\ \gamma & \alpha \end{pmatrix} \tag{2.59}$$

with $|\beta| \neq |\gamma|$. The only possibility for having $\overline{J}_{2\text{D}} = \overline{J}_{\text{o.a.}}$ is to impose $\gamma = \beta = 0$ as well as $A = B = \alpha$ and consequently to have no optical chiral signature whatsoever. This solves the paradox and shows that if gammadion generate nevertheless optical activity with giant gyrotropic factors [24, 28–30] then the system cannot be purely 2D. The third dimension (such as the presence of substrate for example) is enough to break the symmetry between the two directions of transmission through the structure and there is no violation of reciprocity, as the matrix has no anymore reason to satisfy Eq. (2.35) or Eq. (2.37), that is, to belong to $E_{2\text{D}}$.

We also point out that Bunn [53] and later L. Barron [40, 41] already remarked that optical chirality in 3D and also in 2D (see the next section) characterizes not only the structure itself but also the complete illumination protocol including the specific orientation of the incident light relatively to the structure. This idea was recently applied in the context of metamaterials by Zheludev and coworkers by demonstrating specific forms of extrinsic optical chirality in which the individual elements themselves are not chiral, although the complete array of such cells is (due to specific tilts existing between the incident light and the objects [54]). A related scheme has also been discussed at the level of achiral plasmonic nanohole arrays [55].

2.4.1 *Surface Plasmons and Archimedean Spirals: Planar Chirality Gives a Twist to Light*

As mentioned in the Introduction, the first manifestations of optical planar chirality were observed by Zheludev and coworkers [42, 43, 47] using fish-scale periodic metal strips on a dielectric substrate. The chiral structures were first realized at the millimeter scale for the GHz regime in 2006 [42] and soon after scaled down to the nanometer scale for studied in the near-infrared regime in 2008 [47]. Simultaneously with these last studies, we realized planar chiral gratings for SPs on a gold film. These structure shown in Fig. 2.3 (b) are Archimede spirals defined by the parametric equations

$$x(\theta) = \pm\rho(\theta)\cos(\theta),\ y(\theta) = \rho(\theta)\sin(\theta), \tag{2.60}$$

with

$$\rho(\theta) = \frac{P}{2\pi}\theta \tag{2.61}$$

θ varying between $\theta_{\min} = \pi$ and $\theta_{\max} = \theta_{\min} + 18\pi$. The two possible signs $\pm$ define two enantiomers (labelled L and R on Fig. 2.3(b)), which are reciprocal mirror images obtained after reflection across the y-z plane $x = 0$. Such clock wise or anticlockwise spirals were milled on a 310 nm thick gold film using focus ion beam methods. For such Archimede's spirals, the length $P \simeq 760$ nm plays obviously the role of radial period, as at each increment by an angle $\delta\theta = 2\pi$, the radius increases by an amount $\delta\rho = P$. The structure looks like the well known "bull eye's" circular antennas, which are used to resonantly couple monochromatic light with wavelength $\lambda \simeq P$ impinging normally to the structure [56, 57]. We point out that P is actually very closed to the SP wavelength $\lambda_{SP}(\lambda_0) = 760$ nm, which corresponds to the optical wavelength $\lambda_0 \simeq 780$ nm [48]. However,

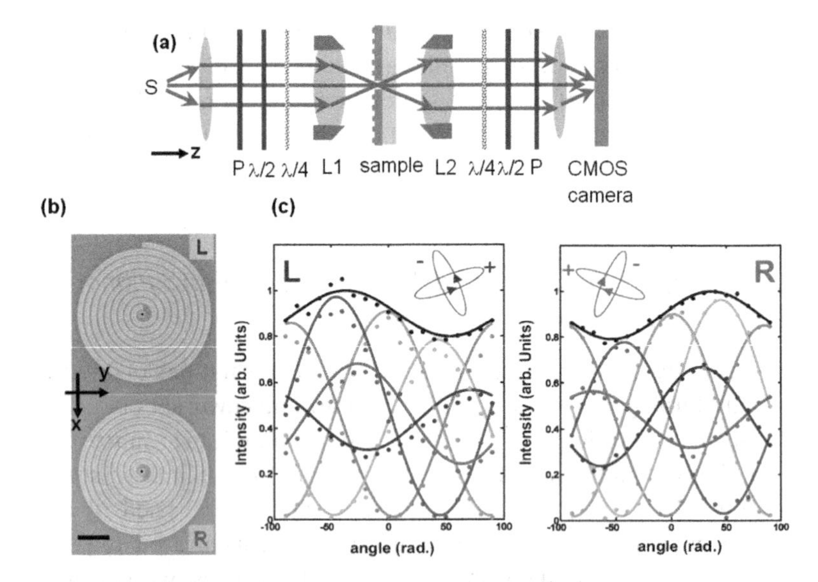

Figure 2.3 Chiral plasmonic metamolecules. (a) Sketch of the polarization tomography set up used in Refs. [48, 60]. (b) Scanning electron micrographs of the left (L) and right (R) handed enantiomer (mirror image) planar chiral structures investigated. The scale bar is 3 μm long. The parameters characterizing the structure are the following: hole diameter $d = 350$ nm, film thickness $h = 310$ nm, grating period $P = 760$ nm, groove width $w = 370$ nm, and groove depth $s = 80$ nm. (c) Analysis of the polarization states for an input light with variable linear polarization for both the left (left panel) and right-handed (right panel) individual chiral structures of (a). The insets in each panel show the ellipses of polarization and the handedness (arrow) associated with the two co-rotating eingenstates associated with the Jones matrix J (blue) and J^{enant} (red). Details are discussed in the text. Figure adapted from [48, 60] with permission.

this small difference is not relevant here. To increase this similarity with a usual bull eye antenna, the groove depth and width were also selected to favor the light coupling to the grating. Importantly, we also milled a 350 diameter hole centered at the origin of the spirals (i.e., $x = y = 0$) in which light can go through. Altogether, the system acted as a chiral bull eye's antenna focussing SPs at the center $x = y = 0$, where they interfere with the incident light before being transmitted through the hole, thanks to a Fano-like mechanism [58–61]. Spectral properties of such antennas showed the typical optical resonance centered at $\lambda \simeq P$ as for their circular or elliptical cousins [45, 56, 57, 60].

Chiral optical properties of the two enantiomorphic structures were studied performing a polarization tomography of the light transmitted through the hole. The method described in Ref. [48, 60] is based on the experimental determination of the 4×4 Mueller matrix [62, 63]. Such a Mueller matrix M characterizes the polarization transformation applied on the incident light beam with Stokes vector

$$\mathbf{S}^{\text{in}} = \begin{pmatrix} S_0^{\text{in}} \\ S_1^{\text{in}} \\ S_2^{\text{in}} \\ S_3^{\text{in}} \end{pmatrix}. \tag{2.62}$$

The resultant Stokes vector $\mathbf{S}^{\text{out}} = M\mathbf{S}^{\text{in}}$ is linked to the electric field $E = [E_x, E_y]$ transmitted through the hole. Furthermore, subsequent theoretical analysis demonstrates a precise connection between the 2×2 Jones matrix J characterizing the system and the 4×4 Mueller matrix M [62]. We used an homemade microscope to focus and control the state of polarization (SOP) of light going through the chiral structures. In order to study experimentally the SOP conversion by our structure, we carried out a complete polarization tomography using the optical setup sketched in Fig. 2.3 (a). A laser beam at $\lambda = 780$ nm is focused normally onto the structure by using an objective L1. The transmitted light is collected by a second objective L2 forming an Airy spot on the camera as expected, as the hole behaves like a point source in an opaque gold film. In our experiments, the intensity is thus defined by taking the maximum of the Airy spot [48, 60]. The SOP of light is prepared and analyzed with half-wave plates, quarter wave plates, and polarizers located before and after the objectives.

To illustrate the polarization conversion induced by the chiral object on the transmitted light, we analyze on Fig. 2.3 (c) the transformation acting on a linearly polarized input light analyzed after transmission in four orthogonal SOP Stokes vectors along the directions: $|x\rangle$ (green), $|y\rangle$ (yellow), $|+45°\rangle$ (cyan), $|-45°\rangle$ (magenta), $|L\rangle$ (red), and $|R\rangle$ (blue). The total transmitted intensity is also shown (black). The symmetries between both panels expected from group theory are observed experimentally in perfect agreement with the theory discussed in Ref. [48]. The insets show in each panel the ellipses of polarization and the handedness (arrow) associated with the two corotating eingenstates associated with the Jones matrix. The good agreement between the measurements and the theoretical predictions deduced from the Jones matrices [48] is clearly seen, together with the mirror symmetries between the two enantiomers. This agreement shows that our theoretical hypothesis about the form of the matrices J (see Eqs. (2.38, 2.41) is experimentally justified. Importantly, the observed symmetries also imply that for unpolarized light, and in complete consistency with spectral studies, the total intensity transmitted by the structures is independent of the chosen enantiomer.

2.4.2 *Chiral Surface Plasmons and Singular Optics: Tailoring Optical Vortices*

Lately, chiral SP modes have also been studied in relation to singular optical effects [64–67]. Indeed, near-field excitations on chiral nanostructures have shown to generate orbital angular momentum (OAM) both in the near field [68–71] and the far field [72, 73]. We have just recently presented a comprehensive analysis of the OAM transfer during plasmonic in-coupling and out-coupling by chiral nanostructures at each side of a suspended metallic membrane, stressing in particular the role of a back-side structure in generating vortex beams as $e^{i\ell\varphi}$ with tunable OAM indices ℓ.

Our device consists of a suspended thin ($h \sim 300$ nm) metallic membrane, fabricated by evaporating a metal film over a poly(vinyl formal) resin supported by a transmission electron microscopy copper grid. After evaporation, the resin is removed using a focused ion beam (FIB), leaving a freely suspended gold membrane. Plasmonic structures are milled, in either concentric (BE) or spiral geometry on both sides of the membrane around a unique central cylindrical aperture acting as the sole transmissive element of the whole device. The general groove radial path is given in the polar $(\hat{\rho}, \hat{\varphi})$ basis, as $\rho_n = (n\lambda_{\text{SP}} + m\varphi\lambda_{\text{SP}}/2\pi)\hat{\rho}$, with n an integer, λ_{SP} the SP wavelength, and m a pitch number. Orientation conventions are chosen with respect to the light propagation direction, so that a right-handed spiral R_m corresponds to $m > 0$ and a left-handed spiral L_m to $m < 0$. As we summarize below, there is an close relation between the spiral pitch m and the topological charge ℓ of the vortex generated in the near field.

Near-field generation of OAM at the front side ($z = 0^+$) of the structured membrane can be understood by considering that each point $\boldsymbol{\rho}_n$ of the groove illuminated by the incoming field is an SP point source, launching an SP wave perpendicular to the groove. With groove widths much smaller than the illumination wavelength, the in-plane component of the generated SP field in the vicinity of the center of the structure is $\mathbf{E}^{\mathrm{SP}}(\boldsymbol{\rho}_0, z = 0^+) \propto \mathrm{G} \cdot [\mathbf{E}^{\mathrm{in}}(\boldsymbol{\rho}_n, z = 0^+) \cdot \hat{\mathbf{n}}_n]\hat{\mathbf{n}}_n$ where $\mathrm{G} = e^{ik_{\mathrm{SP}}|\boldsymbol{\rho}_0 - \boldsymbol{\rho}_n|}/(|\boldsymbol{\rho}_0 - \boldsymbol{\rho}_n|)^{1/2}$ is the Huygens-Fresnel plasmonic propagator and $\hat{\mathbf{n}}_n = \kappa^{-1}(\mathrm{d}^2\boldsymbol{\rho}_n/\mathrm{d}s^2)$ the local unit normal vector determined from the curvature κ and the arc length s of the groove. The resultant SP field is the integral of elementary point sources over the whole groove structure. As indicated by a full evaluation, we can conveniently limit the integration to radial regions $\rho_n \gg \rho_0$ where the grooves become practically annular. This leads to $\hat{\mathbf{n}}_n \sim -\hat{\boldsymbol{\rho}}$ and therefore to a simple expression of the integrated SP field $\mathbf{E}^{\mathrm{SP}} = \mathrm{C}_{\mathrm{in}} \cdot \mathbf{E}^{\mathrm{in}}$, connected to the incoming field by an in-coupling matrix

$$\mathrm{C}_{\mathrm{in}}(m) \propto e^{im\varphi_0} \int_0^{2\pi} \mathrm{d}\varphi e^{im\varphi} e^{-ik_{\mathrm{SP}}\rho_0 \cos\varphi} \hat{\boldsymbol{\rho}} \otimes \hat{\boldsymbol{\rho}}, \tag{2.63}$$

the $\otimes$ symbol denoting a dyadic product.

Contrasting with the recent studies that have been confined to the near field, our suspended membrane opens the possibility to decouple the singular near field into the far field, with an additional structure on the back side of the membrane connected to the front side by the central hole. By symmetry (assuming loss-free unitarity), the out-coupling matrix is simply given as the hermitian conjugate of the in-coupling matrix, that is, $\mathrm{C}_{\mathrm{out}} = \mathrm{C}_{\mathrm{in}}^{\dagger}$, corresponding to a surface field that propagates away from the central hole on the back side. The inout coupling sequence corresponds to the product $\mathrm{T} = \mathrm{C}^{\dagger}(m_{\mathrm{out}}) \cdot \mathrm{C}(m_{\mathrm{in}})$ which, in the circular polarization basis, writes explicitly as

$$\mathrm{T} \propto e^{i(m_{\mathrm{out}} - m_{\mathrm{in}})\varphi} \begin{bmatrix} \mathrm{t}_{++} & \mathrm{t}_{+-}e^{2i\varphi} \\ \mathrm{t}_{-+}e^{-2i\varphi} & \mathrm{t}_{--} \end{bmatrix} \tag{2.64}$$

Table 2.1 Far-field summation rules for OAM generated through the membrane

	+	−
+	$m_{in} - m_{out}$	$m_{in} - m_{out} - 2$
−	$m_{in} - m_{out} + 2$	$m_{in} - m_{out}$

with t_{ij} radial functions based on products of $m_{in,out} \pm 1$-order Bessel functions of the first kind, as detailed in [49]. Note that we use here circular basis conventions that allow associating a positive value to an OAM induced on a right $m > 0$ right spiral. With respect to this chapter, these conventions are such that $+/-$ is associated with $\hat{\mathbf{L}}/\hat{\mathbf{R}}$.

This expression (2.64) reveals two contributions: a polarization-dependent geometric phase, within the matrix, whoch stems from the spin-orbit coupling at the annular groove, and a factorized dynamic phase that arises due to the spiral twist of the structure [69]. In relation to what has just been described above, the structure of Eq. (2.64) shows that, it would belong to the general class of chiral structure, as it is not a mere 2D chiral structure, as seen from the fact that $t_{++} \neq t_{--}$ nor does it belong to the simple optical activity class given that $|t_{+-}| \neq |t_{-+}|$. Obviously, when $m_{in} = m_{out} = 0$, the system is achiral and, in this case, T describes a pure spin–orbit angular momentum transfer, conserving the total angular momentum [69, 74–76].

We have checked experimentally the OAM summation rules that can be drawn from this analysis and that can be gathered in a summation table. A sketch of the experiment is presented in Fig. 2.4. As also shown in panel (c) of the figure, we have been able to generate optical beams in the far field with OAM indices up to $\ell = 8$, in perfect agreement with the expected OAM summation rules. Note that this $\ell = 8$ value is in strict relation with the chosen structures and is not a limit to our device.

Remarkably with suspended membranes, one can actually perform experimentally the path reversal operation described in Section 2.3.5 and in Fig. 2.2. By merely flipping the BE-$(L, R)_1$ structures with respect to the optical axis, one obtains $(R, L)_1$-BE,

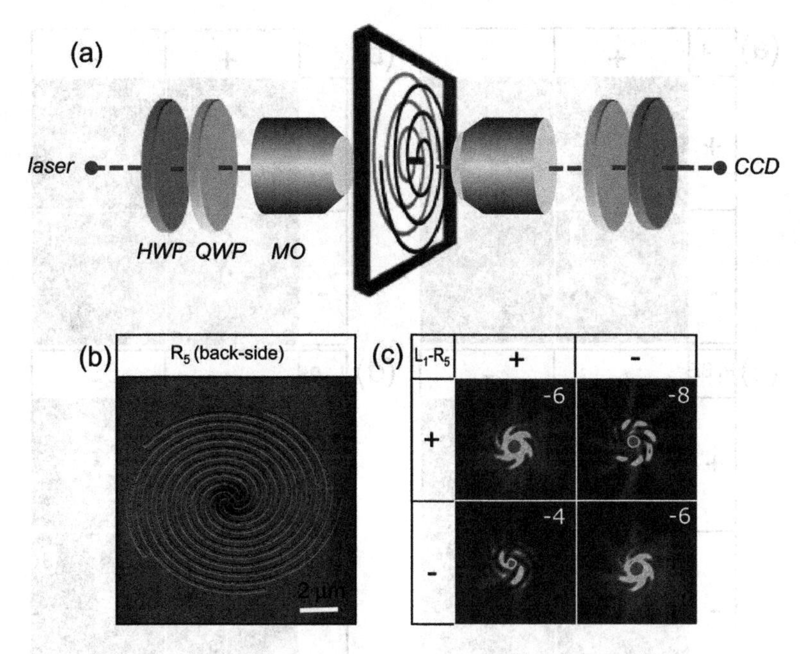

Figure 2.4 (a) Experimental setup: the incoming laser beam is circularly polarized using half (HWP) and quarter-wave (QWP) plates and weakle focused by a microscope objective (5×, NA= 0.13). The transmitted beam is imaged by a second objective (40×, NA = 0.60) and a lens tube ($f = 200$ mm, not shown) on a CCD camera and analyzed in the circular polarization basis by additional HWP and QWP. (b) Scanning electron microscope image of an R_5 spiral milled on the back side of a gold membrane, with $\lambda_{SP} = 768$ nm. (c) Intensity distribution imaged through a $L_1 - R_5$ structure. Labels ($\pm, \pm$) correspond to the combination of circular polarization preparation and analysis. The numbers correspond to the corresponding OAM indices. Figure adapted from [49] with permission.

with enantiomorphic changes $L \leftrightarrow R$ generated by the planar character of the spirals milled on one side of the membrane.

As displayed in Fig. 2.5 when comparing panels (a) and (c) and (b) and (d), the OAM measurements however turn out to be inconsistent with a simple path reversal operation. As we fully explain in [49], this discrepancy points to the pivotal role of the central aperture in the process of OAM conservation, inducing specific OAM selection rules that must be accounted for in the generation process. This discussion is beyond the scope of this

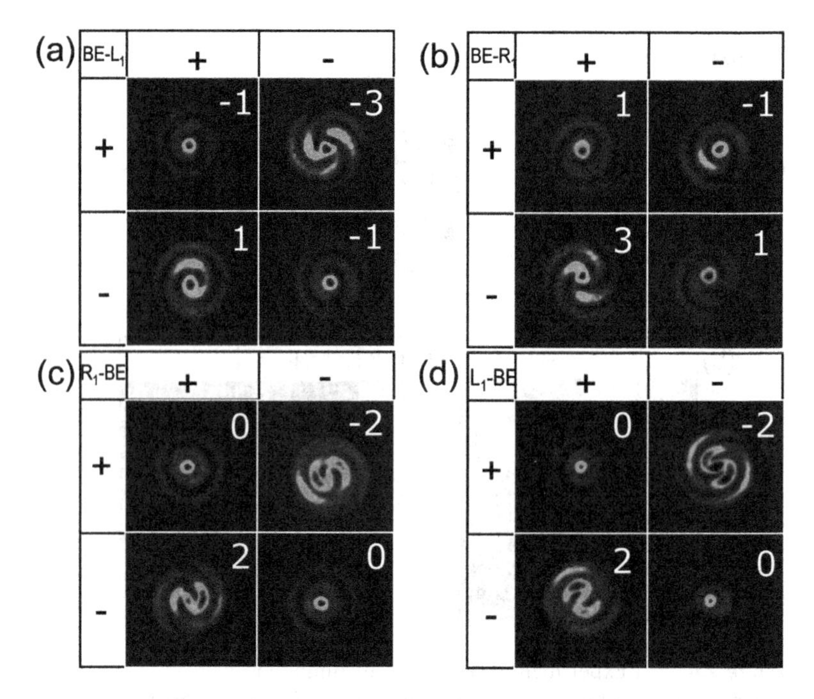

Figure 2.5 Intensity distributions of the beam emerging from (a) BE-L_1, (b) BE-R_1, (c) R_1-BE and (d) L_1-BE structures, respectively. The hole diameter used in all the structures was 400 nm. Same labels as in Fig. 2.4. Figure adapted from [49] with permission.

summary, and we refer the reader to our manuscript for further details.

2.5 Outlook

We have tried in this chapter to give an overview on the close relation between the chiral behavior of optical media and reciprocity. This relation has raised recently interesting issues, particularly salient in the context of metamaterial optics. As we discussed here, an algebraic approach can be useful, as it allows distinguishing different classes of chiral media with respect to reciprocity. Along these lines, one is naturally led to unveil new types of chiroptical behaviors such as *optical planar chirality* a totally original optical signature that contrasts with standard *optical activity*. Doing so, we have also stressed how reciprocity should not be confused with time-reversal invariance. A couple of recent experimental examples have been presented that nicely illustrate these intertwined relations in the realm of nanophotonics, both in terms of polarization dynamics and optical vortices.

Acknowledgments

The authors are grateful to T. W. Ebbesen for his continuous support and Y. Gorodetski, J.-Y. Laluet, and E. Lombard for their participation in the work described here. This work received the financial support of the ERC (grant 227557) and the Agence Nationale de la Recherche (ANR) through the French program Investissement d'Avenir (Equipex Union).

References

1. L. D. Landau, E. M. Lifshitz, L. P. Pitaevskii, *Electrodynamics of continuous media* 2nd *ed.* (Pergamon, New York, 1984).
2. F. I. Fedorov, "On the theory of optical activity in crystals I. The law of conservation of energy and the optical activity tensors," *Opt. Spectrosc.* **6**, 49–53 (1959).
3. E. Hecht, *Optics* 2nd *ed.* (Addison-Wesley, Massachusetts, 1987).
4. S. F. Masson, *Chemical evolution: Origins of the elements, molecules and living systems* pp. 281–284 (Clarendon Press, Oxford, 1991).
5. T. D. Lee and C. N. Yang, "Question of parity conservation in weak interactions," *Phys. Rev.* **104**, 254258 (1956).
6. I. Kant, *Prolegomena to any future metaphysics (1783)* (reissued by Hacket Publishing, 1977).
7. W. Thomson (L. Kelvin), *Baltimore Lectures on Molecular Dynamics and the Wave Theory of Light (1904)* (reissued by Cambridge University Press, 2010).
8. M. Faraday, "Experimental Researches in Electricity. Nineteenth Series", *Phil. Trans. R. Soc. Lond.* **136**, 1–20 (1846).

9. D.-F. M. Arago, "Mémoire sur une modification remarquable qu'éprouvent les rayons lumineux dans leur passage à travers certains corps diaphanes, et sur quelques autres nouveaux phénomènes d'optique," *Mém. Inst. France, Part I* **12** (1811).
10. J. B. Biot, "Sur un nouveau genre d'oscillations que les molécules de la lumière éprouvent en traversant certains cristaux," *Mém. Inst. Nat. Sci. Art.* **13**, 1. (1812).
11. A. Fresnel, "Mémoire sur la double réfraction," presented at the Académie des sciences the 9th December 1822). See also Vol. 1 S. Verdet and L. Fresnel, *œuvres complètes d'Augustin Fresnel* (Imprimerie impériale, Paris, 1866).
12. L. Pasteur, "Mémoire sur la relation qui peut exister entre la forme cristalline et la composition chimique, et sur la cause de la polarization rotatoire," *C. R. Acad. Sci. Paris* **26**, 535–539 (1848).
13. H. D. Flack, "Louis Pasteur discovery of molecular chirality and spontaneous resolution in 1848, together with complete review of his crystallographic and chemical work," *Act. Cryst. A* **65**, 371–389 (2009).
14. J. A. Le Bel, "Sur les relations qui existent entre les formules atomiques des corps organiques et le pouvoir rotatoires de leru dissolutions," *Bull. Soc. Chim. Fr.* **22**, 1874, 337347.
15. J. H. van't Hoff, "Sur les formules de structures dans l'espace," *Arch. Neerl. Sci. Exactes* **9**, 445–454 (1874).
16. A. Cotton, "Absorption inégale des rayons circulaires droit et gauche dans certains corps actifs," Compt. Rend. **120**, 989–991 (1895) and "Dispersion rotatoire anomale des corps absorbants," *Compt. Rend.* **120**, 1044–1046 (1895).
17. J. C. Bose, "On the rotation of plane of polarization of electric waves by a twisted structure." *Proc. R. Soc. London A* **63**, 146–152 (1898).
18. K.F. Lindman, *Ann. Physik* **63**, 621 (1920). For a review see I. V. Lindell, A. H. Sihvola, and J. Kurkijarvi, "Karl F. Lindman: The last Hertzian, and a harbinger of electromagnetic chirality". *IEEE Antennas and Propagation Magazine* **34**, 24–30 (1992).
19. I. Tinoco and M. P. Freeman, "The optical activity of oriented copper helices. I. Experimental," *J. Phys. Chem.* **61**, 1196–1200 (1957).
20. J. B. Pendry, "A chiral route to negative refraction," *Science* **306**, 1353–1355 (2004).
21. A. Papakostas, A. Potts, D. M. Bagnall, S. L. Prosvirnin, H. J. Coles and N. I. Zheludev, "Optical manisfestation of planar chirality," *Phys. Rev. Lett.* **90**, 107404 (2003).

22. A. S. Schwanecke, A. Krasavin, D. M. Bagnall, A. Potts, A. V. Zayats and N. I. Zheludev, "Broken time symmetry of light interaction with planar chiral nanostructures," *Phys. Rev. Lett.* **91**, 247404 (2003).
23. T. Vallius, K. Jefimovs, J. Turunen, P. Vahimaa and Y. Svirko, "Optical activity in subwalength-period arrays of chiral metallic particles," *Appl. Phys. Lett.* **83**, 234–236 (2003).
24. M. Kuwata-Gonokami, N. Saito, Y. Ino, M. Kauranen, K. Jefimovs, T. Vallius, J. Turunen and Y. Svirko, "Giant optical activity in quasi-two-dimensional planar nanostructures," *Phys. Rev. Lett.* **95**, 227401 (2005).
25. B. K. Canfield, S. Kujala, K. Jefimovs, J. Turunen and M. Kauranen, "Linear and nonlinear optical responses influenced by broken symmetry in an array of gold nanoparticles," *Opt. Express* **12**, 5418–5423 (2004).
26. B. K. Canfield, S. Kujala,K. Laiho, K. Jefimovs, J. Turunen and M. Kauranen, "Remarkable polarization sensitivity of gold nanoparticle arrays," *Opt. Express* **12**, 5418–5423 (2004).
27. W. Zhang, A. Potts, A. Papakostas and D. M. Bagnall, "Intensity modulation and polarization rotation of visible light by dielectric planar chiral materials," *Appl. Phys. Lett.* **86**, 231905 (2005).
28. M. Decker, M. W. Klein, M. Wegener and S. Linden, "Circular dichroism of planar chiral magnetic metamaterials," *Opt. Lett.* **32**, 856–858 (2007).
29. E. Plum, V. A. Fedotov, A. S. Schwanecke, N. I. Zheludev and Y. Chen "Giant optical gyrotropy due to electromagnetic coupling," *Appl. Phys. Lett.* **90**, 223113 (2007).
30. A. V. Rogacheva, V. A. Fedotov, A. S. Schwanecke and N. I. Zheludev, "Giant gyrotropy due to electromagnetic-field coupling in a bilayered chiral structure." *Phys. Rev. Lett.* **97**, 177401 (2006).
31. S. L. Prosvirnin and N. I. Zheludev, "Polarization effects in the diffraction of light by planar chiral structure," *Phys. Rev. E* **71**, 037603 (2005).
32. A. V. Krasavin, A. S. Schwanecke and N. I. Zheludev, "Extraordinary properties of light transmission through a small chiral hole in a metallic screen," *J. Opt. A* **8**, S98–S105 (2006).
33. M. Reichelt, S. W. Koch, A. Krasavin, J. V. Moloney, A. S. Schwanecke, T. Stroucken, E. M. Wright and N. I. Zheludev, "Broken enantiomeric symmetry for electromagnetic waves interacting with planar chiral nanostructures", *Appl. Phys. B* **84**, 97–101 (2006).
34. J. K. Gansel, M. Thiel, M. S. Rill, M. Decker, K. Bade, V. Saile, G. von Freymann, S. Linden, M. Wegener, "Gold helix photonic metamaterial as broadband circular polarizer," *Science* **325**, 1513–1515 (2009).

35. C. M. Soukoulis and M. Wegener, "Past achievements and futur challenges in the development of three-dimensional photonic metamaterials," *Nature Photon.* **5**, 523–530 (2011).
36. Y. Zhao, M. A. Belkin, A. Alù, "Twisted optical metamaterials for planarized ultrathin broadband circular polarizers," *Nature Commun.* **3**, 870 (2012).
37. N. I. Zheludev, "The road ahead for metamarials," *Science* **328**, 582–583 (2010).
38. H. A. Lorentz, "The theorem of Poynting concerning the energy in the electromagnetic field and two general propositions concerning the propagation of light," *Amsterdammer Akademie der Wetenschappen* **4**, 176 (1896).
39. R. Carminati, M. Nieto-Vesperinas, J. -J. Greffet, "Reciprocity of evanescent electromagnetic waves," *J. Opt. Soc. Am. A* **15**, 706–712 (1997).
40. L. Hecht, L. D. Barron, "Rayleigh and Raman optical activity from chiral surfaces," *Chem. Phys. Lett.* **225**, 525–530 (1994).
41. L. D. Barron, "Parity and optical activity," *Nature* **238**, 17–19 (1972).
42. V. A. Fedotov, P. L. Mladyonov, S. L. Prosvirnin, A. V. Rogacheva, Y. Chen and N. I. Zheludev, "Asymmetric propagation of electromagnetic waves through a planar chiral structure," *Phys. Rev. Lett.* **97**, 167401 (2006).
43. V. A. Fedotov, A. S. Schwanecke, N. I. Zheludev, V. V. Khardikov and S. L. Prosvirnin, "Asymmetric transmission of light and enantiomerically sensistive plasmon resonance in planar chiral nanostructures," *Nano Lett.* **7**, 1996–1999 (2007).
44. W. L. Barnes, A. Dereux, T. W. Ebbesen, "Surface plasmon subwavelength optics," *Nature* **424**, 824 (2003).
45. C. Genet, T. W. Ebbesen, "Light in tiny holes," *Nature* **445**, 39–46 (2007).
46. L. Novotny, and B. Hecht, *Principles of Nano-Optics* (Cambridge Press, London, 2006).
47. A. S. Schwanecke, V. A. Fedotov, V. V. Khardikov, S. L. Prosvirnin, Y. Chen and N. I. Zheludev, "Asymmetric transmission of light and enantiomerically sensistive plasmon resonance in planar chiral nanostructures," *Nano Lett.* **8**, 2940–2943 (2008).
48. A. Drezet, C. Genet, J.-Y. Laluet, and T. W. Ebbesen, "Optical chirality without optical activity: How surface plasmons give a twist to light," *Opt. Express* **16**, 12559–12570 (2008).
49. Y. Gorodetski, A. Drezet, C. Genet, and T. W. Ebbesen, "Generating far-field orbital angular momenta from near-field chirality," *Phys. Rev. Lett.* **110**, 203906–203910 (2013).

50. H. B. G. Casimir, "On Onsager's Principle of Microscopic Reversibility," *Rev. Mod. Phys.* **17**, 343–350 (1945).
51. L. R. Arnaut, "Chirality in multi-dimensional space with application to electromagnetic characterization of multi-dimensionnal chiral and semi chiral media," *J. Electromagn. Waves Appl.* **11**, 1459–1482 (1997).
52. M. A. Osipov, B. T. Pickup, M. Fehervari, and D. A. Dunmur, "Chirality measure and chiral order parameter for a two dimensionnal system," *Mol. Phys.* **94**, 283–287 (1998).
53. C. W. Bunn, *Chemical Crystallography* (Oxford University Press, New York, 1945).
54. E. Plum, V. A. Fedotov, N. I. Zheludev, "Extrinsic electromagnetic chirality in metamaterials," *J. Opt. A: Pure Appl. Opt.* **11**, 074009 (2009).
55. B. M. Maoz, A. Ben Moshe, D. Vestler, O. Bar-Elli, and G. Markovich, "Chiroptical effects in planar achiral plasmonic oriented nanohole arrays", *Nano Lett.* **12**, 2357–2361 (2012).
56. A. Degiron and T. W. Ebbesen, "Analysis of the transmission process through a single aperture surrounded by periodic corrugations," *Opt. Express* **12**, 3694–3700 (2004).
57. H. J. Lezec, A. Degiron, E. Devaux, R. A. Linke, L. Martin-Moreno, F. J. Garcia-Vidal and T. W. Ebbesen, "Beaming light from a subwavelength aperture," *Science* **297**, 820–822 (2002).
58. C. Genet, M. P. van Exter, and J. P.Woerdman, "Huygens description of resonance phenomena in subwavelength hole arrays," *J. Opt. Soc. Am. A* **22**, 998–1002 (2005).
59. C. Genet, M. P. van Exter, and J. P. Woerdman, "Fano-type interpretation of red shifts and red tails in hole arrays transmission spectra," *Opt. Commun.* **225**, 331–336 (2003).
60. A. Drezet, C. Genet and T. W. Ebbesen, "Miniature plasmonic wave plates," *Phys. Rev. Lett.* **101**, 043902 (2008).
61. Y. Gorodetski, E. Lombard, A. Drezet, C. Genet, T. W. Ebbesen, "A perfect plasmonic quarter-wave plate," *Appl. Phys. Lett.* **101**, 201103 (2012).
62. F. Le Roy-Brehonnet, B. Le Jeune, "Utilization of Mueller matrix formalism to obtain optical targets depolarization and polarization properties," *Prog. Quant. Electr.* **21**, 109–151 (1997).
63. C. Genet, E. Altewischer, M. P. van Exter and J. P. Woerdman, "Optical depolarization induced by arrays of subwavelength metal holes," *Phys. Rev. B* **71**, 033409 (2005).
64. V. E. Lembessis, M. Babiker, and D. L. Andrews, "Surface optical vortices," *Phys. Rev. A* **79**, 011806–011809 (2009).

65. S. Yang, W. Chen, R. L. Nelson, and Q. Zhan, "Miniature circular polarization analyzer with spiral plasmonic lens," *Opt. Lett.* **34**, 3047–3049 (2009).
66. S. Zhang, H. Wei, K. Bao, U. Håkanson, N. J. Halas, P. Nordlander, and H. Xu, "Chiral surface plasmon polaritons on metallic nanowires," *Phys. Rev. Lett.* **107**, 096801–096804 (2011).
67. F. Rüting, A. I. Fernández-Domínguez, L. Martín-Moreno, and F. J. García-Vidal, "Subwavelength chiral surface plasmons that carry tuneable orbital angular momentum," *Phys. Rev. B* **86**, 075437–075441 (2012).
68. T. Ohno and S. Miyanishi, "Study of surface plasmon chirality induced by Archimedes' spiral grooves," *Opt. Express* **14**, 6285–6290 (2007).
69. Y. Gorodetski, A. Niv, V. Kleiner, and E. Hasman, "Observation of the spin-based plasmonic effect in nanoscale structures," *Phys. Rev. Lett.* **101**, 043903–043906 (2008).
70. H. Kim, J. Park , S.-W. Cho , S.-Y. Lee, M. Kang, and B. Lee, "Synthesis and dynamic switching of surface plasmon vortices with plasmonic vortex lens," *Nano Lett.* **10**, 529–536 (2010).
71. S.-W. Cho, J. Park, S.-Y. Lee, H. Kim, and B. Lee, "Coupling of spin and angular momentum of light in plasmonic vortex," *Opt. Express* **20**, 10083–10094 (2012).
72. Y. Gorodetski, N. Shitrit, I. Bretner, V. Kleiner, and E. Hasman, "Observation of optical spin symmetry breaking in nanoapertures," *Nano Lett.* **9**, 3016–3019 (2009).
73. G. Rui, R. L. Nelson, and Q. Zhan, "Circularly polarized unidirectional emission via a coupled plasmonic spiral antenna," *Opt. Lett.* **36**, 4533–4535 (2011).
74. L. Marrucci, C. Manzo, and D. Paparo, "Optical spin-to-orbital angular momentum conversion in inhomogeneous anisotropic media," *Phys. Rev. Lett.* **96**, 163905–163908 (2006).
75. E. Brasselet, N. Murazawa, H. Misawa, and S. Juodkazis, "Optical vortices from liquid crystal droplets," *Phys. Rev. Lett.* **103**, 103903–103906 (2009).
76. E. Lombard, A. Drezet, C. Genet, and T. W. Ebbesen, "Polarization control of non-diffractive helical optical beams through subwavelength metallic apertures," *New J. Phys.* **10**, 023027–023039 (2010).

Chapter 3

Chiral Effects in Plasmonic Metasurfaces and Twisted Metamaterials

Yang Zhao, Amir Nader Askarpour, and Andrea Alù

Department of Electrical and Computer Engineering, The University of Texas at Austin, 1 University Station C0803, Austin, TX 78712, USA
alu@mail.utexas.edu

Chiral effects and optical activity have been a central topic of interest in optics and artificial materials for several decades. Here we discuss and model the physical mechanisms behind extrinsic chiral effects induced over ultrathin plasmonic metasurfaces and in optical metamaterials characterized by controlled arrangements of stacked metasurfaces. We explain these effects using analytical models, supported and validated by full-wave simulations. In this chapter, we provide physical insights into the anomalous optical effects offered by these nanodevices and we envision potential optical applications of these concepts.

3.1 Introduction

The nature of light propagation has fascinated mankind for centuries and has been studied since ancient times by Greek philosophers, who were the first to discover that light travels along straight

Singular and Chiral Nanoplasmonics
Edited by Svetlana V. Boriskina and Nikolay I. Zheludev

ISBN 978-981-4613-17-0 (Hardcover), 978-981-4613-18-7 (eBook)
www.panstanford.com

lines in homogeneous media, can be focused by curved glasses, can be reflected by smooth surfaces, and is refracted when crossing interfaces between media with different density. Mankind has been using these properties to transmit signals, detect, and exchange information for centuries. The physical principles behind these phenomena were established well before 1900 and are based on structures and devices many wavelengths large. However, due to recent advances in science and technology, scientists and researchers have become more interested in controlling the interaction of light with matter in nanostructures of size comparable with the wavelength of operation. These possibilities can largely overcome the conventional limitations of light manipulation and their applications, mostly associated with the diffraction limit. It has been found, in particular, that nanoparticles and their collections may achieve exotic phenomena that are not available in conventional optics. The whole research area of metamaterials, currently one of the most popular in optics, has emerged from these concepts [1–21].

Metamaterials are artificial materials composed of sub-wavelength engineered inclusions, designed to collectively achieve exotic electromagnetic properties at the frequency of interest, beyond those available in nature or in any of their constituents. Although technology and nanofabrication have evolved at a fast pace and reached levels of flexibility that were unimaginable even a few years ago, specific physical and engineering challenges associated with design and realization of optical metamaterials still remain unresolved. In particular, technological challenges limit the realization of fully three-dimensional nanoscale metamaterials, especially in the optical regime [14, 19, 22–27]; in addition, the complex wave matter interaction in large arrays of resonant nanostructures, often with exotic plasmonic properties [28–30], requires parallel advances in the theoretical understanding and modeling of these effects.

Different from metamaterials, metasurfaces or metafilms are effectively the two-dimensional equivalent of metamaterials [31–33]. The interest in their design, realization, and characterization has grown in parallel with the interest in metamaterials, as it was shown that, by tailoring their resonant constituent sub-wavelength inclusions, they may provide analogous exotic electromagnetic phenomena, such as negative index of refraction [34], sub-diffraction imaging [35], nanocircuitry [36–38], and cloaking [39, 40]. Their reduced profile makes them appealing and of easier realization from a practical and technological point of view. In the following, we review our recent efforts in using metasurfaces and their combinations to realize unique chiral effects at the nanoscale, and in modeling these effects using analytical techniques that can provide physical insights into these anomalous optical phenomena.

It is interesting that one of the first forms of exotic artificial materials, realized over a century before the term "metamaterials" was even coined, was produced with the goal of largely increasing optical activity and chirality in natural materials by J. C. Bose in 1898 [41, 42]. In these early experiments, twisted artificial molecules were used to introduce strong chiral effects that were essential to tailor and control the polarization of light and create circular polarization. Today, even more than at those times, control and detection of the polarization state of light is fundamental in several optical and photonic applications, and it is one of the relevant functionalities that distinguish several biological species [43–46] from the human vision system, which cannot detect polarization information. Although linear polarizers are quite easily realized, and they may work over a broad range of frequencies, detection of circular polarization is more challenging, as it is inherently based on phase detection, and it may be usually performed only over a limited range of frequencies using quarter-wave plate technology. However, biological species can detect circular polarization information over much broader bandwidths, and they are able to use this information for orientation, signaling, and defense. Therefore, it would be very relevant to develop analogous functionalities in a man-made photonic system operating at the nanoscale.

As we discuss in the following, the exotic features of plasmonic metasurfaces in the visible domain may provide the necessary tools to induce compact resonances and anomalous optical response for strong chiral effects. In the following sections, we first put forward a general theoretical model for ultrathin planar plasmonic metasurfaces, with a particular attention to their chiral effects, using an averaged transmission-line shunt admittance tensor based on the generalized dipolar polarizability of the inclusions forming the array.

In our model, we fully take into account the dynamic interaction within the array and the cross-polarization coupling. We validate the accuracy of our approach and its limitations by comparing the homogenized metasurface model with full-wave numerical simulations based on the finite integration technique (CST Microwave Studio^{TM} 2011), in order to outline the conditions under which the averaged optical surface impedance may accurately describe the complex wave interaction of planar plasmonic metasurfaces. Then, we apply this analytical model to explore various technological approaches to achieve strong chiral effects using metasurfaces, such as the realization of lithographic periodic arrays of nanoparticles and stacks of plasmonic metasurfaces to form twisted metamaterials, a concept that we have recently introduced to provide strong, broadband optical activity based on a lattice effect [58].

3.2 Wave Interaction with Planar Metasurfaces: Theoretical Basis

One of the major advantages of metasurfaces is the relaxation of complicated fabrication processes required in three-dimensional metamaterials, which is especially challenging at optical frequencies. Recent works based on a three-dimensional lithographic method, direct laser writing (DLW), have demonstrated the realization of three-dimensional metamaterials that exhibit strong, broadband chirality in the near-infrared to mid-infrared regime [26, 47]. DLW is an emerging microfabrication technique based on two-photon absorption to initiate polymerization; the limitation of this method resides in the minimum feature dimensions that may be achieved in the fabricated structure, which is ultimately dominated by the diffraction limit [48–50]. This is due to the fact that light sources commonly utilized in this fabrication method are Ti: Sapphire lasers with an emitting wavelength centered at 800 nm; therefore, the voxel size is severely limited by this length scale. So far, the smallest feature achieved with this method is in the micro-meter range. Planar lithographic methods, such as electron beam lithography, on the contrary, utilize electron beams, implying that, theoretically, the minimum spot size can be as small as a few nanometers, which can provide more possibilities for metamaterial applications in the visible range, including optical activity. Chirality has been demonstrated in either planar [51–55] or three-dimensional metamaterials [56–58] using this fabrication technique. Planar metasurfaces generally exhibit much weaker chiral effects than three-dimensional geometries, because an infinitesimally thin surface is inherently achiral, and excitation at oblique incidence or nonreciprocal responses is required to distinguish between left-handed circular polarization (LCP) and right-handed circular polarization (RCP).

In this chapter, we theoretically discuss how individual and stacked planar metasurfaces may provide strong chiral effects and effectively respond as a bulk three-dimensional chiral metamaterial. We first consider a single, optically thin metasurface located in the $z = 0$ plane, formed by arbitrarily shaped plasmonic nanoparticles embedded in a rectangular lattice with periods d_x and d_y. We assume here and in the following that the lattice constants are much smaller than the wavelength of operation so that only the *zero*-th diffraction order can propagate away from the metasurface plane, and that the inclusions are not too densely packed so that the optical wave interaction can be modeled using the dipolar approximation with good accuracy. We further assume that the inclusions are sufficiently thin in the direction normal to the array to ensure that only an optical displacement current tangential to the surface may be induced.

These assumptions ensure that, when excited by an external plane wave with arbitrary polarization and incidence angle θ, the inclusions are well described by an electric dipole moment parallel to the surface and a magnetic dipole moment normal to it. In the planar array, the local fields impinging on each inclusion are given by the superposition of the impinging fields and the radiation from other dipoles, due to the coupling with the inclusions in the array. Therefore, the metasurface response may be compactly described through a generalized array polarizability tensor $\underline{\underline{\alpha}}_s$ that relates the induced dipole moments at the origin $\mathbf{p}_{00}$, $m_{00}\hat{\mathbf{z}}$ to the impinging fields $\mathbf{E}_{\text{inc}}$ (electric) and $\mathbf{H}_{\text{inc}}$ (magnetic), including the full dynamic coupling among the inclusions [59, 60]:

$$\begin{pmatrix} \mathbf{p}_{00}/\varepsilon_0 \\ \eta_0 m_{00}/\mu_0 \end{pmatrix} = \left(\begin{pmatrix} \underline{\boldsymbol{\alpha}}_{\text{ee}} & \boldsymbol{\alpha}_{\text{em}} \\ \boldsymbol{\alpha}_{\text{me}} & \boldsymbol{\alpha}_{\text{mm}} \end{pmatrix}^{-1} - \begin{pmatrix} \underline{\mathbf{C}}_{tt} & \mathbf{C}_{tz} \\ \mathbf{C}_{zt} & C_{zz} \end{pmatrix} \right)^{-1} \cdot \begin{pmatrix} \mathbf{E}_{\text{inc}} \times \hat{\mathbf{z}} \\ \eta_0 \mathbf{H}_{\text{inc}} \cdot \hat{\mathbf{z}} \end{pmatrix}$$

$$= \underline{\underline{\boldsymbol{\alpha}}}_s \cdot \begin{pmatrix} \mathbf{E}_{\text{inc}} \times \hat{\mathbf{z}} \\ \eta_0 \mathbf{H}_{\text{inc}} \cdot \hat{\mathbf{z}} \end{pmatrix}. \tag{3.1}$$

In this formula, $\underline{\boldsymbol{\alpha}}_{\text{ee}}$ is the transverse electric polarizability tensor (2×2), $\boldsymbol{\alpha}_{\text{mm}}$ is the normal magnetic polarizability coefficient, and $\boldsymbol{\alpha}_{\text{em}}$, $\boldsymbol{\alpha}_{\text{me}}$ are the magnetoelectric polarizability vectors, which take into account the cross-coupling between transverse electric and normal magnetic effects due to artificial magnetism and polarization effects. In addition, the $\mathbf{C}$ elements are the Green's dyadic coefficients taking into account the array coupling [59], $\hat{\mathbf{z}}$ is the unit normal vector to the array, ε_0, μ_0, and $\eta_0 = \sqrt{\mu_0/\varepsilon_0}$ are the free-space permittivity, permeability, and characteristic impedance, respectively. The reciprocity of the metasurface imposes the Onsager constraints [61, 62] on these elements, which may be combined with the reciprocal properties of the coupling dyads:

$$\underline{\boldsymbol{\alpha}}_{\text{ee}} = \underline{\boldsymbol{\alpha}}_{\text{ee}}^{\text{T}}, \quad \boldsymbol{\alpha}_{\text{me}} = -\boldsymbol{\alpha}_{\text{em}}^{\text{T}}, \quad \underline{\mathbf{C}}_{\text{tt}} = \underline{\mathbf{C}}_{\text{tt}}^{\text{T}}, \quad \mathbf{C}_{\text{zt}} = -\mathbf{C}_{\text{tz}}^{\text{T}}, \tag{3.2}$$

where the superscript T indicates the transpose operation, to ensure that the generalized polarizability tensor also has a reciprocal form:

$$\underline{\underline{\boldsymbol{\alpha}}}_s = \begin{pmatrix} \alpha_{xx}^{\text{ee}} & \alpha_{xy}^{\text{ee}} & \alpha_{xz}^{\text{em}} \\ \alpha_{yx}^{\text{ee}} & \alpha_{yy}^{\text{ee}} & \alpha_{yz}^{\text{em}} \\ \alpha_{zx}^{\text{me}} & \alpha_{zy}^{\text{me}} & \alpha_{zz}^{\text{mm}} \end{pmatrix} = \begin{pmatrix} \underline{\boldsymbol{\alpha}}_{\text{s}}^{\text{ee}} & \boldsymbol{\alpha}_{\text{s}}^{\text{em}} \\ \left(-\boldsymbol{\alpha}_{\text{s}}^{\text{em}}\right)^{\text{T}} & \boldsymbol{\alpha}_{\text{s}}^{\text{mm}} \end{pmatrix}. \tag{3.3}$$

In general, the elements of the generalized polarizability tensor are dependent on the incidence angle, as a symptom of spatial dispersion in the array, due to the influence of the coupling dyadics. For smaller periods, however, these effects are usually small. The induced dipole moments on all the other inclusions are related to $\mathbf{p}_{00}$, m_{00} through a simple phase shift associated with the momentum of the impinging excitation, which is conserved.

It follows that, within the dipolar limit, the tensor $\underline{\underline{\boldsymbol{\alpha}}}_s$ compactly describes the metasurface interaction with an arbitrary impinging plane wave, which effectively induces an averaged electric surface current density tangential to the plane

$$\mathbf{K}_{\text{e}} = \frac{j\omega \mathbf{p}_{00}}{d_x d_y} \quad [A/m] \tag{3.4}$$

and a normal magnetic surface current density

$$\mathbf{K}_{\mathrm{m}} = \frac{j\omega m_{00}}{d_x d_y}\widehat{\mathbf{z}}\ [V/m]. \tag{3.5}$$

We have indicated the units of $\mathbf{K}_{\mathrm{e}}$ and $\mathbf{K}_{\mathrm{m}}$ in these formulas for clarity.

3.2.1 *Transverse-Magnetic Excitation*

For transverse-magnetic (TM) polarized excitation, without loss of generality

$$\mathbf{H_{inc}} = \frac{E_0}{\eta_0} e^{-j\sqrt{k_0^2-k_x^2}z} e^{-jk_x x}\widehat{\mathbf{y}}, \tag{3.6}$$

where k_0 is the wave number in free-space. The associated incident electric field is

$$\mathbf{E_{inc}} = e^{-j\sqrt{k_0^2-k_x^2}z} e^{-jk_x x} E_0 \left(\frac{\sqrt{k_0^2 - k_x^2}}{k_0}\widehat{\mathbf{x}} - \frac{k_x}{k_0}\widehat{\mathbf{z}} \right). \tag{3.7}$$

The induced averaged surface currents radiate plane waves described by the vector potentials

$$\mathbf{A} = \frac{-j\mu_0}{2\sqrt{k_0^2 - k_x^2}} \mathrm{e}^{-j\sqrt{k_0^2-k_x^2}z}\mathbf{K}_{\mathrm{e}}, \tag{3.8}$$

and

$$\mathbf{F} = \frac{-j\epsilon_0}{2\sqrt{k_0^2 - k_x^2}} \mathrm{e}^{-j\sqrt{k_0^2-k_x^2}z}\mathbf{K}_{\mathrm{m}}. \tag{3.9}$$

From these expressions, it is easy to calculate the transmitted and reflected fields. First, by combining all the previous equations, we find the total transmitted magnetic fields associated with the magnetic potential **A**:

$$\mathbf{H}_{\mathrm{t}} = \mathbf{H_{inc}} + \frac{jE_0 e^{-j\left(k_x x + \sqrt{k_0^2-k_x^2}z\right)}}{2d_x d_y \eta_0} \times \left(\alpha_{yx}^{ee}\sqrt{k_0^2 - k_x^2}\widehat{\mathbf{x}} - \alpha_{xx}^{ee}\sqrt{k_0^2 - k_x^2}\widehat{\mathbf{y}} - \alpha_{yx}^{ee} k_x\widehat{\mathbf{z}} \right), \tag{3.10}$$

and the corresponding reflected magnetic field

$$\mathbf{H_r} = \frac{jE_0 e^{-j\left(k_x x - \sqrt{k_0^2 - k_x^2} z\right)}}{2d_x d_y \eta_0} \times \left(-\alpha_{yx}^{ee}\sqrt{k_0^2 - k_x^2}\widehat{\mathbf{x}} + \alpha_{xx}^{ee}\sqrt{k_0^2 - k_x^2}\widehat{\mathbf{y}} - \alpha_{yx}^{ee} k_x \widehat{\mathbf{z}}\right). \quad (3.11)$$

Therefore, the transmission and reflection coefficients for electric fields due to the induced electric current density are compactly given by:

$$\begin{aligned} T_{\mathrm{mm}}^{\mathrm{e}} &= 1 - \frac{jk_0 \cos\theta}{2d_x d_y}\alpha_{xx}^{ee}; \quad T_{\mathrm{em}}^{\mathrm{e}} = -\frac{jk_0}{2d_x d_y}\alpha_{yx}^{ee} \\ R_{\mathrm{mm}}^{\mathrm{e}} &= -\frac{jk_0 \cos\theta}{2d_x d_y}\alpha_{xx}^{ee}; \quad R_{\mathrm{em}}^{\mathrm{e}} = -\frac{jk_0}{2d_x d_y}\alpha_{yx}^{ee}, \end{aligned} \quad (3.12)$$

where $\cos\theta = \sqrt{k_0^2 - k_x^2}/k_0$. $T_{\mathrm{mm}}\,(R_{\mathrm{mm}})$ and $T_{\mathrm{em}}\,(R_{\mathrm{em}})$ are the transmission (reflection) coefficients for TM incident polarization radiating TM and TE polarizations, respectively.

The electric potential **F** similarly generates a transmitted electric field

$$\mathbf{E_t} = -\frac{jE_0 e^{-j\left(k_x x + \sqrt{k_0^2 - k_x^2} z\right)}}{2d_x d_y}\alpha_{zx}^{\mathrm{me}} k_x \widehat{\mathbf{y}}, \quad (3.13)$$

and a reflected field

$$\mathbf{E_r} = -\frac{jE_0 e^{-j\left(k_x x - \sqrt{k_0^2 - k_x^2} z\right)}}{2d_x d_y}\alpha_{zx}^{\mathrm{me}} k_x \widehat{\mathbf{y}}. \quad (3.14)$$

Therefore, the transmission and reflection coefficients due to the magnetic current density for TM excitation are

$$\begin{aligned} T_{\mathrm{mm}}^{\mathrm{m}} &= 0\,; \quad T_{\mathrm{em}}^{\mathrm{m}} = -\frac{jk_0 sin\theta}{2d_x d_y}\alpha_{zx}^{\mathrm{me}} \\ R_{\mathrm{mm}}^{\mathrm{m}} &= 0; \quad R_{\mathrm{em}}^{\mathrm{m}} = -\frac{jk_0 sin\theta}{2d_x d_y}\alpha_{zx}^{\mathrm{me}}. \end{aligned} \quad (3.15)$$

Summing Eq. (3.12) and Eq. (3.15), we find the total transmission and reflection coefficients for TM excitation:

$$
\begin{aligned}
T_{\mathrm{mm}} &= T^{\mathrm{e}}_{\mathrm{mm}} + T^{\mathrm{m}}_{\mathrm{mm}} = 1 - \frac{jk_0 \cos\theta}{2d_x d_y}\alpha^{\mathrm{ee}}_{xx} \\
T_{\mathrm{em}} &= T^{\mathrm{e}}_{\mathrm{em}} + T^{\mathrm{m}}_{\mathrm{em}} = -\frac{jk_0}{2d_x d_y}\left(\alpha^{\mathrm{ee}}_{yx} + \sin\theta\alpha^{\mathrm{me}}_{zx}\right) \\
R_{\mathrm{mm}} &= R^{\mathrm{e}}_{\mathrm{mm}} + R^{\mathrm{m}}_{\mathrm{mm}} = -\frac{jk_0 \cos\theta}{2d_x d_y}\alpha^{\mathrm{ee}}_{xx} \\
R_{\mathrm{em}} &= R^{\mathrm{e}}_{\mathrm{em}} + R^{\mathrm{m}}_{\mathrm{em}} = -\frac{jk_0}{2d_x d_y}\left(\alpha^{\mathrm{ee}}_{yx} + \sin\theta\alpha^{\mathrm{me}}_{zx}\right)
\end{aligned}
\tag{3.16}
$$

The above equations compactly describe the interaction of the metasurface with arbitrary TM impinging waves. Notice that the relations $T_{\mathrm{mm}} = 1 + R_{\mathrm{mm}}$ and $T_{\mathrm{em}} = R_{\mathrm{em}}$ hold due to the continuity of the tangential electric field on the surface, which was supposed here to be infinitesimally thin.

3.2.2 *Transverse Electric Plane Wave Incidence*

For transverse electric (TE) excitation, we can assume without losing generality that the impinging electric field is given by

$$
\mathbf{E}_{\mathbf{inc}} = E_0 e^{-j\sqrt{k_0^2-k_x^2}z} e^{-jk_x x}\widehat{\mathbf{y}}. \tag{3.17}
$$

The radiated fields can be similarly calculated as in the previous section, but now, also the normal component of the magnetic field can contribute to induce transverse electric currents on the surface. The transmitted electric field produced by the electric current density is

$$
\mathbf{E}_{\mathbf{t}} = \mathbf{E}_{\mathbf{inc}} - \frac{jE_0 e^{-j\left(k_x x + \sqrt{k_0^2-k_x^2}z\right)}}{2d_x d_y}
\begin{pmatrix}
\frac{\sqrt{k_0^2 - k_x^2}}{k_0}\left(k_x\alpha^{\mathrm{em}}_{xz} + k_0\alpha^{\mathrm{ee}}_{xy}\right)\widehat{\mathbf{x}} \\
+\frac{k_0}{\sqrt{k_0^2 - k_x^2}}\left(k_x\alpha^{\mathrm{em}}_{yz} + k_0\alpha^{\mathrm{ee}}_{yy}\right)\widehat{\mathbf{y}} \\
-\frac{k_x}{k_0}\left(k_x\alpha^{\mathrm{em}}_{xz} + k_0\alpha^{\mathrm{ee}}_{xy}\right)\widehat{\mathbf{z}}
\end{pmatrix},
\tag{3.18}
$$

and the corresponding reflected field is

$$\mathbf{E_r} = -\frac{jE_0 e^{-j\left(k_x x - \sqrt{k_0^2-k_x^2}z\right)}}{2d_x d_y}\left(\begin{array}{l}\frac{\sqrt{k_0^2-k_x^2}}{k_0}\left(k_x\alpha_{xz}^{\mathrm{em}} + k_0\alpha_{xy}^{\mathrm{ee}}\right)\widehat{\mathbf{x}} \\ +\frac{k_0}{\sqrt{k_0^2-k_x^2}}\left(k_x\alpha_{yz}^{\mathrm{em}} + k_0\alpha_{yy}^{\mathrm{ee}}\right)\widehat{\mathbf{y}} \\ +\frac{k_x}{k_0}\left(k_x\alpha_{xz}^{\mathrm{em}} + k_0\alpha_{xy}^{\mathrm{ee}}\right)\widehat{\mathbf{z}}\end{array}\right). \tag{3.19}$$

Therefore, the transmission and reflection coefficients due to electric current density are similarly calculated as

$$\begin{aligned} T_{\mathrm{ee}}^{\mathrm{e}} &= 1 - jk_0\left(\frac{\sec\theta\alpha_{yy}^{\mathrm{ee}} + \tan\theta\alpha_{yz}^{\mathrm{em}}}{2d_x d_y}\right), \quad T_{\mathrm{me}}^{\mathrm{e}} = -jk_0\frac{\alpha_{xy}^{\mathrm{ee}} + \sin\theta\alpha_{xz}^{\mathrm{em}}}{2d_x d_y} \\ R_{\mathrm{ee}}^{\mathrm{e}} &= -jk_0\left(\frac{\sec\theta\alpha_{yy}^{\mathrm{ee}} + \tan\theta\alpha_{zz}^{\mathrm{mm}}}{2d_x d_y}\right), \quad R_{\mathrm{me}}^{\mathrm{e}} = -jk_0\frac{\alpha_{xy}^{\mathrm{ee}} + \sin\theta\alpha_{xz}^{\mathrm{em}}}{2d_x d_y}. \end{aligned} \tag{3.20}$$

Also, the magnetic current density may be induced by both electric and magnetic fields. The transmitted electric field due to its contribution is given by

$$\mathbf{E_t} = -\frac{jE_0 e^{-j\left(k_x x + \sqrt{k_0^2-k_x^2}z\right)}k_x}{2d_x d_y\sqrt{k_0^2-k_x^2}}\left(k_x\alpha_{zz}^{\mathrm{mm}} + k_0\alpha_{zy}^{\mathrm{me}}\right)\widehat{\mathbf{y}}, \tag{3.21}$$

and the reflected field is

$$\mathbf{E_r} = -\frac{jE_0 e^{-j\left(k_x x - \sqrt{k_0^2-k_x^2}z\right)}k_x}{2d_x d_y\sqrt{k_0^2-k_x^2}}\left(k_x\alpha_{zz}^{\mathrm{mm}} + k_0\alpha_{zy}^{\mathrm{me}}\right)\widehat{\mathbf{y}}. \tag{3.22}$$

Therefore, transmission and reflection coefficients for TE excitation induced by the magnetic current density are

$$\begin{aligned} T_{\mathrm{ee}}^{\mathrm{m}} &= -\frac{jk_0}{2d_x d_y}\left(\tan\theta\alpha_{zy}^{\mathrm{me}} + \sin\theta\tan\theta\alpha_{zz}^{\mathrm{mm}}\right); \quad T_{\mathrm{me}}^{\mathrm{m}} = 0 \\ R_{\mathrm{ee}}^{\mathrm{m}} &= -\frac{jk_0}{2d_x d_y}\left(\tan\theta\alpha_{zy}^{\mathrm{me}} + \sin\theta\tan\theta\alpha_{zz}^{\mathrm{mm}}\right); \quad R_{\mathrm{me}}^{\mathrm{m}} = 0. \end{aligned} \tag{3.23}$$

By combining (3.20) and (3.23), we find the total transmission and reflection coefficients for TE excitation:

$$T_{\text{ee}} = T_{\text{ee}}^{\text{e}} + T_{\text{ee}}^{\text{m}} = 1 - jk_0 \left(\frac{\sec\theta \alpha_{yy}^{\text{ee}} + \sin\theta \tan\theta \alpha_{zz}^{\text{mm}}}{2d_x d_y} \right)$$
$$T_{\text{me}} = T_{\text{me}}^{\text{e}} + T_{\text{me}}^{\text{m}} = -jk_0 \left(\frac{\alpha_{xy}^{\text{ee}} + \sin\theta \alpha_{xz}^{\text{em}}}{2d_x d_y} \right) \quad (3.24)$$
$$R_{\text{ee}} = R_{\text{ee}}^{\text{e}} + R_{\text{ee}}^{\text{m}} = -jk_0 \left(\frac{\sec\theta \alpha_{yy}^{\text{ee}} + \sin\theta \tan\theta \alpha_{zz}^{\text{mm}}}{2d_x d_y} \right)$$
$$R_{\text{me}} = R_{\text{me}}^{\text{e}} + R_{\text{me}}^{\text{m}} = -jk_0 \left(\frac{\alpha_{xy}^{\text{ee}} + \sin\theta \alpha_{xz}^{\text{em}}}{2d_x d_y} \right).$$

Also, in this case, $T_{\text{ee(mm)}} = 1 + R_{\text{ee(mm)}}$, $T_{\text{em(me)}} = R_{\text{em(me)}}$, as expected.

3.2.3 *Circularly Polarized Plane Wave Incidence*

As we are interested in the overall chiral response of the metasurface, we can now transform the reflection and transmission tensors obtained in the previous subsections from a linear into a circular basis, assuming circularly polarized inputs. This is easily obtained by considering the following transformation:

$$T_{\text{LL}} = \frac{1}{2}\left(T_{\text{ee}} - jT_{\text{em}} + jT_{\text{me}} + T_{\text{mm}}\right)$$
$$T_{\text{LR}} = \frac{1}{2}\left(-T_{\text{ee}} - jT_{\text{em}} - jT_{\text{me}} + T_{\text{mm}}\right) \quad (3.25)$$
$$T_{\text{RL}} = \frac{1}{2}\left(-T_{\text{ee}} + jT_{\text{em}} + jT_{\text{me}} + T_{\text{mm}}\right)$$
$$T_{\text{RR}} = \frac{1}{2}\left(T_{\text{ee}} + jT_{\text{em}} - jT_{\text{me}} + T_{\text{mm}}\right).$$

which constructs the transmission matrix for circularly polarized inputs

$$\mathbf{T}_{\text{cp}} = \begin{pmatrix} T_{\text{LL}} & T_{\text{LR}} \\ T_{\text{RL}} & T_{\text{RR}} \end{pmatrix}, \quad (3.26)$$

where the second letter in the subscript of each element refers to the polarization of excitation (LCP or RCP) and the first one to the transmitted polarization. We can express the elements of (3.26) in terms of the polarizability elements, obtaining

$$T_{\mathrm{LL}} = \left(1 + \frac{\alpha_{xz}^{\mathrm{em}} k_0 \sin\theta}{2d_x d_y}\right) - \frac{jk_0}{4d_x d_y}\left(\alpha_{xx}^{\mathrm{ee}} \cos\theta + \alpha_{yy}^{\mathrm{ee}} \sec\theta + \alpha_{zz}^{\mathrm{mm}} \sin\theta \tan\theta\right), \tag{3.27}$$

$$T_{\mathrm{LR}} = \frac{jk_0}{4d_x d_y}\left(2j\alpha_{xy}^{\mathrm{ee}} - \alpha_{xx}^{\mathrm{ee}} \cos\theta + \alpha_{yy}^{\mathrm{ee}} \sec\theta + \alpha_{zz}^{\mathrm{mm}} \sin\theta \tan\theta\right), \tag{3.28}$$

$$T_{\mathrm{RL}} = \frac{jk_0}{4d_x d_y}\left(-2j\alpha_{xy}^{\mathrm{ee}} - \alpha_{xx}^{\mathrm{ee}} \cos\theta + \alpha_{yy}^{\mathrm{ee}} \sec\theta + \alpha_{zz}^{\mathrm{mm}} \sin\theta \tan\theta\right), \tag{3.29}$$

$$T_{\mathrm{RR}} = \left(1 - \frac{\alpha_{xz}^{\mathrm{em}} k_0 \sin\theta}{2d_x d_y}\right) - \frac{jk_0}{4d_x d_y}\left(\alpha_{xx}^{\mathrm{ee}} \cos\theta + \alpha_{yy}^{\mathrm{ee}} \sec\theta + \alpha_{zz}^{\mathrm{mm}} \sin\theta \tan\theta\right). \tag{3.30}$$

Similar expressions can be derived for the reflection coefficients, not reported here for brevity.

By inspecting Eqs. (3.27)–(3.30), we find that a nonzero difference between T_{LL} and T_{RR} can only occur at oblique incidence, that is, $\theta \neq 0$. This is due to our assumptions of infinitesimally thin surface and reciprocal response, which makes a planar 2D surface excited at normal incidence inherently achiral for symmetry. In addition, even at oblique incidence, LCP and RCP response may be distinguishable only when the bianisotropy term $\alpha_{xz}^{\mathrm{em}}$ is nonzero, which requires no mirror symmetry along $\hat{\mathbf{y}}$. Analogously, if the plane of incidence is in the yz plane, the polarizability tensor element responsible for the difference between the two handedness is $\alpha_{yz}^{\mathrm{em}}$, which is nonzero for inclusions without mirror symmetry along $\hat{\mathbf{x}}$, consistent with the discussions in [54]. This is shown in Fig. 3.1 for two representative inclusions with asymmetric properties along $\hat{\mathbf{y}}$. When excited at normal incidence, the metasurfaces provide $T_{\mathrm{LL}} = T_{\mathrm{RR}}$, but when excited at 45° in the xz plane, the two transmission coefficients support a different response for the two circular polarizations, with sharp differences around the inclusion resonance. Although this phenomenon is inherently narrowband, it may be exploited to induce chiral response on a single, ultrathin surface, which may be of interest for several optical applications (see [54] and references therein).

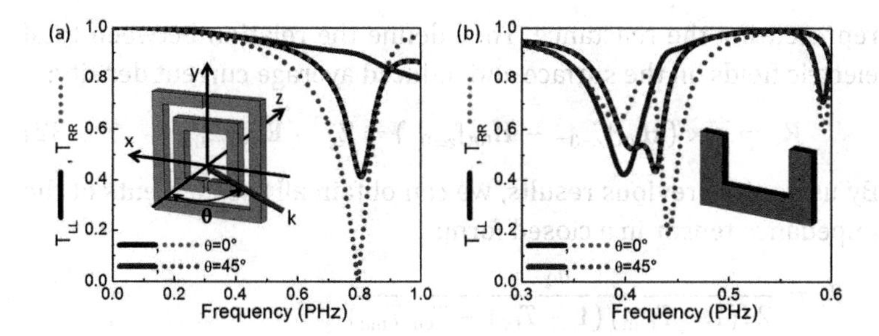

Figure 3.1 Comparison of $T_{\rm LL}$ (solid curves) and $T_{\rm RR}$ (dotted curves) at normal (black) and oblique incidence (45°, red), for (a) PEC and (b) silver split-ring resonator metasurfaces. The unit cell element and direction of propagation is shown in the insets. In both cases, the metasurface lies on the xy plane and lacks mirror symmetry along the y-direction, creating a nonzero $\alpha_{xz}^{\rm em}$ term, which is responsible for the difference between $T_{\rm LL}$ and $T_{\rm RR}$ at oblique incidence.

3.2.4 *Surface Impedance Model*

It is obvious from the previous results that a single, ultrathin metasurface can hardly provide chiral response over a broad bandwidth. For this reason, we aim to apply our results to model stacks of cascaded multilayered structures, which we can analyze using the transmission-line approach based on the results of the previous sections [58, 63]. This approach assumes that the main coupling mechanism among parallel metasurfaces is associated with the *zero*-th order diffraction from the surface, consistent with the previous analysis. This assumption is valid as long as the distance between neighboring surfaces is larger than the period among inclusions in each plane [63]. For simplicity, we will limit our analysis to normal incidence, which ensures that the magnetic current density does not contribute to radiation, consistent with (3.15), (3.23). This ensures that our metasurface can be modeled as a 2 × 2 shunt admittance element with surface impedance $\underline{\mathbf{Z}}_{\rm s}$ or admittance $\underline{\mathbf{Y}}_{\rm s}$:

$$\underline{\mathbf{Z}}_{\rm s} = \begin{pmatrix} Z_{xx} & Z_{xy} \\ Z_{yx} & Z_{yy} \end{pmatrix} = \underline{\mathbf{Y}}_{\rm s}^{-1}. \tag{3.31}$$

Each element in the matrix, $Z_{\rm mn} = r_{\rm mn} + i x_{\rm mn}$ is a complex quantity with real part $r_{\rm mn}$ denoting the resistance and imaginary part $x_{\rm mn}$

representing the reactance. They define the relation between total electric fields on the surface and induced average current density:

$$\mathbf{K}_{\mathrm{e}} = \hat{\mathbf{z}} \times \left(\mathbf{H}_{\mathrm{tot}}|_{z=0^+} - \mathbf{H}_{\mathrm{tot}}|_{z=0^-}\right) = \underline{\mathbf{Z}}_{\mathrm{s}}^{-1} \cdot \mathbf{E}_{\mathrm{tot}}|_{z=0} . \qquad (3.32)$$

By using the previous results, we can obtain all the elements of the impedance tensor in a closed form:

$$\underline{\mathbf{Z}}_{\mathrm{s}} = \frac{\eta_0}{2\left((1-T_{\mathrm{mm}})(1-T_{\mathrm{ee}}) - T_{\mathrm{em}}T_{\mathrm{me}}\right)} \times \begin{pmatrix} T_{\mathrm{mm}}(1-T_{\mathrm{ee}}) + T_{\mathrm{em}}T_{\mathrm{me}} & T_{\mathrm{me}} \\ T_{\mathrm{em}} & T_{\mathrm{ee}}(1-T_{\mathrm{mm}}) + T_{\mathrm{em}}T_{\mathrm{me}} \end{pmatrix}. \qquad (3.33)$$

Equation (3.33) effectively relates the impedance description to the previous analysis, connecting the surface impedance to the generalized polarizability of each metasurface. Conversely, it may also be used to retrieve the effective surface impedance with simple scattering measurements, as the transmission (reflection) coefficients are measurable quantities. Similar to Eqs. (3.16)–(3.24), transmission (reflection) coefficients in linear basis may be expressed in terms of the impedance matrix elements as

$$T_{\mathrm{ee}} = \frac{-4Z_{xy}Z_{yx} + 2Z_{yy}(2Z_{xx}+\eta_0)}{-4Z_{xy}Z_{yx} + (2Z_{yy}+\eta_0)(2Z_{xx}+\eta_0)},$$

$$T_{\mathrm{mm}} = \frac{-4Z_{xy}Z_{yx} + 2Z_{xx}(2Z_{yy}+\eta_0)}{-4Z_{xy}Z_{yx} + (2Z_{yy}+\eta_0)(2Z_{xx}+\eta_0)},$$

$$T_{\mathrm{em}} = \frac{2Z_{yx}\eta_0}{-4Z_{xy}Z_{yx} + (2Z_{yy}+\eta_0)(2Z_{xx}+\eta_0)},$$

$$T_{\mathrm{me}} = \frac{2Z_{xy}\eta_0}{-4Z_{xy}Z_{yx} + (2Z_{yy}+\eta_0)(2Z_{xx}+\eta_0)}, \qquad (3.34)$$

and (3.25) may be used to convert them into a circular basis.

After some manipulations, we find that a nonzero impedance on the off-diagonal term $Z_{xy} = r_{xy} + ix_{xy}$ is responsible for a nonzero value of

$$\Delta_{\mathrm{c}} = T_{\mathrm{LR}} - T_{\mathrm{RL}} = \frac{j4Z_{xy}\eta_0}{4Z_{xy}^2 - (2Z_{xx}+\eta_0)(2Z_{yy}+\eta_0)}. \qquad (3.35)$$

Consistent with the previous results, reciprocity at normal incidence requires $T_{\mathrm{LL}} = T_{\mathrm{RR}}$, and the magnitude of the ratio between cross-polarization terms can be simply written as

$$\left|\frac{T_{\mathrm{LR}}}{T_{\mathrm{RL}}}\right|^2 = \frac{\left[r_{xx} - r_{yy} + \left(x_{xy} + x_{yx}\right)\right]^2 + \left[r_{xy} + r_{yx} - \left(x_{xx} - x_{yy}\right)\right]^2}{\left[r_{xx} - r_{yy} - \left(x_{xy} + x_{yx}\right)\right]^2 + \left[r_{xy} + r_{yx} + \left(x_{xx} - x_{yy}\right)\right]^2}. \tag{3.36}$$

This result, combined with the fact that a reciprocal metasurface satisfies $\underline{\mathbf{Z}}_{\mathrm{s}} = \underline{\mathbf{Z}}_{\mathrm{s}}^{\mathrm{T}}$ and a lossless metasurface requires $\underline{\mathbf{Z}}_{\mathrm{s}} = \underline{\mathbf{Z}}_{\mathrm{s}}^{*T}$, where the asterisk denotes complex conjugate, indicates that losses are inherently required to introduce circular dichroism and a difference in the observed cross-polarization terms for a single metasurface. In the next section, we apply these results to the more exciting possibilities offered by stacked metasurfaces.

3.3 Twisted Metamaterials

In the following, we apply the theoretical model developed in the previous section to analyze stacks of metasurfaces with arbitrary rotation imparted along the stack, as originally proposed in [63]. Before analyzing the chirality response of these structures, we validate the transmission-line model described above with full-wave numerical simulations for specific examples of interest. In particular, we confirm that the retrieved impedance properties obtained for excitation of one single metasurface can be extended to model stacks of two metasurfaces with accuracy, even in the case in which the inclusions of either metasurface are rotated by an arbitrary angle.

3.3.1 *Transmission through a Stack of Two Rotated Metasurfaces*

As an example, in this subsection, we consider a metasurface composed of perfect electric conducting (PEC) nanorod inclusions to evaluate the effect of in-plane rotation along a stack of two identical metasurfaces. We assume that one of the two metasurfaces is obtained by rotating each inclusion in the planar lattice by an arbitrary angle φ. The new inclusion individual polarizability can be obtained from the original one after simply applying the rotation matrix

$$\mathbf{Q} = \begin{pmatrix} \cos\varphi & \sin\varphi \\ -\sin\varphi & \cos\varphi \end{pmatrix}, \tag{3.37}$$

but the generalized polarizability of the surface is in general affected in a more complex way, as the coupling coefficients **C** are also affected by the rotation of the inclusion (notice that this is different than rotating the whole surface, which would not affect the coupling coefficients **C**). We assume here that each inclusion is rotated in a regular planar lattice preserving the original arrangement because this is more of interest in terms of possibilities offered by currently available nanofabrication techniques [58]. As long as the dipolar approximation holds, that is, the coupling among neighboring inclusions is well described by the dipolar fields, the elements of **C** depend only on the lattice arrangement, and therefore within this limit, the surface impedance can be simply calculated as:

$$\underline{\mathbf{Z}}_s^{\mathrm{Rot}} = \mathbf{Q} \cdot \mathbf{Z}_s \cdot \mathbf{Q}^T. \tag{3.38}$$

This assumption is valid as long as the inclusion edges are not too close to each other and is more accurate in the case of nearly symmetric inclusions. Consequently, the transmission coefficients after rotation are modified into

$$\begin{aligned}
T_{\mathrm{mm}}^{\mathrm{Rot}} &= \frac{1}{2}\left[T_{\mathrm{mm}} + T_{\mathrm{ee}} + (T_{\mathrm{mm}} - T_{\mathrm{ee}})\cos(2\varphi) - (T_{\mathrm{me}} + T_{\mathrm{em}})\sin(2\varphi)\right] \\
T_{\mathrm{me}}^{\mathrm{Rot}} &= \frac{1}{2}\left[T_{\mathrm{me}} - T_{\mathrm{em}} + (T_{\mathrm{mm}} - T_{\mathrm{ee}})\sin(2\varphi) + (T_{\mathrm{me}} + T_{\mathrm{em}})\cos(2\varphi)\right] \\
T_{\mathrm{em}}^{\mathrm{Rot}} &= \frac{1}{2}\left[-T_{\mathrm{me}} + T_{\mathrm{em}} + (T_{\mathrm{mm}} - T_{\mathrm{ee}})\sin(2\varphi) + (T_{\mathrm{me}} + T_{\mathrm{em}})\cos(2\varphi)\right] \\
T_{\mathrm{ee}}^{\mathrm{Rot}} &= \frac{1}{2}\left[T_{\mathrm{mm}} + T_{\mathrm{ee}} + (-T_{\mathrm{mm}} + T_{\mathrm{ee}})\cos(2\varphi) + (T_{\mathrm{me}} + T_{\mathrm{em}})\sin(2\varphi)\right].
\end{aligned} \tag{3.39}$$

Figure 3.2 shows the transmission coefficients for a thin metasurface composed of PEC inclusions with 10 nm thickness after rotating the inclusions by 45° and 90°, respectively. We compare the semi-analytical calculations based on Eq. (3.39) using the transmission coefficients calculated before rotation with full-wave simulations. A small discrepancy is only visible near the surface resonance for the 45° angle rotation, which is the one for which the coupling constant is more strongly affected by the relative arrangement of neighboring

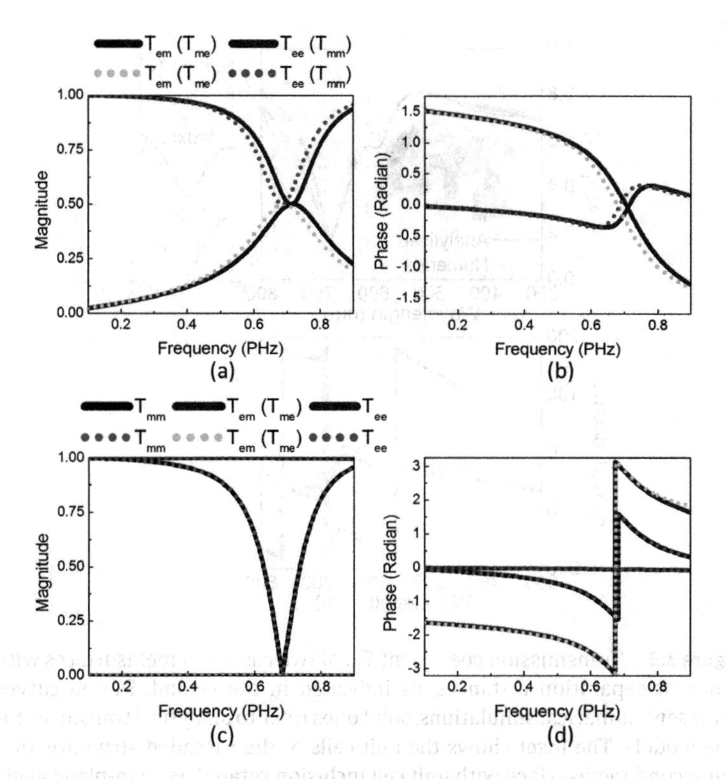

Figure 3.2 Comparison of transmission coefficients for linearly polarized excitation calculated using (3.39) (dotted lines) and with full-wave numerical simulations (solid) for a pair of metasurfaces composed of PEC nanorod inclusions with dimensions 200 nm × 50 nm embedded in a periodic square lattice with period 300 nm. The metasurface thickness is 10 nm, and we consider a rotation angle of (a,b) 45° and (c,d) 90° for the second metasurface.

inclusions. This discrepancy is expected to be further minimized with denser and smaller inclusions.

When two metasurfaces are cascaded, this analytical approach allows us to easily tune and optimize the distance between neighboring metasurfaces and the in-plane rotation angle between consecutive layers for the specific application of interest. The spacing between layers can be readily included in the model by considering a segment of transmission line, under the assumption that the coupling is dominated by the *zero*-th order diffraction

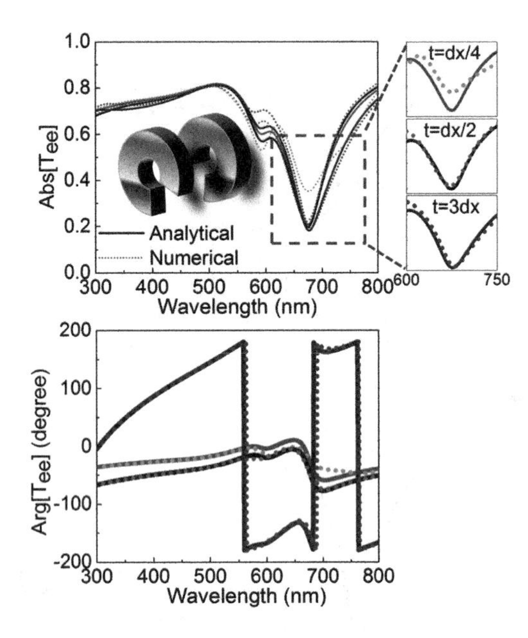

Figure 3.3 Transmission coefficient T_{ee} of two cascaded metasurfaces with different separation distances, as indicated in the legend. Dotted curves represent numerical simulations, solid ones refer to analytical transmission-line models. The inset shows the unit cells of the cascaded structure, that the second metasurface with unit cell inclusion rotated as an in-plane angle with respect to the first metasurface. The outer radius of the split ring is 40 nm, and the inner radius is 20 nm, embedded in a 100 nm by 100 nm square lattice, and an opening of 60° arc; both metasurfaces have thickness of 20 nm.

order, which holds as long as the wavelength is not smaller than $\max\{d_x, d_y\}$. The in-plane rotation is easily modeled using Eq. (3.38), and stacks of two metasurfaces may be analyzed by cascading the ABCD matrices describing the collective interaction of each metasurface and the spacing in between two neighboring surfaces.

Figure 3.3 shows the comparison between analytical results and full-wave simulations for two cascaded metasurfaces made of gold split ring resonator inclusions, for a fixed rotation angle of 60° and a varying distance between the two surfaces. It is seen that for distances smaller than a quarter of the period of the metasurface, the discrepancy becomes noticeable, indicating a noticeable amount of near-field coupling involving higher-order evanescent diffraction orders.

In [63], we have compared more inclusion shapes made of gold using experimentally retrieved permittivity values [64], and excited at optical frequencies, also taken into account plasmonic effects. Our results show that the surface impedance approach can deal with inclusions with arbitrary shapes and rotations, validating this approach as long as the distance between neighboring metasurfaces is larger than a quarter of the transverse metasurface period.

3.3.2 *Optimized Pair of Rotated Metasurfaces for Maximal Chiral Response*

The transmission-line approach using the surface-impedance enables an easy and efficient analysis of arbitrary stacks of metasurfaces, including arbitrary rotation of the inclusions. This model can provide closed-form relations to optimize the design parameters of finite stacks of metasurfaces, in order to maximize the operational bandwidth and extinction ratio for different circular polarization inputs [58].

As an example, Fig. 3.4 describes how the performance of two metasurfaces composed of PEC dipoles, as in Fig. 3.2, can be optimized for operation as a circular polarizer by simultaneously varying the distance and the rotation angle to maximally enhance its overall chiral response. We show the operational bandwidth of this stack, defined as the largest continuous wavelength range over which the ratio of transmittances between opposite circular polarizations is larger than two. Figure 3.4a refers to the range for which $|T_{RR}/T_{LL}|^2 > 2$, Fig. 3.4b to the dual case $|T_{LL}/T_{RR}|^2 > 2$.

Figure 3.4 shows that both bandwidth variations have a periodic nature with respect to the distance, but with a reduced envelope when the separation distance is increased, due to dynamic effects. The results show only positive rotation angles, corresponding to a "right-handed" discrete helix introduced in the cascaded structure and symmetrical results are obtained for the other rotation. Depending on the design parameters, we can select either preferred T_{RR} or T_{LL} for the same rotation. As the plots do not

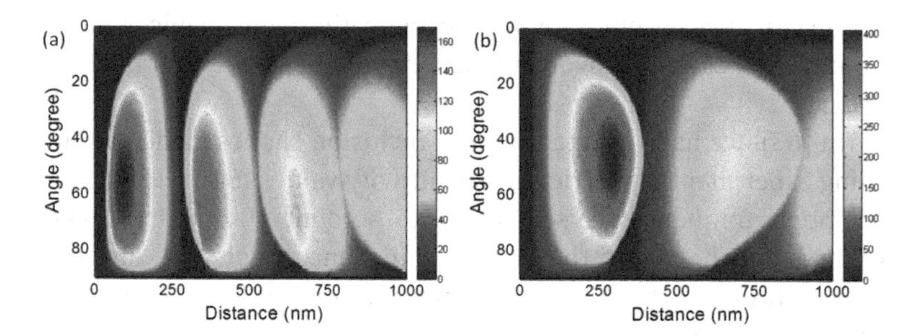

Figure 3.4 Bandwidth [in nm wavelength] of operation as a circular polarizer for a stack of two metasurfaces made of PEC nanorod inclusions, varying their distance and the rotation angle. The inclusion is modeled as a rectangular cuboid with dimensions 50 nm by 275 nm by 50 nm, in a square lattice of period 300 nm. (a) Continuous bandwidth for $|T_{RR}/T_{LL}|^2 > 2$; (b) bandwidth for $|T_{LL}/T_{RR}|^2 > 2$, analyzed within the wavelengths of 300–1000 nm. The space between metasurfaces is assumed to be filled with air.

distinguish between different wavelength ranges, the same design parameters may provide maximum T_{RR} and maximum T_{LL} in different wavelength ranges, explaining the overlap between the two plots. For example, the design with rotation of 60° and distance of 675 nm shows an optimized bandwidth nearly 120 nm in Fig. 3.4a, and a bandwidth nearly 250 nm in Fig. 3.4b, corresponding to two different wavelength ranges.

As extensively discussed in [58], by cascading more layers, we can dramatically increase the bandwidth of operation and the extinction ratio between transmittance of the two circular polarizations, consistent with coupling several resonances and opening a continuous bandgap in a metamaterial array. This transmission-line approach may be applied to analyze an arbitrary number of layers, but in order to analyze the transition toward a continuous bandgap, a generalized Bloch analysis dealing with an infinite stack of rotated structures is more suitable.

3.3.3 *Chiral Effects in Twisted Metamaterials*

When the number of layers becomes very large, cascaded metasurfaces with a consecutive rotation can be viewed as a new class of three-dimensional metamaterials, which we have named "twisted metamaterials" in [58]. We have shown that with seven or eight cascaded metasurfaces, the operational bandwidth as defined in Fig. 3.4 approaches the infinite limit. To analyze the infinite scenario for optimization purposes, we apply a generalized Bloch theorem that can deal with a consecutive rotation applied along the stack, obtaining the generalized eigenmodal dispersion equation:

$$\underline{\mathbf{ABCD}} \cdot \begin{pmatrix} E_y \\ H_x \\ E_x \\ H_y \end{pmatrix} = \underline{\mathbf{R}}(\varphi) \cdot \begin{pmatrix} E_y \\ H_x \\ E_x \\ H_y \end{pmatrix} \cdot \exp^{-j\beta d}. \qquad (3.40)$$

Here, the ABCD matrix includes the transmission through a single metasurface without rotation and the transmission-line segment describing the spacer between two adjacent metasurfaces. The above eigenvalue problem provides β, the propagation constant along the stack, and the corresponding polarization eigenvectors. The rotation matrix $\underline{\mathbf{R}}(\varphi)$ describes the arbitrary in-plane rotation, which for this four-port network taking into account both polarizations is given by

$$\underline{\mathbf{R}}(\varphi) = \begin{pmatrix} \cos\varphi & 0 & -\sin\varphi & 0 \\ 0 & \cos\varphi & 0 & \sin\varphi \\ \sin\varphi & 0 & \cos\varphi & 0 \\ 0 & -\sin\varphi & 0 & \cos\varphi \end{pmatrix}. \qquad (3.41)$$

As an example, Fig. 3.5 shows the result of this generalized Bloch analysis to derive the dispersion relation for eigenmodes supported by the same PEC nanorods as in the previous figure sequentially rotated by 45° with a vacuum spacer of 100 nm between neighboring surfaces. Our analytical results (Fig. 3.5a) are compared with full-wave numerical simulations considering an infinite number of layers (Fig. 3.5b), showing good accuracy, especially at lower frequencies. The red lines indicate eigenmodes with right-handed polarization, and black lines correspond to left-handed polarization. In Fig. 3.6, we compare the calculated analytical eigenmode band diagrams with the transmission coefficients calculated with full-wave simulations for different circularly polarized inputs through a finite stack of nine rotated metasurfaces. The analytical model predicts very well the bandgap for left-handed polarization in the

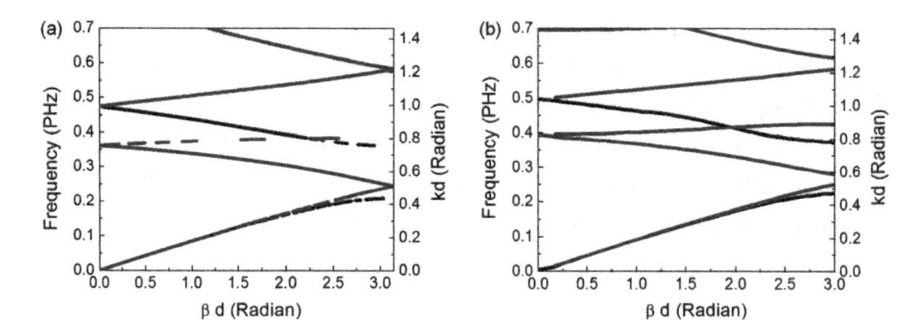

Figure 3.5 (a) Analytical dispersion relation based on our generalized Bloch analysis and (b) full-wave numerical simulations for an infinite number of stacked rotated metasurfaces. The angle of rotation between neighboring metasurfaces is 45° and the separation distance is 100 nm. The metasurfaces are made of PEC nanorod inclusions with dimensions $x \times y \times z = 50$ nm $\times$ 300 nm $\times$ 50 nm in a square lattice 350 nm $\times$ 350 nm.

transmission spectrum, which provides an optimal bandwidth over which large polarization selectivity is achieved through this stack. These results confirm and provide an analytical basis to our designs in [58], showing that this design may form broadband circular polarizers based on stacks of thin planar metasurfaces, which are inherently achiral when isolated.

Another example is shown in Fig. 3.7, keeping the design parameters unchanged from Fig. 3.5, but with a rotation angle of 60°. The analytical results in (a) again show accurate prediction of full-wave simulations (b), especially at low frequencies, with an obvious advantage in terms of computational efficiency. The

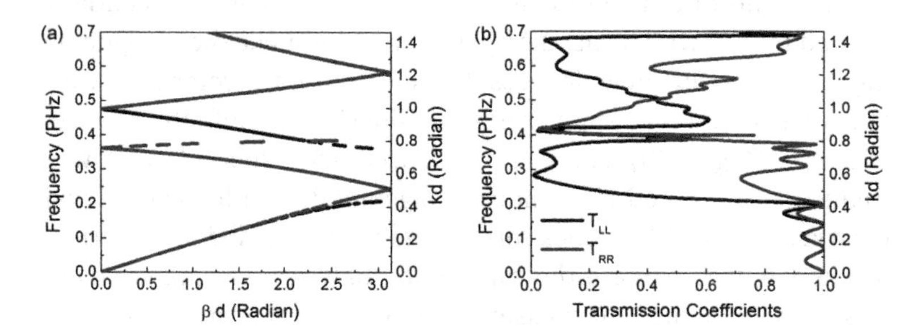

Figure 3.6 (a) Dispersion relation based on the generalized Bloch analysis, similar to Fig. 3.5(a) and (b) transmission spectrum with nine layers of the metasurfaces.

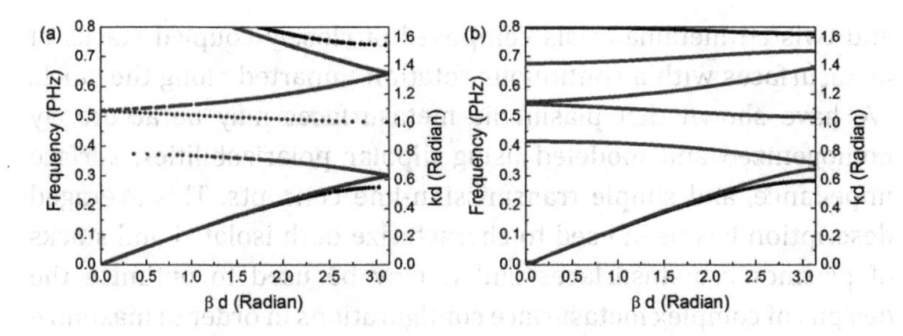

Figure 3.7 Similar to Fig. 3.5, but with a rotation angle of 60°.

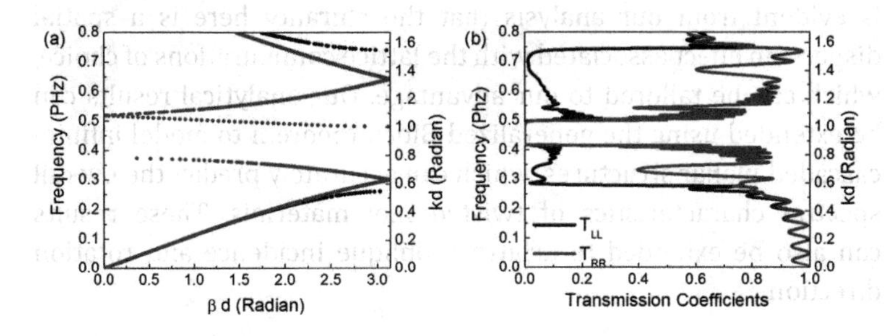

Figure 3.8 Similar to Fig. 3.6, but with a rotation angle of 60°.

power of this generalized Bloch approach consists in the ability of tracking the evolution of the eigenmode band diagram based on the surface impedance as a function of various design parameters, enabling a design optimization in order to realize broadband circular polarizers and large chirality in twisted metamaterials.

Figure 3.8 shows analogous results as in Fig. 3.6 for this design. By changing the rotation angle, a full bandgap can be opened for both polarizations supported by the structure. This is reflected in our full-wave numerical simulations in Fig. 3.8b, which supports total reflection for both input polarizations in a continuous range around 400 THz.

3.4 Conclusion

In this chapter, we have outlined and analytically modeled the physical mechanisms behind chiral effects in isolated metasurfaces and twisted metamaterials composed of closely coupled stacks of metasurfaces with a continuous rotation imparted along the stack. We have shown that plasmonic metasurfaces may be accurately homogenized and modeled using dipolar polarizabilities, surface impedance, and simple transmission-line concepts. This averaged description has been used to characterize both isolated and stacks of plasmonic metasurfaces and it may be used to optimize the designs of complex metasurface configurations in order to maximize extrinsic chiral effects in intrinsically achiral structures, composed of ultrathin metasurfaces with symmetric, achiral inclusions. It is evident from our analysis that the chirality here is a spatial dispersion effect associated with the lattice configurations of choice, which can be tailored to our advantage. Our analytical results can be extended using the generalized Bloch theorem to model infinite cascaded planar structures, which can accurately predict the overall spectral characteristics of twisted metamaterials. These results can also be extended to arbitrary oblique incidence and rotation direction.

Acknowledgments

This work was supported by Office of Naval Research (ONR) Multidisciplinary University Research Initiative (MURI) Grant No. N00014-10-1-0942, by the Norman Hackerman Advanced Research Program (NHARP), and by the Welch Foundation with Grant No. F-1802.

References

1. Shelby, R.A., Smith, D.R., and Schultz, S. (2001). Experimental verification of a negative index of refraction, *Science*, **292**, pp. 77–79.
2. Zharov, A.A., Shadrivov, I.V., and Kivshar, Y.S. (2003). Nonlinear properties of left-handed metamaterials, *Physical Review Letters*, **91**, p. 037401.
3. Linden, S., Enkrich, C., Wegener, M., Zhou, J.F., Koschny, T., and Soukoulis, C.M. (2004). Magnetic response of metamaterials at 100 terahertz, *Science*, **306**, pp. 1351–1353.

4. Smith, D.R., Pendry, J.B., and Wiltshire, M.C.K. (2004). Metamaterials and negative refractive index, *Science*, **305**, pp. 788–792.
5. Alù, A., and Engheta, N. (2005). Achieving transparency with plasmonic and metamaterial coatings, *Physical Review E*, **72**, p. 016623.
6. Fang, N., Lee, H., Sun, C., and Zhang, X. (2005). Sub-diffraction-limited optical imaging with a silver superlens, *Science*, **308**, pp. 534–537.
7. Zhang, S., Fan, W.J., Panoiu, N.C., Malloy, K.J., Osgood, R.M., and Brueck, S.R.J. (2005). Experimental demonstration of near-infrared negative-index metamaterials, *Physical Review Letters*, **95**, p. 137404.
8. Dolling, G., Enkrich, C., Wegener, M., Soukoulis, C.M., and Linden, S. (2006). Simultaneous negative phase and group velocity of light in a metamaterial, *Science*, **312**, pp. 892–894.
9. Leonhardt, U. (2006). Optical conformal mapping, *Science*, **312**, pp. 1777–1780.
10. Schurig, D., Mock, J.J., Justice, B.J., Cummer, S.A., Pendry, J.B., Starr, A.F., and Smith, D.R. (2006). Metamaterial electromagnetic cloak at microwave frequencies, *Science*, **314**, pp. 977–980.
11. Engheta, N. (2007). Circuits with light at nanoscales: Optical nanocircuits inspired by metamaterials, *Science*, **317**, pp. 1698–1702.
12. Lezec, H.J., Dionne, J.A., and Atwater, H.A. (2007). Negative refraction at visible frequencies, *Science*, **316**, pp. 430–432.
13. Rockstuhl, C., Lederer, F., Etrich, C., Pertsch, T., and Scharf, T. (2007). Design of an artificial three-dimensional composite metamaterial with magnetic resonances in the visible range of the electromagnetic spectrum, *Physical Review Letters*, **99**, p. 017401.
14. Shalaev, V.M. (2007). Optical negative-index metamaterials, *Nature Photonics*, **1**, pp. 41–48.
15. Soukoulis, C.M., Linden, S., and Wegener, M. (2007). Negative refractive index at optical wavelengths, *Science*, **315**, pp. 47–49.
16. Tsakmakidis, K.L., Boardman, A.D., and Hess, O. (2007). 'Trapped rainbow' storage of light in metamaterials, *Nature*, **450**, pp. 397–401.
17. Landy, N.I., Sajuyigbe, S., Mock, J.J., Smith, D.R., and Padilla, W.J. (2008). Perfect metamaterial absorber, *Physical Review Letters*, **100**, p. 207402.
18. Papasimakis, N., Fedotov, V.A., Zheludev, N.I., and Prosvirnin, S.L. (2008). Metamaterial analog of electromagnetically induced transparency, *Physical Review Letters*, **101**, p. 253903.

19. Valentine, J., Zhang, S., Zentgraf, T., Ulin-Avila, E., Genov, D.A., Bartal, G., and Zhang, X. (2008). Three-dimensional optical metamaterial with a negative refractive index, *Nature*, **455**, pp. 376–379.
20. Liu, N., Langguth, L., Weiss, T., Kastel, J., Fleischhauer, M., Pfau, T., and Giessen, H. (2009). Plasmonic analogue of electromagnetically induced transparency at the Drude damping limit, *Nature Materials*, **8**, pp. 758–762.
21. Plum, E., Zhou, J., Dong, J., Fedotov, V.A., Koschny, T., Soukoulis, C.M., and Zheludev, N.I. (2009). Metamaterial with negative index due to chirality, *Physical Review B*, **79**, p. 035407.
22. Liu, N., Guo, H.C., Fu, L.W., Kaiser, S., Schweizer, H., and Giessen, H. (2008). Three-dimensional photonic metamaterials at optical frequencies, *Nature Materials*, **7**, pp. 31–37.
23. Soukoulis, C.M., and Wegener, M. (2011). Past achievements and future challenges in the development of three-dimensional photonic metamaterials, *Nature Photonics*, **5**, pp. 523–530.
24. Brueck, S.R.J. (2005). Optical and interferometric lithography - Nanotechnology enablers, *Proceedings of the IEEE*, **93**, pp. 1704–1721.
25. Henzie, J., Lee, M.H., and Odom, T.W. (2007). Multiscale patterning of plasmonic metamaterials, *Nature Nanotechnology*, **2**, pp. 549–554.
26. Gansel, J.K., Thiel, M., Rill, M.S., Decker, M., Bade, K., Saile, V., von Freymann, G., Linden, S., and Wegener, M. (2009). Gold Helix Photonic Metamaterial as Broadband Circular Polarizer, *Science*, **325**, pp. 1513–1515.
27. Liu, N., Liu, H., Zhu, S.N., and Giessen, H. (2009). Stereometamaterials, *Nature Photonics*, **3**, pp. 157–162.
28. Maier, S. A., Brongersma, M. L., Kik, P. G., Meltzer, S., Requicha, A. A. G., Atwater, H. A. (2001). Plasmonics-A route to nanoscale optical devices, *Advanced Materials*, **13**, pp. 1501–1505.
29. Maier, S. A., Atwater, H. A. (2005). Plasmonics: Localization and guiding of electromagnetic energy in metal/dielectric structures, *Journal of Applied Physics*, **98**, p. 011101.
30. Ozbay, E. (2006). Plasmonics: Merging photonics and electronics at nanoscale dimensions, *Science*, **311**, pp. 189–193.
31. Kuester, E.F., Mohamed, M.A., Piket-May, M., and Holloway, C.L. (2003). Averaged transition conditions for electromagnetic fields at a metafilm, *IEEE Transactions on Antennas and Propagation*, **51**, pp. 2641–2651.
32. Holloway, C.L., Mohamed, M.A., Kuester, E.F., and Dienstfrey, A. (2005). Reflection and transmission properties of a metafilm: With an

application to a controllable surface composed of resonant particles, *IEEE Transactions on Electromagnetic Compatibility*, **47**, pp. 853–865.

33. Gordon, J.A., Holloway, C.L., and Dienstfrey, A. (2009). A physical explanation of angle-independent reflection and transmission properties of metafilms/metasurfaces, *IEEE Antennas and Wireless Propagation Letters*, **8**, pp. 1127–1130.
34. Beruete, M., Navarro-Cia, M., Falcone, F., Campillo, I., and Sorolla, M. (2010). Single negative birefringence in stacked spoof plasmon metasurfaces by prism experiment, *Optics Letters*, **35**, pp. 643–645.
35. Chen, P.Y., and Alù, A. (2011). Subwavelength imaging using phase-conjugating nonlinear nanoantenna arrays, *Nano Letters*, **11**, pp. 5514–5518.
36. Alù, A., and Engheta, N. (2008). Tuning the scattering response of optical nanoantennas with nanocircuit loads, *Nature Photonics*, **2**, pp. 307–310.
37. Alù, A., and Engheta, N. (2008). Input impedance, nanocircuit loading, and radiation tuning of optical nanoantennas, *Physical Review Letters*, **101**, pp. 043901.
38. Sun, Y., Edwards, B., Alù, A., and Engheta, N. (2012). Experimental realization of optical lumped nanocircuits at infrared wavelengths, *Nature Materials*, **11**, pp. 208–212.
39. Alù, A. (2009). Mantle cloak: Invisibility induced by a surface, *Physical Review B*, **80**, p. 245115.
40. Burokur, S.N., Daniel, J.P., Ratajczak, P., and de Lustrac, A. (2010). Tunable bilayered metasurface for frequency reconfigurable directive emissions, *Applied Physics Letters*, **97**, p. 064101.
41. J.C. Bose. (1898). On the rotation of plane of polarisation of electric waves by a twisted structure, *Proceedings of the Royal Society*, **63**, pp. 146–152.
42. Sarkar, T.K., and Sengupta, D.L. (1997). An appreciation of J. C. Bose's pioneering work in millimeter waves, *IEEE Antennas and Propagation Magazine*, **39**, pp. 55–63.
43. Chiou, T.H., Kleinlogel, S., Cronin, T., Caldwell, R., Loeffler, B., Siddiqi, A., Goldizen, A., and Marshall, J. (2008). Circular polarization vision in a stomatopod crustacean, *Current Biology*, **18**, pp. 429–434.
44. Roberts, N.W., Chiou, T.H., Marshall, N.J., and Cronin, T.W. (2009). A biological quarter-wave retarder with excellent achromaticity in the visible wavelength region, *Nature Photonics*, **3**, pp. 641–644.

45. Schwind, R. (1991). Polarization vision in water insects and insects living on a moist substrate, *Journal of Comparative Physiology a-Sensory Neural and Behavioral Physiology*, **169**, pp. 531–540.
46. Shashar, N., Rutledge, P.S., and Cronin, T.W. (1996). Polarization vision in cuttlefish: A concealed communication channel, *Journal of Experimental Biology*, **199**, pp. 2077–2084.
47. Seet, K.K., Mizeikis, V., Matsuo, S., Juodkazis, S., and Misawa, H. (2005). Three-dimensional spiral-architecture photonic crystals obtained by direct laser writing, *Advanced Materials*, **17**, pp. 541–545.
48. Maruo, S., Nakamura, O., and Kawata, S. (1997). Three-dimensional microfabrication with two-photon-absorbed photopolymerization, *Optics Letters*, **22**, pp. 132–134.
49. Deubel, M., Von Freymann, G., Wegener, M., Pereira, S., Busch, K., and Soukoulis, C.M. (2004). Direct laser writing of three-dimensional photonic-crystal templates for telecommunications, *Nature Materials*, **3**, pp. 444–447.
50. Lee, K.S., Yang, D.Y., Park, S.H., and Kim, R.H. (2006). Recent developments in the use of two-photon polymerization in precise 2D and 3D microfabrications, *Polymers for Advanced Technologies*, **17**, pp. 72–82.
51. Kuwata-Gonokami, M., Saito, N., Ino, Y., Kauranen, M., Jefimovs, K., Vallius, T., Turunen, J., and Svirko, Y. (2005). Giant optical activity in quasi-two-dimensional planar nanostructures, *Physical Review Letters*, **95**, p. 227401.
52. Rogacheva, A.V., Fedotov, V.A., Schwanecke, A.S., and Zheludev, N.I. (2006). Giant gyrotropy due to electromagnetic-field coupling in a bilayered chiral structure, *Physical Review Letters*, **97**, p. 177401.
53. Decker, M., Klein, M.W., Wegener, M., and Linden, S. (2007). Circular dichroism of planar chiral magnetic metamaterials, *Optics Letters*, **32**, pp. 856–858.
54. Plum, E., Liu, X.X., Fedotov, V.A., Chen, Y., Tsai, D.P., and Zheludev, N.I. (2009). Metamaterials: Optical activity without chirality, *Physical Review Letters*, **102**, p. 113902.
55. Wei, Z., Cao, Y., Fan, Y., Yu, X., and Li, H. (2011). Broadband polarization transformation via enhanced asymmetric transmission through arrays of twisted complementary split-ring resonators, *Applied Physics Letters*, **99**, p. 221907.
56. Demetriadou, A., and Pendry, J.B. (2009). Extreme chirality in Swiss roll metamaterials, *Journal of Physics-Condensed Matter*, **21**, p. 376003.

57. Zhang, S., Zhou, J., Park, Y.-S., Rho, J., Singh, R., Nam, S., Azad, A.K., Chen, H.-T., Yin, X., Taylor, A.J., and Zhang, X. (2012). Photoinduced handedness switching in terahertz chiral metamolecules, *Nature Communications*, **3**, p. 942.
58. Zhao, Y., Belkin, M.A., and Alù, A. (2012). Twisted optical metamaterials for planarized ultrathin broadband circular polarizers, *Nature Communications*, **3**, p. 870.
59. Maradudin, A.A. (2011) Optical wave interaction with two-dimensional arrays of plasmonic nanoparticles, Chapter 3, in *Structured Surfaces as Optical Metamaterials*, Alù, A., and Engheta, N. (eds.), Cambridge University Press, New York, pp. 58–93.
60. Liu, X.-X., and Zhao, Y., Determination of full dynamic polarizability tensor of arbitrary sub-wavelength inclusions in meta-arrays, under review.
61. Sersic, I., Tuambilangana, C., Kampfrath, T., and Koenderink, A.F. (2011). Magnetoelectric point scattering theory for metamaterial scatterers, *Physical Review B*, **83**, p. 245102.
62. Serdyukov, A.N., Semchenko, I.V., Tretyakov, S.A., and Sihvola, A. (2001). *Electromagnetics of Bi-anisotropic Materials: Theory and Applications*, Gordon and Breach, Amsterdam.
63. Zhao, Y., Engheta, N., and Alù, A. (2011). Homogenization of plasmonic metasurfaces modeled as transmission-line loads, *Metamaterials*, **5**, pp. 90–96.
64. Johnson, P.B., and Christy, R.W. (1972). Optical constants of noble metals, *Physical Review B*, **6**, pp. 4370–4379.

Chapter 4

Engineering of Radiation of Optically Active Molecules with Chiral Nano-Meta Particles

Vasily Klimov[a,b] and Dmitry Guzatov[c]

[a] *Lebedev Physical Institute, 53, Leniskij Prospekt, Moscow 119991, Russia*
[b] *National Research Nuclear University "MEPhI", 31, Kashirskoe shosse, Moscow 115409, Russia*
[c] *Yanka Kupala Grodno State University, 22, Ozheshko st., Grodno 230023, Belarus*
klimov256@gmail.com

4.1 Introduction

The word "chirality" is derived from the Greek root "χειρ" meaning "hand". Thus, the term "chirality" denotes such a property of an object that is also a property of the human hand. This term was introduced by lord kelvin in his famous Baltimore Lectures: "I call any geometrical figure, or any group of points, chiral, and say it has chirality, if its image in a plane mirror, ideally realized, cannot be brought to coincide with itself" [34]. This definition implies that, first, "chirality" is the geometric property of an object; second, only spatial, that is, three-dimensional, objects possess this property. Planar (two-dimensional) or linear (one-dimensional) objects do not possess this property in a three-dimensional space.

Singular and Chiral Nanoplasmonics
Edited by Svetlana V. Boriskina and Nikolay I. Zheludev

ISBN 978-981-4613-17-0 (Hardcover), 978-981-4613-18-7 (eBook)
www.panstanford.com

Chiral objects exist in two types: an object and its twin having the form of its mirror reflection, for example, left and right hands, right-hand and left-hand screw, right-hand and left-hand spiral, and so on.

4.1.1 *Chiral (Optically Active) Molecules and Their Importance*

Now due to rapid development of bio-nanotechnologies, any methods of control over enantiomers of chiral molecules are of great importance. Chiral molecules are of great interest because they lie at the base of life [12]. In particular, proteins are constructed almost exclusively from L-amino acids, and nucleic acids from D-sugars (see Fig. 4.1).

Another type of chiral molecules is related with the presence of asymmetric carbon (a chiral center).

It is very remarkable that many biomolecules have so-called optical activity [6], which manifests itself in two related effects: optical rotation and circular dichroism.

The optical rotation is the rotation of the plane of linearly polarized light about the motion direction, as the light travels through a media consisting of chiral molecules. The optical rotation results from a difference in the phase velocities due to different refractive indices n_{left}, n_{right} for right and left-circularly polarized

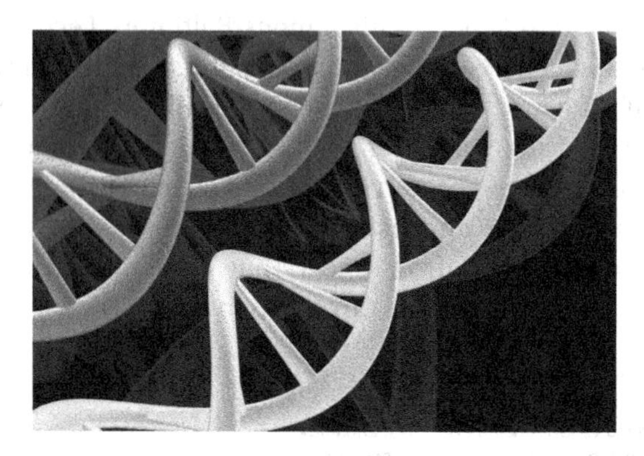

Figure 4.1 A classical right-handed spiral model of DNA, © Depositphotos.com/[Sunagatov Dmitry].

light. As the linearly polarized light can be considered as composed of right- and left-circularly polarized waves with equal amplitudes, this difference in phase velocity creates a phase difference between right- and left-polarized components, which is equivalent to the rotation of the plane of polarization. This effect is characterized by the ratio:

$$\frac{n_{\text{left}} - n_{\text{right}}}{n_{\text{left}} + n_{\text{right}}} \sim \kappa \sim \frac{\text{size of molecule}}{\text{wavelength}} \ll 1, \tag{4.1}$$

which is small for a usual optically active media. The variation of the optical rotation as a function of a wavelength is called "optical rotary dispersion" (ORD).

Another effect of the optical activity—circular dichroism (CD)—is related to the difference in absorption A_{left}, A_{right} of the right and left-polarized light. This effect is characterized by the ratio:

$$\frac{A_{\text{left}} - A_{\text{right}}}{A_{\text{left}} + A_{\text{right}}} \sim \kappa \sim \frac{\text{size of molecule}}{\text{wavelength}} \ll 1, \tag{4.2}$$

which is also small for a usual optically active media.

Optically active molecules are sensitive to light polarization, as they are chiral (and spiral) objects.

The explanation [12] of optical activity effects is based on the fact that in such molecules, both electric dipole moment **d** and magnetic dipole moment **m** of optical transition are different from zero. As a result, the Hamiltonian of light interaction with a "right" molecule, where electric and magnetic dipole momenta are parallel $(\mathbf{d} \parallel \mathbf{m})$, can be written in the form [66]:

$$H_{\text{int}} = -\mathbf{d} \cdot \mathbf{E} - \mathbf{m} \cdot \mathbf{H}, \tag{4.3}$$

while for a "left" molecule (where **d** is antiparallel to **m**), the interaction Hamiltonian looks like

$$H_{\text{int}} = -\mathbf{d} \cdot \mathbf{E} + \mathbf{m} \cdot \mathbf{H} \tag{4.4}$$

As far back as 1930, it was shown [64] that optical activity effects are proportional to the factor

$$\kappa \sim \text{Im}\,(\mathbf{d} \cdot \mathbf{m}^*) \tag{4.5}$$

In (4.5) and further, the asterisk means complex conjugation.

It is important that the chiral parameter κ [and optical rotation (4.1) and CD (4.2)] is proportional to the ratio a/λ, where a is the length of the molecule, and λ is the wavelength. In the optics of natural media, the value of the ratio a/λ turns out to be of the order of 10^{-4}. Indeed, the size of a molecule is several angstroms, that is, of the order of 10^{-8} cm, and the wavelength in the optical region is about 1 micron, that is, of the order of 10^{-4} cm. In organic substances, for example, polymers, the molecules sizes are much greater than the lengths of simple molecules, and therefore, the chiral properties are stronger here.

In spite of their smallness, the chiral effects (or the effects of optical activity) are widely used in biochemical investigation for

1. the detection of small quantities of organic and biological molecules;
2. the determination of whether a protein is folded or not and thus for determination of its secondary and tertiary structure;
3. the comparison of the structure of proteins obtained from different sources or/and the structures of different mutations of the initial protein;
4. the examination of the protein conformational stability relatively different changes of environment (temperature, acidity).

All above-mentioned properties and applications of optical activity are based on light absorption or scattering by biomolecules. Meanwhile, the possibility of making use of radiation of chiral molecules for different biomedical applications (e.g., for resolution of racemic mixtures) was not considered at all. So, in this Chapter (partially based on our recent publications [29–31, 42]), we will try to fill the gaps.

At first sight, it seems rather obvious that radiation of chiral molecules of given handness can be easily controlled because it is well known that it is possible for usual molecules with the help of plasmonic nanoparticles [43]. In particular with the help of island films, the effect of the Raman scattering can be enhanced by 14 orders of magnitude. This allows seeing even single molecules, and now a new type of spectroscopy [surface-enhanced Raman spectroscopy (SERS)] is widely used [45]. Another analogous type of spectroscopy based on surface-enhanced fluorescence (SEF) [21] is also developing fast.

However, a more detailed study shows that for chiral molecules, it is impossible to discriminate radiation of "right" and "left" molecules without a specially arranged nano-environment. The problem is that plasmonic nanoparticles of simple shapes and metal island films do not have noticeable chirality properties, and they can enhance only local electric fields:

$$|\mathbf{E}_{\text{loc}}| \gg |\mathbf{E}_{\text{in}}|, \ |\mathbf{H}_{\text{loc}}| \sim |\mathbf{H}_{\text{in}}| \Rightarrow |\mathbf{E}_{\text{loc}}| \gg |\mathbf{H}_{\text{loc}}| \tag{4.6}$$

As a result, such a plasmonic environment increases the role of an electric dipole momentum in the interaction of chiral molecules with light because the initially large electric dipole momentum ($|\mathbf{d}| \gg |\mathbf{m}|$) will interact with an enhanced electric field:

$$\begin{aligned} H_{\text{R}} &= -\mathbf{d} \cdot \mathbf{E}_{\text{loc}} - \mathbf{m} \cdot \mathbf{H}_{\text{loc}} \approx -\mathbf{d} \cdot \mathbf{E}_{\text{loc}}, \\ H_{\text{L}} &= -\mathbf{d} \cdot \mathbf{E}_{\text{loc}} + \mathbf{m} \cdot \mathbf{H}_{\text{loc}} \approx -\mathbf{d} \cdot \mathbf{E}_{\text{loc}}. \end{aligned} \tag{4.7}$$

Due to this fact, plasmonic nanoparticles of simple shapes and the metal island films will equally enhance radiation of the "right" and "left" molecules. So, for specific control of radiation of enantiomers of chiral molecules, one should arrange such a nano-environment that enhances the magnetic fields and reduces the electric fields. Below, we will show that this is possible with making use of nanoparticles that have both chirality and negative refractive index or negative permeability. Now, such unusual properties are available due to development of metamaterials technologies [11] and chiral metamaterials in particular [46].

4.1.2 *Chiral Metamaterials*

Generally, metamaterials are artificial materials engineered to have properties that cannot be found in nature. The metamaterials consist of nanoparticles of different shapes that are called "meta-atoms." If the meta-atoms have chiral properties, they will form chiral metamaterials.

The examples of the chiral meta-atoms used in metamaterials and their twins are shown in Fig. 4.2. Metal spirals conducting electric current play an important role in development of chiral metamaterials (Fig. 4.2a). The second pair of chiral elements (Fig. 4.2b) is a simplified version of a spiral and is formed by split

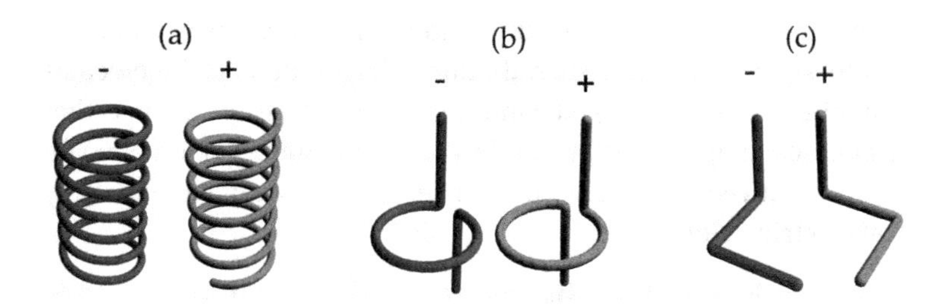

Figure 4.2 The schematic image of chiral meta-atoms used to fabricate chiral metamaterials: (a) the cylindrical spirals, (b) the rings with orthogonal linear ends (Ω particles), and (c) the broken wires with linear parts along the coordinate axes. The signs "+" and "−" denote the right and left chiral elements, correspondingly [43].

rings with linear horns directed normally to the ring plane but in opposite directions. The elements differ by the horns bend direction at the split of the ring. When moving along the wire upwards, in the first element, the right rotation takes place by passing the ring, and in the second element, the left rotation does. Linear parts of the third pair of the chiral particles (Fig. 4.2a) are directed along the axes of the left and right Cartesian coordinate system, correspondingly. These elements are further simplification of previous ones and can have a simpler technological realization.

To understand the optics of chiral particles, let us now consider (see e.g. [19]) what happens if an electromagnetic wave incidents on a small conducting spiral (Fig. 4.2a). Let the field in the area of the spiral has the components E_z and H_z, along the spiral axis. Then, the longitudinal (along the spiral axis) component of the electric dipole moment d_z is formed both by the component E_z of the electric field producing current along the spiral axis and the component H_z of the magnetic field penetrating the spiral rings and inducing the ring current with longitudinal component. Thus,

$$d_z = \alpha_{\mathrm{EE}} E_z + \alpha_{\mathrm{EH}} H_z, \tag{4.8}$$

where α_{EE} is the regular electric polarizability of the spiral, and α_{EH} is so-called electromagnetic cross-polarizability of the particle. Analogously, the longitudinal component of the spiral magnetic dipole moment m_z is formed by the ring electric current produced by the magnetic field penetrating the spiral and the ring current produced by the electric field, as the current runs along the spiral axis only by passing through the spiral rings. As a result, one will have

$$m_z = \alpha_{\text{HH}} H_z + \alpha_{\text{HE}} E_z, \tag{4.9}$$

where α_{HH} is the regular magnetic polarizability of the spiral and α_{HE} is magnetoelectric cross-polarizability of the particle. Moreover, the spiral is characterized by some other (transverse relatively to the spiral axis) components of electric and magnetic dipole moments, which also have a chiral origin. The above arguments become even more illustrative when applied to the second pair of elements shown in Fig. 4.2b.

It is very important that the electromagnetic cross-polarizabilities are of inductive nature and are related to time dependence of the incident fields. Thus, for the harmonic time dependence of the electromagnetic fields, $\exp(-i\omega t)$, the electromagnetic cross-polarizabilities can be presented in the form $\alpha_{\text{EH}} = i\beta$, $\alpha_{\text{HE}} = -i\beta$ where β is a real number, positive for left-handed spirals, and negative for right-handed spirals. So, the longitudinal structure of the spiral dipole moments can be rewritten as:

$$d_z = \alpha_{\text{EE}} E_z + i\beta H_z, \quad m_z = -i\beta E_z + \alpha_{\text{HH}} H_z. \tag{4.10}$$

The properties of spirals (helices) were considered in more details in ([4, 33, 53, 71, 79]). In particular, Jaggard, Mickelson, Papas [33] have shown that for chiral "meta-atom" shown in Fig. 4.2B, the expressions for cross-polarizabilities can be presented in the form:

$$\beta = \alpha_{\text{HH}} \frac{L}{k\left(\pi R^2\right)} = \alpha_{\text{EE}} \frac{k\left(\pi R^2\right)}{L}, \tag{4.11}$$

where L is the total height of the element in Fig. 4.2b and R is the radius of its ring part and K is the wavenumber.

After the polarization properties of a single chiral nanoparticle have been determined, the question of optical properties of metamaterials made of chiral meta-atoms may be considered. Chiral medium can be either an ordered structure in the form of a spatial lattice of chiral meta-atoms or a chaotic mixture of chiral elements. As an exciting example of a regular chiral structure, the planar chiral metamaterial made from 3D gold nanohelices is shown in Fig. 4.3 ([23]). This structure was recently fabricated by laser "writing" of

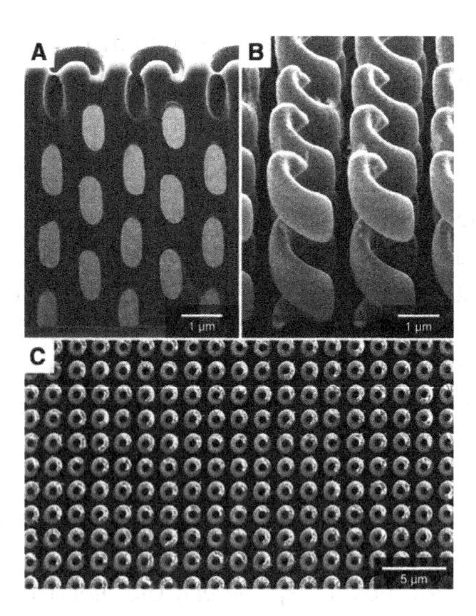

Figure 4.3 Planar chiral metamaterial made of 3D gold nanohelices. (A) Focused-ion-beam (FIB) cut of a polymer structure partially filled with gold by electroplating. (B) Oblique view of a left-handed helix structure after removal of the polymer by plasma etching. (C) Top-view image reveals the circular cross-section of the helices and homogeneity on a larger scale. The lattice constant of the square lattice is $a = 2$ μm. Reprinted from Ref. [23] with permission from AAAS.

helical cavities in a photoresist with subsequent deposition of gold inside these holes.

In any case, known mixing rules (see, e.g., [53, 71]) can be used to elaborate the macroscopic description from the microscopic one.

The application of this procedure to chiral metamaterials with chaotically situated nanoparticles, with taking (4.10) into account, results in the following constitutive equations:

$$\mathbf{D} = \varepsilon\mathbf{E} + i\kappa\mathbf{H}, \quad \mathbf{B} = \mu\mathbf{H} - i\kappa\mathbf{E}, \tag{4.12}$$

where κ is the chirality parameter, ε is the permittivity, and μ is the permeability of a chiral medium. Often, the constitutive Eqs. (4.12) are written as

$$\mathbf{D} = \varepsilon\left(\mathbf{E} + \tilde{\beta}\mathrm{rot}\mathbf{E}\right), \quad \mathbf{B} = \mu\left(\mathbf{H} + \tilde{\beta}\mathrm{rot}\mathbf{H}\right), \tag{4.13}$$

where $\tilde{\beta} = \beta/(\varepsilon\mu\omega/c)$, ω is the frequency, and c is the velocity of light in a vacuum. Using the Maxwell's equations, the equivalence of these representations can be shown easily. Further, we will use the constitutive equations in the form (4.13). This symmetrical form of the notation of constitutive equation in chiral media is called the form of Drude–Born–Fedorov [10].

Thus, the chiral metamaterials resemble natural optically active media, and one can expect that with the help of the chiral metamaterials, one can improve or even create new technologies of dealing with enantiomers of optically active biomolecules.

These hopes are based on the fact that the chirality parameter κ for the chiral metamaterials can be increased substantially in comparison with chirality parameter for usual optically active media. This increase is due to two effects.

1. Meta-atom length L is substantially greater than the molecule size.
2. The interaction of light with meta-atom can be additionally increased due to plasmon resonance effects.

For example, one can use the resonance of the current along the spiral element. The external linear spiral sizes can be small relative to the wavelength, but the total length of the wire forming the spiral can be of the order of the wavelength providing the resonance condition. In this case, the chirality is not a small effect at all, and the properties of chiral metamaterial will differ substantially from the properties of usual optically active media even with making use of the accumulation of small effect by means of large length of optical device.

Thus, one can expect that, making use of resonant chiral nano-meta-particles can increase the effectiveness of investigation of the chiral molecules (proteins and DNA) properties substantially. In particular, one can expect that a chiral nano-metaparticle can change radiation of right and left enantiomers of optically active biomolecules specifically. It is this problem that will be considered in the present chapter.

The remaining part of this chapter is as follows. In Section 4.2, we will consider the radiation of optically active molecule near a chiral half-space with arbitrary parameters. In Sections 4.3 and 4.4, we will examine the radiation of optically active molecules near a single chiral metasphere (i.e., a sphere made of an arbitrary metamaterial) and near a cluster of two metaspheres, which forms a chiral nanoantenna. In Section 4.5, we will discuss possible applications of our results.

4.2 Radiation of Chiral Molecule Near Chiral Half-Space

The study of an electric dipole, that is, usual (nonchiral) molecule, radiation, which is located near the conducting half-space, has a century-long history starting from the work by A. Sommerfeld in 1909 [68]. Until now, the electric dipole radiation near layered structures with arbitrary parameters is considered in details, including DNG layers and even chiral layers [2, 13, 17, 20, 39, 41, 48, 58, 61–63, 65, 74–76, 78].

However, as far as we know, at the present time, there are no works devoted to the radiation of an optically active (chiral) molecule near half-space made of chiral metamaterials [23, 57, 60, 72], which may have a number of unique properties. In this section, we will consider the spontaneous emission of an optically active molecule located near the chiral half-space. The geometry of the problem under consideration is shown in Fig. 4.4.

4.2.1 *Spontaneous Radiation of a Chiral Molecule Placed near a Chiral Half-Space*

In the description of the electric and magnetic fields in the chiral medium, we will use the material equations in the Drude–Born–Fedorov form (4.13). Substituting these relations into the Maxwell's equations and using the Bohren's transformation [8], one can present electric and magnetic fields as a linear combination of the right and left-polarized waves:

$$\mathbf{E} = \mathbf{Q}_L - iZ\mathbf{Q}_R, \quad \mathbf{H} = -iZ^{-1}\mathbf{Q}_L + \mathbf{Q}_R, \tag{4.14}$$

In (4.14) and further, $Z = \sqrt{\mu}/\sqrt{\varepsilon}$ is the impedance of the chiral medium. $\mathbf{Q}_L$, $\mathbf{Q}_R$ in (4.14) are eigenmodes of the Maxwell equations

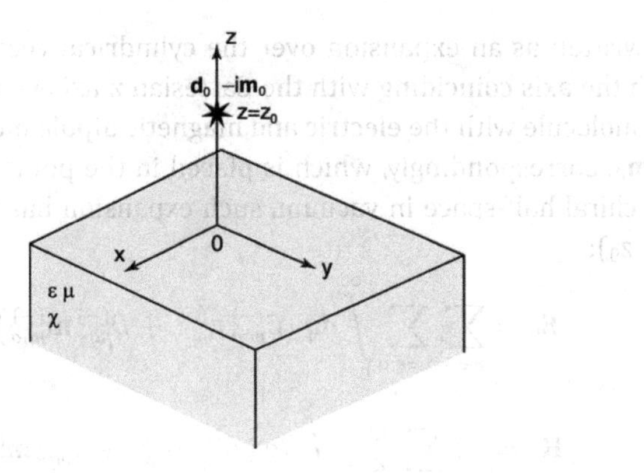

Figure 4.4 The geometry of the problem of spontaneous radiation of a chiral molecule located near a chiral half-space. The molecule is placed in a vacuum.

and satisfy the following equations:

$$\mathrm{rot}\mathbf{Q}_L = +k_L\mathbf{Q}_L, \quad \mathrm{rot}\mathbf{Q}_R = -k_R\mathbf{Q}_R, \tag{4.15}$$

where

$$k_L = \frac{k_0 n}{1-\chi n}, \quad k_R = \frac{k_0 n}{1+\chi n}, \tag{4.16}$$

are wavenumbers of the left (L) and right-polarized (R) waves; $n = \sqrt{\varepsilon\mu}$ is the refractive index; $k_0 = \omega/c$ is the wavenumber in a vacuum; $\chi = k_0\tilde{\beta}$ is the dimensionless chirality parameter [see (4.13)].

As we assume that the half-space is made of a metamaterial with arbitrary values of permittivity and permeability, it is of great importance to determine branches of square roots in (4.14) and (4.16) correctly. Here and below, we assume that $\sqrt{\varepsilon\mu} = \sqrt{\varepsilon}\sqrt{\mu}$. This definition provides a negative index of refraction $\mathrm{Re}\left(\sqrt{\varepsilon\mu}\right) < 0$ for DNG-metamaterial ($\varepsilon < 0$ and $\mu < 0$) and a positive index of refraction $\mathrm{Re}\left(\sqrt{\varepsilon\mu}\right) > 0$ for dielectric ($\varepsilon > 0$ and $\mu > 0$), and also with this definition $\mathrm{Im}\left(\sqrt{\varepsilon\mu}\right) > 0$ for metal ($\varepsilon < 0$ and $\mu > 0$) and for MNG-metamaterial ($\varepsilon > 0$ and $\mu < 0$). In addition, in the case of a material with losses ($\mathrm{Im}(\varepsilon) \geq 0$, $\mathrm{Im}(\mu) \geq 0$), the accepted definition provides a positive coefficient of extinction, that is, $\mathrm{Im}\left(\sqrt{\varepsilon\mu}\right) > 0$.

To solve the problem of the field induced by a chiral molecule, the electric and magnetic fields of the molecule in free space can be written as an expansion over the cylindrical vector harmonics, with the axis coinciding with the Cartesian z-axis (see Fig. 4.4). For the molecule with the electric and magnetic dipole moments $\mathbf{d}_0$ and $-i\mathbf{m}_0$, correspondingly, which is placed in the point $z_0 > 0$ above the chiral half-space in vacuum, such expansion has the form ($0 < z < z_0$):

$$\mathbf{E}_0 = \sum_{n=0}^{1} \sum_{\sigma=e,o} \int_0^\infty dq \left(C_{nq\sigma}^{(-)} \mathbf{n}_{nq\sigma}^{(-)} + D_{nq\sigma}^{(-)} \mathbf{m}_{nq\sigma}^{(-)} \right),$$

$$\mathbf{H}_0 = -i \sum_{n=0}^{1} \sum_{\sigma=e,o} \int_0^\infty dq \left(D_{nq\sigma}^{(-)} \mathbf{n}_{nq\sigma}^{(-)} + C_{nq\sigma}^{(-)} \mathbf{m}_{nq\sigma}^{(-)} \right), \qquad (4.17)$$

where q is the radial wavenumber; the vector cylindrical harmonics $\mathbf{n}_{nq\sigma}^{(-)}$ and $\mathbf{m}_{nq\sigma}^{(-)}$, and the coefficients $C_{nq\sigma}^{(-)}$ and $D_{nq\sigma}^{(-)}$ (where $\sigma = e,\ o$ corresponds to "even" and "odd" waves) are presented in the explicit form in the Appendix 4.A.

The right or left eigenmodes $\mathbf{Q}_R$ and $\mathbf{Q}_L$ of the chiral half-space [see (4.14)] can also be presented in the form of the expansion over the vector cylindrical harmonics:

$$\mathbf{Q}_L = \sum_{n=0}^{1} \sum_{\sigma=e,o} \int_0^\infty dq\, A_{nq\sigma} \left(\mathbf{n}_{nq\sigma}^{(L)} + \mathbf{m}_{nq\sigma}^{(L)} \right),$$

$$\mathbf{Q}_R = \sum_{n=0}^{1} \sum_{\sigma=e,o} \int_0^\infty dq\, B_{nq\sigma} \left(\mathbf{n}_{nq\sigma}^{(R)} - \mathbf{m}_{nq\sigma}^{(R)} \right). \qquad (4.18)$$

The explicit expression for the vector cylindrical harmonics $\mathbf{n}_{nq\sigma}^{(J)}$ and $\mathbf{m}_{nq\sigma}^{(J)}$ (where $\sigma = o,\ e;\ J = L,\ R$) are presented in Appendix 4.A. Note that the flow of energy corresponding to these cylindrical harmonics is directed inside the chiral half-space for any choice of parameters.

Electric and magnetic fields of the reflected wave can be written in the following form:

$$\mathbf{E}^{sc} = \sum_{n=0}^{1} \sum_{\sigma=e,o} \int_0^\infty dq \left(C_{nq\sigma} \mathbf{n}_{nq\sigma}^{(+)} + D_{nq\sigma} \mathbf{m}_{nq\sigma}^{(+)} \right),$$

$$\mathbf{H}^{sc} = -i \sum_{n=0}^{1} \sum_{\sigma=e,o} \int_0^\infty dq \left(D_{nq\sigma} \mathbf{n}_{nq\sigma}^{(+)} + C_{nq\sigma} \mathbf{m}_{nq\sigma}^{(+)} \right), \qquad (4.19)$$

where the vector cylindrical harmonics $\mathbf{n}_{nq\sigma}^{(+)}$ and $\mathbf{m}_{nq\sigma}^{(+)}$ are defined in Appendix 4.A.

To find the unknown coefficients of expansions (4.18) and (4.19), one can use the continuity of the tangential components of the electric and magnetic fields at the interface [17]. As the molecule is placed out of the chiral half-space, then to find the spontaneous emission decay rate, we need to know only the coefficients C_{nqe} and D_{nqe} for which one can obtain the following expressions ($n = 0, 1$; $\sigma = e, o$):

$$C_{nq\sigma} = -b_q C_{nq\sigma}^{(-)} - c_q D_{nq\sigma}^{(-)}, \quad D_{nq\sigma} = -a_q D_{nq\sigma}^{(-)} - c_q C_{nq\sigma}^{(-)}, \tag{4.20}$$

where the functions were introduced:

$$\begin{aligned} a_q &= \left(A_q(L)\,W_q(R) + A_q(R)\,W_q(L)\right)/\Delta, \\ b_q &= \left(V_q(L)\,B_q(R) + V_q(R)\,B_q(L)\right)/\Delta, \\ c_q &= \left(B_q(L)\,W_q(R) - B_q(R)\,W_q(L)\right)/\Delta, \\ \Delta &= V_q(L)\,W_q(R) + V_q(R)\,W_q(L), \end{aligned} \tag{4.21}$$

in which ($J = L, R$)

$$\begin{aligned} A_q(J) &= S(k_J)/Z - S(k_0), \quad & B_q(J) &= S(k_J) - S(k_0)/Z, \\ V_q(J) &= S(k_J)/Z + S(k_0), \quad & W_q(J) &= S(k_J) + S(k_0)/Z, \end{aligned} \tag{4.22}$$

where $S(k_J) = \sqrt{k_J^2 - q^2}/k_J$.

In the present work, we assume that $\sqrt{k_0^2 - q^2} = i\sqrt{q^2 - k_0^2}$ for $q > k_0$, and $\sqrt{k_J^2 - q^2} = \sqrt{k_J - q}\sqrt{k_J + q}$. For dielectric ($k_J > 0$), this definition provides fulfillment of the relations $\mathrm{Re}\left(\sqrt{k_J^2 - q^2}\right) > 0$, when $k_J > q$, and also $\mathrm{Im}\left(\sqrt{k_J^2 - q^2}\right) > 0$, when $k_J < q$. For DNG-metamaterial, for which the refractive index is negative ($k_J = -|k_J|$), we obtain $\mathrm{Re}\left(\sqrt{k_J^2 - q^2}\right) < 0$, when $|k_J| > q$, and also $\mathrm{Im}\left(\sqrt{k_J^2 - q^2}\right) > 0$, when $|k_J| < q$. For metal and MNG-metamaterial, we have $\mathrm{Im}\left(\sqrt{k_J^2 - q^2}\right) > 0$. In addition, in the case of materials with losses (i.e., $\mathrm{Im}(k_J) > 0$), this definition provides $\mathrm{Im}\left(\sqrt{k_J^2 - q^2}\right) > 0$.

Equating the denominator Δ in (4.21) to zero, we obtain the dispersion equation for the wavenumbers q of the chiral-plasmon modes excited on the surface the chiral half-space

$$\begin{aligned} &(S(k_L)/Z + S(k_0))(S(k_R) + S(k_0)/Z) \\ &+ (S(k_R)/Z + S(k_0))(S(k_L) + S(k_0)/Z) = 0. \end{aligned} \tag{4.23}$$

Assuming $\chi = 0$, from (4.23), we get two independent equations: $\varepsilon\sqrt{k_0^2 - q^2} + \sqrt{k_0^2\varepsilon\mu - q^2} = 0$ and $\mu\sqrt{k_0^2 - q^2} + \sqrt{k_0^2\varepsilon\mu - q^2} = 0$, which define wavenumbers of the TM and TE modes, respectively, [3]. In the case $\mu = 1$, the first equation is reduced to the usual dispersion equation for the surface plasmon $q = k_0\sqrt{\varepsilon/(\varepsilon + 1)}$ (where $\varepsilon < -1$) [43, 59].

In the case of a small chirality ($\chi \to 0$, $\chi > 0$), one can find an asymptotic solution of (4.23). In the most interesting cases $\mu = \pm 1$ and $\varepsilon = -|\varepsilon|$, one can obtain

$$\begin{aligned} q &= k_0\sqrt{\frac{|\varepsilon|}{|\varepsilon| - 1}}\left(1 + \frac{|\varepsilon|\chi^2}{(|\varepsilon| + 1)^2(|\varepsilon| - 1)}\right), \\ \mu &= 1, \quad |\varepsilon| > 1, \quad q \geq k_0, \\ q &= k_0\frac{|\varepsilon| - 1}{\chi\sqrt{2|\varepsilon|}}, \quad \mu = -1, \quad |\varepsilon| > 1, \quad q \geq k_0\sqrt{|\varepsilon|}, \\ q &= k_0\frac{1 - |\varepsilon|}{\chi\sqrt{2|\varepsilon|}}, \quad \mu = -1, \quad 0 < |\varepsilon| < 1, \quad q \geq k_0. \end{aligned} \tag{4.24}$$

In Fig. 4.5, the dispersion curves corresponding to (4.23) and (4.24) are shown. From this figure, one can see that in the case of a metal half-space, which has a small addition of chirality ($\varepsilon < 0$, $\mu = 1$, $\chi = 0.1$), the wavenumbers of the surface chiral-plasmon modes almost coincide with the wavenumbers of the surface plasmon modes in a nonchiral metal (see Fig. 4.5a). In the case of chiral DNG-metamaterial ($\varepsilon < 0$, $\mu = -1$, $\chi = 0.1$), there are two branches of surface modes (see Fig. 4.5b). For nonchiral DNG half-space ($\mu = -1$, $\chi = 0$), these branches coincide and transformed to $\varepsilon = -1$ for all $q \geq k_0$ (see Fig. 4.5b). It is this degenerate plasmon mode that leads to a number of paradoxical properties of the dipole radiation near the DNG half-space [39, 41, 56, 58]. From Fig. 4.5, it also follows that the asymptotic expressions (4.24) describe the solutions of the dispersion equation well.

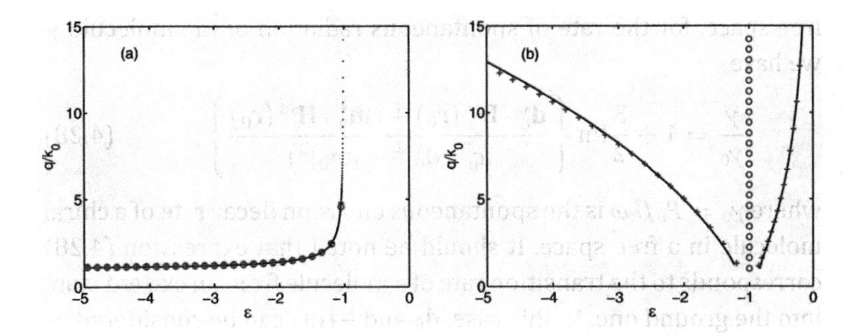

Figure 4.5 The solution of the dispersion equation (4.23) at $\chi = 0.1$. (a) $\mu = 1$; (b) $\mu = -1$. The crosses show the asymptotic solution (4.24). The circles correspond to the solution of the equation $\varepsilon\sqrt{k_0^2 - q^2} + \sqrt{k_0^2\varepsilon\mu - q^2} = 0$. The points show the numerical solution of Eq. (4.23).

The relative rate of spontaneous emission of a chiral molecule near a chiral (or any other) material body can be defined from the classical point of view as the ratio of the power lost by the molecule to drive the electromagnetic field and the analogous power of the molecule in free unbounded space.

The total rate of doing work by the fields in a finite volume V is [73]:

$$P = \frac{1}{2}\mathrm{Re}\int_V dV\left(\mathbf{j}_E^*(\mathbf{r})\cdot\mathbf{E}(\mathbf{r}) + \mathbf{j}_H^*(\mathbf{r})\cdot\mathbf{H}(\mathbf{r})\right), \tag{4.25}$$

Here, $\mathbf{E} = \mathbf{E}_0 + \mathbf{E}^{sc}$ and $\mathbf{H} = \mathbf{H}_0 + \mathbf{H}^{sc}$ are the total electric and magnetic fields near the chiral molecule. Densities of the external currents in this case are defined by the molecular dipole moments:

$$\left\{\begin{array}{c}\mathbf{j}_E\\ \mathbf{j}_H\end{array}\right\} = -i\omega\left\{\begin{array}{c}\mathbf{d}_0\\ -i\mathbf{m}_0\end{array}\right\}\delta(\mathbf{r}-\mathbf{r}_0), \tag{4.26}$$

where $\delta(\mathbf{r}-\mathbf{r}_0)$ is the Dirac delta function; $\mathbf{r}_0$ is the radius vector of the position of a chiral molecule.

The power P_0, which is radiated by the dipole sources in the empty space, is

$$P_0 = \frac{ck_0^4}{3}\left(|\mathbf{d}_0|^2 + |\mathbf{m}_0|^2\right). \tag{4.27}$$

Substituting currents (4.26) into the power (4.25) and normalizing the resulting expression to the power lost by the chiral molecule in free space, for the rate of spontaneous radiation of the molecule γ we have

$$\frac{\gamma}{\gamma_0} = 1 + \frac{3}{2}\mathrm{Im}\left\{\frac{\mathbf{d}_0^* \cdot \mathbf{E}^{sc}(\mathbf{r}_0) + i\mathbf{m}_0^* \cdot \mathbf{H}^{sc}(\mathbf{r}_0)}{k_0^3\left(|\mathbf{d}_0|^2 + |\mathbf{m}_0|^2\right)}\right\}, \tag{4.28}$$

where $\gamma_0 = P_0/\hbar\omega$ is the spontaneous emission decay rate of a chiral molecule in a free space. It should be noted that expression (4.28) corresponds to the transition rate of a molecule from an excited state into the ground one. In this case, $\mathbf{d}_0$ and $-i\mathbf{m}_0$ can be considered as matrix elements of the electric and magnetic dipole moments of the selected transition. To take into account the possibility of transition into several states, one should sum corresponding partial radiation rates.

Substituting expressions (4.19) into (4.28) and changing variables $q = \sqrt{k_0^2 - h^2}$ and $q = \sqrt{k_0^2 + p^2}$, one can find

$$\frac{\gamma}{\gamma_0} = 1 + \frac{3}{4}\mathrm{Re}\left\{\frac{\int_0^{k_0} dhU(h)\exp(2ihz_0) - i\int_0^{\infty} dpU(ip)\exp(-2pz_0)}{k_0^3\left(|\mathbf{d}_0|^2 + |\mathbf{m}_0|^2\right)}\right\}, \tag{4.29}$$

where

$$\begin{aligned} U(h) = {} & 2\left(h^2 - k_0^2\right)\left\{\beta_h |d_{0z}|^2 + \alpha_h |m_{0z}|^2 + 2\delta_h Re\left(d_{0z}m_{0z}^*\right)\right\} \\ & + \left(d_{0x}^*h + im_{0y}^*k_0\right)\left\{\beta_h\left(d_{0x}h + im_{0y}k_0\right) + \delta_h\left(m_{0x}h + id_{0y}k_0\right)\right\} \\ & + \left(d_{0y}^*h - im_{0x}^*k_0\right)\left\{\beta_h\left(d_{0y}h - im_{0x}k_0\right) + \delta_h\left(m_{0y}h - id_{0x}k_0\right)\right\} \\ & + \left(m_{0x}^*h + id_{0y}^*k_0\right)\left\{\alpha_h\left(m_{0x}h + id_{0y}k_0\right) + \delta_h\left(d_{0x}h + im_{0y}k_0\right)\right\} \\ & + \left(m_{0y}^*h - id_{0x}^*k_0\right)\left\{\alpha_h\left(m_{0y}h - id_{0x}k_0\right) + \delta_h\left(d_{0y}h - im_{0x}k_0\right)\right\}. \end{aligned} \tag{4.30}$$

The coefficients α_h, β_h, and δ_h can be obtained from the coefficients a_q, b_q, and c_q [see (4.21)], correspondingly, with the help of substitution $q = \sqrt{k_0^2 - h^2}$; expression for $U(ip)$ one can derive from (4.30) by substituting $h \to ip$; the coefficients α_{ip}, β_{ip}, and δ_{ip} are obtained from the coefficients a_q, b_q and c_q, respectively, by substitution $q = \sqrt{k_0^2 + p^2}$.

The physical reason of the separation of (4.29) into two parts is due to the presence of both propagating [the first integral in (4.29)] and evanescent (the second integral) waves in radiation of the dipole [54].

In the case $k_0 \to 0$, the expression for the rate of spontaneous emission of a chiral molecule placed near the chiral half-space becomes

$$\frac{\gamma}{\gamma_0} = \frac{3}{2\left(2k_0 z_0\right)^3} \left\{ \begin{array}{l} \left(\dfrac{|\mathbf{d}_0|^2 + |d_{0z}|^2}{|\mathbf{d}_0|^2 + |\mathbf{m}_0|^2} \right) \mathrm{Im}\left(\dfrac{(\varepsilon - 1)(\mu + 1) + \varepsilon\mu\chi^2}{(\varepsilon + 1)(\mu + 1) - \varepsilon\mu\chi^2} \right) + \\ \left(\dfrac{|\mathbf{m}_0|^2 + |m_{0z}|^2}{|\mathbf{d}_0|^2 + |\mathbf{m}_0|^2} \right) \mathrm{Im}\left(\dfrac{(\varepsilon + 1)(\mu - 1) + \varepsilon\mu\chi^2}{(\varepsilon + 1)(\mu + 1) - \varepsilon\mu\chi^2} \right) + \\ \left(\dfrac{2\mathrm{Re}\left(\mathbf{d}_0 \cdot \mathbf{m}_0^* + d_{0z} m_{0z}^*\right)}{|\mathbf{d}_0|^2 + |\mathbf{m}_0|^2} \right) \mathrm{Im}\left(\dfrac{2\varepsilon\mu\chi}{(\varepsilon + 1)(\mu + 1) - \varepsilon\mu\chi^2} \right) \end{array} \right\}. \tag{4.31}$$

In the case of a chiral medium without losses, the expression (4.31) is equal to zero. Thus, the quasistatic limit (4.31) should be associated with nonradiative (Joule) losses in chiral media.

The vanishing of the denominator in (4.31) determines the values of ε and μ, which correspond to the excitation of chiral-plasmon resonances in chiral half-space in the quasistatic approximation:

$$(\varepsilon + 1)(\mu + 1) = \varepsilon\mu\chi^2. \tag{4.32}$$

In the case of a nonchiral medium ($\chi = 0$), from (4.32) we obtain two independent equations: $\varepsilon + 1 = 0$ and $\mu + 1 = 0$ [59].

4.2.2 *Graphical Illustrations and Discussion of the Results Obtained*

Analytical properties of the expression (4.29) [which are defined by the poles corresponding to (4.23) and by branching points at $q = k_0,\ k_L,\ k_R$] are much more complicated than analogous expressions for a nonchiral case [17], and we shall not investigate them here. Instead of this, we will calculate the integrals numerically, although that is relatively easy in the case of nonzero imaginary parts of ε and μ, which lead to a shift of singularities from the real axis (the integration path) into the complex plane of the integration variable.

Now, to be more specific, we consider only the typical case of a chiral molecule with an equal Cartesian projection of the electric dipole moment ($\mathbf{d}_0 = \{d_0, d_0, d_0\}$) and with an equal Cartesian projection of the magnetic dipole moment ($\mathbf{m}_0 = \{m_0, m_0, m_0\}$). In this special case, the integrand expression (4.30) can be written as follows:

$$U(h) = 2\left[\left(2h^2 - k_0^2\right)\beta_h - k_0^2\alpha_h\right]|d_0|^2 + 2\left[\left(2h^2 - k_0^2\right)\alpha_h - k_0^2\beta_h\right]|m_0|^2 + 8\left(h^2 - k_0^2\right)\delta_h Re\left(d_0 m_0^*\right). \quad (4.33)$$

Here, as before, we will call the molecules with parallel $\mathbf{d}_0$ and $\mathbf{m}_0$ (d_0 and m_0 have the same signs) "right" molecules, and the molecules with antiparallel $\mathbf{d}_0$ and $\mathbf{m}_0$ (d_0 and m_0 have the opposite signs) we will call "left" molecules.

In Fig. 4.6, the dependence of the spontaneous decay rate of a chiral molecule on its distance to the interface is shown. As it follows from the figure, there is a difference between the decay rates of "right" and "left" molecule enantiomers. In this case, the difference is biggest for chiral metamaterial with $\varepsilon > 0$ and $\mu < 0$ (MNG-metamaterial) (see Fig. 4.6c) and for chiral metamaterial with $\varepsilon < 0$ and $\mu < 0$ (DNG-metamaterial) (see Fig. 4.6d). In the case of a chiral dielectric, this difference is smaller (see Fig. 4.6a). For a metal with chiral properties, the curves for "left" and "right" molecules have almost no difference (see Fig. 4.6b). This is due to a rather large imaginary part of the wavenumbers in this case. In addition, from Fig. 4.6, it follows that the asymptotic relation (4.31) describes well the nonradiative part of the total rate of spontaneous decay for small values of $k_0 z_0$ and for different values of permittivity and permeability.

Figure 4.7 shows the ratio of the spontaneous decay rate of a "left" molecule and the spontaneous decay rate of a "right" molecule (γ_L/γ_R) and, vice versa, the ratio of the spontaneous radiation rate of a "right" molecule and the spontaneous radiation rate of a "left" molecule (γ_R/γ_L), as a function of permittivity and permeability of the chiral half-space. From this figure, it follows that for certain values ε and μ, the radiation rates of the "right" and "left" molecules differ by 10–15 times. It is seen that a chiral metamaterial with $-1 < \varepsilon < 0$ and $\mu \approx -1$ (DNG-metamaterial) will increase radiation of "left" molecule and suppress radiation of "right" molecules. At the same time, a chiral metamaterial with $\varepsilon > 0$ and $\mu \approx -1$ (MNG-metamaterial) and also a metamaterial with $\varepsilon < -1$ and $\mu \approx -1$

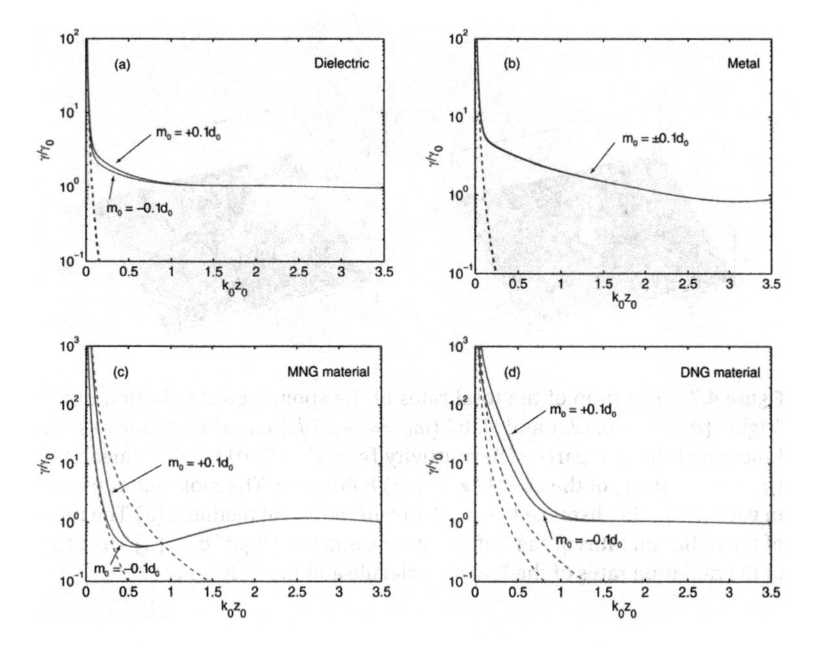

Figure 4.6 The total rate of spontaneous radiation of the "right" ($m_0 = +0.1d_0$) and "left" ($m_0 = -0.1d_0$) chiral molecules located in vacuum near a chiral ($\chi = 0.2$) half-space as the function of the normalized distance $k_0 z_0$ between the molecule and the half-space. The permittivity and permeability of the chiral medium are $\varepsilon = \varepsilon' + i0.01$ and $\mu = \mu' + i0.01$, respectively. (a) Dielectric, $\varepsilon' = 3$ and $\mu' = 1$. (b) Metal, $\varepsilon' = -3$ and $\mu' = 1$. (c) MNG-metamaterial, $\varepsilon' = 3$ and $\mu' = -1$. (d) DNG-metamaterial, $\varepsilon' = -3$ and $\mu' = -1$. The dashed lines correspond to the asymptotic solution (4.31).

(DNG-metamaterial) will increase the radiation of a "right" molecule and slow down the radiation of "left" molecules. More clearly, the difference in the spontaneous radiation rate of the enantiomers of chiral molecules is shown in Fig. 4.8.

Note, that from a practical point of view, for the experimental study of the effect of differences in the rates of spontaneous radiation of the "right" and "left" molecule enantiomers, the chiral MNG-metamaterial is more promising, because the technology of its implementation (technology of the split ring resonators) is now well established [55, 67]. On the contrary, the chiral DNG-metamaterials can be also synthesized [46].

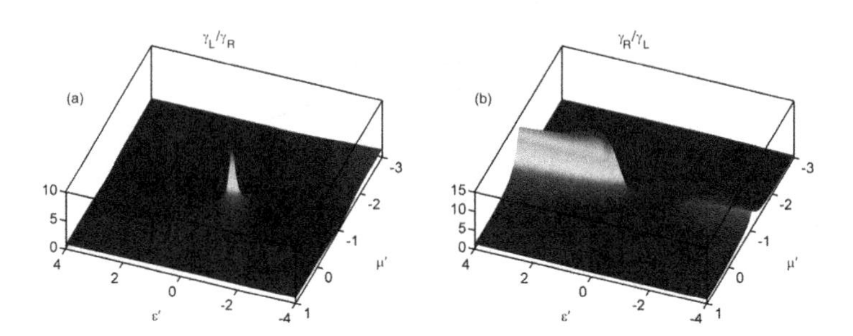

Figure 4.7 The ratio of the total rates of the spontaneous radiation of the "right" ($m_0 = +0.1d_0$) and "left" ($m_0 = -0.1d_0$) chiral molecules as the function of the real parts of permittivity ($\varepsilon = \varepsilon' + i0.01$) and permeability ($\mu = \mu' + i0.01$) of the chiral ($\chi = 0.2$) half-space. The molecule is placed in vacuum on the distance $k_0 z_0 = 0.1$ from the chiral medium. (a) The ratio of the radiation rates of the "left" molecule and the "right" one. (b) The ratio of the radiation rates of the "right" molecule and the "left" one.

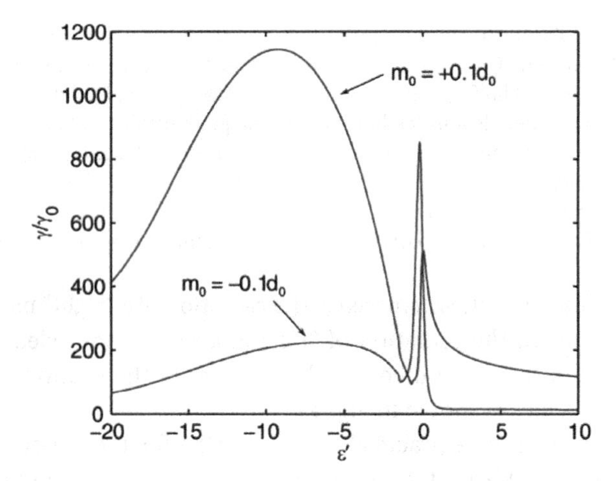

Figure 4.8 The total rate of the spontaneous emission of "left" ($m_0 = +0.1d_0$) and "right" ($m_0 = -0.1d_0$) chiral molecules as the function of the real part of the permittivity $\varepsilon = \varepsilon' + i0.01$ of a chiral ($\chi = 0.2$) half-space at the fixed permeability $\mu = -1 + i0.01$. The molecule is placed in vacuum at the distance $k_0 z_0 = 0.1$ from the chiral medium.

Thus, even a relatively simple geometry, like a half-space from a chiral metamaterial, can have different effects on the spontaneous radiation of the enantiomers of chiral molecules. In practice, this effect can be used to create devices for detection and selection of the "right" and "left" molecule enantiomers. Despite the fact that the chiral half-space does not affect the "right" and "left" molecules so much as the chiral nanoparticle do (see the next section), this geometry is simpler from a technological point of view, and even such a simple system can serve as a basis for the development of devices for the selection of the enantiomers in racemic mixtures of molecules.

4.3 Radiation of a Chiral Molecule Near a Chiral Spherical Particle

Optical properties of chiral spherical particles have been studied quite extensively. Detailed investigation of Eigen oscillations of chiral sphere was presented in [44], where an interesting interplay between chirality and spatial helicity of Eigen modes was found. The scattering of a plane electromagnetic wave by homogeneous [9, 53] and inhomogeneous [50] chiral microspheres is investigated as well as the scattering by chiral spherical shells [49]; the scattering of the Hermite–Gaussian beam by a chiral microsphere is also considered [77].

A significant number of works is also devoted to the study of the effect of spherical particles on the radiation of an electric dipole. The influence of dielectric microspheres on the spontaneous emission of atoms was considered in [15, 16, 37]; the spontaneous decay of an atom located near the microsphere made of DNG-metamaterial was analyzed in [35]. The radiation of an electric dipole located inside the chiral spherical particle is considered in [47].

At the same time, the problem of the influence of chiral spherical particles on the radiation of chiral molecules was not widely discussed so far. In this section, we present the results of investigation of the effect of the chiral spherical particle on the spontaneous emission of chiral (optically active) molecule, which is located near the particle (see also [29, 42]). All analytical results will

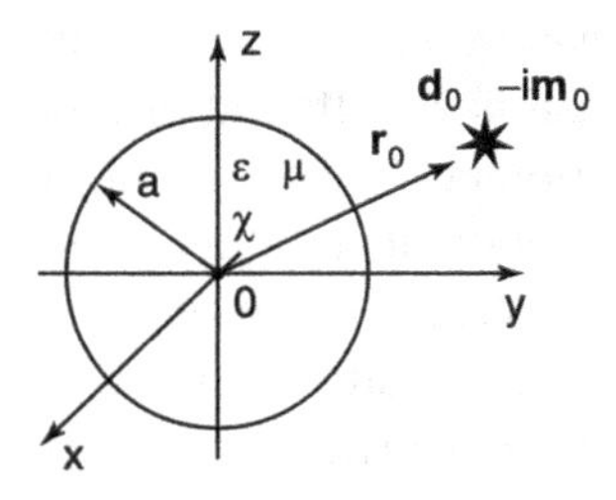

Figure 4.9 The geometry of the problem of the spontaneous emission of a chiral molecule located near a chiral spherical particle. The molecule is placed in vacuum.

be obtained for an arbitrary particle size, its material composition, and arbitrary relations between the electric and magnetic dipole moments of the chiral molecule. The geometry of the problem is shown in Fig. 4.9. Note that in this case, we use the classical formulation of the problem of spontaneous radiation. However, the full quantum electrodynamics approach gives the same results [29].

4.3.1 *Spontaneous Emission of a Chiral Molecule placed near a Chiral Spherical Particle*

In the case of a spherical particle, it is convenient to express the left and right-polarized eigenwaves in (4.15) through the vector spherical harmonics $\mathbf{N}\psi_{mn}^{(J)}$ and $\mathbf{M}\psi_{mn}^{(J)}$ (where $J = L, R$) [8]

$$\mathbf{Q}_L = \sum_{n=1}^{\infty}\sum_{m=-n}^{n} A_{mn}\left(\mathbf{N}\psi_{mn}^{(L)} + \mathbf{M}\psi_{mn}^{(L)}\right),$$

$$\mathbf{Q}_R = \sum_{n=1}^{\infty}\sum_{m=-n}^{n} B_{mn}\left(\mathbf{N}\psi_{mn}^{(R)} - \mathbf{M}\psi_{mn}^{(R)}\right), \tag{4.34}$$

the explicit expressions for which are given in Appendix 4.B.

From (4.34), one can see that $\mathbf{Q}_L$ and $\mathbf{Q}_R$ are expressed in terms of fixed combinations of the vector spherical harmonics $\mathbf{N}\psi_{mn}^{(L)} + \mathbf{M}\psi_{mn}^{(L)}$ and $\mathbf{N}\psi_{mn}^{(R)} - \mathbf{M}\psi_{mn}^{(R)}$, which always have nonzero component along the radius (see Appendix 4.B) and, therefore, can never be reduced to the usual TM or TE fields.

In accordance with the Bohren's transformation (4.14) and equations (4.34), the electric and magnetic fields inside a chiral spherical particle take the form:

$$\mathbf{E} = \sum_{n=1}^{\infty}\sum_{m=-n}^{n}\left[A_{mn}\left(\mathbf{N}\psi_{mn}^{(L)} + \mathbf{M}\psi_{mn}^{(L)}\right) - iZ\,B_{mn}\left(\mathbf{N}\psi_{mn}^{(R)} - \mathbf{M}\psi_{mn}^{(R)}\right)\right],$$

$$\mathbf{H} = \sum_{n=1}^{\infty}\sum_{m=-n}^{n}\left[-iZ^{-1}A_{mn}\left(\mathbf{N}\psi_{mn}^{(L)} + \mathbf{M}\psi_{mn}^{(L)}\right) + B_{mn}\left(\mathbf{N}\psi_{mn}^{(R)} - \mathbf{M}\psi_{mn}^{(R)}\right)\right]. \quad (4.35)$$

Electric and magnetic fields of the chiral molecule with the electric and magnetic dipole moments $\mathbf{d}_0$ and $-i\mathbf{m}_0$, respectively, which is located in the point $\mathbf{r}_0$ in vacuum, can be presented as an expansion over the vector spherical harmonics in the following form ($r_0 > r$):

$$\mathbf{E}_0 = \sum_{n=1}^{\infty}\sum_{m=-n}^{n}\left(A_{mn}^{(0)}\mathbf{N}\psi_{mn}^{(0)} + B_{mn}^{(0)}\mathbf{M}\psi_{mn}^{(0)}\right),$$

$$\mathbf{H}_0 = -i\sum_{n=1}^{\infty}\sum_{m=-n}^{n}\left(B_{mn}^{(0)}\mathbf{N}\psi_{mn}^{(0)} + A_{mn}^{(0)}\mathbf{M}\psi_{mn}^{(0)}\right), \quad (4.36)$$

where $\mathbf{N}\psi_{mn}^{(0)}$ and $\mathbf{M}\psi_{mn}^{(0)}$, and $A_{mn}^{(0)}$ and $B_{mn}^{(0)}$ are defined in Appendix 4.B.

Electric and magnetic fields of the reflected wave can be written as:

$$\mathbf{E}^{sc} = \sum_{n=1}^{\infty}\sum_{m=-n}^{n}\left(C_{mn}\mathbf{N}\zeta_{mn} + D_{mn}\mathbf{M}\zeta_{mn}\right),$$

$$\mathbf{H}^{sc} = -i\sum_{n=1}^{\infty}\sum_{m=-n}^{n}\left(D_{mn}\mathbf{N}\zeta_{mn} + C_{mn}\mathbf{M}\zeta_{mn}\right), \quad (4.37)$$

where $\mathbf{N}\zeta_{mn}$ and $\mathbf{M}\zeta_{mn}$ are defined in Appendix 4.B.

To find the coefficients in expansions (4.35) and (4.37), one can use the continuity of the tangential components of the electric and magnetic fields on the surface of the sphere [69].

For the coefficients C_{mn} and D_{mn} that define the reflected fields in (4.28), one can obtain the following expressions:

$$C_{mn} = -\alpha_n A_{mn}^{(0)} + i\delta_n B_{mn}^{(0)}, \quad D_{mn} = -\beta_n B_{mn}^{(0)} + i\delta_n A_{mn}^{(0)}, \quad (4.38)$$

where functions α_n, β_n, and δ_n have the form

$$\alpha_n = \frac{A_n^{(L)}W_n^{(R)} + A_n^{(R)}W_n^{(L)}}{V_n^{(L)}W_n^{(R)} + V_n^{(R)}W_n^{(L)}}, \quad \beta_n = \frac{V_n^{(L)}B_n^{(R)} + V_n^{(R)}B_n^{(L)}}{V_n^{(L)}W_n^{(R)} + V_n^{(R)}W_n^{(L)}},$$

$$\delta_n = i\frac{B_n^{(L)}W_n^{(R)} - B_n^{(R)}W_n^{(L)}}{V_n^{(L)}W_n^{(R)} + V_n^{(R)}W_n^{(L)}}. \quad (4.39)$$

In (4.39), the abbreviations ($J = L, R$) were introduced:

$$A_n^{(J)} = Z^{-1}\psi_n(k_J a)\,\psi_n'(k_0 a) - \psi_n'(k_J a)\,\psi_n(k_0 a),$$
$$B_n^{(J)} = \psi_n(k_J a)\,\psi_n'(k_0 a) - Z^{-1}\psi_n'(k_J a)\,\psi_n(k_0 a),$$
$$V_n^{(J)} = Z^{-1}\psi_n(k_J a)\,\zeta_n'(k_0 a) - \psi_n'(k_J a)\,\zeta_n(k_0 a),$$
$$W_n^{(J)} = \psi_n(k_J a)\,\zeta_n'(k_0 a) - Z^{-1}\psi_n'(k_J a)\,\zeta_n(k_0 a), \quad (4.40)$$

where $\psi_n(k_J a)$ and $\zeta_n(k_0 a)$ are Riccati-Bessel functions [1] (see also Appendix 4.B); the prime near the function [in (4.40) and below] means its derivative; a is the radius of the spherical particle (see Fig. 4.9); k_J is the wavenumber (4.16).

The relative radiative decay rate of the spontaneous emission (which is proportional to the power of the spontaneous emission) of a chiral molecule located near the chiral spherical particle can be calculated as the ratio of the radiation power of the system "molecule and particle" (P_{rad}) and the radiation power of the molecule in a free space ($P_{0,rad}$) [36]. To find the time-averaged power of radiation of the molecule and particle system, it is necessary to calculate power flow across a sphere of an infinite radius [69]:

$$P_{\mathrm{rad}} = \frac{c}{8\pi}\mathrm{Re}\int\limits_{S_\infty}\left[\left(\mathbf{E}^{sc} + \mathbf{E}_0\right), \left(\mathbf{H}^{sc*} + \mathbf{H}_0^*\right)\right]_n dS, \quad (4.41)$$

where the electric and magnetic fields of a chiral molecule in a free space $\mathbf{E}_0$, $\mathbf{H}_0$ have form ($r > r_0$):

$$\mathbf{E}_0 = \sum_{n=1}^{\infty}\sum_{m=-n}^{n}\left(C_{mn}^{(0)}\mathbf{N}\zeta_{mn} + D_{mn}^{(0)}\mathbf{M}\zeta_{mn}\right),$$
$$\mathbf{H}_0 = -i\sum_{n=1}^{\infty}\sum_{m=-n}^{n}\left(D_{mn}^{(0)}\mathbf{N}\zeta_{mn} + C_{mn}^{(0)}\mathbf{M}\zeta_{mn}\right), \quad (4.42)$$

with $C_{mn}^{(0)}$ and $D_{mn}^{(0)}$ defined in Appendix 4.B.

Evaluating the integral (4.41) and normalizing the resulting expression on the time-averaged power of the radiation of a chiral molecule located in a free space $P_{0,\mathrm{rad}} = P_0$ [see (4.27)], for the relative radiative decay rate γ_{rad} we obtain the following expression:

$$\frac{\gamma_{\mathrm{rad}}}{\gamma_0} = \frac{3}{2k_0^6\left(|\mathbf{d}_0|^2 + |\mathbf{m}_0|^2\right)}\sum_{n=1}^{\infty}\sum_{m=-n}^{n}\frac{n(n+1)}{2n+1}\frac{(n+m)!}{(n-m)!}$$
$$\times\left(\left|C_{mn}^{(0)} + C_{mn}\right|^2 + \left|D_{mn}^{(0)} + D_{mn}\right|^2\right), \quad (4.43)$$

where γ_0 is the spontaneous emission decay rate in a free space. The expression (4.43) corresponds to a transition in the molecule from one of the excited states to the ground state, and $\mathbf{d}_0$ and $-i\mathbf{m}_0$ are the electric and magnetic dipole moments of this transition.

In the case of a molecule located on the positive side of the z axis at the point r_0, the expression (4.43) can be written in the following explicit form:

$$\begin{aligned}
\frac{\gamma_{\text{rad}}}{\gamma_0} &= \frac{3}{8\left(k_0r_0\right)^2\left(|\mathbf{d}_0|^2+|\mathbf{m}_0|^2\right)}\sum_{n=1}^{\infty}(2n+1) \\
&\times \Big(\big|\left(d_{0x}-id_{0y}\right)\left(\psi_n'\left(k_0r_0\right)-\alpha_n\zeta_n'\left(k_0r_0\right)\right)+i\delta_n\left(m_{0x}-im_{0y}\right)\zeta_n'\left(k_0r_0\right) \\
&-\left(m_{0y}+im_{0x}\right)\left(\psi_n\left(k_0r_0\right)-\alpha_n\zeta_n\left(k_0r_0\right)\right)-i\delta_n\left(d_{0y}+id_{0x}\right)\zeta_n\left(k_0r_0\right)\big|^2 \\
&+\big|\left(d_{0x}+id_{0y}\right)\left(\psi_n'\left(k_0r_0\right)-\alpha_n\zeta_n'\left(k_0r_0\right)\right)+i\delta_n\left(m_{0x}+im_{0y}\right)\zeta_n'\left(k_0r_0\right) \\
&-\left(m_{0y}-im_{0x}\right)\left(\psi_n\left(k_0r_0\right)-\alpha_n\zeta_n\left(k_0r_0\right)\right)-i\delta_n\left(d_{0y}-id_{0x}\right)\zeta_n\left(k_0r_0\right)\big|^2 \\
&+\big|\left(m_{0x}-im_{0y}\right)\left(\psi_n'\left(k_0r_0\right)-\beta_n\zeta_n'\left(k_0r_0\right)\right)+i\delta_n\left(d_{0x}-id_{0y}\right)\zeta_n'\left(k_0r_0\right) \\
&-\left(d_{0y}+id_{0x}\right)\left(\psi_n\left(k_0r_0\right)-\beta_n\zeta_n\left(k_0r_0\right)\right)-i\delta_n\left(m_{0y}+im_{0x}\right)\zeta_n\left(k_0r_0\right)\big|^2 \\
&+\big|\left(m_{0x}+im_{0y}\right)\left(\psi_n'\left(k_0r_0\right)-\beta_n\zeta_n'\left(k_0r_0\right)\right)+i\delta_n\left(d_{0x}+id_{0y}\right)\zeta_n'\left(k_0r_0\right) \\
&-\left(d_{0y}-id_{0x}\right)\left(\psi_n\left(k_0r_0\right)-\beta_n\zeta_n\left(k_0r_0\right)\right)-i\delta_n\left(m_{0y}-im_{0x}\right)\zeta_n\left(k_0r_0\right)\big|^2\Big) \\
&+\frac{3}{2\left(k_0r_0\right)^4\left(|\mathbf{d}_0|^2+|\mathbf{m}_0|^2\right)}\sum_{n=1}^{\infty}(2n+1)\,n\,(n+1) \\
&\times\Big(|d_{0z}\left(\psi_n\left(k_0r_0\right)-\alpha_n\zeta_n\left(k_0r_0\right)\right)+i\delta_n m_{0z}\zeta_n\left(k_0r_0\right)|^2 \\
&+|m_{0z}\left(\psi_n\left(k_0r_0\right)-\beta_n\zeta_n\left(k_0r_0\right)\right)+i\delta_n d_{0z}\zeta_n\left(k_0r_0\right)|^2\Big).
\end{aligned} \tag{4.44}$$

In a special case of a nonchiral particle ($\chi = 0$, $\delta_n = 0$), the well-known expressions [37] for the spontaneous emission radiative decay rate of an atom with a nonzero electric or magnetic dipole moment of the transition can be derived from (4.43).

If there are losses in the chiral particle, there is a channel of nonradiative transition from the excited state to the ground one. The general expression for the total (radiative and nonradiative) rate of spontaneous emission was obtained earlier [see (4.28)]. Subtracting the radiative component from the total rate, we obtain the nonradiative decay rate of the spontaneous emission decay. In the case of chiral particles without losses, the expression (4.28) gives the same results as the expression (4.43). In this Chapter, we will not consider the nonradiative channel of the spontaneous decay. A detailed investigation of this case is performed in [44].

4.3.2 *Analysis of the Results obtained and Graphical Illustrations*

In a very important case of a chiral nanoparticle, that is, a particle having a size much smaller than the wavelength of the radiation, the expression (4.43) can be expanded in a series over the small parameter $k_0 a \to 0$. Taking into account only the main terms of the expansion, we find the following expression for the spontaneous emission relative radiative decay rate [42]:

$$\frac{\gamma_{\mathrm{rad}}}{\gamma_0} = \frac{1}{|\mathbf{d}_0|^2 + |\mathbf{m}_0|^2} \left\{ \begin{array}{l} \left| \mathbf{d}_0 + \frac{\alpha_{\mathrm{EE}}}{r_0^3} \left(3\mathbf{n}\left(\mathbf{n}\cdot\mathbf{d}_0\right) - \mathbf{d}_0\right) - \frac{i\alpha_{\mathrm{EH}}}{r_0^3} \left(3\mathbf{n}\left(\mathbf{n}\cdot\mathbf{m}_0\right) - \mathbf{m}_0\right) \right|^2 \\ + \left| \mathbf{m}_0 + \frac{\alpha_{\mathrm{HH}}}{r_0^3} \left(3\mathbf{n}\left(\mathbf{n}\cdot\mathbf{m}_0\right) - \mathbf{m}_0\right) + \frac{i\alpha_{\mathrm{HE}}}{r_0^3} \left(3\mathbf{n}\left(\mathbf{n}\cdot\mathbf{d}_0\right) - \mathbf{d}_0\right) \right|^2 \end{array} \right\}, \tag{4.45}$$

where electromagnetic polarizabilities of a chiral spherical particle in the uniform field have the form:

$$\begin{aligned} \alpha_{\mathrm{EE}} &= a^3 \frac{(\varepsilon - 1)(\mu + 2) + 2\varepsilon\mu\chi^2}{(\varepsilon + 2)(\mu + 2) - 4\varepsilon\mu\chi^2}, \\ \alpha_{\mathrm{HH}} &= a^3 \frac{(\varepsilon + 2)(\mu - 1) + 2\varepsilon\mu\chi^2}{(\varepsilon + 2)(\mu + 2) - 4\varepsilon\mu\chi^2}, \\ \alpha_{\mathrm{EH}} &= -\alpha_{\mathrm{HE}} = a^3 \frac{3i\varepsilon\mu\chi}{(\varepsilon + 2)(\mu + 2) - 4\varepsilon\mu\chi^2}; \end{aligned} \tag{4.46}$$

n is the unit vector directed from the center of the particle to the molecule position.

The vanishing of the denominator (4.46) determines the values of ε and μ, which correspond to excitation of the chiral-plasmon resonances in chiral spherical nanoparticle:

$$(\varepsilon + 2)(\mu + 2) = 4\varepsilon\mu\chi^2. \tag{4.47}$$

In a special case of a nonchiral material ($\chi = 0$), one can obtain two independent equations from (4.47): $\varepsilon + 2 = 0$ and $\mu + 2 = 0$, which correspond to the dipole plasmon resonances in a nonchiral spherical nanoparticle [43].

In Fig. 4.10, the structure of plasmonic resonances for a spherical nanoparticle without chirality and with the chirality factor $\chi = 0.1$

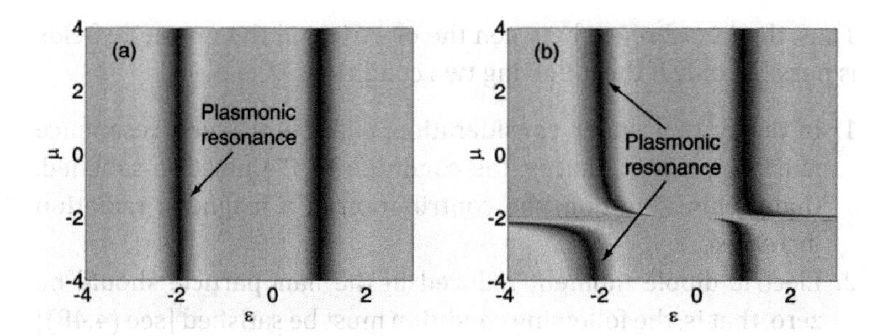

Figure 4.10 The dependence of the coefficient α_1 (see (4.39)) as the function of permittivity (ε) and permeability (μ). (a) $\chi = 0$. (b) $\chi = 0.1$. The particle is placed in vacuum, $k_0 a = 0.1$.

is shown. From this figure, it is seen that TM modes in a nonchiral nanoparticle (see Fig. 4.10a) have electric dipole plasmon resonance at $\varepsilon \approx -2$, nearly independent on permeability.

At the same time, any chirality (even small) of a nanoparticle (see Fig. 4.10b) leads to a radical change in the shape of the resonance line, which becomes dependent in a nontrivial way on both the permittivity and permeability of the material of the nanoparticle. Below, we will see that these "hybrid" chiral-plasmon modes in the nanoparticle provide the effective interference between radiation of electric and magnetic dipole moments of the chiral molecule.

In practice, the orientation of the molecule can be arbitrary with respect to the nanoparticle surface. Therefore in such cases, one should average (4.45) over the orientations of the electric and magnetic dipole moments. For spiral molecules with collinear electric and magnetic dipole moments (i.e., for those with $\mathbf{m}_0 = \xi \mathbf{d}_0$), the averaging leads to the following expression:

$$\frac{\gamma_{\text{rad}}}{\gamma_0} = 1 + \frac{2\left|\alpha_{\text{EE}} - i\xi\alpha_{\text{EH}}\right|^2}{r_0^6\left(1+|\xi|^2\right)} + \frac{2\left|\alpha_{\text{HE}} - i\xi\alpha_{\text{HH}}\right|^2}{r_0^6\left(1+|\xi|^2\right)}. \qquad (4.48)$$

As a rule, the magnetic dipole moment of the molecule is much smaller than the electric dipole moment $\xi << 1$. The chirality parameter is also usually small ($\chi << 1$) (see, however, [25]). So, the second term in (4.48) corresponding to the induced electric dipole moment is usually greater than the term corresponding to the induced magnetic dipole moment [the third term in (4.48)].

Thus, the interference between the electric and the magnetic fields is possible only if the following two conditions take place.

1. In the system under consideration, a chiral-plasmon resonance must be present, that is, the condition (4.47) must be satisfied. Under this condition, the contribution of a magnetic radiation increases.
2. Electric dipole moment induced in the nanoparticle should be zero, that is, the following condition must be satisfied [see (4.48)]

$$\alpha_{\mathrm{EE}} - i\xi\alpha_{\mathrm{EH}} = 0. \tag{4.49}$$

The solution of the system of Eqs. (4.47) and (4.49) determines values of permittivity and permeability of the nanoparticle, which correspond to minimal values of the radiative decay rate of a chiral molecule. It means that the interference between the electric dipole and magnetic dipole radiation becomes maximal and destructive if

$$\mu \to -\frac{2d_0}{d_0 + 2\chi m_0}, \quad \varepsilon \to -\frac{2m_0}{m_0 + 2\chi d_0}. \tag{4.50}$$

From (4.50), it follows that by changing the sign of m_0, that is, when the chirality of the molecule is changed, the resonant permeability μ varies only slightly ($\chi << 1$, $m_0 << d_0$) and approximately equals to -2. On the contrary, the resonant permittivity ε may have different signs for molecules with different chirality. This means that both the nanoparticles with simultaneously negative ε and μ (DNG-metamaterials), and nanoparticles with negative μ and positive ε (MNG-metamaterials) are suitable for the effective control of radiation of chiral molecules.

One should note that all the results obtained [including (4.50)] are valid for any values of the chirality parameter and the relation between electric and magnetic dipole moments, not only for the small ones. For a larger chirality parameter (see, e.g., [25]), one can expect even more pronounced effects.

Figure 4.11 shows the radiative decay rate of spontaneous emission of a chiral molecule located near the DNG chiral spherical nanoparticles as a function of the permittivity for a fixed value of permeability. As it is well seen in the figure, for molecules that differ only in the orientation of the magnetic dipole moment of the transition (for the "left" and the "right" molecules), the radiative

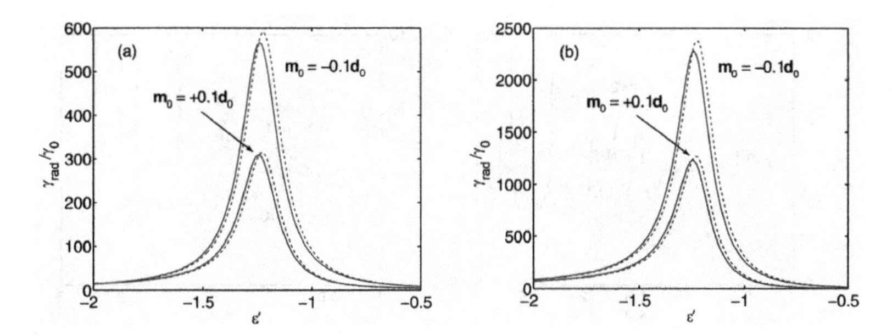

Figure 4.11 Radiative decay rate (4.43) of the spontaneous emission of a chiral molecule located in the z-axis in close vicinity to the surface of the chiral DNG spherical nanoparticle ($r_0 \to a$) with $\varepsilon = \varepsilon' + i0.1$, $\mu = -1.6$, $\chi = 0.2$, and $k_0 a = 0.1$. (a) Electric and magnetic dipole moments of the transition of the molecule are oriented along the x-axis (tangentially to the particle surface). (b) Electric and magnetic dipole moments of the transition of the molecule are oriented along the z-axis (normally to the particle surface). The dashed line shows the asymptotic expression (4.45). Nanoparticle is placed in a vacuum.

decay rates take substantially different values, and thus, under such conditions, it is possible to control effectively the emission properties of enantiomers.

From Fig. 4.11, one can see that the quasistatic asymptotic expression (4.45) gives a good description of the properties of the arbitrary chiral molecules radiation near nanoparticles indeed. In the case when retardation effects are significant, it is necessary to make use of the full expression for the radiative decay rate (4.43).

In Fig. 4.12, the radiative decay rate of spontaneous emission of a chiral molecule placed near the chiral spherical particle with positive values of permittivity and permeability is shown as a function of the size parameter. As it is seen from this figure, the presence of chirality in a spherical particle leads to a shift of the maxima of the spontaneous emission radiative decay rate in comparison with the case of nonchiral particle. Indeed, as it is known, a condition that allows to estimate the position of these resonance is given by the relation $k_0 a\sqrt{\varepsilon\mu} \approx n + 1/2$. In the case of chiral spherical particles, left and right-polarized waves exist simultaneously, and $k_L a > k_0 a\sqrt{\varepsilon\mu}$ if $\chi > 0$. Hence, the above condition should be changed to $k_L a \approx n + 1/2$. As a result, this leads

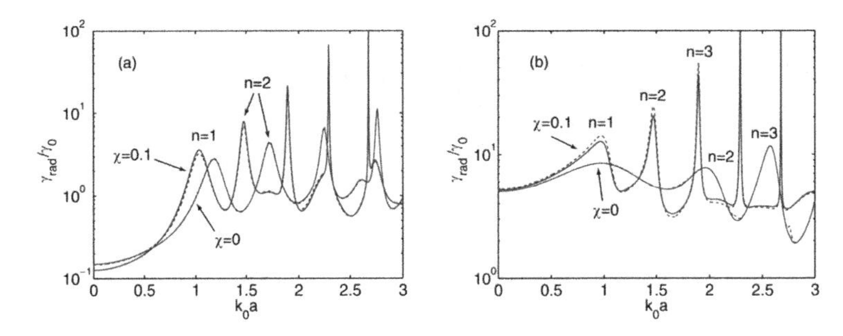

Figure 4.12 Radiative decay rate of the spontaneous emission of a chiral molecule located in close vicinity to the surface of a chiral dielectric spherical particle ($r_0 \to a$) with $\varepsilon = 6$ and $\mu = 1$ as a function of its size, $k_0 a$. (a) The transition electric dipole moment of the molecule is oriented tangentially to the surface of the particle. (b) The transition electric dipole moment of the molecule is normal to the surface of the particle. The solid line corresponds to the transition magnetic dipole moment of a molecule oriented along the surface ($m_{0x} = 0.1 d_0$), while the dashed line corresponds to the transition magnetic dipole moment of a molecule oriented normally to the sphere surface ($m_{0z} = 0.1 d_0$). The particle is placed in vacuum.

to smaller resonant values of $k_0 a$ than those of a nonchiral particle. Even more interesting feature of ordinary dielectric particles with a small admixture of chirality is a substantial increase in the quality factor of whispering gallery modes. This figure clearly shows that the corresponding linewidth can be increased by factor 9 or even more.

Note also that in Fig. 4.12 only a slight difference between the radiative decay rates of spontaneous emission corresponding to the different orientations of the transition magnetic dipole moment of a chiral molecule at a fixed electric dipole moment is observed. However, one should expect an increase of this difference in the case of increasing of absolute values of the magnetic dipole momentum.

Figure 4.13 shows the radiative decay rate of spontaneous emission of a chiral molecule located near chiral spherical nanoparticles as a function of the chiralityχ. As it is clearly seen in this figure, changing the chirality of the sphere has the greatest impact on the rate of spontaneous decay of the molecules near a dielectric and left-handed spherical particles, in which high-Q modes can be excited. When the chirality parameter approaches a critical value $\chi_{\text{crit}} = 1/\sqrt{\varepsilon\mu}$, the number of oscillations in the radiative

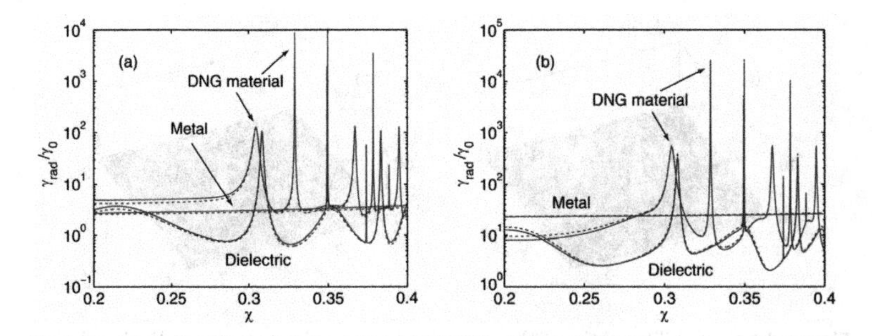

Figure 4.13 Radiative decay rate of the spontaneous emission of a chiral molecule ($m_{0x} = m_{0z} = 0.1d_0$) located in close vicinity to surface of a chiral spherical particle ($r_0 \to a$) with the size $k_0a = 1$, as a function of the chirality parameter χ. Dielectric, metal, and DNG spheres have $\{\varepsilon = 4, \mu = 1\}$, $\{\varepsilon = -4, \mu = 1\}$, and $\{\varepsilon = -4, \mu = -1.11\}$, correspondingly. (a) Tangential orientation of electric dipole momentum; (b) normal orientation of electric dipole momentum. The solid line corresponds to the transition magnetic dipole moment of the molecule oriented along the surface, and the dashed line corresponds to the transition magnetic dipole moment of the molecule oriented normally to the surface. The particle is placed in vacuum.

decay rate increases rapidly, and $k_L \to \infty$ [see (4.16)]. In the case of metal particles, the dependence of the radiative decay rate of spontaneous emission on the parameter χ is weak due to the absence of propagating waves.

Figure 4.14 shows the ratio of the radiative decay rate of spontaneous emission of the "left" molecule and the radiative decay rate of the "right" molecule ($\gamma_{\text{rad}}^L/\gamma_{\text{rad}}^R$), and vice versa ($\gamma_{\text{rad}}^R/\gamma_{\text{rad}}^L$). From this figure, it follows that if the condition (4.50) is satisfied, then the decay rates of the "left" and "right" molecules differ by factor 15 or 60 or even more, depending on the chirality of the molecule considered as a reference one. In other words, nanoparticles with a parameter given by (4.50) will enhance the radiation of the "right" molecules and slow down the radiation of the "left" molecules, and vice versa. Let us stress that negative μ is of crucial importance for such discrimination.

From this figure, it follows also that the effect of discrimination of radiation of different enantiomers in the presence of chiral spherical nano-metaparticle is greater than the analogous values for the case of a chiral molecule near the chiral half-space (cf. Fig. 4.7 and

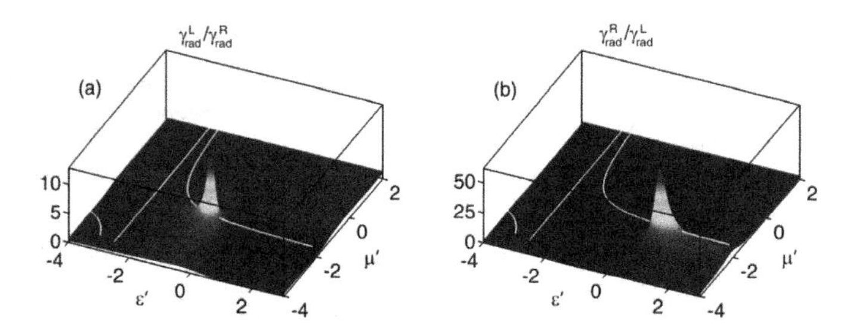

Figure 4.14 (a) The ratio of the averaged over orientations radiative decay rate of spontaneous emission of the "left" ($m_0 = -0.1d_0$) chiral molecule and the radiative decay rate of spontaneous emission of the "right" ($m_0 = +0.1d_0$) chiral molecule and (b) vice versa as a function of the real part of permittivity ($\varepsilon = \varepsilon' + i0.1$) and the real part of permeability ($\mu = \mu' + i0.1$) of a sphere. The nanoparticle size $k_0a = 0.1$, the chirality parameter $\chi = 0.2$. The molecule is placed in close vicinity to the surface of the spherical nanoparticle ($r_0 \to a$). The white line shows the position of the chiral-plasmon resonance in the spherical nanoparticle [see (4.47)]. The nanoparticle is placed in vacuum.

Fig. 4.14). We will present possible applications of this effect of discrimination of the radiation of "right" and "left" enantiomers in Section 4.5.

4.4 Radiation of a Chiral Molecule Near a Cluster of Two Chiral Spherical Particles

Results obtained in the previous section do not exhaust all possible geometries with chiral spherical nanoparticles. In practice, very often, a chiral molecule can be located near a cluster of several chiral particles. This problem is very complex and, despite of its actuality, there are no works on this subject yet. Only the case of an electric dipole radiation near a cluster of usual (nonchiral) spherical or spheroidal particles is more or less studied [7, 26, 28, 38, 52].

In this section, we will partially fill this gap and consider the influence of a cluster of two equal chiral spherical particles (i.e., chiral nanoantenna) on the radiation of a chiral molecule (see also [31]). All analytical results will be obtained for an arbitrary

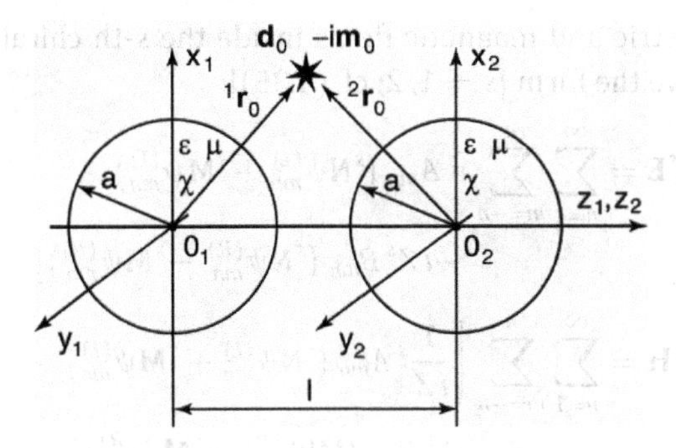

Figure 4.15 The geometry of the problem of spontaneous emission of a chiral molecule situated near a cluster of two chiral spherical particles. The molecule is placed in vacuum.

particles size and the distance between them, for an arbitrary composition of the particles, and for arbitrary relations between the electric and magnetic dipole moments of a chiral molecule. The geometry of the problem under consideration is shown in Fig. 4.15.

4.4.1 *Spontaneous Radiation of a Chiral Molecule Located near a Cluster of Two Chiral Spherical Particles*

In the case of a cluster of two chiral spherical particles, it is convenient to present the electromagnetic field inside each particle in local spherical coordinates associated with this particle. For definiteness, let us assume that the local Cartesian coordinate systems have parallel axes and a common z-axis (see Fig. 4.15). Local coordinates and all other quantities related to the s-th (s = 1, 2) spherical particle will be labeled by an additional index "1" or "2." For simplicity, we restrict ourselves to a cluster of two identical spherical particles with radii a, which are made from chiral material with the material parameters ε, μ, and χ. We also assume that the cluster is placed in vacuum. The case of the cluster of two different chiral particles can be considered in an analogous way with the help of our approach.

The electric and magnetic fields inside the s-th chiral spherical particle have the form [$s = 1, 2$; cf. (4.35)]:

$$
\begin{aligned}
{}^s\mathbf{E} = \sum_{n=1}^{\infty}\sum_{m=-n}^{n} \Big[{}^s A_{mn} \left({}^s\mathbf{N}\psi_{mn}^{(L)} + {}^s\mathbf{M}\psi_{mn}^{(L)} \right) \\
- iZ\, {}^s B_{mn} \left({}^s\mathbf{N}\psi_{mn}^{(R)} - {}^s\mathbf{M}\psi_{mn}^{(R)} \right) \Big], \qquad (4.51) \\
{}^s\mathbf{H} = \sum_{n=1}^{\infty}\sum_{m=-n}^{n} \left[\frac{1}{iZ}\, {}^s A_{mn} \left({}^s\mathbf{N}\psi_{mn}^{(L)} + {}^s\mathbf{M}\psi_{mn}^{(L)} \right) \right. \\
\left. + {}^s B_{mn} \left({}^s\mathbf{N}\psi_{mn}^{(R)} - {}^s\mathbf{M}\psi_{mn}^{(R)} \right) \right],
\end{aligned}
$$

where the vector spherical harmonics ${}^s\mathbf{N}\psi_{mn}^{(J)}$ and ${}^s\mathbf{M}\psi_{mn}^{(J)}$ (where $J = L, R$) in local coordinates ${}^s\mathbf{r}$ of the s-th spherical particle are defined in Appendix 4.B. Z again stands for impedance of particles.

The electric and magnetic fields of a chiral molecule with the electric and magnetic dipole moments $\mathbf{d}_0$ and $-i\mathbf{m}_0$, placed at the point ${}^s\mathbf{r}_0$ of the s-th local system of coordinates, can be presented as an expansion over the vector spherical harmonics in the following form [${}^s r_0 > {}^s r$; $s = 1, 2$; cf. (4.36)]:

$$
\begin{aligned}
{}^s\mathbf{E}_0 = \sum_{n=1}^{\infty}\sum_{m=-n}^{n} \left({}^s A_{mn}^{(0)}\, {}^s\mathbf{N}\psi_{mn}^{(0)} + {}^s B_{mn}^{(0)}\, {}^s\mathbf{M}\psi_{mn}^{(0)} \right), \qquad (4.52) \\
{}^s\mathbf{H}_0 = -i \sum_{n=1}^{\infty}\sum_{m=-n}^{n} \left({}^s B_{mn}^{(0)}\, {}^s\mathbf{N}\psi_{mn}^{(0)} + {}^s A_{mn}^{(0)}\, {}^s\mathbf{M}\psi_{mn}^{(0)} \right),
\end{aligned}
$$

where the vector spherical harmonics ${}^s\mathbf{N}\psi_{mn}^{(0)}$ and ${}^s\mathbf{M}\psi_{mn}^{(0)}$, and the coefficients ${}^s A_{mn}^{(0)}$ and ${}^s B_{mn}^{(0)}$ can be found in Appendix 4.B.

The electric and magnetic field of the reflected wave can be expressed as the sum of the partial fields from each of the particles [22]:

$$
\mathbf{E}^{sc} = {}^1\mathbf{E}^{sc} + {}^2\mathbf{E}^{sc}, \quad \mathbf{H}^{sc} = {}^1\mathbf{H}^{sc} + {}^2\mathbf{H}^{sc}, \qquad (4.53)
$$

where [$s = 1, 2$; cf. (4.37)]

$$
\begin{aligned}
{}^s\mathbf{E}^{sc} = \sum_{n=1}^{\infty}\sum_{m=-n}^{n} \left({}^s C_{mn}\, {}^s\mathbf{N}\zeta_{mn} + {}^s D_{mn}\, {}^s\mathbf{M}\zeta_{mn} \right), \qquad (4.54) \\
{}^s\mathbf{H}^{sc} = -i \sum_{n=1}^{\infty}\sum_{m=-n}^{n} \left({}^s D_{mn}\, {}^s\mathbf{N}\zeta_{mn} + {}^s C_{mn}\, {}^s\mathbf{M}\zeta_{mn} \right),
\end{aligned}
$$

and the vector spherical harmonics ${}^{s}\mathbf{N}\zeta_{mn}$ and ${}^{s}\mathbf{M}\zeta_{mn}$ are defined in Appendix 4.B.

To find the unknown coefficients of expansions (4.51) and (4.54), it is necessary to use the continuity of the tangential components of the electric and magnetic fields on the surface of each of the sphere of a cluster [32]. In doing so, one should also use the addition-translation theorem [18, 22, 24] that allows one to express vector spherical harmonics in the local coordinates of one particle through the vector spherical harmonics in local coordinates of the another particle.

As the molecule is outside the spherical particles, then, to find the radiative decay rate, we need only the coefficients ${}^{1}C_{mn}$, ${}^{1}D_{mn}$, ${}^{2}C_{mn}$, and ${}^{2}D_{mn}$ that satisfy the following equations:

$$\begin{aligned}
{}^{1}C_{mn} &+ \sum_{q=|m|}^{\infty}\left(V_{mnq}\alpha_n - i\,W_{mnq}\delta_n\right){}^{2}C_{mq} \\
&+ \sum_{q=|m|}^{\infty}\left(W_{mnq}\alpha_n - i\,V_{mnq}\delta_n\right){}^{2}D_{mq} = -\alpha_n{}^{1}A_{mn}^{(0)} + i\delta_n{}^{1}B_{mn}^{(0)}, \\
{}^{1}D_{mn} &+ \sum_{q=|m|}^{\infty}\left(V_{mnq}\beta_n - i\,W_{mnq}\delta_n\right){}^{2}D_{mq} \\
&+ \sum_{q=|m|}^{\infty}\left(W_{mnq}\beta_n - i\,V_{mnq}\delta_n\right){}^{2}C_{mq} = -\beta_n{}^{1}B_{mn}^{(0)} + i\delta_n{}^{1}A_{mn}^{(0)},
\end{aligned} \tag{4.55}$$

$$\begin{aligned}
{}^{2}C_{mn} &+ \sum_{q=|m|}^{\infty}(-1)^{q+n}\left(V_{mnq}\alpha_n + i\,W_{mnq}\delta_n\right){}^{1}C_{mq} \\
&- \sum_{q=|m|}^{\infty}(-1)^{q+n}\left(W_{mnq}\alpha_n + i\,V_{mnq}\delta_n\right){}^{1}D_{mq} \\
&= -\alpha_n{}^{2}A_{mn}^{(0)} + i\delta_n{}^{2}B_{mn}^{(0)}, \\
{}^{2}D_{mn} &+ \sum_{q=|m|}^{\infty}(-1)^{q+n}\left(V_{mnq}\beta_n + i\,W_{mnq}\delta_n\right){}^{1}D_{mq} \\
&- \sum_{q=|m|}^{\infty}(-1)^{q+n}\left(W_{mnq}\beta_n + i\,V_{mnq}\delta_n\right){}^{1}C_{mq} \\
&= -\beta_n{}^{2}B_{mn}^{(0)} + i\delta_n{}^{2}A_{mn}^{(0)},
\end{aligned} \tag{4.56}$$

where the lower limit of the summation over q must be equal to 1 in the case $m = 0$, and equal to $|m|$ in other cases ($m \neq 0$); the functions V_{mnq} and W_{mnq} are defined in Appendix 4.B; the functions α_n, β_n, and δ_n are defined by (4.39).

To find the radiative decay rate of the spontaneous emission of the chiral molecule located near a cluster of two spherical particles, it is necessary to calculate the total power of the radiation of the system "molecule and cluster" [see (4.41)]. If we use local coordinates related to the first particle, the radiative part of the spontaneous emission decay rate [cf. (4.43)] will have the following form:

$$\frac{\gamma_{\text{rad}}}{\gamma_0} = \frac{3}{2k_0^6\left(|\mathbf{d}_0|^2 + |\mathbf{m}_0|^2\right)} \sum_{n=1}^{\infty} \sum_{m=-n}^{n} \frac{n(n+1)}{2n+1} \frac{(n+m)!}{(n-m)!} \times \left(\left|{}^1C_{mn}^{(0)} + {}^1\tilde{C}_{mn}\right|^2 + \left|{}^1D_{mn}^{(0)} + {}^1\tilde{D}_{mn}\right|^2\right), \tag{4.57}$$

$${}^1\tilde{C}_{mn} = {}^1C_{mn} + \sum_{q=|m|}^{\infty} \left(\tilde{V}_{mnq}\,{}^2C_{mq} + \tilde{W}_{mnq}\,{}^2D_{mq}\right),$$

$${}^1\tilde{D}_{mn} = {}^1D_{mn} + \sum_{q=|m|}^{\infty} \left(\tilde{V}_{mnq}\,{}^2D_{mq} + \tilde{W}_{mnq}\,{}^2C_{mq}\right). \tag{4.58}$$

In (4.58), the lower limit of the summation over q must be equal to 1 in the case $m = 0$, and equal to $|m|$ in another case with $m \neq 0$; the coefficients ${}^1C_{mn}^{(0)}$, ${}^1D_{mn}^{(0)}$, $\tilde{V}_{mnq}$, and $\tilde{W}_{mnq}$ can be found in Appendix 4.B If there are losses in the chiral particle, there is a probability of the nonradiative transition from the excited state to the ground one. The general expression for the total (radiative + nonradiative) decay rate of the spontaneous emission in the case of a cluster of two spherical particles has the form [cf. (4.28)]:

$$\frac{\gamma}{\gamma_0} = 1 + \frac{3}{2}\text{Im}\left\{\frac{\sum_{s=1}^{2}\left(\mathbf{d}_0^* \cdot {}^s\mathbf{E}^{sc}\left({}^s\mathbf{r}_0\right) + i\mathbf{m}_0^* \cdot {}^s\mathbf{H}^{sc}\left({}^s\mathbf{r}_0\right)\right)}{k_0^3\left(|\mathbf{d}_0|^2 + |\mathbf{m}_0|^2\right)}\right\}, \tag{4.59}$$

where the reflected fields ${}^s\mathbf{E}^{sc}\left({}^s\mathbf{r}_0\right)$ and ${}^s\mathbf{H}^{sc}\left({}^s\mathbf{r}_0\right)$ [see (4.54)] should be calculated at the position of the chiral molecule ${}^s\mathbf{r}_0$. Note that in the case of chiral spherical particles without losses, the expression (4.59) gives the same results as the expression (4.57). In what follows, for simplicity, we will not study the nonradiative channel of the spontaneous decay.

4.4.2 *The Analysis of the Results Obtained and Graphical Illustrations*

The process of the spontaneous decay of chiral molecules placed near a cluster of two chiral spherical particles is very complicated, and for clarity, we will present graphical illustrations only for the most interesting case of a chiral molecule placed in the gap between chiral particles on the symmetry axis (the z-axis, see Fig. 4.15). In this case, ${}^1\theta_0 = 0$ and ${}^2\theta_0 = \pi$; the expressions for the coefficients ${}^sA^{(0)}_{mn}$, ${}^sB^{(0)}_{mn}$, ${}^sC^{(0)}_{mn}$, and ${}^sD^{(0)}_{mn}$ (where $s = 1, 2$) become simpler (see Appendix 4.B), and nonzero coefficients are only ${}^sC_{mn}$ and ${}^sD_{mn}$ with $m = 0, 1, -1$; therefore, the lower limit of the summation over q in (4.55), (4.56), and (4.58) must be equal to 1.

For simplicity, we will consider the radiative decay rate of spontaneous emission only for chiral molecules with equal Cartesian projection of the electric dipole moment ($\mathbf{d}_0 = \{d_0, d_0, d_0\}$) and with equal Cartesian projection of the magnetic dipole moment ($\mathbf{m}_0 = \{m_0, m_0, m_0\}$).

Figure 4.16a shows the radiative decay rate of spontaneous emission of a chiral molecule located in the gap between two nonchiral dielectric spherical particles near the surface of the first sphere as the function of the parameter k_0a for different values of the distance $l/(2a)$ between particles. One can see that for large distances between the spheres, the decay rate behavior is similar to a decay rate in the case of a single sphere. However, for a small gap between the nanoparticles, one can observe substantial changes in spectra. The most interesting feature is that now the maximum of the decay rate appears for smaller size of nanoparticles (or for larger wavelength).

More interesting effects arise in the case of a cluster of chiral dielectric particles. Figure 4.16b shows the dependencies analogous to those shown in Fig. 4.16a, but for chiral particles. As it is seen in Fig. 4.16b, the nonzero chirality leads to the redistribution of the maxima of the spontaneous emission decay rate and to the appearance of new high-Q resonances in comparison with the case of a nonchiral cluster. A more important observation is that for chiral particles, one can observe the difference between the radiative decay rates of spontaneous emission for "right" and "left" molecules.

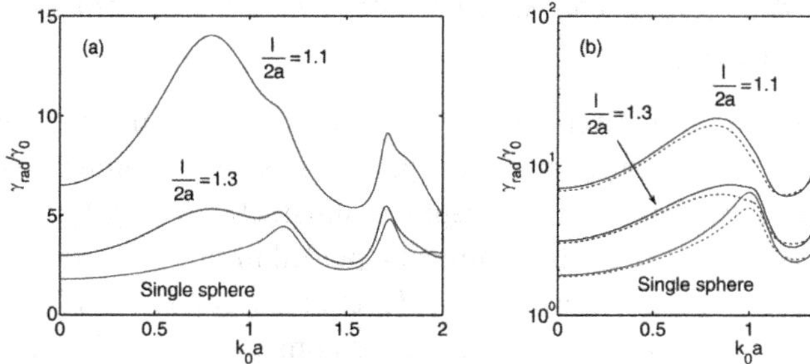

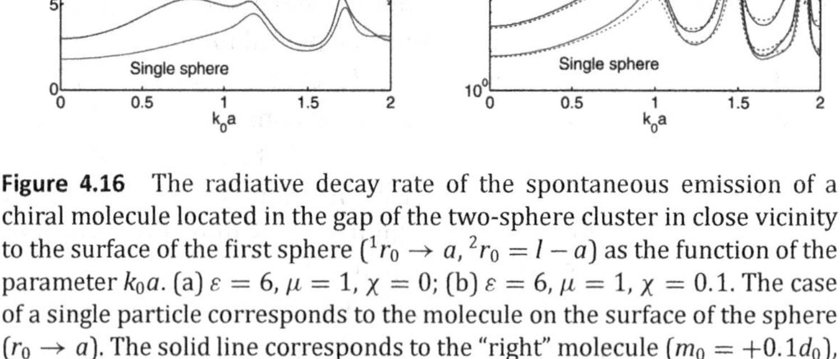

Figure 4.16 The radiative decay rate of the spontaneous emission of a chiral molecule located in the gap of the two-sphere cluster in close vicinity to the surface of the first sphere (${}^1r_0 \to a, {}^2r_0 = l - a$) as the function of the parameter k_0a. (a) $\varepsilon = 6, \mu = 1, \chi = 0$; (b) $\varepsilon = 6, \mu = 1, \chi = 0.1$. The case of a single particle corresponds to the molecule on the surface of the sphere ($r_0 \to a$). The solid line corresponds to the "right" molecule ($m_0 = +0.1d_0$), and the dashed line corresponds to the "left" molecule ($m_0 = -0.1d_0$). In the case of nonchiral particles, the dependencies for "right" and "left" molecules coincide. The particles are placed in vacuum.

When the distance between the particles of the cluster is increased, the difference remains, and the dependence of the spontaneous emission radiative decay rate on k_0a tends to the case of a single particle.

Figure 4.17 shows the spontaneous radiative decay rate of chiral molecules located near single chiral spherical particles (Fig. 4.17a) and in the gap between chiral spherical particles of a cluster (Fig. 4.17b) as the functions of chirality χ. The figure shows that the dependence of the radiative decay rate on the chirality parameter can vary substantially in comparison with the dependence for a single particle. It is well seen that the most significant difference arises in the case of DNG-metamaterial particles. Indeed, in the case of closely spaced DNG particles [$l/(2a) = 1.1$], there are a number of high-Q resonances in the dependence of the radiative decay rate on χ, and the dependence itself becomes rather complicated. In the case of a cluster of two chiral dielectric particles, there is also a substantial difference from the case of a single particle. In particular, it is seen that there is an increase of numbers of the radiative decay maxima, but it is not as significant as in the case of DNG particles.

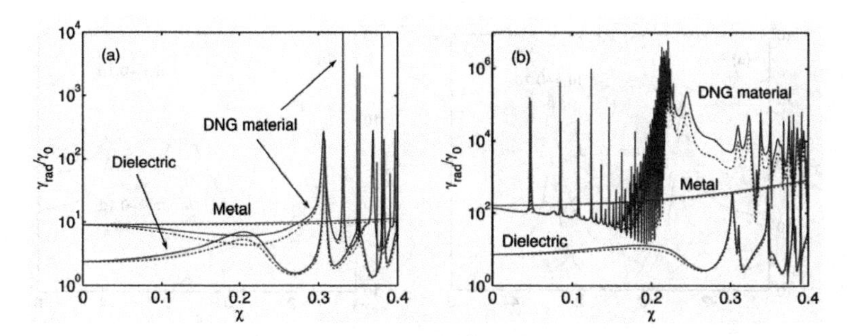

Figure 4.17 The dependence of the radiative decay rate of the spontaneous emission of a chiral molecule on the chirality parameter χ. (a) The molecule is placed in close vicinity to the surface of a single sphere ($r_0 \to a$). (b) The molecule is placed in the gap of the two-sphere cluster in close vicinity to the surface of the first sphere ($^1r_0 \to a$, $^2r_0 = l - a$). Metal, dielectric, and DNG spheres have permittivity and permeability: $\varepsilon = -4$, $\mu = 1$; $\varepsilon = 4$, $\mu = 1$, and $\varepsilon = -4$, $\mu = -1.1$, correspondingly. The solid line corresponds to the "right" molecule ($m_0 = +0.1d_0$), and the dashed line corresponds to the "left" molecule ($m_0 = -0.1d_0$). The relative distance between the particles centers is $l/(2a) = 1.1$. The particles are placed in vacuum.

As it can be seen in Fig. 4.17, in the case of molecule in the gap of chiral nanoantenna, the number of maximums increases in comparison with the case of a single nanoparticle. It is a consequence of the fact that a cluster has a larger number of nondegenerate chiral-plasmon modes than a single nanoparticle. Analogous phenomena for the plasmon modes take place in the case of a cluster of two metal nanoparticles [38]. Increasing of the number of excited modes allows controlling the spontaneous emission of a chiral molecule more precisely located in the gap of a cluster of two chiral nanoparticles by varying the distance between the particles.

In Fig. 4.18, the radiative decay rates of spontaneous emission of "right" and "left" molecules located both near the single chiral spherical nanoparticle (Fig. 4.18a) and in the gap between chiral spherical nanoparticles (Fig. 4.18b) is shown as a function of the real part of permittivity, ε, at the fixed permeability, $\mu = -1.6$. From comparison of Fig. 4.18a and Fig. 4.18b, it follows that the radiative decay rate of spontaneous emission of chiral molecules in the gap of

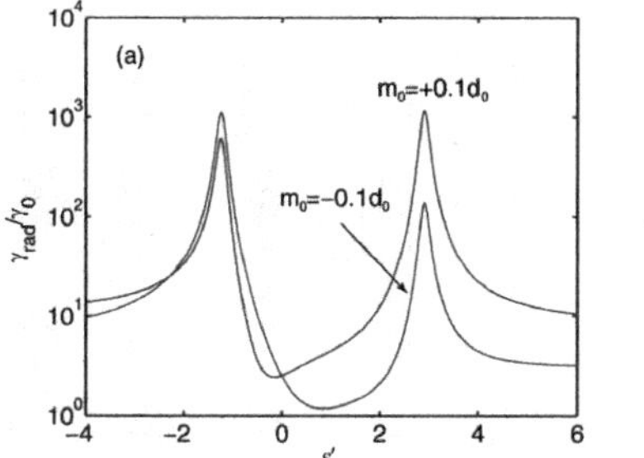

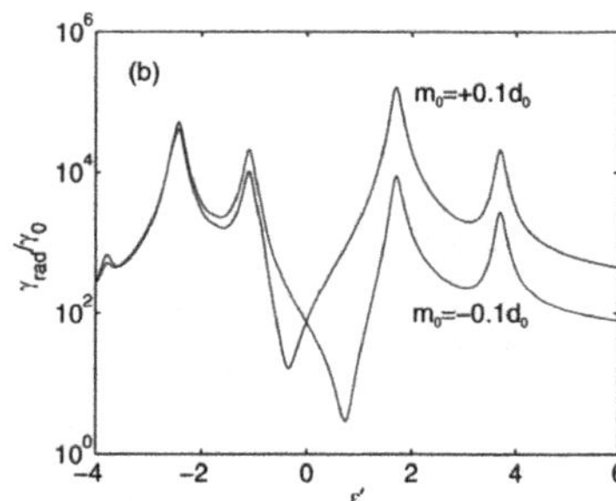

Figure 4.18 The dependence of the radiative decay rate of the spontaneous emission of the "right" ($m_0 = +0.1d_0$) and "left" ($m_0 = -0.1d_0$) chiral molecules on the real part of permittivity ($\varepsilon = \varepsilon' + i0.1$) at $\mu = -1.6$, $k_0 a = 0.1$, and $\chi = 0.2$. (a) The molecule is placed in close vicinity to the surface of a single sphere ($r_0 \to a$). (b) The molecule is placed in the gap of the two-sphere cluster in close vicinity to the surface of the first sphere (${}^1r_0 \to a, {}^2r_0 = l - a$). The relative distance between the particles centers is $l/(2a) = 1.1$. The particles are placed in vacuum.

the chiral nanoantenna has substantially greater values than in the case of chiral molecule near a single chiral nanoparticle.

In Fig. 4.19, the ratio of the radiative decay rate of spontaneous emission of the "left" molecule and the radiative decay rate of the "right" molecule ($\gamma^L_{\text{rad}}/\gamma^R_{\text{rad}}$), and vice versa ($\gamma^R_{\text{rad}}/\gamma^L_{\text{rad}}$), is shown as functions of the permittivity and permeability of the chiral cluster. The molecule is located in the gap between the particles near the surface of the first spherical particles of the cluster.

From this figure, it follows that for certain values ε and μ, there is a very significant difference in the radiative decay rates of the "right" and "left" molecules in 40 or even in 300 (and more) times, which can exceed significantly the analogous results for a single chiral nanoparticle (cf. Fig. 4.14 and Fig. 4.19).

As it is seen in Fig. 4.19, as well as for a single chiral nanoparticle, a chiral nanoantenna with $\varepsilon < 0$ and $\mu < 0$ (DNG-metamaterial) will enhance the radiation of the "left" molecules and slow down the radiation of the "right" molecules, while the cluster with $\varepsilon > 0$ and $\mu < 0$ (MNG-metamaterial) will enhance the radiation of the "right" molecules and suppress the radiation of the "left" molecules.

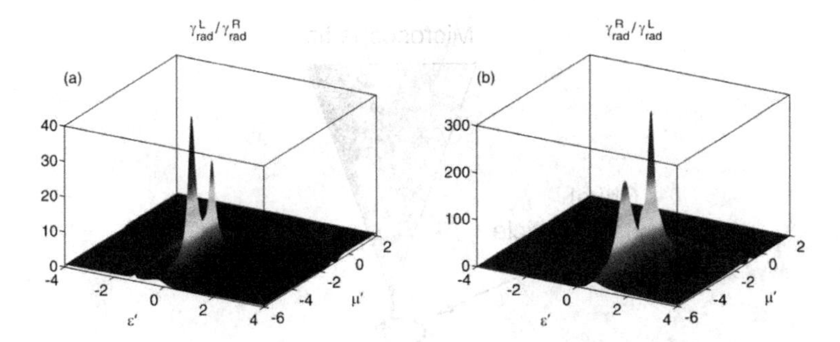

Figure 4.19 (a) The ratio of the radiative decay rate of spontaneous emission of the "left" ($m_0 = -0.1d_0$) chiral molecule and the radiative decay rate of spontaneous emission of the "right" ($m_0 = +0.1d_0$) chiral molecule, and (b) vice versa, as a function of the real part of the permittivity ($\varepsilon = \varepsilon' + i0.1$) and the real part of the permeability ($\mu = \mu' + i0.1$) of the material of cluster of two chiral nanoparticles with $k_0a = 0.1$ and $\chi = 0.2$. The molecule is placed in the gap of the two-sphere cluster in close vicinity to the surface of the first sphere ($^1r_0 \to a$, $^2r_0 = l - a$) in the common z-axis. The relative distance between the particles centers is $l/(2a) = 1.1$. The particles are placed in vacuum.

The significant difference between the radiative decay rates of "right" and "left" enantiomers of molecules located near a cluster of two chiral nanoparticles made of metamaterials allows using such clusters for creation of effective devices for detection and selection of enantiomers. Note that due to increase of local magnetic fields, the effect of discrimination for a chiral nanoantenna is much greater than for a single chiral nanoparticle. Possible variants of the application of these effects are discussed in the following section.

4.5 Applications

The predicted effect of influence of chiral nanoparticles on the radiation of optically active molecules can be used in many applications [42]. Here, we mention only two, the most obvious ones. The first application is the observation of "right" or "left" molecules separately with a scanning microscope (see Fig. 4.20).

To do so, specially prepared nanoparticles [see (4.50)] are attached to the tip of a scanning microscope to increase the

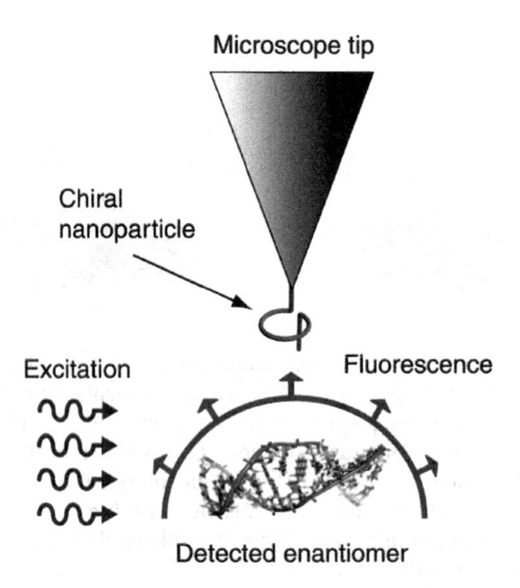

Figure 4.20 The operation principle of a chiral nanoscope that allows distinguishing "right" and "left" enantiomers. A chiral nanoparticle (or a short nanospiral) is placed on the tip of the microscope. Such a nanoparticle is specially tuned to satisfy the conditions (4.50). Thus, this nanoscope is tuned to accelerate the radiation of the molecule with selected ("left" or "right") chirality. As a result, the molecules with selected chirality appear as bright spots while molecules with opposite chirality remain invisible. Reprinted from Ref. 42, with permission from IOP Publishing.

spontaneous emission of a molecule with a selected chirality. A short metal spiral can be used as such a nanoparticle, because it can have chirality and negative magnetic response. Indeed, the intensity of the fluorescence of a molecule is defined by the relation:

$$I_{\mathrm{fluor}} = \hbar\omega \frac{\gamma_{\mathrm{pump}}\gamma_{\mathrm{rad}}}{\gamma_{\mathrm{pump}} + \gamma_{\mathrm{nonrad}} + \gamma_{\mathrm{rad}}}, \tag{4.60}$$

where γ_{pump}, γ_{rad}, and γ_{nonrad} are the rate of excitation, the radiative, and nonradiative decays of the molecule, respectively. In the presence of nanoparticles with the parameters (4.50), the radiative decay rate of one sort of molecules (e.g., "right" ones) increases, while the radiative decay rate of molecules with the opposite chirality ("left") is inhibited. Due to this fact, the contrast between the brightness of the "right" and "left" molecules can reach the value

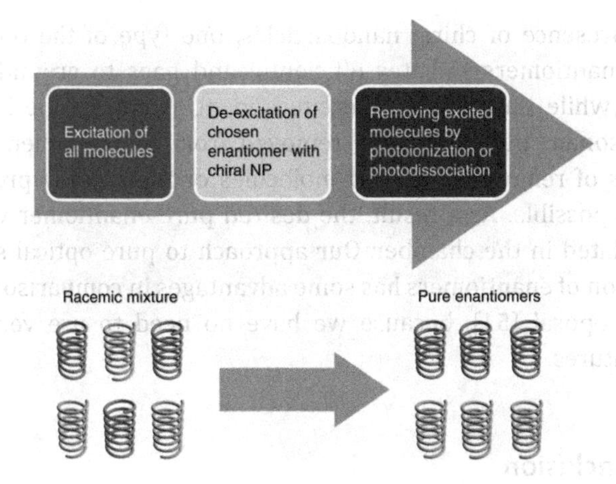

Figure 4.21 The operation principle of a device allowing one to resolve "right" and "left" enantiomers. The racemic mixture with excited enantiomers of both types of chirality is placed into a reaction chamber. The effective interaction of specially tuned chiral nanoparticles [according to the conditions (4.50)] takes place only for molecules with selected chirality. It allows them to become unexcited quickly, whereas molecules with another type of chirality remain excited. Such molecules can be removed from the chamber by different methods. As a result, the desired pure enantiomer will be accumulated in the chamber. Reprinted from Ref. 42, with permission from IOP Publishing.

of 10–100 or more times and, consequently, one will see images of molecules with the chosen chirality only.

Chiral molecules play especially important role in biology and pharmacy. Therefore, it is extremely important to arrange the separation of the "right" and "left" enantiomers of molecules in the racemic mixtures. This can be done in various ways, for example, by using radiation pressure forces in the electromagnetic field of left or right-handed circularly polarized electromagnetic waves [27], or by using spiral optical beams [40]. Our results pave the way to new devices for the separation of enantiomers in racemic mixtures.

The operation principle of such a device is shown in Fig. 4.21. The key element of this scheme is a reaction chamber containing nanoparticles prepared in accordance with the conditions (4.50). The racemic mixture of enantiomers is placed in this chamber and then excited by one or another way (e.g., photoexcitation). Due to the presence of chiral nanoparticles, one type of the optically active enantiomers radiates efficiently and goes to ground state quickly, while the remaining excited enantiomers can be ionized by a resonant field, and then removed from the chamber. Other methods of removal of excited molecules or their decay products are also possible. As a result, the desired pure enantiomer will be accumulated in the chamber. Our approach to pure optical spatial separation of enantiomers has some advantages in comparison with other proposal [51], because we have no need to use very low temperatures.

4.6 Conclusion

Thus, in this Chapter, the radiation of optically active molecules is considered for different chiral meta-environment such as a half-space, single nanoparticle, and a nanoantenna made of two metaparticles. We have shown that all these geometries allow one to control radiation of different enantiomers by his will. The degree of this control is increased from a chiral half-space to a nanoantenna, where the enhancement of local fields and chiral discrimination is maximal. It is very important to note that such a control is impossible without making use of modern μ negative (MNG) and/or double-negative (DNG) metamaterials.

The possibility to arrange a system in which "right" and "left" molecules have different decay rates pave the way to different applications starting from the detection of a single "right" DNA molecule among millions of usual "left" DNA molecules to develop devices for separation of enantiomers during drug mass production.

Acknowledgments

The authors thank the Russian Foundation for Basic Research (VK, grants numbers 11-02-91065, 11-02-92002, 11-02-01272, and 12-02-90014) and the Belarusian Republican Foundation for Fundamental Research (DG, grant number F12R-006) for partial financial support of this work.

Appendix 4.A

Chiral Molecule near a Chiral Half-Space

In Appendix 4.A, mathematical expressions necessary to solve the problem of radiation of a chiral molecule near a chiral half-space will be presented.

4.A.1 Vector Cylindrical Harmonics

Vector cylindrical harmonics describing electric and magnetic fields inside a chiral half-space ($z < 0$, see Fig. 4.4) have the following form ($n = 0,\ 1,\ 2,\ \ldots; q > 0$):

$$\mathbf{m}_{nqe}^{(J)} = \mathrm{rot}\left(\mathbf{e}_z J_n(q\rho)\cos(n\varphi)\exp\left(-i\sqrt{k_J^2 - q^2}z\right)\right),$$

$$\mathbf{m}_{nqo}^{(J)} = \mathrm{rot}\left(\mathbf{e}_z J_n(q\rho)\sin(n\varphi)\exp\left(-i\sqrt{k_J^2 - q^2}z\right)\right),$$

$$\mathbf{n}_{nqe}^{(J)} = \frac{1}{k_J}\mathrm{rot}\mathbf{m}_{nqe}^{(J)}, \quad \mathbf{n}_{nqo}^{(J)} = \frac{1}{k_J}\mathrm{rot}\mathbf{m}_{nqo}^{(J)}, \tag{4.A.1}$$

where $0 \le \rho < \infty$, $0 \le \varphi < 2\pi$ are polar coordinates; $\mathbf{e}_z$ is the unit vector directed along the Cartesian z axis; k_J is the wavenumber of left ($J = L$) and right ($J = R$) polarized waves in a chiral medium, and $J_n(q\rho)$ is the Bessel function [1].

Vector cylindrical harmonics describing electric and magnetic fields outside the chiral medium ($n = 0,\ 1,\ 2, \ldots; q > 0$) have the following form:

$$\mathbf{m}_{nqe}^{(\pm)} = \mathrm{rot}\left(\mathbf{e}_z J_n(q\rho)\cos(n\varphi)\exp\left(\pm i\sqrt{k_0^2 - q^2}z\right)\right),$$

$$\mathbf{m}_{nqo}^{(\pm)} = \mathrm{rot}\left(\mathbf{e}_z J_n(q\rho)\sin(n\varphi)\exp\left(\pm i\sqrt{k_0^2 - q^2}z\right)\right),$$

$$\mathbf{n}_{nqe}^{(\pm)} = \frac{1}{k_0}\mathrm{rot}\mathbf{m}_{nqe}^{(\pm)}, \quad \mathbf{n}_{nqo}^{(\pm)} = \frac{1}{k_0}\mathrm{rot}\mathbf{m}_{nqo}^{(\pm)}, \tag{4.A.2}$$

where the "+" and "−" signs correspond to outgoing and incoming waves and k_0 is the wavenumber outside the chiral medium (in vacuum).

Note that the analytic branches of the square roots $\sqrt{k_J^2 - q^2}$ and $\sqrt{k_0^2 - q^2}$ in (4.A.1) and (4.A.2) are defined in the same way as in (4.22) and ensure proper directions of energy flows.

More information about the properties of the vector cylindrical harmonics can be found, for example, in [69].

4.A.2 Electromagnetic Field of a Chiral Molecule in Cylindrical Coordinates

The expressions for the electric and magnetic fields of a chiral molecule placed at the point z_0 ($z_0 > 0$) of the Cartesian z axis in vacuum have the following forms:

$$\mathbf{E}_0 = \left\{ \begin{matrix} \mathbf{E}_0^{(+)}, & z > z_0 \\ \mathbf{E}_0^{(-)}, & z < z_0 \end{matrix} \right\}, \quad \mathbf{H}_0 = \left\{ \begin{matrix} \mathbf{H}_0^{(+)}, & z > z_0 \\ \mathbf{H}_0^{(-)}, & z < z_0 \end{matrix} \right\}, \tag{4.A.3}$$

where

$$\mathbf{E}_0^{(\pm)} = \sum_{n=0}^{1} \sum_{\sigma=e,o} \int_0^{\infty} dq \left(C_{nq\sigma}^{(\pm)} \mathbf{n}_{nq\sigma}^{(\pm)} + D_{nq\sigma}^{(\pm)} \mathbf{m}_{nq\sigma}^{(\pm)} \right),$$

$$\mathbf{H}_0^{(\pm)} = -i \sum_{n=0}^{1} \sum_{\sigma=e,o} \int_0^{\infty} dq \left(D_{nq\sigma}^{(\pm)} \mathbf{n}_{nq\sigma}^{(\pm)} + C_{nq\sigma}^{(\pm)} \mathbf{m}_{nq\sigma}^{(\pm)} \right). \tag{4.A.4}$$

The coefficients of the expansion in (4.A.4) have the following form:

$$C_{nqe}^{(\pm)} = \left(\delta_{n0} q \frac{i d_{0z} k_0}{\sqrt{k_0^2 - q^2}} + \delta_{n1} k_0 \left(\pm d_{0x} - \frac{i m_{0y} k_0}{\sqrt{k_0^2 - q^2}} \right) \right) \exp\left(\mp i \sqrt{k_0^2 - q^2} z_0 \right),$$

$$C_{nqo}^{(\pm)} = \delta_{n1} k_0 \left(\pm d_{0y} + \frac{i m_{0x} k_0}{\sqrt{k_0^2 - q^2}} \right) \exp\left(\mp i \sqrt{k_0^2 - q^2} z_0 \right),$$

$$D_{nqe}^{(\pm)} = \left(\delta_{n0} q \frac{i m_{0z} k_0}{\sqrt{k_0^2 - q^2}} + \delta_{n1} k_0 \left(\pm m_{0x} - \frac{i d_{0y} k_0}{\sqrt{k_0^2 - q^2}}\right)\right)$$
$$\exp\left(\mp i \sqrt{k_0^2 - q^2 z_0}\right),$$
$$D_{nqo}^{(\pm)} = \delta_{n1} k_0 \left(\pm m_{0y} + \frac{i d_{0x} k_0}{\sqrt{k_0^2 - q^2}}\right) \exp\left(\mp i \sqrt{k_0^2 - q^2 z_0}\right), \tag{4.A.5}$$

where δ_{np} is the Kronecker's delta. Note that the condition of applicability of the above expression is $\mathrm{Im}\left(\sqrt{k_0^2 - q^2}\right) \geq 0$ [17].

Appendix 4.B

A Chiral Molecule Near Chiral Microspheres

In Appendix 4.B, the mathematical expressions necessary to solve the problem of radiation of a chiral molecule near a cluster of two chiral spherical particles will be presented.

4.B.1 Vector Spherical Harmonics

Vector spherical harmonics describing electric and magnetic fields inside the chiral spherical particle have the following form ($n = 1, 2, 3, \ldots; m = 0, \pm 1, \pm 2, \ldots, \pm n$):

$$\mathbf{M}\psi_{mn}^{(J)} = \frac{1}{k_J r} \mathrm{rot}\left(\mathbf{r} \psi_n (k_J r) P_n^m (\cos\theta) e^{im\varphi}\right), \; \mathbf{N}\psi_{mn}^{(J)} = \frac{1}{k_J} \mathrm{rot} \mathbf{M}\psi_{mn}^{(J)}, \tag{4.B.1}$$

where $0 \leq r < \infty$, $0 \leq \theta \leq \pi$, and $0 \leq \varphi < 2\pi$ are spherical coordinates; $\mathbf{r}$ is the radius vector; k_J is the wavenumber of the left ($J = L$) and the right ($J = R$) polarized waves in chiral medium; $\psi_n (k_J r) = k_J r j_n (k_J r)$; $j_n (k_J r)$ is the spherical Bessel function [1], and $P_n^m (\cos\theta)$ is the associated Legendre function [1].

Vector spherical harmonics describing electric and magnetic fields outside the chiral spherical particle (in vacuum) have the following form ($n = 1,\ 2,\ 3,\ \ldots;\ m = 0,\ \pm 1,\ \pm 2,\ \ldots,\ \pm n$):

$$\mathbf{M}\zeta_{mn} = \frac{1}{k_0 r}\mathrm{rot}\left(\mathbf{r}\zeta_n\left(k_0 r\right) P_n^m\left(\cos\theta\right) e^{im\varphi}\right), \quad \mathbf{N}\zeta_{mn} = \frac{1}{k_0}\mathrm{rot}\mathbf{M}\zeta_{mn}, \tag{4.B.2}$$

where $\zeta_n\left(k_0 r\right) = k_0 r h_n^{(1)}\left(k_0 r\right)$; $h_n^{(1)}\left(k_0 r\right)$ is the spherical Hankel function of the first kind [1], and k_0 is the wavenumber in vacuum.

Information about the properties of vector cylindrical harmonics can be also found, for example, in [69].

To obtain the spherical vector harmonics ${}^s\mathbf{N}\psi_{mn}^{(J)}$, ${}^s\mathbf{M}\psi_{mn}^{(J)}$, ${}^s\mathbf{N}\zeta_{mn}$, and ${}^s\mathbf{M}\zeta_{mn}$ in the local coordinates of the s-th particle ($s = 1, 2$), it is necessary to replace the coordinates r, θ, φ by the local coordinates ${}^s r$, ${}^s\theta$, ${}^s\varphi$ in the expressions (4.B.1) and (4.B.2).

4.B.2 Electromagnetic Field of a Chiral Molecule in Spherical Coordinates

In (4.36), the vector spherical harmonics $\mathbf{N}\psi_{mn}^{(0)}$ and $\mathbf{M}\psi_{mn}^{(0)}$ can be obtained from the expressions for harmonics $\mathbf{N}\psi_{mn}^{(J)}$ and $\mathbf{M}\psi_{mn}^{(J)}$ [see (4.B.1)] by changing the index $J \to 0$. The coefficients of expansion in (4.36) can be written down as

$$\begin{aligned} A_{mn}^{(0)}\left(\mathbf{r}_0\right) &= ik_0^3\left[F_{mn}\left(\mathbf{r}_0, \mathbf{d}_0\right) + iG_{mn}\left(\mathbf{r}_0, -i\mathbf{m}_0\right)\right], \\ B_{mn}^{(0)}\left(\mathbf{r}_0\right) &= ik_0^3\left[G_{mn}\left(\mathbf{r}_0, \mathbf{d}_0\right) + iF_{mn}\left(\mathbf{r}_0, -i\mathbf{m}_0\right)\right], \end{aligned} \tag{4.B.3}$$

where $F_{mn}\left(\mathbf{r}_0, \mathbf{d}_0\right)$ and $G_{mn}\left(\mathbf{r}_0, \mathbf{d}_0\right)$ have the form:

$$\begin{aligned} F_{mn}\left(\mathbf{r}_0, \mathbf{d}_0\right) &= -\frac{1}{2}\left(d_{0x} - id_{0y}\right)\frac{1}{n}b_{m-1,n-1}\left(\mathbf{r}_0\right) \\ &+\frac{1}{2}\left(d_{0x} - id_{0y}\right)\frac{1}{n+1}b_{m-1,n+1}\left(\mathbf{r}_0\right) \\ &+\frac{1}{2}\left(d_{0x} + id_{0y}\right)\frac{1}{n}\left(n-m-1\right)\left(n-m\right)b_{m+1,n-1}\left(\mathbf{r}_0\right) \\ &-\frac{1}{2}\left(d_{0x} + id_{0y}\right)\frac{1}{n+1}\left(n+m+1\right)\left(n+m+2\right)b_{m+1,n+1}\left(\mathbf{r}_0\right) \\ &+d_{0z}\frac{1}{n}\left(n-m\right)b_{m,n-1}\left(\mathbf{r}_0\right) + d_{0z}\frac{1}{n+1}\left(n+m+1\right)b_{m,n+1}\left(\mathbf{r}_0\right), \end{aligned} \tag{4.B.4}$$

$$G_{mn}(\mathbf{r}_0, \mathbf{d}_0) = \frac{1}{2}\left(d_{0y} + id_{0x}\right)\frac{(2n+1)}{n(n+1)}b_{m-1,n}(\mathbf{r}_0)$$
$$-\frac{1}{2}\left(d_{0y} - id_{0x}\right)\frac{(2n+1)}{n(n+1)}(n-m)(n+m+1)\,b_{m+1,n}(\mathbf{r}_0)$$
$$+id_{0z}\frac{(2n+1)}{n(n+1)}mb_{mn}(\mathbf{r}_0), \qquad (4.B.5)$$

where

$$b_{mn}(\mathbf{r}_0) = \frac{(n-m)!}{(n+m)!}\frac{\zeta_n(k_0r_0)}{k_0r_0}P_n^m(\cos\theta_0)\,e^{-im\varphi_0}. \qquad (4.B.6)$$

In (4.B.4), (4.B.5), and (4.B.6), r_0, θ_0, φ_0 are the spherical coordinates of the position of a chiral molecule, and the function $\zeta_n(k_0r_0)$ is defined above [see (4.B.2)]. In the case of $\mathbf{m}_0 = 0$, the expressions (4.B.3) coincide with the known results [14]. Explicit expressions for the functions $F_{mn}(\mathbf{r}_0, -i\mathbf{m}_0)$ and $G_{mn}(\mathbf{r}_0, -i\mathbf{m}_0)$ can be obtained from (4.B.4) and (4.B.5) by changing all the Cartesian components of the vector $\mathbf{d}_0$ to corresponding components of the vector $-i\mathbf{m}_0$. In a special case of a chiral molecule located on the z–axis, one can obtain ($\theta_0 = 0$ or $\theta_0 = \pi$)

$$F_{mn}(\mathbf{r}_0, \mathbf{d}_0) = -\frac{1}{2}\left[\delta_{m1}\left(d_{0x} - id_{0y}\right) - \delta_{m,-1}\left(d_{0x} + id_{0y}\right)n(n+1)\right]$$
$$\times\frac{(2n+1)}{n(n+1)}\frac{\zeta_n'(k_0r_0)}{k_0r_0}P_{n+1}(\cos\theta_0)$$
$$+\delta_{m0}d_{0z}(2n+1)\frac{\zeta_n(k_0r_0)}{(k_0r_0)^2}P_{n+1}(\cos\theta_0), \qquad (4.B.7)$$

$$G_{mn}(\mathbf{r}_0, \mathbf{d}_0) = \frac{1}{2}\left[\delta_{m1}\left(d_{0y} + id_{0x}\right) - \delta_{m,-1}\left(d_{0y} - id_{0x}\right)n(n+1)\right]$$
$$\times\frac{(2n+1)}{n(n+1)}\frac{\zeta_n(k_0r_0)}{k_0r_0}P_n(\cos\theta_0), \qquad (4.B.8)$$

where the prime near the function means its derivative, and δ_{mp} is the Kronecker's delta.

The vector spherical harmonics $\mathbf{N}\zeta_{mn}$ and $\mathbf{M}\zeta_{mn}$ are defined in (4.B.2), and the coefficients $C_{mn}^{(0)}$ and $D_{mn}^{(0)}$ can be obtained from the expressions for coefficients $A_{mn}^{(0)}$ and $B_{mn}^{(0)}$ [see (4.B.3)], correspondingly, by changing b_{mn} to

$$c_{mn}(\mathbf{r}_0) = \frac{(n-m)!}{(n+m)!}\frac{\psi_n(k_0r_0)}{k_0r_0}P_n^m(\cos\theta_0)\,e^{-im\varphi_0}, \qquad (4.B.9)$$

where the function $\psi_n(k_0r_0)$ is defined above [(see (4.B.1)].

To write the expressions for a chiral molecule fields in a free space in local spherical coordinates of the s-th particle ($s = 1, 2$), it is necessary to replace in (4.36) and (4.42) spherical harmonics $\mathbf{N}\psi_{mn}^{(0)}$, $\mathbf{M}\psi_{mn}^{(0)}$, $\mathbf{N}\zeta_{mn}$, and $\mathbf{M}\zeta_{mn}$ with their analogues in the local system of coordinates ${}^s\mathbf{N}\psi_{mn}^{(0)}$, ${}^s\mathbf{M}\psi_{mn}^{(0)}$, ${}^s\mathbf{N}\zeta_{mn}$, and ${}^s\mathbf{M}\zeta_{mn}$; the coordinates of the molecule location r_0, θ_0, φ_0 should be replaced with the local coordinates ${}^sr_0, {}^s\theta_0, {}^s\varphi_0$ defining the molecule position ${}^s\mathbf{r}_0$ in the local s-th system of coordinates. As a result, we will have ${}^sA_{mn}^{(0)} = A_{mn}^{(0)}({}^s\mathbf{r}_0)$, ${}^sB_{mn}^{(0)} = B_{mn}^{(0)}({}^s\mathbf{r}_0)$, ${}^sC_{mn}^{(0)} = C_{mn}^{(0)}({}^s\mathbf{r}_0)$, and ${}^sD_{mn}^{(0)} = D_{mn}^{(0)}({}^s\mathbf{r}_0)$. The Cartesian components of the vectors $\mathbf{d}_0$ and $\mathbf{m}_0$ remain unchanged when one local system of coordinates is changed onto another system (see Fig. 4.15).

4.B.3 Elements of the Translational Addition Theorem for Vector Spherical Harmonics

Vector spherical harmonics written in one local system of coordinates can be expressed through the vector spherical harmonics written in another local system of coordinates with the help of the translational addition theorems [18, 22, 24]. By omitting details, for the local systems of coordinates under consideration (see Fig. 4.15), this theorem allows us to find coefficients used in the Section 4.4.1. For V_{mnq} and W_{mnq}, we have

$$V_{mnq} = U_{mnq} - k_0 l \left[\frac{n-m}{n(2n-1)} U_{m,n-1,q} + \frac{n+m+1}{(n+1)(2n+3)} U_{m,n+1,q} \right],$$
$$W_{mnq} = -ik_0 l \frac{m}{n(n+1)} U_{mnq}, \quad U_{mnq} = \sum_{\sigma=|n-q|}^{n+q} Q_{\sigma mnq} \frac{\zeta_\sigma(k_0 l)}{k_0 l}, \tag{4.B.10}$$

where l is the distance between origins of local systems of coordinates (see Fig. 4.15); the function $\zeta_\sigma(k_0 l)$ is defined above [see (4.B.2)], and

$$Q_{\sigma mnq} = (-1)^m i^{n-q-\sigma} (2n+1) \left\{ \frac{(n-m)!(q+m)!}{(n+m)!(q-m)!} \right\}^{1/2} C_{qmn,-m}^{\sigma 0} C_{q0n0}^{\sigma 0}, \tag{4.B.11}$$

where $C^{\sigma 0}_{qmn,-m}$ and $C^{\sigma 0}_{q0n0}$ are the Clebsch-Gordan coefficients [70]. By calculating (4.B.11), it is necessary to take into account that $C^{\sigma 0}_{q0n0}$ is nonzero only if the sum $n + q + \sigma$ is even [70].

For $\tilde{V}_{mnq}$ and $\tilde{W}_{mnq}$, we have the following expressions:

$$\tilde{V}_{mnq} = \tilde{U}_{mnq} - k_0 l \left[\frac{n-m}{n(2n-1)} \tilde{U}_{m,n-1,q} + \frac{n+m+1}{(n+1)(2n+3)} \tilde{U}_{m,n+1,q} \right],$$

$$\tilde{W}_{mnq} = -ik_0 l \frac{m}{n(n+1)} \tilde{U}_{mnq}, \quad \tilde{U}_{mqn} = \sum_{\sigma=|n-q|}^{n+q} O_{\sigma mnq} \frac{\psi_\sigma (k_0 l)}{k_0 l}, \tag{4.B.12}$$

where $\psi_\sigma (k_0 l)$ is defined above [see (4.B.1)], and

$$O_{\sigma mnq} = i^{n-q-\sigma} (2\sigma + 1) \left\{ \frac{(n-m)!\,(q+m)!}{(n+m)!\,(q-m)!} \right\}^{1/2} C^{nm}_{qm\sigma 0} C^{n0}_{q0\sigma 0}. \tag{4.B.13}$$

References

1. Abramowitz, M. and Stegun I. A. (Eds.) (1965). *Handbook of Mathematical Functions*, Dover, New York.
2. Ali, S. M., Habashy, T. M., and Kong, J. A. (1992). Spectral-domain dyadic Green's function in layered chiral media, *J. Opt. Soc. Am. A*, 9, pp. 413–423.
3. Alu, A. and Engheta, N. (2004). Guided modes in a waveguide filled with a pair of Single-Negative (SNG), Double-Negative (DNG), and/or Double-Positive (DPS) layers, *IEEE Trans. Microw. Theory Tech.*, 52, pp. 199–210.
4. Arnaut, L. R. and Davis, L. E. (1993). Chiral properties of lossy n-turn helices in the quasi-stationary approximation using a transmission-line model, 23rd European Microwave Conference, Sept. 6–10, pp. 176–178 (in Madrid, Spain).
5. Barron, L. D. (1997). From cosmic chirality to protein structure and function: Lord Kelvin's legacy, QJM-Month *J. Assoc. Phys.*, 90, pp. 793–800.
6. Barron, L. D. (2004). *Molecular Light Scattering and Optical Activity*, 2nd Ed., Cambridge University Press, UK.
7. Blanco, L. A. and Garcia de Abajo F. J. (2004). Spontaneous light emission in complex nanostructures, *Phys. Rev. B*, 69, pp. 205414-1–205414-12.

8. Bohren, C. F. (1974). Light scattering by an optically active sphere, *Chem. Phys. Lett.*, 29, pp. 458–462.
9. Bohren, C. F. and Huffman, D. R. (1983). *Absorption and Scattering of Light by Small Particles*, John Wiley-Interscience, New York.
10. Bokut', B. V., Serdyukov, A. N., and Fedorov, F. I. (1971). Phenomenological theory of optically active crystals, *Sov. Phys. Crystallog.*, 15, pp. 871–874.
11. Cai, W. and Shalaev, V. (2009). *Optical Metamaterials: Fundamentals and Applications*, Springer, Berlin.
12. Cantor, C. R. and Schimmel, P. R. (1980). *Biophysical Chemistry, Part 2: Techniques for the Study of Biological Structure and Function*, W. H. Freeman, Oxford.
13. Chance, R. R., Prock, A., and Silbey, A. (1978). Molecular fluorescence and energy transfer near interfaces, *Adv. Chem. Phys.*, 37, pp. 1–65.
14. Chew, H., McNulty, P. J., and Kerker, M. (1976). Model for Raman and fluorescent scattering by molecules embedded in small particles, *Phys. Rev. A*, 13, pp. 396–404; Erratum, *Phys. Rev. A*, 14, p. 2379.
15. Chew, H. (1987). Transition rates of atoms near spherical surfaces, *J. Chem. Phys.*, 87, pp. 1355–1360.
16. Chew, H. (1988). Radiation and lifetimes of atoms inside dielectric particles, *Phys. Rev. A*, 38, pp. 3410–3416.
17. Chew, W. C. (1995). *Waves and Fields in Inhomogeneous Media* (IEEE Press, New York).
18. Dufva, T. J., Sarvas, J., and Sten, J. C.-E. (2008). Unified derivation of the translational addition theorems for the spherical scalar and vector wave functions, *Prog. Electromagn. Res. B*, 4, pp. 79–99.
19. Feynman, R. P., Leighton, R. B., and Sands, M. (2005). *The Feynman Lectures on Physics including Feynman's Tips on Physics: The Definitive and Extended Edition*, 2nd Ed., Vol. 1, Ch. 33, §5, Addison–Wesley, USA.
20. Ford, G. W. and Weber, W. H. (1984). Electromagnetic interactions of molecules with metal surfaces, *Phys. Rep.*, 113, pp. 195–287.
21. Fort, E. and Grésillon, S. (2008). Surface enhanced fluorescence, *J. Phys. D: Appl. Phys.*, 41, pp. 013001-1–013001-31.
22. Fuller, K. A. (1994). Scattering and absorption cross sections of compounded spheres. I. Theory for external aggregation, *J. Opt. Soc. Am. A*, 11, pp. 3251–3260.
23. Gansel, J. K., Thiel, M., Rill, M. S., Decker, M., Bade, K., Saile, V., von Freymann, G., Linden, S., and Wegener, M. (2009). Gold helix photonic metamaterial as broadband circular polarizer, *Science*, 325, pp. 1513–1515.

24. Gerardy, J. M. and Ausloos, M. (1982). Absorption spectrum of clusters of spheres from the general solution of Maxwell's equations. II. Optical properties of aggregated metal spheres, *Phys. Rev. B*, 25, pp. 4204–4229.
25. Gomez, A., Lakhtakia, A., Margineda, J., Molina-Cuberos, G. J., Nunez, M. J., Saiz Ipina, J. A., Vegas, A., and Solano, M. A. (2008). Full-wave hybrid technique for 3-d isotropic-chiral-material discontinuities in rectangular waveguides: Theory and experiment, *IEEE Trans. Microw. Theory Tech.*, 56, pp. 2815–2825.
26. Grishina, N. V., Eremin, Yu. A., and Sveshnikov, A. G. (2012). Analysis of plasmon resonances of closely located particles by the discrete sources method, *Opt. Spect.*, 113, pp. 440–445.
27. Guzatov, D. V. and Klimov, V. V. (2011). Chiral particles in a circularly polarised light field: new effects and applications, *Quant. Electron.*, 41, pp. 526–533.
28. Guzatov, D. V. and Klimov, V. V. (2011). Optical properties of a plasmonic nano-antenna: an analytical approach, *New. J. Phys.*, 13, pp. 053034-1–053034-26.
29. Guzatov, D. V. and Klimov, V. V. (2012). The influence of chiral spherical particles on the radiation of optically active molecules, *New J. Phys.*, 14, pp. 123009-1–123009-19.
30. Guzatov, D. V., Klimov, V. V., and Poprukailo, N. S. (2013). Spontaneous radiation of a chiral molecule located near a half-space of a bi-isotropic material, *J. Exp. Theor. Phys.*, 116, pp. 531–540.
31. Guzatov, D. V., and Klimov, V. V. (2014). Spontaneous radiation of a chiral molecule placed near a cluster of two chiral spherical particles, *Quant. Electron.*, (in print).
32. Ivanov, Ye. A. (1968). *Diffraction of Electromagnetic Waves on Two Bodies*, Nauka i Tekhnika, Minsk (in Russian).
33. Jaggard, D. L., Michelson, A. R., and Papas, C. H. (1979). On electromagnetic waves in chiral media, *Appl. Phys.*, 18, pp. 211–216.
34. Kelvin, Lord (1904). *Baltimore Lectures on Molecular Dynamics and the Wave Theory of Light*, C. J. Clay, London).
35. Klimov, V. V. (2002). Spontaneous emission of an excited atom placed near a "left-handed" sphere, *Opt. Commun.*, 211, pp. 183–196.
36. Klimov, V. V. and Ducloy, M. (2004). Spontaneous emission rate of an excited atom placed near a nanofiber, *Phys. Rev A*, 69, pp. 013812-1–013812-17.

37. Klimov, V. V. and Letokhov, V. S. (2005). Electric and magnetic dipole transitions of an atom in the presence of spherical dielectric interface, *Laser Phys.*, 15, pp. 61–73.

38. Klimov, V. V. and Guzatov, D. V. (2007). Strongly localized plasmon oscillations in a cluster of two metallic nanospheres and their influence on spontaneous emission of an atom, *Phys. Rev. B*, 75, pp. 024303-1–024303-7.

39. Klimov, V. (2009). Novel approach to a perfect lens, *JETP Lett.*, 89, pp. 270–273.

40. Klimov, V. V., Bloch, D., Ducloy, M., and Rios Leite J. R. (2009). Detecting photons in the dark region of Laguerre-Gauss beams, *Opt. Express*, 17, pp. 9718–9723.

41. Klimov, V. V., Baudon, J., and Ducloy, M. (2011). Comparative focusing of Maxwell and Dirac fields by negative-refraction half-space, *Europhys. Lett.*, 94, pp. 20006-1–20006-6.

42. Klimov, V. V., Guzatov, D. V., and Ducloy, M. (2012). Engineering of radiation of optically active molecules with chiral nano-meta-particles, *Europhys. Lett.*, 97, pp. 47004-1–47004-6.

43. Klimov, V. (2014). *Nanoplasmonics: Fundamentals and Applications*, Pan Stanford Publishing, Singapore.

44. Klimov, V. V., Zabkov, I. V., Pavlov, A. A., and Guzatov, D. V. (2014). Eigen oscillations of a chiral sphere and their influence on radiation of chiral molecules, *Opt. Exp.* (submitted).

45. Kneipp, K., Moskovits, M., and Kneipp, H. (Eds.) (2006). *Surface-Enhanced Raman Scattering: Physics and Applications* (Topics in Applied Physics), Springer Verlag, Berlin.

46. Kwon, D.-H., Werner, D. H., Kildishev, A. V., and Shalaev, V. M. (2008). Material parameter retrieval procedure for general bi-isotropic metamaterials and its application to optical chiral negative-index metamaterial design, *Opt. Express*, 16, pp. 11822–11829.

47. Lakhtakia, A., Varadan, V. K., and Varadan, V. V. (1990). Radiation by a point electric dipole embedded in a chiral sphere, *J. Phys. D: Appl. Phys.*, 23, pp. 481–485.

48. Li, K. (2009). *Electromagnetic Fields in Stratified Media*, Zhejiang University Press, China.

49. Li, L.-W., You D., Leong, M.-S., Yeo, T.-S., and Kong, J. A. (2000). Electromagnetic scattering by multilayered chiral-media structures: A scattering-to-radiation transform, *Prog. Electromagn. Res.*, 26, pp. 249–291.

50. Li, L.-W., Dan, Y., Leong, M.-S., and Kong, J. A. (2009). Electromagnetic scattering by an inhomogeneous chiral sphere of varying permittivity: A discrete analysis using multilayered model, *Prog. Electromagn. Res.*, 23, pp. 239–263.
51. Li, X. and Shapiro, M. (2010). Spatial separation of enantiomers by coherent optical means, *J. Chem. Phys.*, 132, pp. 041101-1–041101-3.
52. Liaw, J.-W., Chen, C.-S., and Chen, J.-H. (2010). Enhancement or quenching effect of metallic nanodimer on spontaneous emission, *J. Quant. Spectrosc. Radiat. Transf.*, 111, pp. 454–465 (2010).
53. Lindell, I. V., Sihvola, A. H., Tretyakov, S. A., and Viitanen, A. J. (1994). *Electromagnetic Waves in Chiral and Bi-isotropic Media*, Artech House, Boston.
54. Lukosz, W. and Kunz, R. E. (1977). Light emission by magnetic and electric dipoles close to a plane interface. I. Total radiated power, *J. Opt. Soc. Am.*, 67, pp. 1607–1615.
55. Pendry, J. B., Holden, A. J., Robbins, D. J., and Stewart, W. J. (1999). Magnetism from conductors and enhanced nonlinear phenomena, *IEEE Trans. Microw. Theory Tech.*, 47, pp. 2075–2084.
56. Pendry, J. B. (2000). Negative refraction makes a perfect lens, *Phys. Rev. Lett.*, 85, pp. 3966–3969.
57. Pendry, J. B. (2004). A chiral route to negative refraction, *Science*, 306, pp. 1353–1355.
58. Petrin, A. B. (2008). Electromagnetic wave propagation from a point source in air through a medium with a negative refractive index, *JETP Lett.*, 87, pp. 464–469.
59. Pitarke, J. M., Silkin, V. M., Chulkov, E. V., and Echenique, P. M. (2007). Theory of surface plasmons and surface-plasmon polaritons, *Rep. Prog. Phys.*, 70, pp. 1–87.
60. Plum, E., Zhou, J., Dong, J., Fedotov, V. A., Koschny, T., Soukoulis, C. M., and Zheludev, N. I. (2009). Metamaterial with negative index due to chirality, *Phys. Rev. B*, 79, pp. 035407-1–035407-6.
61. Qiu, C.-W., Yao, H.-Y., Li, L.-W., Zouhdi, S., and Yeo, T.-S. (2007). Eigenfuctional representation of dyadic Green's functions in planarly multilayered general Faraday chiral media, *J. Phys. A: Math. Theor.*, 40, pp. 5751–5766.
62. Radi, Y., Nikmehr, S., and Hosseinzadeh, S. (2011). A rigorous treatment of vertical dipole impedance located above lossy DPS, MNG, ENG, and DNG half-space, *Prog. Electromagn. Res.*, 116, pp. 107–121.

63. Ren, W. (1994). Dyadic Green's functions and dipole radiations in layered chiral media, *J. Appl. Phys.*, 75, pp. 30–35.
64. Rosenfeld, L. (1928) Quantenmechanische Theorie der natürlichen optischen Aktivität von Flüssigkeiten und Gasen, Z. Physik, 52, pp. 161–174 (in German).
65. Ruppin, R. and Martin, O. J. F. (2004). Lifetime of an emitting dipole near various types of interfaces including magnetic and negative refractive materials, *J. Chem. Phys.*, 121, pp. 11358–11361.
66. Schellman, J. A. (1975). Circular dichroism and optical rotation, *Chem. Rev.*, 75, pp. 323–331.
67. Smith, D. R., Padilla, W. J., Vier, D. C., Nemat-Nasser, S. C., and Schultz, S. (2000). Composite medium with simultaneously negative permeability and permittivity, *Phys. Rev. Lett.*, 84, pp. 4184–4187.
68. Sommerfeld, A. (1909). Über die Ausbreitung der Wellen in der drahtlosen Telegraphie, *Ann. d. Phys.*, 28, pp. 665–736 (in German).
69. Stratton, J. A. (1941). *Electromagnetic Theory*, McGraw-Hill, New York.
70. Varshalovich, D. A., Moskalev, A. N., and Khersonskii, V. K. (1975). *Quantum Theory of Angular Momemtum*, Nauka, Leningrad (in Russian).
71. Vinogradov, A. P. (2001). *Electrodynamics of Composite Materials*, URSS, Moscow (in Russian).
72. Wang, B., Zhou, J., Koschny, T., and Soukoulis, C. M. (2009). Nonplanar chiral metamaterials with negative index, *Appl. Phys. Lett.*, 94, pp. 151112-1–151112-3.
73. Weinstein, L. A. (1988). *Electromagnetic Waves*, Radio i Svyaz, Moscow (in Russian).
74. Wylie, J. M. and Sipe, J. E. (1984). Quantum electrodynamics near an interface, *Phys. Rev. A*, 30, pp. 1185–1193.
75. Xu, J.-P., Yang, Y.-P., Lin, Q., and Zhu, S.-Y. (2009). Spontaneous decay of a two-level system near the left-handed slab, *Phys. Rev. A*, 79, pp. 043812-1–043812-9.
76. Yao, P., Van Vlack, C., Reza, M., Patterson, M., Dignam, M. M., and Hughes, S. (2009). Ultrahigh Purcell factors and Lamb shifts in slow light metamaterial waveguides, *Phys. Rev. B*, 80, pp. 195106-1–195106-11.
77. Yokota, M., He, S., and Takenaka T. (2001). Scattering of a Hermite-Gaussian beam field by a chiral sphere, *J. Opt. Soc. Am. A*, 18, pp. 1681–1689.

78. Zhu, X., Pan, W.-Y., and Guan, B.-R. (2009). Electromagnetic field generated by a horizontal electric dipole on a double negative medium half space, *Prog. Electromagn. Res.*, 6, pp. 123–137.
79. Zouhdi, S., Fourrier-Lamer, A., and Mariotte, F. (1992). On the relationships between constitutive parameters of chiral materials and dimensions of chiral objects (helices), *J. Phys. III* France, 2, pp. 337–342.

Chapter 5

Unusual Optical Properties of Helical Metallic Photonic Crystals and Chiral Channels in Dielectric Photonic Crystals

Hongqiang Li,[a] Jian Wen Dong,[b] and Che Ting Chan[c]

[a]*Key Laboratory of Advanced Micro-structure Materials, MOE, Department of Physics, Tongji University, Shanghai 200092, China*
[b]*State Key Laboratory of Optoelectronic Materials and Technologies, School of Physics and Engineering, Sun Yat-Sen University, Guangzhou 510275, China*
[c]*Department of Physics, The Hong Kong University of Science and Technology, Clear Water Bay, Hong Kong, China*
phchan@ust.hk

In this chapter, we will show that a photonic crystal comprising an array of subwavelength metallic helices has a unique photonic dispersion that allows for the realization of negative refraction for frequencies both above and below the polarization gap for electromagnetic waves propagating along the helical axis. The helical photonic crystal can also serve as a broadband wave plate for waves propagating perpendicular to the helical axis of the metallic photonic crystal. In addition, we show that robust transport of light can be achieved using chiral photonic guided modes in a channel of

Singular and Chiral Nanoplasmonics
Edited by Svetlana V. Boriskina and Nikolay I. Zheludev

ISBN 978-981-4613-17-0 (Hardcover), 978-981-4613-18-7 (eBook)
www.panstanford.com

a dielectric photonic crystal. The light transport is immune to the scattering of isotropic homogenous impurities and the phenomenon bears some phenomenological similarity to robust transport of electrons in topological insulators, although the mechanism is not the same. In particular, the system is time-reversal invariant and the robust one-way transport does not require an external field, which is distinct from the previous strategies employing two-dimensional magnetic photonic crystals with explicit time reversal breaking by an external field.

5.1 Introduction

This chapter is concerned with some rather unusual optical properties derived from the helical symmetry in metallic photonic crystals (PCs) and the chirality of some defect modes in dielectric PCs. Chirality has attracted much attention recently in the field of metamaterials [8, 20, 23, 26, 28, 32, 36–38, 41, 44, 52, 54]. A crystal of metallic helices [11, 12] is a very special member of the chiral family. Although the physics of many chiral media [8, 11, 12, 20, 23, 26, 36–38, 41, 52, 54] can be attributed to the lack of mirror symmetry and the local resonance of discrete chiral entities, the continuous helical symmetry of a helix results in a photonic dispersion that has many unique features not found in other chiral systems [48–50] and we will review these special properties in the following sections.

The prediction of negative refraction in chiral media [32] has fueled the interest to achieve negative refractive index [38, 47, 52, 54], strong optical activity [23, 26, 36], and circular dichroism [8, 20, 37, 41]. Chiral metamaterials are typically realized as layers of discrete chiral resonators, and the optical properties are interpreted using effective medium theory [5, 7, 27, 28, 44]. In this review, we will start from the micro-structure of the building block (without the effective medium approximation) and this can give us a deeper understanding of the relation between structure and functionality. We will see that metallic spiral arrays can operate as a broadband wave plate for the incidence wave propagating along directions perpendicular to the helix axis. This is different from conventional wave plates with a certain thickness of birefringent crystals that can only operate in a narrow frequency range [3, 24].

Helical structures are useful for optical activity [24, 25, 43], broad-band antenna [21, 22], and traveling wave tube [33–35, 42]. Helices are potent building blocks for metamaterials [11, 12, 32] and it is known that chirality can lead to a negative refractive index [5, 7, 27, 28, 32, 44]. In addition, an analytical model for helicoidal spirals [2] predicted the elliptical polarization of eigenstates and bandgap along the directions orthogonal to the spiral axis. Polarization gaps were demonstrated in a gold helix metamaterial in the THz regime [12]. However, the underlying physics of such kind of helicoidal metamaterials is still under exploration. In the following, we will present a band theory for the helix array by combining multiple scattering theory (MST) [6, 29] with the semi-analytical solution for a single helix [42]. We will also review some experimental results that directly demonstrate the negative refraction in the helical systems and we will review the wave propagation along directions perpendicular to the helix axis [51].

We will then switch gear to dielectric PCs and show that robust transport of light can be achieved using chiral defect modes. The robust transport of electrons in quantum Hall systems is well known [19, 40, 53]. In such systems, electrons in the chiral edge states propagate in one direction and the transport is robust against backscattering from impurities. Recently, chiral edge states have been predicted [1, 13, 45] and experimentally realized [9, 39, 46] in magneto-optical PCs with a large external magnetic field. These one-way states rely on breaking time-reversal symmetry. On the contrary, a new class of topological states has been found in topological insulators [4, 10, 14, 16, 18] that possesses time-reversal symmetry. Robust transport in the form of counter-propagating currents carrying spin-up or spin-down electrons come as a consequence of Kramer degeneracy and strong spin-orbit coupling. These topologically protected electronic states do not require an applied magnetic field. However, both Kramers' degeneracy and spin-orbit coupling are specific to electronic systems; it is not

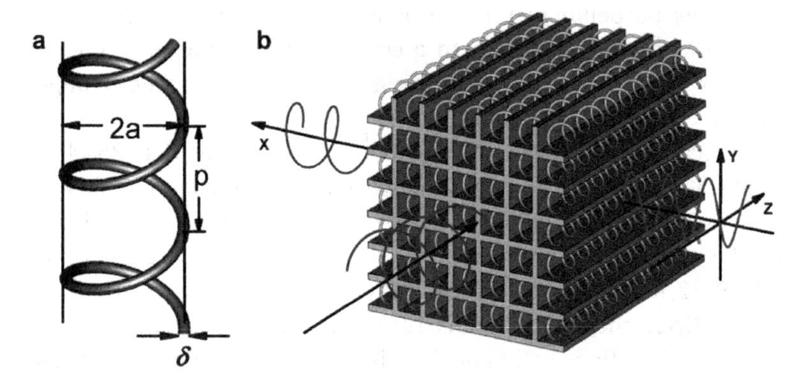

Figure 5.1 Schematic picture of (a) a metallic helix unit and (b) a square array of metallic helices. Wave propagations along the axis direction and transverse plane of helices are considered.

obvious that robust transport of light can be realized in the photonic system without time-reversal symmetry breaking.

We will show that chirality can enable robust transport in TRS invariant photonic system and the phenomenon can be observed simply by drilling a channel inside a layer-by-layer 3D PC. High transmittance will be observed in such a chiral channel even if the channel is blocked by a perfect electric conductor (PEC) obstacle with size larger than the channel's cross-section.

5.2 Metallic Helix Array

5.2.1 *Band Theory of Metallic Helix Array*

We will now consider the band dispersion of a PC composed of metallic helices and examine its implications. Exact solution for the electromagnetic (EM) modes in a metal helix is difficult, but there are elegant approximate solutions [33–35, 42]. A helix remains invariant if it is rotated and simultaneously displaced along the helical axis and this special rotational/translational symmetry imposes an associated phase factor for the EM wave propagation [33, 42]. Figure 5.1 shows a model of the square array of right-handed

(RH) metallic helices. As an example, we assume that a single helix has a pitch of $p = 4.4$ mm, radius $a = 3.3$ mm, and the diameter of metallic wires $\delta = 0.8$ mm. The helices form a square array in the xy plane with a lattice constant of $d = 11$ mm. We define a pitch angle ψ by $\cot\psi = 2\pi a/p$. We note that a helix comes back to itself after being translated by a distance of Δz and being rotated simultaneously by an angle of $2\pi\Delta z/p$ (for RH helix) or $-2\pi\Delta z/p$ [for left-handed (LH) helix], and thus, any physical quantity should satisfy the symmetry condition

$$U(\rho, \phi, z) = U(\rho, \phi \pm 2\pi\Delta z/p, z + \Delta z), \quad (5.1)$$

with $+/-$ sign for the RH/LH helix, respectively [33, 42]. The periodicity along the helical (z) axis implies that the field components for an RH helix system can be expanded by functions of the form

$$\psi_n(\rho, \phi, z) = \exp(ik_z z)F_n(\rho)\exp(-in\phi + 2n\pi z/p), \quad (5.2)$$

where k_z is the Bloch wavevector along the z axis. The angular term should be $\exp(in\phi)$ if the helix is LH. The radial function $F_n(\rho)$ can be expressed in terms of modified Bessel functions.

5.2.2 *MST Approach to Solve for the Band Structure of Helix Crystals*

We employ cylindrical coordinates (ρ, ϕ, z) with the z axis parallel to axes of helices. The EM fields inside (region I) and outside (region II) the helices are expanded using modified Bessel and Hankel functions I_n and K_n, and the I_n and K_n become conventional Bessel functions for a state above the light line in free space as the arguments change from real to imaginary values.

For a periodic array of RH helices, the EM waves at a point (ρ, ϕ, z) with components of the electric field (E_ρ, E_φ, E_z) and magnetic field (H_ρ, H_φ, H_z) can be expressed as the linear combination of Bloch state series $e^{i(k_z+2\pi n/p)z}e^{-in\varphi}$ defined in Eqs. (1, 2). If we are dealing with an LH helix, the angular function shall be $e^{in\varphi}$. For a point at (ρ, ϕ, z) falling inside a certain RH helix (region I), the electric field and magnetic field read

$$
\begin{aligned}
E_{z1} &= \sum_n -\tau_n^2 A_n I_n(\tau_n\rho)\, e^{-in\varphi} e^{i(k_z+2\pi n/p)z} \\
E_{\rho 1} &= \sum_n \left[i(k_z + 2\pi n/p)\tau_n A_n I_n'(\tau_n\rho) \right. \\
&\quad \left. + \frac{\omega\mu n}{\rho} B_n I_n(\tau_n\rho) \right] e^{-in\varphi} e^{i(k_z+2\pi n/p)z} \\
E_{\varphi 1} &= \sum_n \left[\frac{(k_z + 2\pi n/p)n}{\rho} A_n I_n(\tau_n\rho) \right. \\
&\quad \left. - i\omega\mu\tau_n B_n I_n'(\tau_n\rho) \right] e^{-in\varphi} e^{i(k_z+2\pi n/p)z} \\
H_{z1} &= \sum_n -\tau_n^2 B_n I_n(\tau_n\rho)\, e^{-in\varphi} e^{i(k_z+2\pi n/p)z} \\
H_{\rho 1} &= \sum_n \left[i(k_z + 2\pi n/p)\tau_n B_n I_n'(\tau_n\rho) \right. \\
&\quad \left. - \frac{\omega\varepsilon n}{\rho} A_n I_n(\tau_n\rho) \right] e^{-in\varphi} e^{i(k_z+2\pi n/p)z} \\
H_{\varphi 1} &= \sum_n \left[\frac{(k_z + 2\pi n/p)n}{\rho} B_n I_n(\tau_n\rho) \right. \\
&\quad \left. + i\omega\varepsilon\tau_n A_n I_n'(\tau_n\rho) \right] e^{-in\varphi} e^{i(k_z+2\pi n/p)z}.
\end{aligned} \tag{5.3}
$$

If the point is outside the helices (region II), the fields read

$$
\begin{aligned}
E_{z2} &= \sum_n -\tau_n^2 C_n K_n(\tau_n\rho)\, e^{-in\phi} e^{i(k_z+2\pi n/p)z} \\
&\quad + \sum_{l,n} -\tau_l^2 C_l (-1)^l S_{n-l}(\tau_l) I_n(\tau_l\rho)\, e^{-in\phi} e^{i(k_z+2\pi l/p)z} \\
E_{\rho 2} &= \sum_n \left[i(k_z + 2\pi n/p)\tau_n C_n K_n'(\tau_n\rho) \right. \\
&\quad \left. + \frac{\omega\mu n}{\rho} D_n K_n(\tau_n\rho) \right] e^{-in\phi} e^{i(k_z+2\pi n/p)z} \\
&\quad + \sum_{l,n} \left[i(k_z + 2\pi l/p)\tau_l C_l (-1)^l S_{n-l}(\tau_l) I_n'(\tau_l\rho) \right. \\
&\quad \left. + \frac{\omega\mu n}{\rho} D_l (-1)^l S_{n-l}(\tau_l) I_n(\tau_l\rho) \right] e^{-in\phi} e^{i(k_z+2\pi n/p)z}
\end{aligned}
$$

$$E_{\phi 2} = \sum_n \left[\frac{(k_z + 2\pi n/p)n}{\rho} C_n K_n(\tau_n \rho) \right.$$
$$\left. - i\omega\mu\tau_n D_n K_n'(\tau_n \rho) \right] e^{-in\phi} e^{i(k_z + 2\pi n/p)z}$$
$$+ \sum_{l,n} \left[\frac{(k_z + 2\pi n/p)n}{\rho} C_l (-1)^l S_{n-l}(\tau_l) I_n(\tau_l \rho) \right.$$
$$\left. - i\omega\mu\tau_l D_l (-1)^l S_{n-l}(\tau_l) I_n'(\tau_l \rho) \right] e^{-in\phi} e^{i(k_z + 2\pi n/p)z}$$
$$H_{z2} = \sum_n -\tau_n^2 D_n K_n(\tau_n \rho) e^{-in\phi} e^{i(k_z + 2\pi n/p)z}$$
$$+ \sum_{l,n} -\tau_l^2 D_l (-1)^l S_{n-l}(\tau_l) I_n(\tau_l \rho) e^{-in\phi} e^{i(k_z + 2\pi n/p)z}$$
$$H_{\rho 2} = \sum_n \left[i(k_z + 2\pi n/p)\tau_n D_n K_n'(\tau_n \rho) \right.$$
$$\left. - \frac{\omega\varepsilon n}{\rho} C_n K_n(\tau_n \rho) \right] e^{-in\phi} e^{i(k_z + 2\pi n/p)z}$$
$$+ \sum_{l,n} \left[i(k_z + 2\pi l/p)\tau_l D_l (-1)^l S_{n-l}(\tau_l) I_n'(\tau_l \rho) \right.$$
$$\left. - \frac{\omega\varepsilon n}{\rho} C_l (-1)^l S_{n-l}(\tau_l) I_n(\tau_l \rho) \right] e^{-in\phi} e^{i(k_z + 2\pi l/p)z}$$
$$H_{\phi 2} = \sum_n \left[\frac{(k_z + 2\pi n/p)n}{\rho} D_n K_n(\tau_n \rho) \right.$$
$$\left. + i\omega\varepsilon\tau_n C_n K_n'(\tau_n \rho) \right] e^{-in\phi} e^{i(k_z + 2\pi n/p)z}$$
$$+ \sum_{l,n} \left[\frac{(k_z + 2\pi n/p)n}{\rho} D_l (-1)^l S_{n-l}(\tau_l) I_n(\tau_l \rho) \right.$$
$$\left. + i\omega\varepsilon\tau_l C_l (-1)^l S_{n-l}(\tau_l) I_n'(\tau_l \rho) \right] e^{-in\phi} e^{i(k_z + 2\pi l/p)z}, \quad (5.4)$$

where ω is the circular frequency, ε and μ are permittivity and permeability in vacuum, respectively. $\tau_n = -iT_n$ and T_n is the transverse component of wavevector k satisfying to $T_n^2 = k^2 - (k_z + 2\pi n/p)^2$. $S_l(\tau) = \sum_{q \neq 0} K_l(\tau R_q) e^{il\varphi_q} e^{i\mathbf{k}_t \cdot \mathbf{R}_q}$ is a lattice sum running over the nodes (R_q, φ_q) of the square lattice in cylindrical coordinates where k_i is the transverse component of wavevector k in the vacuum. $I_n'(x) = dI_n(x)/dx$ and $K_n'(x) = dK_n(x)/dx$. A_n, B_n, C_n, and D_n are the coefficients to be determined. We follow Sensiper to assume uniformly distributed surface current flow along the metal wires [42], and the metal wires are treated as thin conducting tapes in the so-called tape-helix model. Under that assumption, the boundary conditions require that the local electric field on metal wires must be perpendicular to the line of metal wire, so that we can assign the surface current component $J_\perp$ or alternatively $E_\perp(a)$ to be zero where $J_\perp = J_z \cos\psi - J_\phi \sin\psi$, $E_{//} = E_t(a, \phi, p\phi/2\pi) = E_\phi \cos\psi + E_z \sin\psi$. The coefficients A_n, B_n, C_n, and D_n can be determined by considering the boundary conditions that link the wave fields at interface of regions I and II. At the interface (the surface of the conducting tapes), we have

$$\begin{aligned} E_{t1}(a) &= E_{t2}(a)\, H_{t2}(a) - H_{t1}(a) = \alpha_f \\ E_{z1}(a) &= E_{z2}(a)\, H_{z2}(a) - H_{z1}(a) = -J_{s\phi}(a) \\ E_{\phi 1}(a) &= E_{\phi 2}(a)\, H_{\phi 2}(a) - H_{\phi 1}(a) = J_{sz}(a), \end{aligned} \tag{5.5}$$

where $J_{s\phi}(a)$ and $J_{sz}(a)$ refers to the angular and radial components of surface currents on the interface, respectively. These components, flowing on the conducting tapes (or metallic wires), must propagate along the z direction for the sake of boundary requirements so that they shall be decomposed in form of harmonic waves in space

$$J_{s\phi}(a) = \sum_n J_{\phi n} e^{-in\phi} e^{i\beta_n z} \quad J_{sz}(a) = \sum_n J_{zn} e^{-in\phi} e^{i\beta_n z}, \tag{5.6}$$

where $\beta_n = \beta_0 + 2\pi n/p$ is the Bloch wavevector in the nth order. $J_{s\phi}(a)$ and $J_{sz}(a)$ can also be expressed as the superposition of the components $J_{//}$ and $J_\perp$ that are along and perpendicular to the metallic wire, as

$$J_{s\phi} = J_{//} \cos\psi - J_\perp \sin\psi \quad J_{sz} = J_{//} \sin\psi + J_\perp \cos\psi. \tag{5.7}$$

Since $J_\perp = 0$, for the Fourier components of $J_{s\phi}(a)$ and $J_{sz}(a)$, we always have

$$J_{\phi n} = J_{//n} \cos\psi \quad J_{zn} = J_{//n} \sin\psi\,. \tag{5.8}$$

As the surface currents oscillate along the line of metallic wires in uniform distribution for the incidence waves along the direction of helix axis, $J_{\phi n}$ as a function of z reads as

$$J_{\phi n} = \begin{cases} \dfrac{\frac{p}{\delta} J\, e^{i\left[\beta_0 \frac{p\phi}{2\pi} + \beta_n\left(z - \frac{p\phi}{2\pi}\right)\right]}}{\sqrt{1 - \xi\left[2\left(z + \frac{p\phi}{2\pi}\right)/\delta\right]^2}} & \left(\frac{p\phi}{2\pi} - \frac{\delta}{2} < z < \frac{p\phi}{2\pi} + \frac{\delta}{2}\right) \\ 0 & \left(z < \frac{p\phi}{2\pi} - \frac{\delta}{2},\, z > \frac{p\phi}{2\pi} + \frac{\delta}{2}\right) \end{cases} . \tag{5.9}$$

When the metallic tapes are narrow compared with the operational wavelength so that $\xi \to 0$, we have

$$J_{\phi n} = JR_n \quad R_n = \frac{\sin \dfrac{n\pi\delta}{p}}{\dfrac{n\pi\delta}{p}} = \sin c \frac{n\pi\delta}{p}. \tag{5.10}$$

By substituting Eq. (5.10), Eq. (5.8), and Eq. (5.6) into Eq. (5.5), we obtain the following equations for A_n, B_n, C_n, and D_n

$$A_n I_n(\tau_n a) = C_n K_n(\tau_n a) + \sum_l C_n (-1)^n S_{l-n}(\tau_n, \mathrm{k_i})\, I_l(\tau_n a)$$

$$B_n I_n'(\tau_n a) = D_n K_n'(\tau_n a) + \sum_l D_n (-1)^n S_{l-n}(\tau_n, \mathrm{k_i})\, I_l'(\tau_n a)$$

$$D_n K_n(\tau_n a) + \sum_l D_n (-1)^n S_{l-n}(\tau_n, \mathrm{k_i})\, I_l(\tau_n a) - B_n I_n(\tau_n a)$$
$$= \frac{JR_n}{\tau_n^2} \cos\psi$$

$$C_n K_n'(\tau_n a) + \sum_l C_n (-1)^n S_{l-n}(\tau_n, \mathrm{k_i})\, I_l'(\tau_n a) - A_n I_n'(\tau_n a)$$
$$= \frac{JR_n}{i\omega\varepsilon\tau_n}\left(\sin\psi - \frac{n\beta_n}{a\tau_n^2}\cos\psi\right), \tag{5.11}$$

which can be further reduced to

$$A_n = \frac{JR_n y_n}{i x_n \omega\varepsilon\tau_n}\left(-\sin\psi + \frac{n\beta_n}{a\tau_n^2}\cos\psi\right)$$

$$B_n = \frac{JR_n z_n}{x_n \tau_n^2}\cos\psi$$

$$C_n = \frac{JR_n}{i x_n \omega\varepsilon\tau_n}\left(-\sin\psi + \frac{n\beta_n}{a\tau_n^2}\cos\psi\right) I_n(\tau_n a)$$

$$D_n = \frac{JR_n}{x_n \tau_n^2}\cos\psi\, I_n'(T_n a), \tag{5.12}$$

and the intermediate parameters x_n, y_n, and z_n read

$$x_n = \frac{1}{\tau_n a} + (-1)^n \sum_l S_{l-n}(\tau_n, \mathrm{k_i}) \left[I_l(\tau_n a) I_n'(\tau_n a) - I_l'(\tau_n a) I_n(\tau_n a) \right]$$

$$y_n = K_n(\tau_n a) + (-1)^n \sum_l S_{l-n}(\tau_n, \mathrm{k_i}) I_l(\tau_n a)$$

$$z_n = K_n'(\tau_n a) + (-1)^n \sum_l S_{l-n}(\tau_n, \mathrm{k_i}) I_l'(\tau_n a), \tag{5.13}$$

where $S_l(\tau, \mathrm{k_i}) = \sum_{p \neq 0} K_l(\tau R_p) e^{il\varphi_p} e^{i\mathrm{k_i}.\mathrm{R_p}}$ is a lattice sum. These intermediate parameters satisfy with the Wronskian relation $x_n = I_n'(\tau_n a) y_n - I_n(\tau_n a) z_n$ of modified Bessel functions $I_n(\tau_n a)$ and its first order derivative $I_n'(\tau_n a)$. Now, we can obtain the analytical forms of all the components of EM waves propagating inside the helical system, and the tangential component $E_t(a)$ of electric field on conducting tape can be expressed by E_ϕ and E_z in a form of $E_t(a) = E_\phi \cos\psi + E_z \sin\psi$. If we adopt all the representations of Eq. (5.3) in region I or those of Eq. (5.4) in region II, at the interface of the two regions, $E_t(a)$ has the form

$$E_t(a) = \cos\psi \sum_n \left[\frac{\beta_n n}{a} A_n I_n(\tau_n a) - i\omega\mu\tau_n B_n I_n'(\tau_n a) \right] e^{-in\phi} e^{i\beta_n z}$$
$$+ \sin\psi \sum_n -\tau_n^2 A_n I_n(\tau_n a) e^{-in\phi} e^{i\beta_n z}. \tag{5.14}$$

After some further reduction, Eq. (5.14) can be expressed as

$$E_t(a) = \frac{\sin^2\psi}{i\omega\varepsilon a} \sum_n \frac{1}{x_n \tau_n a} \Bigg[\Bigg(\tau_n^2 a^2 - 2na\beta_n \cot\psi$$
$$+ \frac{n^2 \beta_n^2}{\tau_n^2} \cot^2\psi \Bigg) y_n I_n(\tau_n a)$$
$$+ k^2 a^2 \cot^2\psi z_n I_n'(\tau_n a) \Bigg] J R_n e^{-in\phi} e^{i\beta_n z}. \tag{5.15}$$

As the surface currents flow along the line of metallic wires without swirling, a reasonable approximation is that the tangential electric field components shall be zero along the center line of conducting tapes, so that we have $E_t(a, \phi, p\phi/2\pi) = 0$, and the eigen-equation

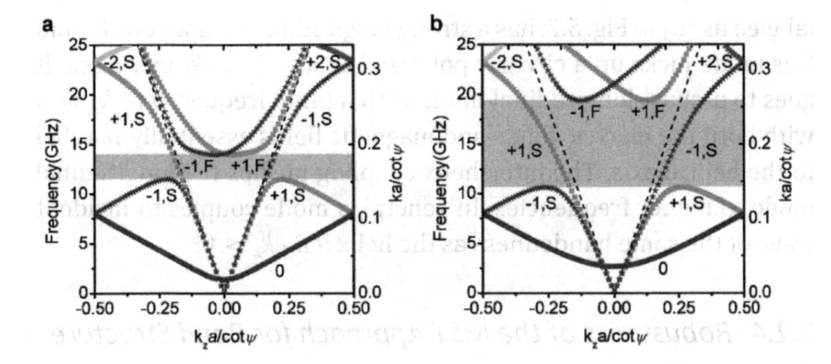

Figure 5.2 Photonic band structures of helix crystals along the helix axis for three lattice constants (a) $d = 20$ mm, (b) $d = 11$ mm. The pitch $p = 4.4$ mm, the radius $a = 3.3$ mm.

for our helical system

$$\sum_n \left[\left(k_z^2 a^2 - k^2 a^2 + \frac{n^2 k^2}{\tau_n^2} \cot^2 \psi \right) y_n I_n (\tau_n a) \right.$$

$$\left. + k^2 a^2 \cot^2 \psi z_n I_n' (\tau_n a) \right] \frac{R_n}{x_n \tau_n} = 0. \tag{5.16}$$

We can see that the Eq. (5.16) regresses back to the eigen-equation of a single tape helix if we remove the lattice term in x_n, y_n, and z_n.

5.2.3 *Band Structure of Metallic Helix Array*

The computed photonic band dispersions give us an intuitive understanding of the optical property of helix arrays. Figure 5.2a,b show the band structure of the helix array along the helix axis for different lattice constants. We label the eigenmodes in Fig. 5.2 by their dominant term in Eq. (5.2). For example, the $(-1,S)$ modes (blue stars in Fig. 5.2) have $n = -1$ term as the dominant term and "S" stands for a "slow mode" below the light line, and we use the subscript "F" for a mode inside the light cone. Equations (5.3, 5.4) show that there is a $\pm/2$ phase difference between the radial and the angular components for both the electric field and magnetic field, implying that the eigenmodes are LH or RH circularly polarized (LCP or RCP). Eigenmode analysis indicates that the $n = \pm 1$ modes are indeed either LCP or RCP. The lowest frequency branch (red), labeled as (0) in Fig. 5.2, has a strong longitudinal component E_z and this mode picks up a circular polarized character as k_z increases. It goes to a quasi-longitudinal mode with a finite frequency at $k_z = 0$ with both the electric fields and magnetic fields essentially parallel to the helical axis. The inter-helix coupling pushes the longitudinal mode to higher frequencies. In general, a mode couples to incident wave of the same handedness as the helix if $n \cdot k_z > 0$.

5.2.4 *Robustness of the MST Approach for Band Structure*

We test our semi-analytic MST approach by comparing with brute force numerical simulations. For comparison, the total length $L = (p^2 + (2\pi a)^2)^{1/2}$ of a unit is fixed. Figure 5.3 shows that the band structure calculated by our analytical method [colored dots in Fig. 1.3(a), (c), and (e)] is basically the same as those computed by the commercial package CST Microwave Studio [colored dots in Fig. 1.3(b), (d), and (f)].

5.2.5 *Wide Polarization Gap and Negative Refraction at Low Frequencies*

An important feature of the band structure is a wide polarization gap [shaded in grey in Fig. 5.2(c)] that only allows the passage of incident waves of the opposite handedness. For example, a RH helix array has an RCP gap. The gap becomes wider for a higher helix filling ratio (smaller d). Another interesting feature of the band structure is the emergence of negative group velocity bands at both sides of the polarization gap, which is different from the previous theoretical prediction that the negative refraction only happens above the resonant gap [32]. Both the high frequency (+1,F) and the lower frequency (+1,S) branch exhibit negative group velocities. The (+1,S) branch exhibits negative group velocity after reaching a maximum frequency that pins the lower edge of the polarization gap. Concomitant with the negative refraction bands in the slow mode, one can see from Figs. 5.2 and 5.3 that there are band crossings at the BZ boundary ($k = \pi/p$) and the degenerate modes are pinned at frequencies $f_{0,1}$ and $f_{1,2}$ that are nearly independent of the lattice constant p. A comparative study shows that the genuine longitudinal

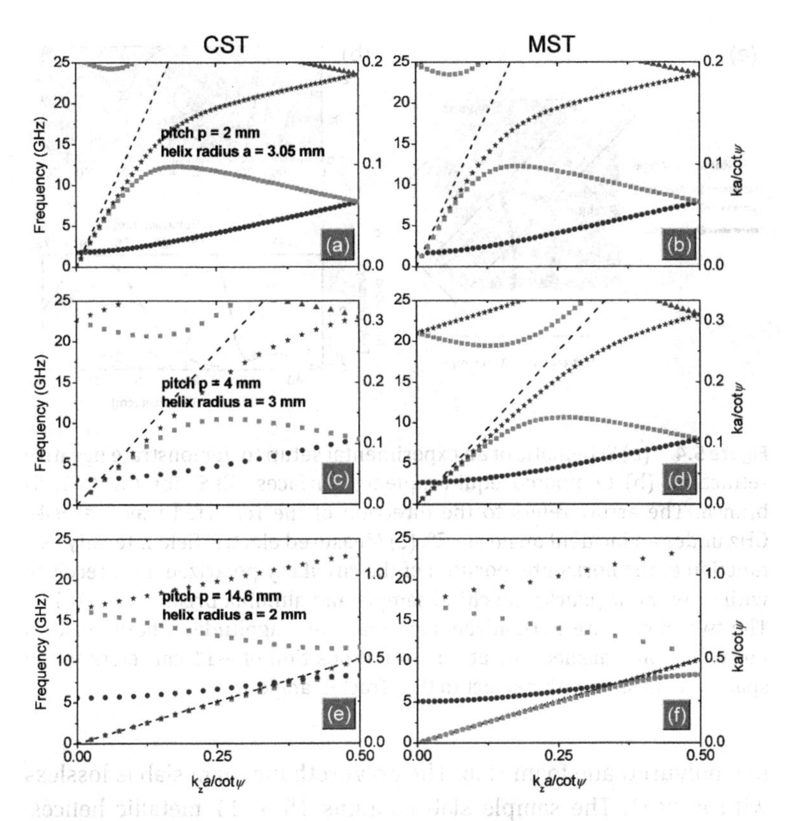

Figure 5.3 Comparing band structures computed with MST (right panels) and by full wave numerical solver (left panels). The total length of a metallic helix unit is fixed.

and/or circularly polarized eigenmodes and the negative dispersion bands in the slow mode and band crossing will all disappear when the helices are cut into "discrete" spirals [50]. The slope of the negative dispersion band above the polarization gap becomes very small without the helical symmetry. Thus, we conclude that many salient features of the band structure are direct consequences of the helical symmetry requirement [Eq. (5.1)]. The negative refraction is demonstrated by measurements inside an anechoic chamber using a slab of the helix array with geometric parameters corresponding to the band structure shown in Fig. 5.2(b). Another helix sample is fabricated by periodically embedding the clockwise metallic helices

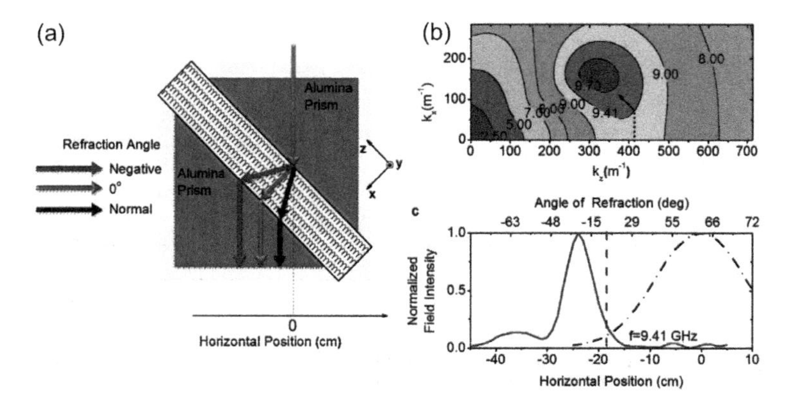

Figure 5.4 (a) Schematic of an experimental setup to demonstrate negative refraction. (b) Computed equi-frequency surfaces (EFS) for the (+1, S) branch. The arrow refers to the direction of the refracted waves at 9.41 GHz under an incident angle at 45°. (c) Measured electric field intensity as a function of the horizontal position of the circularly polarized horn receiver, with or without (black) the chiral sample and alumina prisms at 9.41 GHz. The two curves are normalized such that the magnitude of both peaks is unity. The blue dashed line at horizontal position of −11 cm refers to the spatial beam shift with respect to 0° refracted angle.

in a polyurethane foam slab. The polyurethane foam slab is lossless with $\varepsilon \approx 1$. The sample slab contains 15×11 metallic helices, each having 140 periods along helical axis (z-axis). Computed EFS show that negative refraction can be achieved at both sides of the polarization gap and we try to realize the negative refraction at the lower edge of the gap. As the (±1,S) modes lie below the light line, we excite the (±1,S) modes by prism coupling techniques [see Fig. 5.4(a)], and the refractive angle is found by measuring the spatial beam shift. Two isosceles right-angled triangular alumina prisms ($\varepsilon_r = 8.9$) are placed so that they touch the sample slab at both sides and a Gaussian beam is normally incident in xz-plane to the air-prism interface from a linearly polarized horn emitter (operating at 8.2–12.4 GHz with a gain factor of 24.6 dB), ensuring an incident angle of 45° from alumina to sample. The local field intensity is measured by the LCP/RCP horn receiver as a function of the horizontal position. The spatial shift of the outgoing beam is found by measuring the peak position at the interface of prism. The coordinate origin in the horizontal position is aligned with the position of the horn emitter, marked by the dashed vertical line in Fig. 5.4(a). Negative refraction is observed from 9.18 GHz to 9.48 GHz with a refraction angle from -17.44° to -50.11°, in good agreement with the computed EFS shown in Fig. 5.4 (b). The solid line in Fig. 5.4(c) presents the spatial profile of local field intensity measured at 9.41 GHz. A peak value is measured at the horizontal position of -24 cm, corresponding to a refraction angle of -46.5°, roughly equal to -45.8° estimated by EFS analysis. Thus, negative refraction below the polarization gap is demonstrated experimentally.

5.2.6 *Transverse Propagation in Helix Array: A Functionality of Broadband Wave Plate*

For EM waves propagating along the transverse plane that is perpendicular to the axis of helices, the ellipticity and difference between wavevectors of the two states can be fixed in a wide frequency range by choosing appropriate geometric parameters, and this feature can be exploited to make highly transparent broadband wave plate. We will review proof-of-principle microwave experiments that verify the thickness-dependent polarization character of transmitted waves.

Figure 5.5 shows the band structure $\omega(k_x, k_y = k_z = 0)$ in the transverse plane of a wave plate made with RH metallic helix array [see the photo in the inset]. The comparative results by varying critical geometric parameters associated with helical symmetry and Bragg scattering are shown in Fig. 1.5(b) and (c). The sample shown in Fig. 5.5(a) has pitch $p = 4$ mm, helix radius $a = 3$ mm, wire diameter $\delta = 0.6$ mm, lattice constant $d = 11$ mm. Our calculations show that the lowest branch B_R [black line in Fig. 5.5(b)] is dictated by the degenerate $\pm$ 1st orders of helical Bloch states with E-fields along the y direction, while the second lowest branch B_L [red dashed line in Fig. 5.5(b)] is dictated by the 0th order of helical Bloch state with E-fields along the helix axis [50]. Consequently, a B_R/B_L state shall pick up an REP/LEP character with long axis along the y /z direction. We see from Fig. 1.5 (b) and (c) that, the branch B_R, nearly linear at small k_x, opens up a Bragg gap at Brillouin zone (BZ) boundary with the fourth branch (blue dash-dot line) with the same

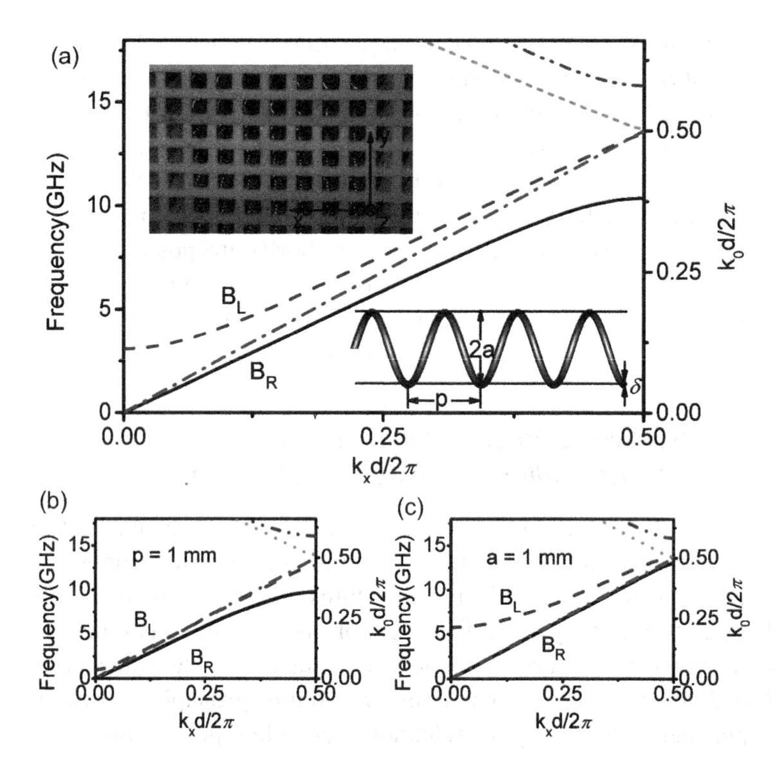

Figure 5.5 A slab of helices as a wave plate for transversely propagating waves. (a) The band structure $\omega(k_x, k_y = k_z = 0)$ for transverse propagation through helices arranged in a square array; (The photo in the inset presents a photo of the seven-layered sample.) The geometric parameters are the pitch $p = 4$ mm, helix radius $a = 3$ mm, wire diameter $\delta = 0.6$ mm, the lattice constant $d = 11$ mm; Band structures by (b) varying pitch only to $p = 1$ mm, and (c) varying helix radius only to $a = 1$ mm, all other parameters are fixed.

handedness. On the contrary, the branch B_L has a locally resonant character with a cut-off frequency at $k_x = 0$, but becomes linear quickly at a small value of k_x, and joins the third lowest branch of the same handedness at the BZ boundary without opening a noticeable gap. The dispersions of the B_L and B_R branches can be engineered so that they run parallel to each other in a wide frequency range. This is because they are controlled by the EM coupling along the helix (z) axis and the Bragg scattering (in the *xy* plane) among the helices, respectively, and the two different mechanisms can be independently adjusted by different sets of structural parameters. Such property is not likely to be found in other systems.

The E-field of two orthogonal elliptical states $|\phi_{LEP}\rangle$ and $|\phi_{REP}\rangle$ propagating along x axis can be expressed as

$$E_{REP} = (\alpha \mathrm{u}_y + i\beta \mathrm{u}_z)\, e^{i(k_{REP}x-\omega t)}, \tag{5.17}$$

$$E_{LEP} = (\beta \mathrm{u}_y - i\alpha \mathrm{u}_z)\, e^{i(k_{LEP}x-\omega t)} \quad , \tag{5.18}$$

where α and β are two different positive numbers that determine the axis ratio of elliptical states, u_y and u_z denote the unitary vectors along the y and z coordinate axes. Figure 5.5(a) shows that the B_L and B_R branches are essentially linear and parallel to each other in a wide frequency range of 3.9 GHz $\sim$ 9.6 GHz. Within this type of band dispersion, two orthogonal eigenstates $|\phi_{LEP}\rangle$ and $|\phi_{REP}\rangle$ of the B_L and B_R branches pick up the same $\Delta k = k_{REP} - k_{LEP}$ between their wavevectors. The analysis of EFS in Fig. 5.6 shows that this property holds for all directions in the transverse plane, as if the helix array behaves as an isotropic medium for the transversely propagating waves. Within the same frequency range, the axial ratio of the in-plane field components for the LEP/REP branch (Fig. 5.5) is roughly fixed as well. These properties are useful for producing a broadband wave plate.

Consider, for example, a linearly polarized plane wave propagating along the x direction with a polarization angle of θ with respect to y-axis (see Fig. 5.1). The polarization of the outgoing wave [24]

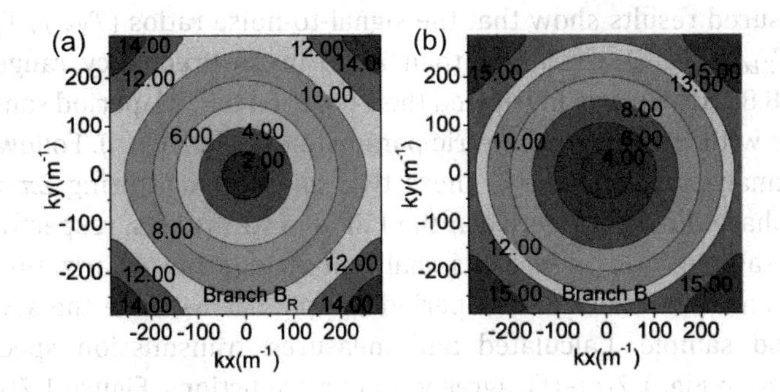

Figure 5.6 Equi-frequency surfaces of the BR and BL branches for the helix sample shown in the inset of Fig. 6(a).

can be identified by its axial ratio $ar = E_y/iE_z$ as

$$ar = \frac{(i\alpha/\beta + \tan\theta)\, e^{-i\Delta k \cdot L} + (i\beta/\alpha - \tan\theta)}{-(i + \beta/\alpha \tan\theta)\, e^{-i\Delta k \cdot L} + (i - \alpha/\beta \tan\theta)}, \tag{5.19}$$

where L is the slab thickness. The transmitted waves will be transformed to RH or LH circularly polarized (RCP or LCP) at $\theta = 0$ or $\theta = \pi/2$ provided that the axial ratios of REP and LEP eigenstates and the thickness of helix slab satisfy the conditions $\alpha/\beta = \sqrt{2} + 1$ and $\Delta k \cdot L = \pi$. The sample shown in the inset of Fig. 5.5(a), which has seven periods along x direction, can realize such a wave plate with appropriately geometric parameters. Such a helix slab can also rotate linear polarization in a wide frequency range. If $\Delta k \cdot L = \pi$ and $\theta = \pm\pi/4$, the transmitted wave is still linearly polarized but with rotated polarization direction along $\mp\pi/4$. This property is independent of the axial ratio of elliptical eigenstates. In contrast, a conventional wave plate cannot implement two different kinds of polarization transformation simultaneously.

Figure 5.7(b) presents measured and simulated transmission spectra. According to the calculated band structure in Fig. 5.5(a), the thinnest wave plate for linear-to-circular polarization transformation (or vice versa) only requires seven periods along the x direction. Both the calculations and measurements demonstrate that the transmitted waves are transformed to RCP or LCP, respectively, under y-polarized or z-polarized incidence as shown by the solid lines and dashed lines in Fig. 1.7(a,b). The calculated/measured transmittance is above 95%/85% in the range of $3.9 \sim 9.6$ GHz. Measured results show that the signal-to-noise ratios ($T_{Y,RCP}/T_{Y,LCP}$ or $T_{Z,LCP}/T_{Z,RCP}$) are larger than 20dB in the frequency range of 4.1–8.8GHz. We also fabricated the 14-period and 21-period sample slabs with the same geometric parameters in Fig. 5.5(a). Following the analysis stated above, these two samples shall bring 2π and 3π phase difference between the LEP and REP states, respectively. The sample with 14 periods shall not change the polarization of incident waves, and the 21-period sample behaves like the seven-period sample. Calculated and measured transmission spectra, shown in Fig. 1.7(c)–(f), agree with our predictions. Figure 1.7(g,h) show that the seven-layered sample can rotate the polarization direction of a linear polarized wave. The polarization angle of linear

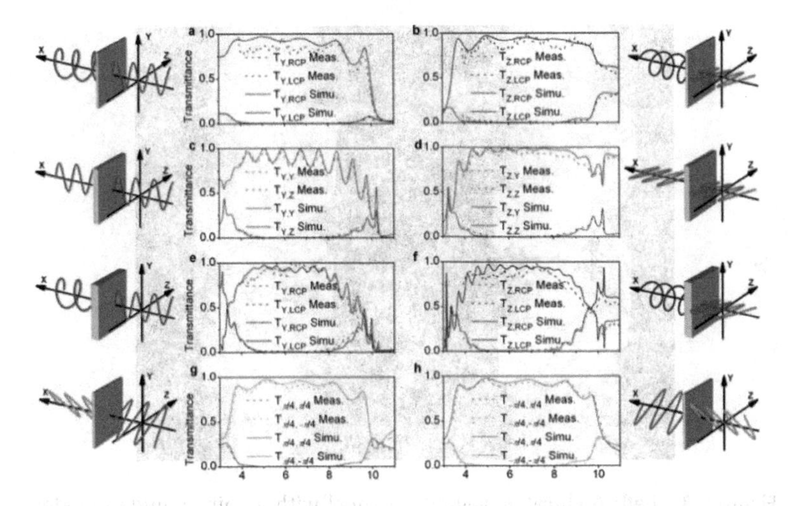

Figure 5.7 Transmission spectra of helix samples as a wave plate. Transmission spectra are calculated and measured for three samples with 7, 14, and 21 periods along the propagating direction (x-direction). (a) and (b), (g) and (h) for seven-period sample, (c) and (d), (e) and (f) for 14, 21-period samples, respectively. The first and the second subscripts, i, and j, of the transmission spectra $T_{i,j}$ refer to the polarized state of the incident and the transmitted waves, respectively. The letter Y or Z denotes a linear polarization along y or z direction. The symbol $\pi/4$ or $-\pi/4$ denotes linear polarization with a polarization angle at $\pi/4$ or $-\pi/4$ about y-axis. The transmission spectra are normalized to the total power of the incidence.

polarized wave is transformed from $\theta = \pm\pi/4$ to $\theta = \mp\pi/4$ as predicted.

5.3 Dielectric Photonic Crystals

5.3.1 *Robust Transport of Light in Chiral Channels in a Dielectric Photonic Crystal*

In the following, we will consider 3D dielectric PCs and examine the wave propagation inside chiral channels. Let us first consider a dielectrics waveguide with both inversion and mirror symmetries. The waveguide modes are always linearly polarized and do not differentiate between LH and RH circularly polarized incident wave.

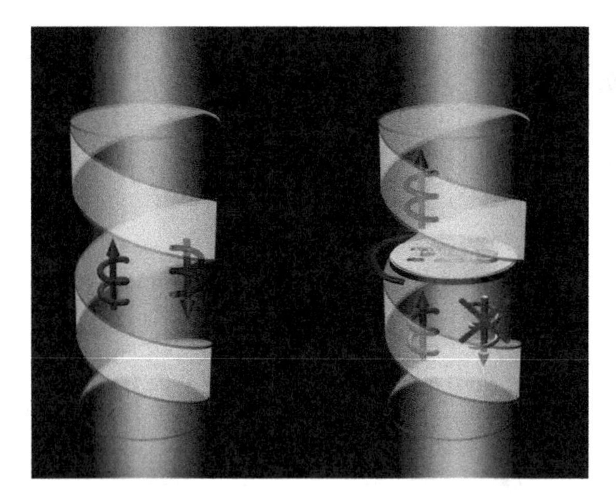

Figure 5.8 Left: A chiral waveguide channel with a pair of guided modes of only one chirality. Right: Robust transport of light when the chiral mode encounters an obstacle inside the channel.

Here, we define an LH (RH) circularly polarized wave such that the tip of the electric fields trace out an LH (RH) helix in space at a frozen time. When an LH circularly polarized wave propagates in the positive z-direction, the representation of the electric field is $(1, i, 0)/\sqrt{2}$. When the LH wave propagates in the negative z-direction, the representation becomes $(1, -i, 0)/\sqrt{2}$.

In a chiral channel with neither inversion nor mirror symmetry, the waveguide mode will have chirality due to the lifting of LH/RH degeneracy. For a pair of counter-propagating chiral modes with the same handedness, their coupling is suppressed as the temporal rotation of the electric field at a given point in space has opposite directions. Moreover, it is possible that in a certain frequency range, the channel only allow modes of a specific handedness to propagate (say an LH upward mode and an LH downward mode, see the left panel in Fig. 5.8). When a PEC slab is inserted into such a chiral waveguide, incident LH upward wave will be reflected by the slab and become an RH downward wave. But the chiral waveguide supports no RH propagating mode because of a RH polarization gap. Consequently, the wave has to go around the slab (see right panel in Fig. 5.8), resulting in backscattering-immune transport.

To understand the robust transport, let us consider an isotropic homogenous plate lying on the x-y plane. An LH circularly polarized wave from the negative z-direction is incident on this plate. The representation of the electric field of this polarization is $(1, i, 0)/\sqrt{2}$. The reflection phase change in the electric field will be the same for both the x and y components due to the isotropy of the surface. As a result, no relative phase change occurs, and the representation of the reflected wave is still $(1, i, 0)/\sqrt{2}$. However, the reflected wave is propagating to the negative z-direction, and the wave propagating in the negative z-direction with a polarization of $(1, i, 0)/\sqrt{2}$ is an RH circularly polarized wave. Therefore, an LH circularly polarized wave will change to RH polarization upon reflection from an isotropic homogenous surface, which is not allowed.

5.3.2 *Sample Construction*

In this section, we will show that the backscattering-immune transport discussed in the previous section can be realized easily in a prototypical 3D PC. Consider a 3D layer-by-layer PC [15, 31], built by stacking square bars as shown in Fig. 5.9(a). The cross-section of the square bars is $b \times b$, and the periodicity in the z direction and on the x-y plane is a and d, respectively. There are four chiral ladders hidden in the layer-by-layer structure, although the PC itself is non-chiral (NC). To highlight this point pictorially, we shaded a segment of building blocks using blue color in Fig. 5.9(a), showing that a discrete chiral ladder is twisting upwards. In Fig. 5.9(b), we employed different colors to identify the four chiral ladders in the PC. In particular, blue and cyan denote two LH ladders, of which the phase difference between neighboring ladders is π. And red and pink denote two RH chiral ladders with a π phase difference. The handedness of each chiral ladder is also plotted by colored square arrows in the bottom plane of Fig. 5.9(b).

If we drill a void channel with a square pattern along the z direction, the channel can either be NC or chiral, depending on the channel's central axis. We can construct a chiral channel by removing one of the blue-colored LH ladder with its central axis passing through the star in the bottom plane of Fig. 5.9(d). Figure 5.9(f)

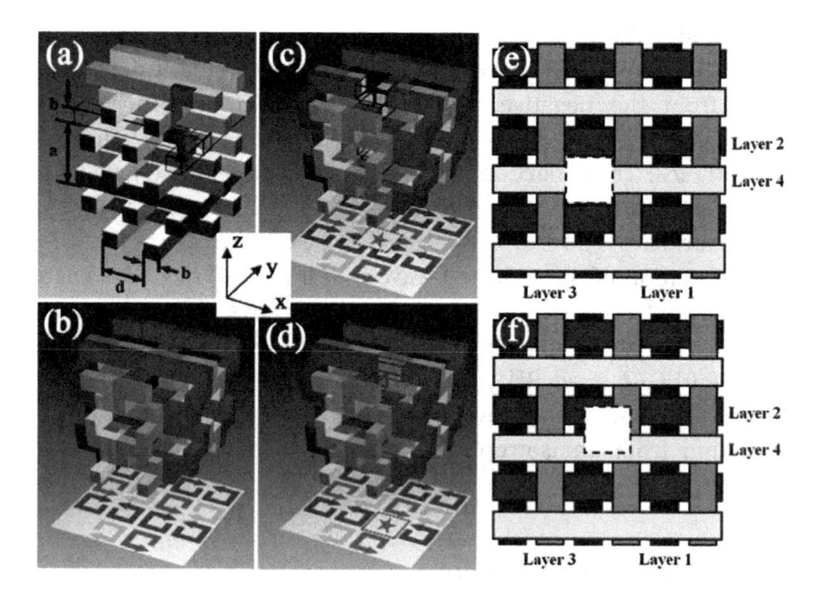

Figure 5.9 (a) PC consisting of square bars (yellow). A segment of one chiral ladder is marked in blue to guide the eye. (b) PC comprising two LH ladders (blue and cyan) and two RH ladders (red and pink). (c)/(d) NC/chiral channel constructed by removing material along the z direction. Colored arrows in the bottom plane of (b–d) show the handedness of the ladders. Stars in (c) and (d) show the position of the channel's central axis. (e) and (f) are top view of the NC and chiral channel created in the PC.

shows the top view of the chiral channel that has neither mirror nor inversion symmetry, but has four-fold helical symmetry. For the NC channel, its central axis is shifted by $(d/4, d/4)$ relative to the chiral channel, which is located at the intersection of two neighboring perpendicular bars. The NC channel can be constructed by removing square bar segments in two neighboring layers of a period along the z-direction [see Fig. 5.9(c,e)].

5.3.3 *Calculation Method and Experimental Setup*

The robust transport can be demonstrated by simulations and microwave experiments. We calculated the band structures of both NC and chiral channels using a plane wave method [17] using a supercell with the size of $5d \times 5d \times a$. The transmission spectra are calculated using the finite-difference-time-domain (FDTD) method [30]. Perfectly matched layer boundary conditions are used. An EM source with x-linear polarization is placed with the distance of a lattice constant below the entrance of the channel. The source has the cross-section of $0.5625a \times 1.4375a$, which is the same size as the X11644A microwave guided-port in the experimental measurement. A monitor with the cross-section of $a \times a$ is positioned with the distance of a lattice constant above the exit of the channel to collect the transmission energy.

Figure 5.10(a) shows the illustration of the experimental setup. The microwave is emitted through an X11644A waveguide from 7 to

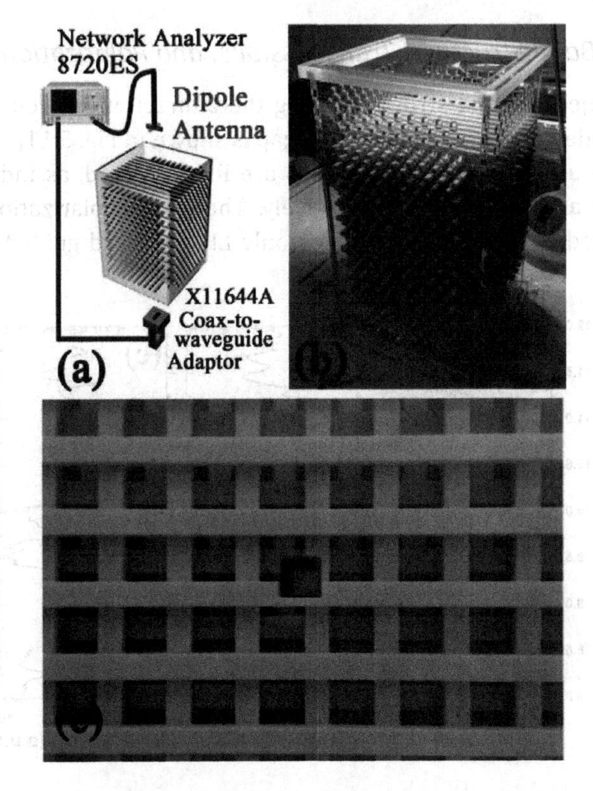

Figure 5.10 (a) Experimental setup for transport properties measurement. (b) Photograph of the experimental sample. Surrounding transparent plexiglass walls are used for supporting the sample. (c) The exits of the chiral channel.

13 GHz with x-linear polarization. The source is placed 16 mm below the channel entrance. A 15 mm long dipole antenna is placed at the channel exit to measure the transmitted amplitude and phase. By aligning the dipole antenna along either x- or y-direction, the transmitted amplitude of either x- or y-polarization as well as the relative phase between the x- and y-polarization can be measured. A photograph of the sample is shown in Fig. 5.10(b). The thickness of the sample is 160 mm, and we design $a = 16$ mm, $b = 4$ mm, $d = 11.32$ mm so that the working frequency of robust transport behavior is around 10 GHz. Photographs of the chiral channel are shown in Fig. 5.10(c).

5.3.4 *Band Structure, Transmissions, and Polarization*

The projected band structure along the z-direction as well as the four guided modes in the full bandgap is shown in Fig. 5.11(a). Two of them are LH polarized and two are RH polarized, as indicated by blue and red colors, respectively. There is a polarization gap (indicated by light blue) in which only LH-polarized guided mode

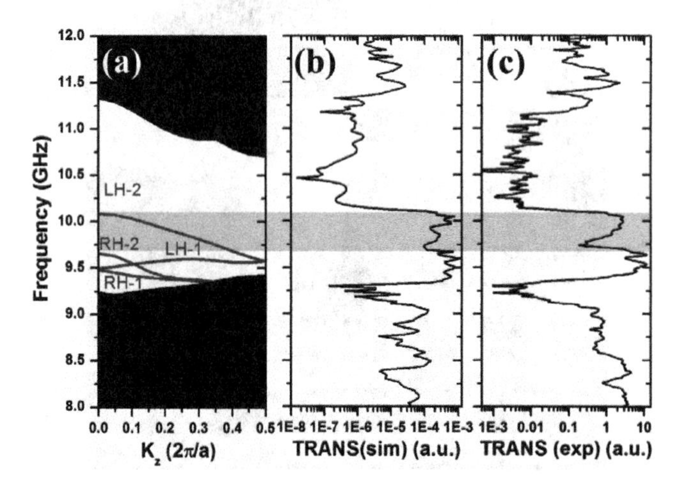

Figure 5.11 (a) Band structure, (b) simulated transmission spectrum, and (c) measured transmission spectrum for the chiral channel. In (a), blue solid lines stand for LH-guided modes, and red solid lines stand for RH-guided modes. Black colors stand for projected pass bands.

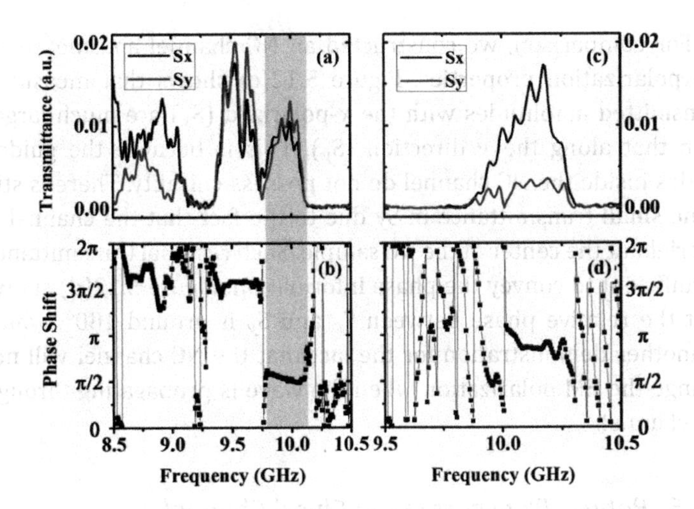

Figure 5.12 (a) Measured transmitted amplitudes of S_x (black) and S_y (red). (b) Relative phase between S_x and S_y. The backscattering-immune state is highlighted by light blue. The nearly equaling amplitudes of S_x and S_y and nearly 90° phase shift show that the output EM wave is nearly LH circularly polarized. (c) Measured transmitted amplitudes of S_x (black) and S_y (red). (d) Relative phase shift between S_x and S_y. The dominant amplitude in the component of S_x and the nearly 180° phase shift between S_x and S_y show that the output EM wave is nearly x-linear polarized.

can propagate. This implies robust transport in this frequency region. Both simulation and microwave total transmission spectra are plotted in Fig. 1.1(b,c), which agree well with the band structure.

In principle, the transmitted wave should be LH polarized inside the polarization gap, even if the incidence wave is linearly polarized. This is also verified by the experimental results shown in light blue region of Fig. 1.12(a,b). The spectra of the transmitted amplitudes S_x and S_y are almost the same, and the relative phase is nearly 90°. However, outside the polarization gap from 9.36 to 9.78 GHz, the measured transmitted amplitudes S_y are almost twice as the amplitudes S_x from 9.59 to 9.78 GHz, while the relative phases are nearly 270°, indicating that the transmitted waves are of elliptical polarization. The amplitudes S_x and S_y are almost the same from 9.36 to 9.59 GHz, while the relative phase with the value of around 270°, showing that the transmitted waves are of RH polarization.

For comparison, we constructed an NC channel and measured the polarization properties. Figure 5.12(c) shows that measured transmitted amplitudes with the x-polarized (S_x) are much larger than that along the y direction (S_y). This is because the guided modes inside the NC channel do not possess chirality. There is still some small transmittance in Sy due to the fact that the channel is not right at the center of the NC sample. Such residual transmittance is sufficient to convey the phase information. Figure 5.12(d) shows that the relative phase between S_x and S_y is around 180°, which is another demonstration for the fact that the NC channel will not change the EM polarization when the wave is propagating through the channel.

5.3.5 *Robust Transport in the Chiral Channel*

Now, we come to demonstrate the robust transport property of the chiral channel states. Figure 5.13 shows the measured transmissions

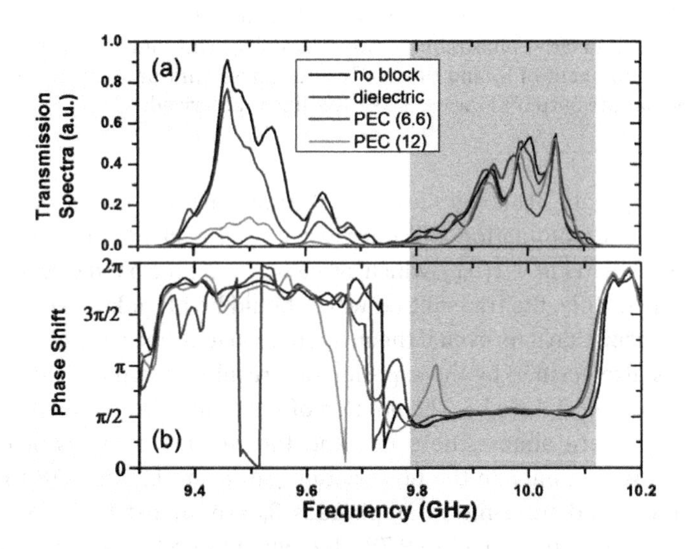

Figure 5.13 (a) Measured total transmittance and (b) measured phase shift for the backscattering-immune transport in the chiral channel. Black solid line: no obstacle in channel. Blue/green solid line: small/large PEC plate in channel. Red solid line: dielectric plate in channel. Light blue highlights the region for the polarization gap.

with different obstacles inserted inside the chiral channels. Light blue regions of Fig. 5.13 highlight the presence of robust transport in the chiral channel for the frequency region inside the polarization gap. The first obstacle is an aluminum block (PEC block) measuring 6.6 mm × 6.6 mm × 2.5 mm, approximately the same as the channel's cross-section. We can see that the forward transmission (blue) is almost the same as that of the unblocked channel (black). If a larger PEC block with nearly double the size is inserted, the transmission (green) is only slightly decreased. The relative phase of the PEC blocked channel shown in Fig. 5.13(b) is nearly the same as that of the un-blocked channel, indicating that the polarizations of the output wave do not change even the obstacle is present.

We did some FDTD simulations to demonstrate the robust transport property of the chiral channel states. The simulated geometry parameters are the same as those in the experiment and an E_x-polarized source at 9.9 GHz is used. The patterns within a period of $2a$ are shown to highlight the salient features. Poynting vector patterns without and with the small-size PEC slab are shown in Fig. 1.14(a) and (b), respectively. Energy flux propagates up the unblocked channel in a helical shape inside the LH chiral channel. When the LH polarized wave encounters a slab inserted inside the chiral channel, the reflection will change LH to RH polarization. However, the reflected wave cannot propagate because of the absence of backward RH-guided mode. The only way for the flux is to continue moving upward, resulting in the robust transport phenomenon and hence high transmittance.

The behaviors of robust transport cannot be sustained outside the band gap. Figure 5.13(b) illustrates that low transmission is recorded from 9.4 to 9.7 GHz (blue and green curves), as the EM wave is strongly reflected by the PEC blocks. The results of relative phase between Sx and Sy, plotted in Fig. 5.13(b), have larger fluctuations than those inside the polarization gap, showing that the polarization character is changed a lot after the obstacle is inserted.

For the case of a dielectric slab, the robust transport effect can also be observed in the chiral channel. When an alumina slab with the same size as the first PEC slab is inserted, the transmissions remain high inside the polarization gap (see red curves in Fig. 5.13). Note that the spectrum is slightly different from

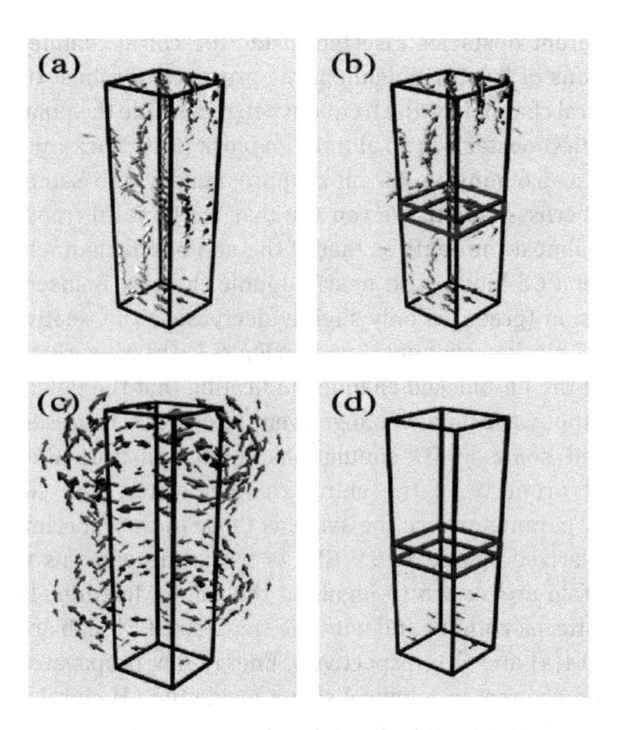

Figure 5.14 Poynting vector patterns (arrows) for (a)/(b) the unblocked/PEC-blocked chiral channel, and (c)/(d) unblocked/PEC-blocked NC channel. Red boundary represents the PEC plate. Black rectangles guide eyes to see the boundary of waveguide channel.

that of the unblocked channel due to the Fabry–Pérot interference effect. We have calculated Poynting vector patterns in the chiral channel blocked by the square dielectric slab at the frequency of 9.9 GHz. We find that part of flux propagates through the dielectric slab instead of going around it. This will change the optical path of the channel so that the Fabry–Pérot resonant peaks are different from the unblocked channel.

We also did some simulations to demonstrate the robust transport property of the chiral channel states by utilizing different kinds of obstacles. Three representative examples, including PEC sphere, PEC rectangular, and PMC square obstacles, are illustrated in Fig. 5.15. One can clearly see that the forward transmissions (red) are almost the same as those of the unblocked channel

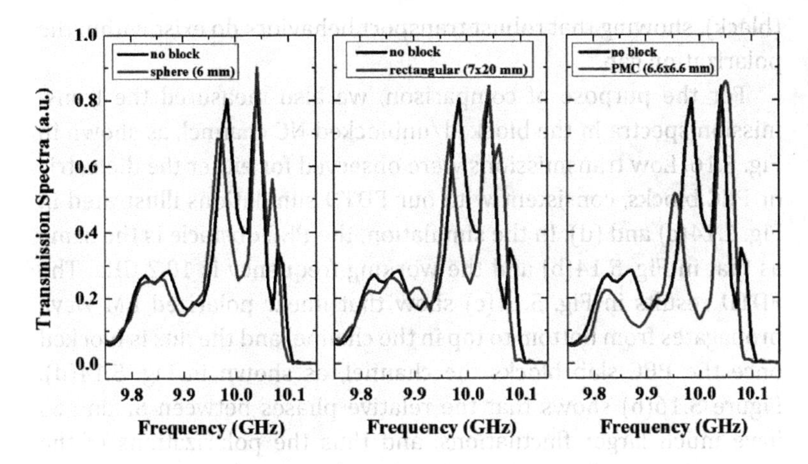

Figure 5.15 Simulated total transmittance in the frequency within the LH polarization gap in the chiral channel blocked by (a) PEC sphere, (b) PEC rectangular, and (c) PMC square slab.

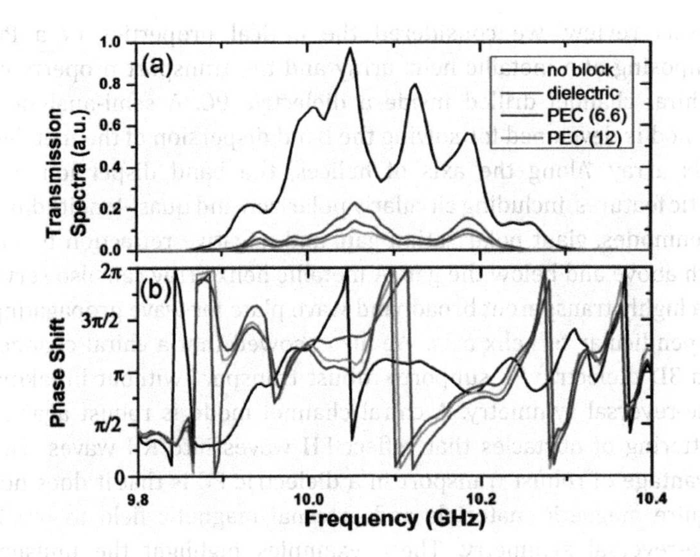

Figure 5.16 (a) Measured total transmittance and (b) measured phase shift for the EM transport in the NC channel. Black solid line: no obstacle in channel. Blue/green solid line: small/large PEC plate in channel. Red solid line: dielectric plate in channel.

(black), showing that robust transport behaviors do exist within the polarization gap.

For the purpose of comparison, we also measured the transmission spectra in the blocked/unblocked NC channel, as shown in Fig. 5.16. Low transmissions were observed for either the dielectric or PEC blocks, consistent with our FDTD simulations illustrated in Fig. 1.14(c) and (d). In the simulation, the PEC obstacle is the same as that in Fig. 5.14(b) and the working frequency is 10.2 GHz. The FDTD results in Fig. 5.14(c) show that linear polarized EM wave propagates from bottom to top in the channel and the flux is blocked once the PEC slab blocks the channel, as shown in Fig. 5.14(d). Figure 5.16(b) shows that the relative phases between S_x and S_y have much larger fluctuations, and thus the polarizations of the output wave do change when the obstacle is present.

5.4 Summary

In this review, we considered the optical properties of a PC composing of a metallic helix array and the transport property of a chiral channel drilled inside a dielectric PC. A semi-analytical method is developed for solving the band dispersion of the metallic helix array. Along the axis of helices, the band dispersion has exotic features, including circularly polarized and quasi-longitudinal eigenmodes, giant polarization gap, and negative refraction bands both above and below the gap. A metallic helix array can also serve as a highly transparent broadband wave plate for wave propagating perpendicular to helix axis. We also showed that a chiral channel in a 3D dielectric PC supports robust transport without breaking time-reversal symmetry. A chiral channel mode is robust against scattering of obstacles that reflect LH waves into RH waves. The advantage of robust transport in a dielectric PC is that it does not require magnetic materials and external magnetic field to break time-reversal symmetry. These examples highlight the unusual properties of helical and chiral photonic systems and some of these properties may find useful applications.

Acknowledgment

This work was supported by Hong Kong CRF grant HKUST2/CRF/11G.

HQL is supported by NSFC (No. 11174221, 10974144), the Fundamental Research Funds for the Central Universities, and SHEDF (No. 06SG24). JWD is supported by the NSFC (11274396, 11074311, 61235002), the Guangdong Natural Science Foundation (S2012010010537), and the Fundamental Research Funds for the Central Universities (2012300003162498).

References

1. Ao, X., Lin, Z. and Chan, C. T. (2009) One-way edge mode in a magneto-optical honeycomb photonic crystal, *Physical Review B,* **80**, p. 033105.
2. Belov, P. A., Simovski, C. R. and Tretyakov, S. A. (2003) Example of bianisotropic electromagnetic crystals: The spiral medium, *Physical Review E,* **67**, p. 056622.
3. Born, M. and Wolf, E. (1999) *Principles of Optics: Electromagnetic Theory of Propagation, Interference and Diffraction of Light,* 7th Ed. (Cambridge University Press, Cambridge, UK).
4. Chen, Y. L., Analytis, J. G., Chu, J.-H., Liu, Z. K., Mo, S.-K., Qi, X. L., Zhang, H. J., Lu, D. H., Dai, X., Fang, Z., Zhang, S. C., Fisher, I. R., Hussain, Z. and Shen, Z.-X. (2009) Experimental Realization of a three-dimensional topological insulator, Bi2Te3, *Science,* **325**, pp. 178–181.
5. Cheng, Q. and Cui, T. J. (2006) Negative refractions in uniaxially anisotropic chiral media, *Physical Review B,* **73**, p. 113104.
6. Chin, S. K., Nicorovici, N. A. and McPhedran, R. C. (1994) Green's function and lattice sums for electromagnetic scattering by a square array of cylinders, *Physical Review E,* **49**, p. 4590.
7. Engheta, N., Jaggard, D. L. and Kowarz, M. W. (1992) Electromagnetic waves in Faraday chiral media, *Antennas and Propagation, IEEE Transactions on,* **40**, pp. 367–374.
8. Fedotov, V. A., Mladyonov, P. L., Prosvirnin, S. L., Rogacheva, A. V., Chen, Y. and Zheludev, N. I. (2006) Asymmetric propagation of electromagnetic waves through a planar chiral structure, *Physical Review Letters,* **97**, pp. 167401–167404.

9. Fu, J.-X., Liu, R.-J. and Li, Z.-Y. (2010) Robust one-way modes in gyromagnetic photonic crystal waveguides with different interfaces, *Applied Physics Letters,* **97**, pp. 041112–041113.
10. Fu, L., Kane, C. L. and Mele, E. J. (2007) Topological insulators in three dimensions, *Physical Review Letters,* **98**, p. 106803.
11. Gansel, J. K., Wegener, M., Burger, S. and Linden, S. (2010) Gold helix photonic metamaterials: A numerical parameter study, *Optics Express,* **18**, pp. 1059–1069.
12. Gansel, J. K., Thiel, M., Rill, M. S., Decker, M., Bade, K., Saile, V., von Freymann, G., Linden, S. and Wegener, M. (2009) Gold helix photonic metamaterial as broadband circular polarizer *Science,* **325**, pp. 1513–1515.
13. Haldane, F. D. M. and Raghu, S. (2008) Possible realization of directional optical waveguides in photonic crystals with broken time-reversal symmetry, *Physical Review Letters,* **100**, p. 013904.
14. Hasan, M. Z. and Kane, C. L. (2010) Colloquium: Topological insulators, *Reviews of Modern Physics,* **82**, pp. 3045–3067.
15. Ho, K. M., Chan, C. T., Soukoulis, C. M., Biswas, R. and Sigalas, M. (1994) Photonic band gaps in three dimensions: New layer-by-layer periodic structures, *Solid State Communications,* **89**, pp. 413–416.
16. Hsieh, D., Qian, D., Wray, L., Xia, Y., Hor, Y. S., Cava, R. J. and Hasan, M. Z. (2008) A topological Dirac insulator in a quantum spin Hall phase, *Nature,* **452**, pp. 970–974.
17. Johnson, S. and Joannopoulos, J. (2001) Block-iterative frequency-domain methods for Maxwell's equations in a planewave basis, *Opt. Express,* **8**, pp. 173–190.
18. Kane, C. L. and Mele, E. J. (2005) Quantum spin Hall effect in graphene, *Physical Review Letters,* **95** p. 226801.
19. Klitzing, K. v., Dorda, G. and Pepper, M. (1980) New method for high-accuracy determination of the fine-structure constant based on quantized hall resistance, *Physical Review Letters,* **45**, pp. 494–497.
20. Krasavin, A. V., Schwanecke, A. S., Zheludev, N. I., Reichelt, M., Stroucken, T., Koch, S. W. and Wright, E. M. (2005) Polarization conversion and "focusing" of light propagating through a small chiral hole in a metallic screen, *Applied Physics Letters,* **86**, pp. 201105–201103.
21. Kraus, J. D. (1947) Helical beam antennas, *Electronics,* **20**, pp. 109–111.
22. Kraus, J. D. (1950) *Antennas* (McGraw-Hill Book Co Inc, New York).
23. Li, T. Q., Liu, H., Li, T., Wang, S. M., Wang, F. M., Wu, R. X., Chen, P., Zhu, S. N. and Zhang, X. (2008) Magnetic resonance hybridization and optical

activity of microwaves in a chiral metamaterial, *Applied Physics Letters,* **92**, pp. 131111–131113.

24. Lindell, I. V., Shivola, A. H., Tretyakov, S. A. and Viitanen, A. J. (1994) *Electromagnetic Waves in Chiral and BI-Isotropic Media* (Artech House, Norwood, MA).
25. Lindeman, K. F. (1920) Über eine durch ein isotropes System von spiralförmigen Resonatoren erzeugte Rotationspolarization der elektromagnetischen Wellen, *Annalen der Physik,* **63**, pp. 621–644.
26. Liu, H., Genov, D. A., Wu, D. M., Liu, Y. M., Liu, Z. W., Sun, C., Zhu, S. N. and Zhang, X. (2007) Magnetic plasmon hybridization and optical activity at optical frequencies in metallic nanostructures, *Physical Review B,* **76**, p. 073101.
27. Mackay, T. G. and Lakhtakia, A. (2004) Plane waves with negative phase velocity in Faraday chiral mediums, *Physical Review E,* **69**, p. 026602.
28. Monzon, C. and Forester, D. W. (2005) Negative refraction and focusing of circularly polarized waves in optically active media, *Physical Review Letters,* **95**, p. 123904.
29. Nicorovici, N. A., McPhedran, R. C. and Botten, L. C. (1995) Photonic band gaps for arrays of perfectly conducting cylinders, *Physical Review E,* **52**, p. 1135.
30. Oskooi, A. F., Roundy, D., Ibanescu, M., Bermel, P., Joannopoulos, J. D. and Johnson, S. G. (2010) Meep: A flexible free-software package for electromagnetic simulations by the FDTD method, *Computer Physics Communications,* **181**, pp. 687–702.
31. Özbay, E., Abeyta, A., Tuttle, G., Tringides, M., Biswas, R., Chan, C. T., Soukoulis, C. M. and Ho, K. M. (1994) Measurement of a three-dimensional photonic band gap in a crystal structure made of dielectric rods, *Physical Review B,* **50**, pp. 1945–1948.
32. Pendry, J. B. (2004) A chiral route to negative refraction, *Science,* **306**, pp. 1353–1355.
33. Pierce, J. R. (1947) Theory of the beam-type traveling-wave tube, *Proceedings of the IRE,* **35**, pp. 111–123.
34. Pierce, J. R. and Field, L. M. (1947) Traveling-wave tubes, *Proceedings of the IRE,* **35**, pp. 108–111.
35. Pierce, J. R. and Tien, P. K. (1954) Coupling of modes in helixes, *Proceedings of the IRE,* **42**, pp. 1389–1396.
36. Plum, E., Fedotov, V. A., Schwanecke, A. S., Zheludev, N. I. and Chen, Y. (2007) Giant optical gyrotropy due to electromagnetic coupling, *Applied Physics Letters,* **90**, pp. 223113–223113.

37. Plum, E., Liu, X. X., Fedotov, V. A., Chen, Y., Tsai, D. P. and Zheludev, N. I. (2009) Metamaterials: Optical activity without chirality, *Physical Review Letters,* **102**, pp. 113902–113904.
38. Plum, E., Zhou, J., Dong, J., Fedotov, V. A., Koschny, T., Soukoulis, C. M. and Zheludev, N. I. (2009) Metamaterial with negative index due to chirality, *Physical Review B (Condensed Matter and Materials Physics),* **79**, pp. 035407–035406.
39. Poo, Y., Wu, R.-x., Lin, Z., Yang, Y. and Chan, C. T. (2011) Experimental realization of self-guiding unidirectional electromagnetic edge states, *Physical Review Letters,* **106**, p. 093903.
40. Prange, R. E. and Girvin, S. M. (1987) *The Quantum Hall Effect* (Springer, New York).
41. Prosvirnin, S. L. and Zheludev, N. I. (2005) Polarization effects in the diffraction of light by a planar chiral structure, *Physical Review E,* **71**, p. 037603.
42. Sensiper, S. (1955) Electromagnetic Wave propagation on helical structures (A review and survey of recent progress), *Proceedings of the IRE,* **43**, pp. 149–161.
43. Tinoco, I. and Freeman, M. P. (1957) The optical activity of oriented copper helices. I. Experimental, *The Journal of Physical Chemistry,* **61**, pp. 1196–1200.
44. Tretyakov, S., Nefedov, I., Sihvola, A., Maslovski, S. and Simovski, C. (2003) Waves and energy in chiral nihility, *Journal of Electromagnetic Waves and Applications,* **17**, pp. 695–706.
45. Wang, Z., Chong, Y. D., Joannopoulos, J. D. and Soljačić, M. (2008) Reflection-free one-way edge modes in a gyromagnetic photonic crystal, *Physical Review Letters,* **100**, p. 013905.
46. Wang, Z., Chong, Y., Joannopoulos, J. D. and Soljacic, M. (2009) Observation of unidirectional backscattering-immune topological electromagnetic states, *Nature,* **461**, pp. 772–775.
47. Wiltshire, M. C. K. and et al. (2009) Chiral Swiss rolls show a negative refractive index, *Journal of Physics: Condensed Matter,* **21**, p. 292201.
48. Wu, C., Li, H. and Chan, C. T. A multiple-scattering approach for metallic helix array, *unpublished.*
49. Wu, C., Li, H., Yu, X., Li, F. and Chan, C. T. Asymmetric signatures of axial propagation through metallic helix array, *unpublished.*
50. Wu, C., Li, H., Wei, Z., Yu, X. and Chan, C. T. (2010) Theory and experimental realization of negative refraction in a metallic helix array, *Physical Review Letters,* **105**, p. 247401.

51. Wu, C., Li, H., Yu, X., Li, F., Hong, C. and Chan, C. T. (2011) Metallic helix array as a broadband wave plate, *Physical Review Letters,* **107**, p. 177401
52. Zhang, S., Park, Y.-S., Li, J., Lu, X., Zhang, W. and Zhang, X. (2009) Negative refractive index in chiral metamaterials, *Physical Review Letters,* **102**, p. 023901.
53. Zhang, Y., Tan, Y.-W., Stormer, H. L. and Kim, P. (2005) Experimental observation of the quantum Hall effect and Berry's phase in graphene, *Nature,* **438**, pp. 201–204.
54. Zhou, J., Dong, J., Wang, B., Koschny, T., Kafesaki, M. and Soukoulis, C. M. (2009) Negative refractive index due to chirality, *Physical Review B,* **79**, p. 121104.

Chapter 6

Chiral Surface Plasmon Polaritons on One-Dimension Nanowires

Shunping Zhang[a] and Hongxing Xu[a,b]

[a]*Center for Nanoscience and Nanotechnology, and School of Physics and Technology, Wuhan University, Wuhan 430072, China*
[b]*Beijing National Laboratory for Condensed Matter Physics and Institute of Physics, Chinese Academy of Sciences, Beijing, Box 603-146, 100190, China*
hxxu@iphy.ac.cn

6.1 Introduction

An object is chiral when it is not identical to its mirror image. For electromagnetic waves, circularly polarized plane waves are chiral in the sense that each "enantiomer" carries an opposite spin angular momentum. The interactions of chiral objects with left- or right-handed chiral electromagnetic waves are different, giving rise to the circular dichroism (CD) that can serve as a measurement of the chirality of the object. For most natural molecules or biomolecules, CD is typically about 10^{-2} (electronic) or $10^{-4} \sim 10^{-6}$ (vibrational) compared with their parent signals [1]. Although the CD signals of most natural molecules occur in the ultraviolet (UV) range, rationally designed chiral "meta-atoms" and "meta-molecules" have

Singular and Chiral Nanoplasmonics
Edited by Svetlana V. Boriskina and Nikolay I. Zheludev

ISBN 978-981-4613-17-0 (Hardcover), 978-981-4613-18-7 (eBook)
www.panstanford.com

shown greatly enhanced CD response with a large tunable spectral window. Depending on their feature sizes, shape and compositions, and so on, those artificial molecules can work in a frequency domain ranging from microwave to optical frequency. The enhanced CD response and the large tunability are usually originated from the metal components, which carry a large amount of free electrons that react strongly to external electromagnetic radiation. These collective excitations of the free electrons on metal surfaces, known as surface plasmons (SPs), have been a subject of intense academic interest in the past decades [2, 3]. By exciting localized SPs, electromagnetic field can be confined into a subwavelength volume around the metal structures. The use of SPs enables the scaling down of the artificial chiral objects into the nanometer scale.

Two categories of chiral structures are intensively investigated. One is two-dimensional (2D) planar chiral structures that usually sit on a substrate, and the other is three-dimensional (3D) chiral structures fabricated by either well-controlled top-down approach [4] or bottom-up assembling of small metallic nanoparticles (NPs) [5–7]. A 2D object, for example a spiral, is chiral in the sense that it cannot be superimposed on its mirror image by in-plane rotation or translations [8]. Asymmetric light propagation and optical rotations can be observed in analogues to the Faraday effect [9, 10]. On the contrary, the chirality of a 3D chiral object, such as a helix, is apparent. The propagation of right-handed and left-handed circularly polarized light through such helices is significantly different. An array of such helices can function as an ideal polarization-dependent filter [11]. For achiral structures, the asymmetric transmission of light with different handedness can also be externally accomplished by taking into account the incident direction of light [12]. But such kind of CD is not an intrinsic property of the structures and will vanish when the incident excitation is averaged over the whole space.

Besides the structural chirality, chiral nanoplasmonics also concerns the chirality of the local field associated with SPs, with emphases on the local optical chirality and its influence on a nearby chiral molecule [13, 14]. For a chiral structure, such as a spiral, the near-fields of its SP resonances are chiral, which have been confirmed by both experimental observations [15] and electromagnetic simulations [16]. For achiral nanostructures, the local field chirality can be transferred from the incident circularly polarized light. For example, Biagioni et al. proposed to use a cross Au dipole antenna to generate a tiny hot spot with highly circular polarization [17]. Very recently, however, theoretical calculations predict that even for a single isotropic metal sphere excited by linearly polarized light, the local field can be chiral [18]. The local optical chirality enhancement accelerates the excitation rate of a chiral molecule, opening up a new degree for surface enhanced light-matter interactions.

In this chapter, we will discuss the generation and the properties of chiral propagating SPs on an achiral one-dimensional (1D) metallic nanowire [19]. Propagating SPs are usually termed surface plasmon polaritons (SPPs) owing to their coupled nature of the collective motions of electrons and the associated electromagnetic wave. Different from localized SPs in non-extended structures, chiral SPPs are surface waves that can propagate over a significant distance, enabling their potential application as photonic information carriers beyond the diffraction limitation. This chapter is organized as follows. In Section 2, we briefly introduce the SPP modes on a cylindrical metal nanowire. The effective refractive indices and the attenuation constants are compared. In Section 3, the generation of chiral SPPs under local optical excitation is discussed. In Section 4, the influences of the external excitation and the structural parameters on chiral SPPs are summarized. These include the in-coupling efficiency of each nanowire SPP and the period of plasmon helices. Finally, we discuss the applications of chiral SPPs in Section 5 and summarize the chapter in Section 6.

6.2 Surface Plasmon Polaritons on Metallic Nanowires

The SPP modes on a cylindrical metal wire can be described by analytical solutions that can be found in many textbooks [20, 21]. Briefly, in cylindrical coordinates (ρ, ϕ, x), the radial and axial dependence of the SPP modes can be described by the Bessel functions and an exponential function, respectively. For region inside the cylinder, Bessel functions of the first kind behave well at the origin while the Hankel functions satisfy the outgoing wave requirement outside the cylinder. In the azimuthal direction, the exponential functions $e^{\pm im\phi}$ ($m = 0, 1, 2, \ldots$) fulfill the periodic condition. Equivalently, the linear combination of them, that is, the trigonometric functions, $cos(m\phi)$ and $sin(m\phi)$, can be used as eigen functions as well. These trigonometric functions present clear pictures of specific collective oscillations that propagate along the nanowire surface. Therefore, it is convenient to adapt the trigonometric functions when discussing plasmonic waveguides. The wave vectors of the SPPs characterize several important features of the modes, including the propagation constant and the penetration depth away from the metal-dielectric interface. The propagation constant $k_{m,\parallel}$ is complex, with its real and imaginary part defining the phase constant and attenuation constant, respectively. By solving the transcendental equation involving Bessel functions and their derivatives, the complex propagation constants of four low-order SPPs on a Ag nanowire can be obtained, as shown in Fig. 6.1. The permittivity of Ag at the vacuum wavelength $\lambda_0 = 632.8$ nm, $\varepsilon^{\mathrm{M}} = -18.36 + 0.4786i$, is interpolated from the experimental data by Johnson and Christy [22]. The surrounding matrix is assumed to be oil with $\varepsilon^{\mathrm{D}} = 2.25$. As having been discussed, for example, by D. E. Chang et al. [23], the $m = 0$ mode exists in a nanowire of arbitrary size. On the contrary, the high-order modes ($m \geq 2$) have distinct cutoff sizes below which the modes cannot sustain. The $m = 1$ mode, however, is not strictly cutoff but suffers from a rapid expansion of its mode area at the thin wire limit. As shown in Fig. 6.1(a), when the radius of the nanowire is smaller than 50 nm, the real part of its propagation constant approaches to that of the light in the surrounding medium. This means that on a very thin wire, the $m = 1$ mode becomes similar to a plane wave propagating parallel to the wire axis. Therefore, this mode becomes difficult to excite under normal (to the wire axis) incidence. This property is important in the formation of chiral SPPs as will be discussed below. As the radius of the nanowire increases, the real part of the propagation constant of the $m = 0$ mode decreases, whereas those for the other modes increase. Notice that for very thick wire, the cylinder surface is similar to a flat interface so that all the modes will eventually approach the SPP on a single metal-oil interface.

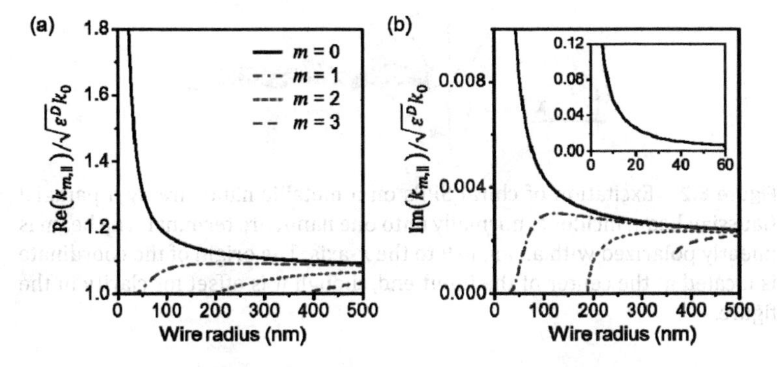

Figure 6.1 The real (a) and imaginary (b) part of the propagation constant of the m-th order SPP on a Ag nanowire as a function of the wire radius. Inset to (b) shows an enlarged region for the $m = 0$ mode for small wire radius. The permittivity of the surrounding medium is $\varepsilon^D = 2.25$. $k_0 = 2\pi/\lambda_0$ and $\lambda_0 = 632.8$ nm.

Figure 6.1(b) shows the attenuation constants of different SPP modes. As the radius of the nanowire decreases, the attenuation constant of the $m = 0$ mode increases, meaning that the mode is getting lossy. Therefore, this mode can be regarded as a short-range SPP for thin wires. On the contrary, the $m = 1$ mode has much smaller loss in the thin wire limit due to the extension of its mode profile into the non-absorbing dielectric region. Accordingly, this mode can be regarded as a long-ranged SPP. The higher order modes ($m = 2$ and 3) have even smaller attenuation constants than the $m = 1$ mode, and also show reduced losses as the radius of the metal wire decreases.

6.3 Generation of Chiral Surface Plasmon Polaritons

In analogy with circularly polarized light, the superposition of two orthogonal degenerate modes with a $\pi/2$ phase shift can result in a circularly polarized state, which is chiral. Further interference with the coexcited fundamental mode ($m = 0$) generates a time-averaged helically distributed near-field energy flow along the metal surface (see Fig. 6.2), which is chiral as well. If not specified, chiral SPPs refer to the helically distributed SPPs in the following discussions.

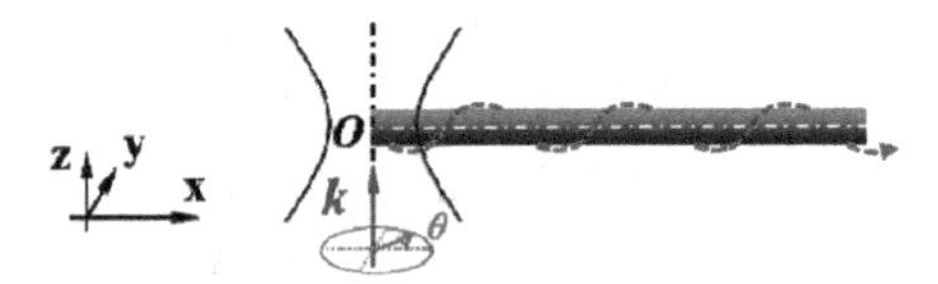

Figure 6.2 Excitation of chiral SPPs on a metallic nanowire by a paraxial Gaussian beam incidents normally onto one nanowire terminal. The beam is linearly polarized with an angle θ to the x-axis. The origin of the coordinate is located at the center of the input end, though it is offset for clarity in the figure.

It is suggested that the high-order SPPs on cylindrical wire can be viewed as being propagating spirally over the wire surface [24]. Actually, this interpretation comes from the decomposition of the wave vector into an axial and an azimuthal component and it is in principle different from the chiral SPPs discussed here.

Under excitation at the nanowire end, those modes with profiles overlapping with the incident light can be excited simultaneously. As shown in Fig. 6.3, for incident polarization parallel to the nanowire axis, the $m = 0$ and $m = 1$ (z-direction) modes can be excited. For perpendicular polarization, the $m = 1$ (y-direction) and $m = 2$ modes are excited. If the radius of the nanowire is small enough so that higher order modes ($m \geq 2$) are cutoff (see Fig. 6.1), we have the $m = 0$ and two degenerate $m = 1$ modes excited on the nanowire for oblique polarization. Comparing Fig. 6.3(b) and

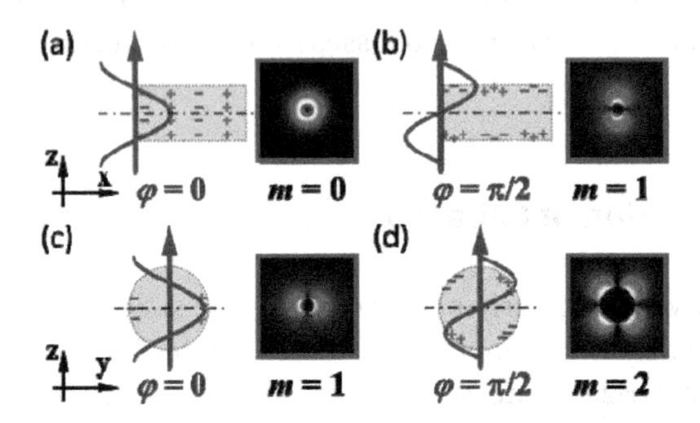

Figure 6.3 Optical excitation of different SPP modes for incident polarization parallel (a, b) or perpendicular (c, d) to the wire axis. ϕ denotes the incident phase of the excitation beam. Reproduced from Ref. [19], with permission of the American Physical Society

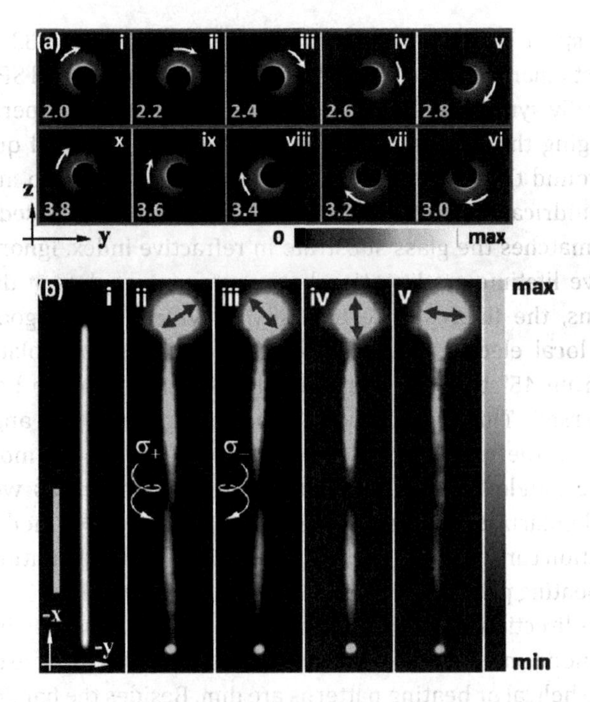

Figure 6.4 (a) Electromagnetic simulation showing the helical distribution of energy flow along a Ag nanowire ($R = 60$ nm, $\varepsilon^{\mathrm{D}} = 2.25$). The number in each panel, 2.0 ~ 3.8 μm, indicates the distance between the plane of the plot and the input-end of the nanowire. (b) Experimental verifications of chiral SPPs by fluorescence imaging. Scale bar, 5 μm. Reproduced from Ref. [19], with permission of the American Physical Society.

(c) indicates that the excitation of the $m = 1$ mode in the z- and y-directions corresponds to an incident instantaneous profile with a phase delay of $\pi/2$. Therefore, the condition for the excitation of circularly polarized state is naturally satisfied. The coherent interference between the coexistence axially polarized $m = 0$ mode and the circularly polarized state results in a helical pattern.

Figure 6.4 shows the calculated and measured local field distribution of the chiral SPP around an Ag nanowire. As shown in Fig. 6.4(a), finite element method (FEM) simulation of a 5 μm long Ag nanowire (radius $R = 60$ nm) embedded in oil reveals the helical distribution of the energy flow along the nanowire. The incident paraxial Gaussian beam is polarized at an angle $\theta = 45°$

with respect to the wire axis. The period of helix is 1.83 μm for this particular case. Experimental verification of the chiral SPPs on a chemically synthesized Ag nanowire ($R \sim 150$ nm) was performed by imaging the fluorescence from uniformly spin-coated quantum dots around the wire, as shown in Fig. 6.4(b). In order to maintain the cylindrical symmetry, the Ag nanowire was immersed in oil, which matches the glass substrate in refractive index. Ignoring the radiative lifetime modification between quantum dots at different locations, the fluorescence intensity provides a fairly good map of the local electric field intensity. By changing the polarization angle from 45° to −45°, the handedness of the plasmon helix can be reversed. This is because the incident polarization angle can determine the relative phase between the two $m = 1$ modes, so that the handedness of the chiral SPP is controlled as well. For parallel polarization shown in Fig. 6.4(b-iv), the $m = 1$ mode in the y-direction can be hardly excited. Then, the helical distribution turns into a beating pattern between the $m = 0$ mode and the $m = 1$ mode in the z-direction. For perpendicular polarization, however, both the $m = 0$ mode and the $m = 1$ mode in the z-direction are not excited so that the helical or beating patterns are dim. Besides the handedness, the incident polarization angle can also alter the relative amplitude of the $m = 1$ mode in the y- or z-direction, which further determines the degree of circular polarization of the chiral SPP.

6.4 Influence of External or Structual Parameters

As the chiral SPPs depend on the relative amplitude of each excited mode, the effect of the external excitation can have a large impact on the properties of the chiral SPPs. Besides the polarization angle mentioned above, the incident direction of the Gaussian beam can affect the excitation efficiency of the different SPP modes as well [25]. The emission of SPP into light at nanowire terminals has been shown to be maximized at a certain angle to the wire axis [26]. Although it has not been pointed out in the original study, this emission characteristic is actually related to the $m = 0$ mode. As it can be anticipated from the optical reciprocity, the coupling of an incident light beam into the $m = 0$ mode will reach its maximum when the direction of the incident beam is oriented by the same angle to the wire axis. However, the end emission of the $m = 1$ mode is mainly in the direction along the wire axis [27] so that end-fire excitation can be a more optimal configuration for this mode. Therefore, if the incident angle is varied, the efficiency of in-coupling of different SPP modes changes so that the ellipticity of the chiral SPP is modified.

In the excitation configuration depicted in Fig. 6.2, the retardation effect is a critical issue to excite the $m = 1$ mode in the z-direction. Thus, the ratio between the incident wavelength and the nanowire radius is critical in the formation of chiral SPPs. Figure 6.5 shows the calculated transmission spectra of a Ag nanowire for inclined ($\theta = 45°$) and perpendicular ($\theta = 90°$) incident polarization. For perpendicular excitation, the transmission spectrum shows Fabry–Pérot resonances due to the interference of the $m = 1$ mode (y-direction) in short wavelength range. For longer wavelengths, the excitation efficiency is reduced due to the smaller overlap between the mode profile and the incident light. For inclined polarization, the excitation of the $m = 0$ mode in the long wavelength region dominates the transmission spectrum. The

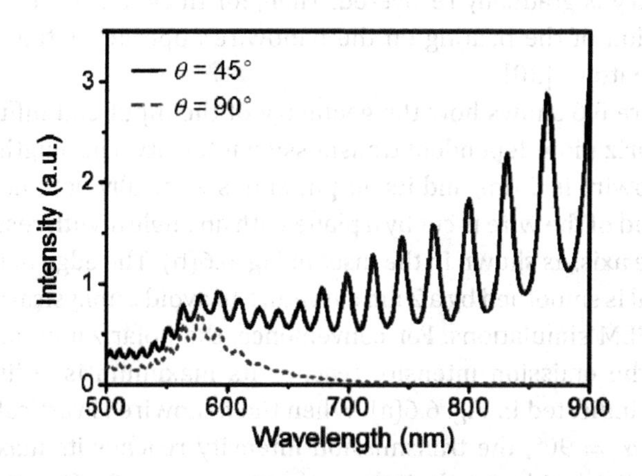

Figure 6.5 FEM calculated transmission spectra of an Ag nanowire normally excited by a paraxial Gaussian beam with inclined ($\theta = 45°$) or perpendicular ($\theta = 90°$) polarization. The length of the nanowire is 5.0 μm, and $R = 60$ nm, $\varepsilon^{D} = 2.25$. The data for $\theta = 45°$ are taken from Ref. [19], with permission of the American Physical Society.

increasing transmission intensity in the long wavelength region is mainly due to the higher excitation efficiency. The clearer Fabry–Pérot resonances indicate a smaller cavity loss for the SPP. In the short wavelength region, however, the emission of the $m = 1$ mode (excited by the perpendicular component) dominates the transmission spectrum so that the Fabry–Pérot resonances are correlated for $\theta = 45^\circ$ and 90°.

Unlike the external excitation parameters, the structural parameters can determine the intrinsic properties of the nanowire SPPs, a general property of plasmonic structures [3, 28]. For example, the shape of the nanowire cross-section, the diameter, and the refractive index of the surrounding medium can modify the properties of each wire SPP and the chiral SPPs. The presence of a dielectric substrate can mediate the coupling of different plasmon modes, giving birth to a new set of hybridized modes [29]. When illuminated by a laser beam at one terminal, the interference of several coexcited hybridized modes results in a beating pattern rather than the helical pattern mentioned above. Further adding of a dielectric coating layer onto the substrate-supported nanowire can reduce the asymmetry of the dielectric environment so that the cylindrical symmetry is gradually recovered. Then, for thick dielectric coating, the period of the beating on the nanowire approaches that of the helical pattern [30].

Figure 6.6 shows how the geometry of the input end influences the polarization-dependent transmission intensity. The length of the Ag nanowire is 5 μm, and its output end is vertically flattened. The input end of the wire is cut by a plane with an angle α with respect to the wire axis, as shown in the inset in Fig. 6.6(b). The edge of the cut terminal is smoothed by a 2 nm curvature to avoid unphysical results in the FEM simulations. For convenience, the polarization angle at which the emission intensity reaches its maximum is defined as $\theta_{\max}$, as indicated in Fig. 6.6(a). When the nanowire is vertically cut, that is, $\alpha = 90^\circ$, the transmission intensity reaches its maximum for parallel incident polarization, that is, $\theta_{\max} = 0^\circ$. However, for $\alpha = 45^\circ$, the transmission intensity is maximized when $\theta_{\max} = 69^\circ$. Figure 6.6(b) summarizes the dependence of $\theta_{\max}$ and the maximal transmission intensity on the cutting angle. The $\theta_{\max}$ goes up monotonously from 0° to 69° as α is decreased from 90° to 45°.

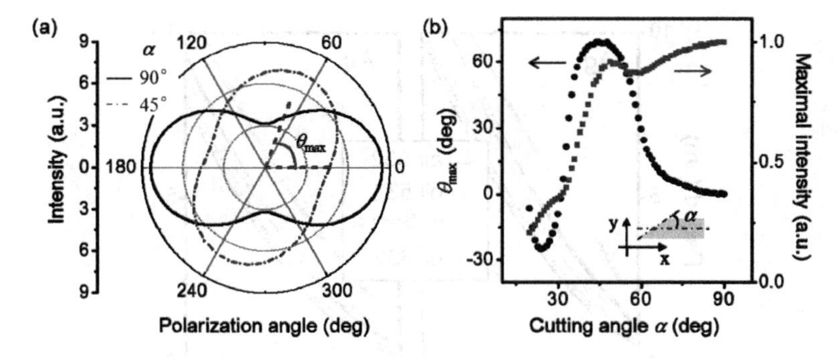

Figure 6.6 Effect of the nanowire input end geometry. (a) Calculated emission intensity at the distal end as a function of the incident polarization angle θ, for cutting angle $\alpha = 90^\circ$ (blue) and 45° (red). The polarization angle at which the emission intensity reaches maximum is defined as θ_{max}. (b) The dependence of θ_{max} (black dot) and the maximal emission power (red square) on the cutting angle α. Inset: The geometry of the cut input end with the sharp tip edge smoothed by a 5 nm curvature. The radius of the nanowire is $R = 60$ nm and the incident wavelength is $\lambda_0 = 632.8$ nm.

Then, θ_{max} decreases to minus values and goes up again when α is further decreased. As the tip becomes sharper, especially when α is smaller than $\sim 45^\circ$, the coupling into the nanowire becomes weaker so that the maximal transmission intensity is reduced. Detailed correlation of the in-coupling and out-coupling polarization state can be found in Ref. [31], in which the influence of the geometry of the tip is emphasized.

The period of the helical pattern is determined by $\Lambda = 2\pi\left[\mathrm{Re}\left(k_{0,\parallel} - k_{1,\parallel}\right)\right]^{-1}$, where $k_{0,\parallel}$ and $k_{1,\parallel}$ are the propagation constants of the $m = 0$ and $m = 1$ mode. According to Fig. 6.1(a), the real part of the propagation constant of the $m = 0$ mode decreases as the radius of nanowire increases, whereas the propagation constant of the $m = 1$ mode increases. Therefore, the pitch of the helical pattern is getting larger for thick wire. This behavior applies for nanowires of different metals, surrounding media with different refractive indexes and different wavelengths, as shown in Fig. 6.7. For a nanowire in the same matrix, the helix period is larger for 632.8 nm vacuum wavelength than that for 532 nm. At a given wavelength, the helix period is larger in air than that

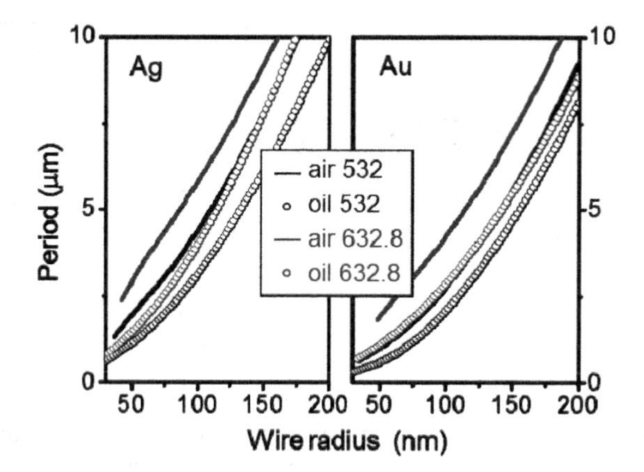

Figure 6.7 Period of the plasmon helix on a Ag or Au nanowire, embedded in air (lines) or oil (open circles). The calculations correspond to a vacuum wavelength $\lambda_0 = 532$ nm (black) or 632.8 nm (red).

in oil. Comparing Ag and Au wires, the helix period is longer on Ag wires under the same circumstance.

So far, we have discussed the chiral SPPs involving low-order SPP modes while all the higher order ($m \geq 2$) modes are neglected. Generally, the superposition of different SPP modes of arbitrary order can generate beating or more complex patterns. If only two modes are involved, the period of the resulting beating pattern is given by $\Lambda_{mm'} = 2\pi \left(\mathrm{Re}\left(k_{m,\parallel} - k_{m',\parallel}\right)\right)^{-1}$, where $m < m'$. Figure 6.8 shows the beating period as a function of the nanowire radius at a vacuum wavelength of 632.8 nm. Apparently, the periods of the plasmon beating grow monotonously as the radius of the nanowire increases. Further, due to the big difference between the propagation constants of the $m = 0$ and $m = 3$ modes, Λ_{03} has the smallest value while Λ_{01} is the largest for a given nanowire radius.

6.5 Applications of Chiral Surface Plasmon Polaritons

As originally proposed in Ref. [19], chiral SPPs on metallic nanowires can be used to generate localized chiral near field for enantiomeric

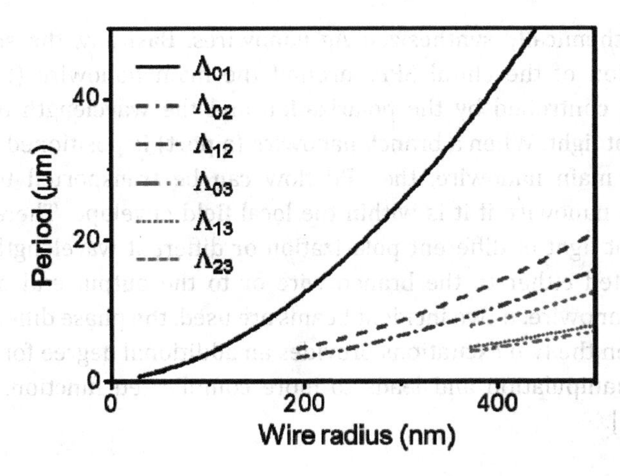

Figure 6.8 Periods of plasmon beat constituted by different SPP modes ($m = 0$, 1, 2, and 3) as a function of the Ag wire radius. The permittivity of the surrounding medium is $\varepsilon^{\mathrm{D}} = 2.25$.

sensing. The degree of circular polarization of the emitted photons from the output end can be tuned by the polarization angle of the incident light. Then, the excitation rate of a chiral object, such as a chiral molecule, can be controlled by adjusting the local optical chirality. For this application, the $m = 0$ mode usually serves as a background excitation and should be eliminated. For example, using shorter excitation wavelength or changing the incident direction may facilitate the excitation of the $m = 1$ mode while suppress the $m = 0$ mode (see the discussions following Fig. 6.5). The emission of SPPs at the output end of the nanowire can be modulated by the helical pattern, thanks to the coherent interference nature of chiral SPPs. A rainbow effect, that is, an angle-dependent emission, is expected. This requires further experimental confirms. With the above-mentioned properties, a metallic nanowire can serve as a subwavelength light source that connects many fancy physical phenomena.

Another attractive application of the chiral SPPs is to use their helically distributed local field for subwavelength plasmonic circuits. Successful examples of positive plasmonic devices, such as splitters, demultiplexers, and modulators [32, 33], have been demonstrated using chemically synthesized Ag nanowires. Basically, the spatial extension of the chiral SPPs around the main nanowire (trunk) can be controlled by the polarization and the wavelength of the incident light. When a branch nanowire (a port) is positioned close to the main nanowire, the SPP flow can be transported to the branch nanowire if it is within the local field envelope. Therefore, incident light of different polarization or different wavelength can be routed either to the branch wire or to the output end of the main nanowire. If two incident beams are used, the phase difference between the two excitations provides an additional degree for local field manipulation and leads to more complicated functionalities [34, 35].

6.6 Summary and Perspective

We have introduced a kind of chiral collective electron motions associated with electromagnetic wave sustained on the metal-dielectric interface of 1D nanowire. The excitation of chiral SPPs by external light beam, and their handedness, period, and the influence of the structural parameters are discussed. Possible applications of chiral SPPs are expected, including enhanced light-chiral objects interactions and functionalized plasmonic devices.

Another aspect of chiral SPPs is their angular momentum. An optical beam is known to carry orbital angular momentum (OAM) when it has a spiral phase front [36]. Exploiting the OAM can largely improve the information transmission capacity of the devices [37]. It is straightforward to think of using chiral SPPs to transfer similar techniques into plasmonic waveguides. Benefiting from the subwavelength feature of the SPPs, such OAM-encoded information may be transferred in an integrated on-chip circuit. On a helically grooved metal nanowire, chiral SPPs are shown to carry fractal OAM depending on the mode wave vector of the SPP [38]. In addition, recent calculations show that there are spin flows associated with the SPP at the planar metal-dielectric interface [39]. However, a systematic description of the spin and orbital angular momentum properties of SPPs and the related phenomena is still urgently demanded.

References

1. Berova, N., Di Bari, L., and Pescitelli, G. (2007). Application of electronic circular dichroism in configurational and conformational analysis of organic compounds, *Chem. Soc. Rev.*, **36**, p. 914.
2. Barnes, W. L., Dereux, A., and Ebbesen, T. W. (2003). Surface plasmon subwavelength optics, *Nature (London)*, **424**, p. 824.
3. Halas, N. J., Lal, S., Chang, W. S., Link, S., and Nordlander, P. (2011). Plasmons in strongly coupled metallic nanostructures, *Chem. Rev.*, **111**, p. 3913.
4. Hentschel, M., Schaeferling, M., Weiss, T., Liu, N., and Giessen, H. (2012). Three-dimensional chiral plasmonic oligomers, *Nano Lett.*, **12**, p. 2542.
5. Guerrero-Martinez, A., Lorenzo Alonso-Gomez, J., Auguie, B., Magdalena Cid, M., and Liz-Marzan, L. M. (2011). From individual to collective chirality in metal nanoparticles, *Nano Today*, **6**, p. 381.
6. Kuzyk, A., Schreiber, R., Fan, Z., Pardatscher, G., Roller, E. M., Hogele, A., Simmel, F. C., Govorov, A. O., and Liedl, T. (2012). DNA-based self-assembly of chiral plasmonic nanostructures with tailored optical response, *Nature*, **483**, p. 311.
7. Fan, Z. Y., and Govorov, A. O. (2010). Plasmonic circular dichroism of chiral metal nanoparticle assemblies, *Nano Lett.*, **10**, p. 2580.
8. Kuwata-Gonokami, M., Saito, N., Ino, Y., Kauranen, M., Jefimovs, K., Vallius, T., Turunen, J., and Svirko, Y. (2005). Giant optical activity in quasi-two-dimensional planar nanostructures, *Phys. Rev. Lett.*, **95**, 227401.
9. Papakostas, A., Potts, A., Bagnall, D. M., Prosvirnin, S. L., Coles, H. J., and Zheludev, N. I. (2003). Optical manifestations of planar chirality, *Phys. Rev. Lett.*, **90**, 107404.
10. Fedotov, V. A., Mladyonov, P. L., Prosvirnin, S. L., Rogacheva, A. V., Chen, Y., and Zheludev, N. I. (2006). Asymmetric propagation of electromagnetic waves through a planar chiral structure, *Phys. Rev. Lett.*, **97**, 167401.
11. Gansel, J. K., Thiel, M., Rill, M. S., Decker, M., Bade, K., Saile, V., von Freymann, G., Linden, S., and Wegener, M. (2009). Gold helix photonic metamaterial as broadband circular polarizer, *Science*, **325**, p. 1513.
12. Plum, E., Fedotov, V. A., and Zheludev, N. I. (2008). Optical activity in extrinsically chiral metamaterial, *Appl. Phys. Lett.*, **93**, 191911.
13. Tang, Y., and Cohen, A. E. (2010). Optical chirality and its interaction with matter, *Phys. Rev. Lett.*, **104**, 163901.

14. Tang, Y., and Cohen, A. E. (2011). Enhanced enantioselectivity in excitation of chiral molecules by superchiral light, *Science*, **332**, p. 333.

15. Hendry, E., Carpy, T., Johnston, J., Popland, M., Mikhaylovskiy, R. V., Lapthorn, A. J., Kelly, S. M., Barron, L. D., Gadegaard, N., and Kadodwala, M. (2010). Ultrasensitive detection and characterization of biomolecules using superchiral fields, *Nature Nanotech.*, **5**, p. 783.

16. Schäeferling, M., Dregely, D., Hentschel, M., and Giessen, H. (2012). Tailoring enhanced optical chirality: Design principles for chiral plasmonic nanostructures, *Phys. Rev. X*, **2**, 031010.

17. Biagioni, P., Huang, J., Duò, L., Finazzi, M., and Hecht, B. (2009). Cross resonant optical antenna, *Phys. Rev. Lett.*, **102**, 256801.

18. García-Etxarri, A., and Dionne, J. A. (2012). Surface enhanced circular dichroism spectroscopy mediated by non-chiral nanoantennas, *Phys. Rev. B*, **87**, 235409.

19. Zhang, S. P., Wei, H., Bao, K., Håkanson, U., Halas, N. J., Nordlander, P., and Xu, H. X. (2011). Chiral surface plasmon polaritons on metallic nanowires, *Phys. Rev. Lett.*, **107**, 096801.

20. Stratton, J. A. (1941) *Electromagnetic Theory* (McGraw-Hill, New York).

21. Jackson, J. D. (1998) *Classical Electrodynamics* (Wiley, New York).

22. Johnson, P. B., and Christy, R. W. (1972). Optical constants of noble metals, *Phys. Rev. B*, **6**, p. 4370.

23. Chang, D. E., Sørensen, A. S., Hemmer, P. R., and Lukin, M. D. (2007). Strong coupling of single emitters to surface plasmons, *Phys. Rev. B*, **76**, 035420.

24. Schmidt, M. A., and Russell, P. S. J. (2008). Long-range spiralling surface plasmon modes on metallic nanowires, *Opt. Express*, **16**, p. 13617.

25. Fang, Z., Fan, L., Huang, S., and Zhu, X. (2010). Applications of the surface plasmon polariton in the au nanocircuit, *J. Korean Phys. Soc.*, **56**, p. L1725.

26. Li, Z. P., Hao, F., Huang, Y. Z., Fang, Y. R., Nordlander, P., and Xu, H. X. (2009). Directional light emission from propagating surface plasmons of silver nanowires, *Nano Lett.*, **9**, p. 4383.

27. Al-Bader, S. J., and Jamid, H. A. (2007). Diffraction of surface plasmon modes on abruptly terminated metallic nanowires, *Phys. Rev. B*, **76**, 235410.

28. Kelly, K. L., Coronado, E., Zhao, L. L., and Schatz, G. C. (2003). The optical properties of metal nanoparticles: The influence of size, shape, and dielectric environment, *J. Phys. Chem. B*, **107**, p. 668.

29. Zhang, S. P., and Xu, H. X. (2012). Optimizing substrate-mediated plasmon coupling toward high-performance plasmonic nanowire waveguides, *ACS Nano*, **6**, p. 8128.
30. Wei, H., Zhang, S., Tian, X., and Xu, H. (2013). Highly tunable propagating surface plasmons on supported silver nanowires, *Proc. Natl. Acad. Sci. USA*, **110**, p. 4494.
31. Li, Z., Bao, K., Fang, Y., Huang, Y., Nordlander, P., and Xu, H. (2010). Correlation between incident and emission polarization in nanowire surface plasmon waveguides, *Nano Lett.*, **10**, p. 1831.
32. Fang, Y. R., Li, Z. P., Huang, Y. Z., Zhang, S. P., Nordlander, P., Halas, N. J., and Xu, H. X. (2010). Branched silver nanowires as controllable plasmon routers, *Nano Lett.*, **10**, p. 1950.
33. Li, Z. P., Zhang, S. P., Halas, N. J., Nordlander, P., and Xu, H. X. (2011). Coherent modulation of propagating plasmons in silver-nanowire-based structures, *Small*, **7**, p. 593.
34. Wei, H., Li, Z., Tian, X., Wang, Z., Cong, F., Liu, N., Zhang, S., Nordlander, P., Halas, N. J., and Xu, H. (2011). Quantum dot-based local field imaging reveals plasmon-based interferometric logic in silver nanowire networks, *Nano Lett.*, **11**, p. 471.
35. Wei, H., Wang, Z., Tian, X., Käll, M., and Xu, H. (2011). Cascaded Logic Gates in Nanophotonic plasmon networks, *Nat. Commun.*, **2**, 387.
36. Allen, L., Beijersbergen, M. W., Spreeuw, R. J. C., and Woerdman, J. P. (1992). Orbital angular momentum of light and the transformation of Laguerre–Gaussian laser modes, *Phys. Rev. A*, **45**, p. 8185.
37. Wang, J., Yang, J.-Y., Fazal, I. M., Ahmed, N., Yan, Y., Huang, H., Ren, Y., Yue, Y., Dolinar, S., Tur, M., and Willner, A. E. (2012). Terabit free-space data transmission employing orbital angular momentum multiplexing, *Nat. Photon.*, **6**, p. 488.
38. Rüeting, F., Fernández-Domínguez, A. I., Martín-Moreno, L., and García-Vidal, F. J. (2012). Subwavelength chiral surface plasmons that carry tuneable orbital angular momentum, *Phys. Rev. B*, **86**, 075437.
39. Bliokh, K. Y., and Nori, F. (2012). Transverse spin of a surface polariton, *Phys. Rev. A*, **85**, 061801.

Chapter 7

Manipulation of Surface Plasmon Patterns with Chirality of Metallic Structure

Il-Min Lee, Seung-Yeol Lee, Yohan Lee, and Byoungho Lee
National Creative Research Center for Active Plasmonics Application Systems, Inter-University Semiconductor Research Center and School of Electrical Engineering, Seoul National University, Gwanak-Gu Gwanakro 1, Seoul 151-744, Korea
byoungho@snu.ac.kr

In this chapter, we discuss the surface plasmon related variation in chirality for some illustrative metallic apertures with chiral shapes or distributions. When light is transmitted through a metallic plate with a chiral aperture pattern, provided that only plasmonic-mediated transmission exists, it is possible to clearly observe that the spin angular momentum of the incident light is converted into the orbital angular momentum of the transmitted light. The scope of this chapter is to discuss the geometrical dependence of such plasmonic spin-to-orbital momentum conversion in complex geometrical structures. Patterns of gammadion with several bent angles and quasiperiodic apertures with complex chiral patterns are discussed.

Singular and Chiral Nanoplasmonics
Edited by Svetlana V. Boriskina and Nikolay I. Zheludev

ISBN 978-981-4613-17-0 (Hardcover), 978-981-4613-18-7 (eBook)
www.panstanford.com

7.1 Introduction: Optical Chirality and Plasmonic Chiral Patterns

Light beams possessing or carrying angular momentum (AM) have been intensively studied from diverse areas such as studies of cosmic objects in rotation [1], rotating atoms with photons [2], and many other applications [3]. AMs that are usually carried by an optical field can be classified into two types, namely, spin AM (SAM) and orbital AM (OAM) [3, 4]. Although the separation of these two AMs is not based on rigorous physical ground [4–6], for paraxial beams, it is generally accepted that SAM and OAM are strongly associated with the polarization and spatial distribution of the optical field, respectively. In other words, for the case of a paraxial beam of light, the SAM is in the relation with the rotation of the polarization of light in the basis of left- or right-hand circularity and the OAM is related on the dependence of the light field to the azimuth angle in its complex amplitude. The conversion between these two AMs is the preferred choice in many studies for explaining optical singularity or vortices in focused beams or scattered light [7–9].

Recently, great interest has developed in light interactions with materials having chirality [8–12]. The word chirality refers to the asymmetry or handedness of a configuration that cannot be superimposed over its mirror image. More precisely, a chiral structure which contains helix, screw-like circular geometry is often called helical structure. When an optical field passes through a chiral, optically thick material, the polarization of the transmitted field is rotated from its original state. This phenomenon is called optical activity [11]. Optical activity is based on a type of birefringence: polarization-dependent phase retardations enforced by the structure. Therefore, such optical activity usually occurs in a three-dimensional object and is rarely observed in the case of a planar or an optically thin pattern. However, recent studies have proven that such phenomena can be associated with planar plasmonic chiral structures [8–10, 12]. Surface plasmon polaritons (SPPs) are light-coupled collective oscillations of electrons in a metal surface [13]. When a light field is transmitted through a planar structure that contains a chiral metallic pattern, the SPPs are excited along the metallic pattern. As the excitations of the SPPs are generally sensitive to the polarization state of the incident light, the excited SPPs can be the source of the polarization-dependent phase retardation of the transmitted or reflected light. If the topological pattern of the metallic structure is asymmetric or has handedness, and on which the SPPs can be efficiently excited by illuminating light, we refer to such a structure as a chiral plasmonic structure. There is a long list of publications dealing with the subject of chirality in plasmonics and some of those are listed in the references at the end of this chapter [8–10, 12, 14–24]. In short, the chirality in a metallic pattern can modify an AM in a light field.

The focus of this chapter is on the dependence of chirality in an optical field on the characteristics of a metallic chiral pattern. For this objective, we briefly review the basic concepts related to plasmonic chiral optics. Either rods or apertures can be used to make up metallic chiral patterns for plasmonics. For analytical simplicity, we will consider the simpler case: the aperture. For this, we will consider two representative examples of metallic patterns with chirality: gammadion and quasiperiodic spirals. For the gammadion, we change the bent angles of the wings to examine the dependence of chirality on the transmitted light. For the second case, we investigate the effect of the spatially varying chirality on transmitted light.

7.2 Plasmonic Apertures with Chirality: A Simple Analogical Approach

Before we consider the specific examples for this chapter, let us begin with a simpler structure: a circular slit pattern on a planar metallic plate. Assume that circularly polarized light is incident from the backside of the circular slit, as shown in Fig. 7.1.

Throughout this chapter, the field rotational directions for left-circular polarization (LCP) and right-circular polarization (RCP) will refer to the clockwise (CW) and counter-clockwise (CCW) rotational directions when observed from the front side of the beam propagation direction as in Fig. 7.1. In addition, we assume that the critical dimension of the apertures (the width of the slit or the diameter of the circular aperture) is narrower than the cutoff to prohibit any propagation of the conventional photonic mode through it. We also assume that the thickness of the metal plate is thicker than the skin depth of the SPPs to exclude any transmissions via tunneling through the metal plate. Under these assumptions, the transmitted light after the metal plate can be regarded as radiated waves arising from the coupling of the transferred plasmonic-guided modes through the aperture(s) [25]. Therefore, the relative phase in the transmitted light is strongly associated with that of the transmitted plasmonic mode.

The excitations of the plasmonic mode via a metallic slit or aperture are polarization-dependent: only the electric field component that is perpendicular to the slit can contribute to the excitation of SPPs when the light is normally incident on the metallic plate [26]. In consequence, the SPPs propagating on the metal plate follow the phase of the incident light as depicted in Fig. 7.1. In this figure, the gray arrows approaching from four directions schematically represent the phase distributions of the SPPs at the exit surface of the metal plate at a given fixed time. As shown in the figure, near the center of the slit pattern, the azimuth distribution of the phase of the SPPs follows that of the incident field with constant phase retardation. Consequently, the phase distribution in

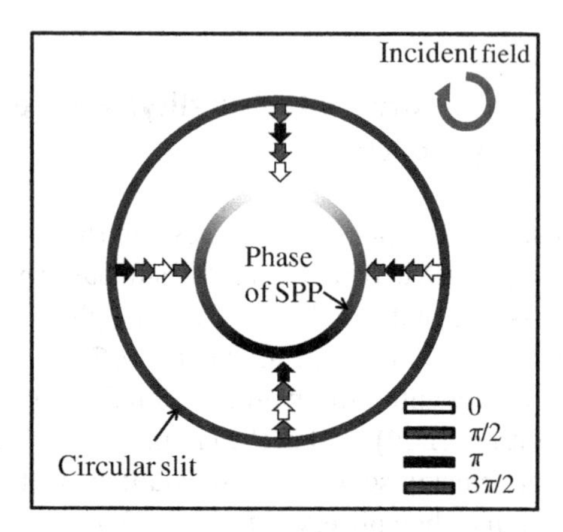

Figure 7.1 Schematic diagram of excited SPPs from circularly polarized light (LCP) illuminated from the backside of a circular metallic slit.

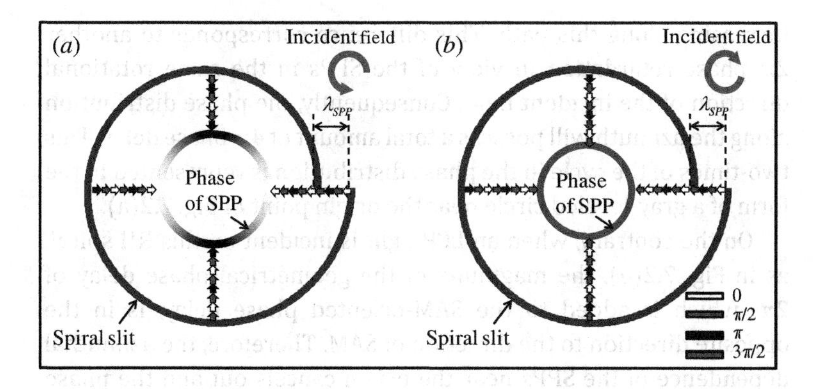

Figure 7.2 Schematic diagram of excited SPPs from a circularly polarized light illuminated from the backside of a spiral metallic slit. The polarization of the incident light is (a) RCP and (b) LCP.

the transmitted light from this slit pattern also follows that of the SPPs at the slit. Therefore, the diffracted light from this slit pattern will have an OAM that coincides with the SAM of the incident field. This is a simple illustration of the interaction between SAM and OAM for a plasmonic structure.

Now, instead of the circularly symmetric slit, let us move on to a chiral slit pattern as shown in Fig. 7.2. In this case, the distance between the slit and the rotational center (the origin of the pattern) linearly increases, ending with a distance of λ_{SPP} to its starting point after one cycle of rotation. As the distance increases in the direction of the CCW rotation, we will refer to this type of helical pattern as a right-helical (RH) spiral. For this structure, the phase distributions near the center of the slit pattern do not simply follow the phase of the incident light. The spatial distributions of the field magnitude of the responses for the incidences of RCP and LCP light are very different.

Let us first consider the case in which the RCP is incident on an RH spiral in Fig. 7.2(a). In this case, the SPPs excited from the slit will have phase distributions following the SAM of the incident field as in the circular slit pattern; this corresponds to the amount of 2π phase delay along one cycle of rotation in the azimuthal direction. However, the propagation distances of the SPPs from the slit to the origin of the pattern, or the points at equi-distances from the origin, are also increasing along this path. This difference corresponds to another 2π phase retardation in view of the SPPs in the same rotational direction of the incident field. Consequently, the phase distribution along the azimuth will possess a total amount of 4π phase delay. This two-times of the cycle in the phase distribution is represented in the form of a gray-leveled circle near the origin point of Fig. 7.2(a).

On the contrary, when an LCP light is incident on this RH spiral, as in Fig. 7.2(b), the magniude of the geometrical phase delay of 2π, which is added to the SAM-oriented phase delay, is in the opposite direction to the direction of SAM. Therefore, the azimuthal dependence of the SPPs near the origin cancels out and the phase distribution near the origin is constant as shown in Fig. 7.2(b).

To appropriately describe such a SAM-OAM conversion, it is convenient to adopt the convention of a topological charge number [26, 27]. The topological charge number is defined as the number corresponding to the phase modulation along a circular path near the singular point of the field divided by 2π. When a light field is scattered by matter with a geometrical helicity, it is known that the topological charge number m_{T} can be expressed as

$$m_{\mathrm{T}} = m_{\mathrm{O}} + m_{\mathrm{SO}} + m_{\mathrm{G}}, \tag{7.1}$$

where m_{O}, m_{SO}, and m_{G} are the AM numbers coming from OAM carried by the incident light, SAM-OAM conversion, and the geometrical helicity, respectively. As these numbers can be positive, zero, or negative, m_{T} can be larger or smaller than m_{O} according to the relation between the geometry of the scatterer and the spin state of the incident field.

Using the convention of topological charge number, it is easy to describe the cases in Figs. 7.1 and 7.2. For these examples, the results given by Eq. (7.1) are as follows:

$$m_{\mathrm{T}} = m_{\mathrm{O}} + m_{\mathrm{SO}} + m_{\mathrm{G}} = 0 - 1 + 0 = -1 \text{ (Fig. 7.1)}, \tag{7.2}$$

$$m_{\mathrm{T}} = m_{\mathrm{O}} + m_{\mathrm{SO}} + m_{\mathrm{G}} = 0 + 1 + 1 = 2 \text{ [Fig. 7.2(a)]}, \tag{7.3}$$

$$m_{\mathrm{T}} = m_{\mathrm{O}} + m_{\mathrm{SO}} + m_{\mathrm{G}} = 0 - 1 + 1 = 0 \text{ [Fig. 7.2(b)]}. \tag{7.4}$$

The AM numbers in Eq. (7.1) are not necessarily integers. They can be rational or irrational numbers. Even worse, for some complicated geometries, the AM number m_{G} cannot be identified in a simple manner. In the following sections, we will examine such examples and investigate the effect arising in cases of geometrically complicated patterns.

7.3 Dependence of Chirality on the Bent Angle of a Gammadion Metal Slit

Gammadion shapes are one of the representative chiral patterns that have been studied by several researchers in investigating plasmonic chirality: for example, see [19, 20, 23]. In a gammadion shape, there are four bent wings that are bent in the same rotational direction. This rotational arrangement of the bent wings confers chirality to the gammadion. If we define the bent angle θ_r in a gammadion as the angle between the inner and the outer segments of a wing as in Fig. 7.3(a), there are several possible variations in the shape as shown in Fig. 7.3(b–e). When the bent angle is $\theta_r = 180°$ as in Fig. 7.3(b), the shape becomes a cross-pattern that has no chiral property.

To examine the chirality, we conducted numerical simulations for the apertures in Fig. 7.3(b–e) with a Fourier modal method (FMM) based technique involving a rigorous coupled wave analysis (RCWA) [27, 28].

Throughout this chapter, we assumed that light with a wavelength of 750 nm is incident from the backside ($z < 0$) of the metallic

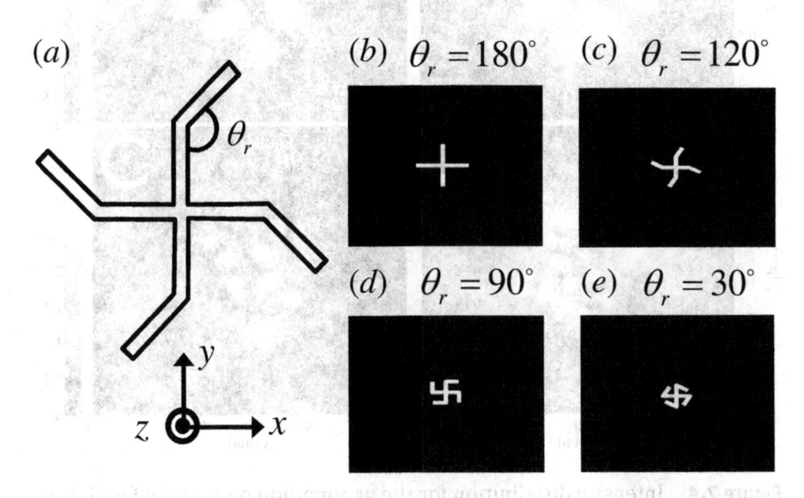

Figure 7.3 (a) Schematic diagram of a gammadion with the definition of the bent angle θ_r and the convention of coordinate axis. Several examples of a gammadion depending on the bent angle are presented in (b–e). The bent angles in (b–e) are shown at the upper sides of each figure.

aperture located in a silver plate with a thickness of 200 nm. At this wavelength, the permittivity of the silver is assumed to have an approximate value of $-23.95 + 1.44i$ [29]. For the gammadion patterns in Fig. 7.3, the widths of each slit (wing) are 150 nm and the lengths of each segment of the wing are equally 400 nm. The joints at each wing bent are assumed to be smoothly connected via a circular arc. In this section, all the transmitted field or intensity patterns are obtained at the x–y plane 100 nm above the aperture ($z = 100$ nm).

Let us consider the responses for the apertures in Fig. 7.3(b) and (d). The intensity patterns for the RCP and LCP incidences on these apertures are presented in Fig. 7.4. Although the aperture in Fig. 7.3(b) ($\theta_r = 180°$) is achiral, slight difference in the intensity patterns in Fig. 7.4(a) and (b) can be observable: the handedness

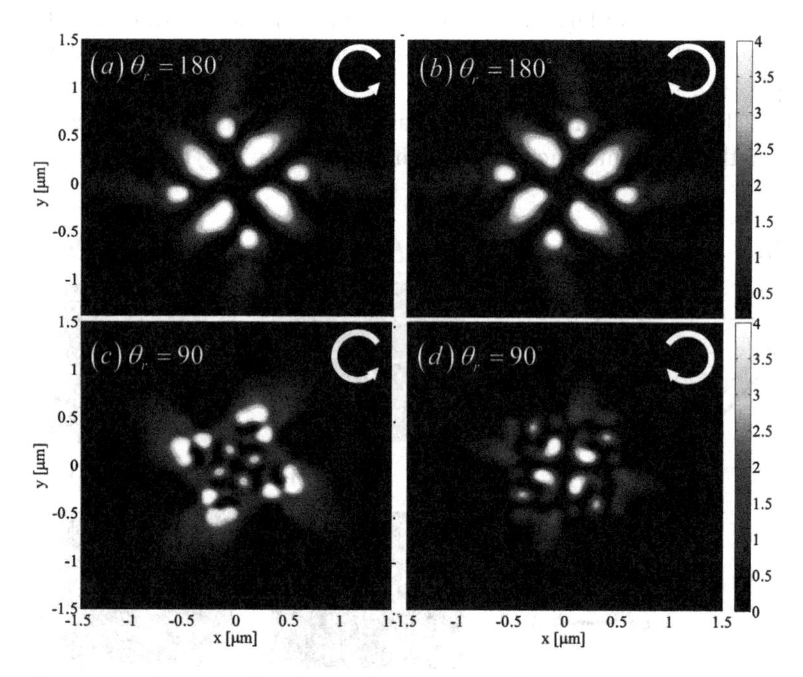

Figure 7.4 Intensity distribution for the gammadion patterns in Fig. 7.3(b) and (d). In the first row, (a) and (b) are the results for the aperture in Fig. 3(b) ($\theta_r = 180°$). In the second row, (c) and (d) are for the aperture in Fig. 7.3(d) ($\theta_r = 90°$). In the first column, (a) and (c) are for the incidence of RCP. In the second column, (b) and (d) are for the incidence of LCP.

of the intensity patterns in these two images is opposite to each other. Since m_O and m_G in Eq. (7.1) are zero, the handedness in these intensity distributions obviously originates from the SAM-OAM conversion: m_{SO}. Therefore, the intensity distributions shown in Fig. 7.4(a) and (b) are mirror images of one another. Another feature in Fig. 4(a) and (b) is that the intensity patterns have four-fold rotational symmetries. Note that the rotational symmetry axes for the RCP and LCP incidences are also mirror images of one another.

The distinct difference between the two polarization states can be seen in the responses of the bent structure of Fig. 7.3(d) ($\theta_r = 90°$). The responses are shown in Fig. 7.4(c) and (d). For this aperture, the strong polarization dependence arises not only from the SAM of the incident light (m_{SO}) but also from the geometrical factor m_G. As in Fig. 7.4(a) and (b), the intensity distributions in Fig. 7.4(c) and (d) also have four-fold rotational symmetries. However, the clearest difference between the intensity distributions in Fig. 7.4(c) and (d) compared with those in Fig. 7.4(a) and (b) is that the intensity of the light incidence of LCP [Fig. 7.4(d)] forms a strong focal area near the center of the aperture transmittance. On the contrary, the intensity for RCP [Fig. 7.4(c)] is distributed toward the end of gammadion's wings. Such a polarization-dependent 'focusing' effect is known to occur when the handedness of the incident beam and the direction of the twist in the pattern are in the same direction [19]. We will see whether such relationship holds for all the various shapes of gammadions presented in Fig. 7.3 later in this section.

Before we discuss on reason for such polarization-dependent characteristics further, let us move onto the remaining two cases in Fig. 7.3: the apertures in Fig. 7.3(c) ($\theta_r = 120°$) and (e) ($\theta_r = 30°$). The bent angles for Fig. 7.3(c) and (e) are obtuse and acute. We will see that these differences in bent angle cause a great contrast in the plasmonic responses between these patterns. We present the intensity distributions for these two patterns under RCP and LCP incidences in Fig. 7.5.

For the case of the obtuse angle in Fig. 7.5(a) and (b), we can see the concentrated field at the center for the LCP case in Fig. 7.5(b) but a dark center can be seen for the RCP case in Fig. 7.5(a). This relation is similar to the responses considered in Fig. 7.4(c) and (d). However,

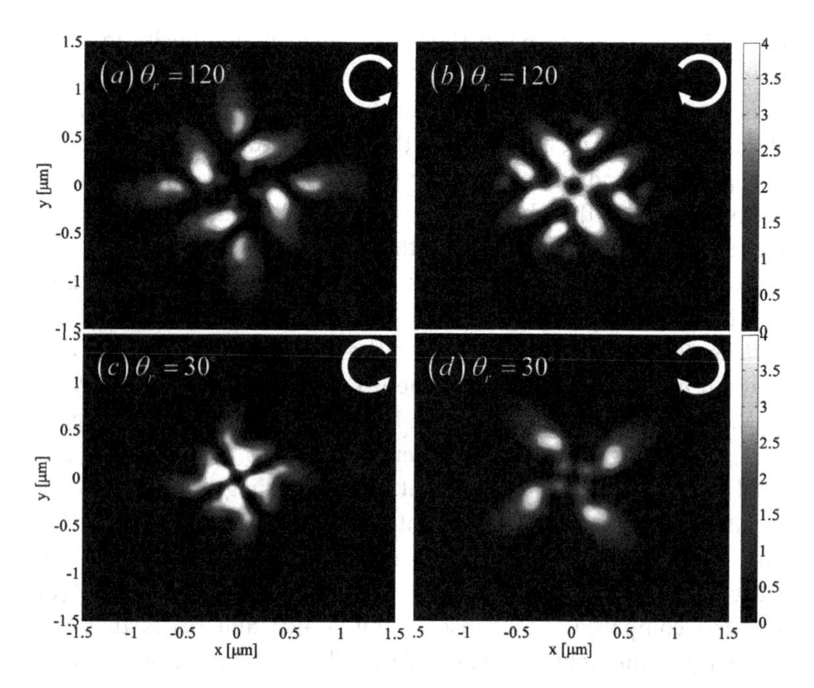

Figure 7.5 Intensity distributions for the gammadion pattern for Fig. 7.3(c) ($\theta_r = 120°$) are shown in (a–b), and for Fig. 7.3(e) ($\theta_r = 30°$) are in (c–d). All other conditions are the same as in Fig. 7.4.

for the case of the acute angle in Fig. 7.5(c) and (d), a brighter central intensity can be observed in the case of the RCP incidence as in Fig. 7.5(c). On the contrary, distributed brighter regions at the ends of the wings for the LCP incidence are observed in Fig. 7.5(d). Therefore, the sense of co-directionality between the SAM and the helicity of geometry to 'focus' the field to the center [19] is reversed in this acute structure.

For the limitation in length, we do not show the time-dependent evolution of the field distributions in this chapter. However, if we observe the time-dependent responses for the gammadion apertures in Fig. 7.3, the differences in the responses between obtuse and acute wings become more clear. The most distinct contrast between these patterns is that the time-dependent locations of the peaks in the field transmissions evolve in different directions for these bent angles.

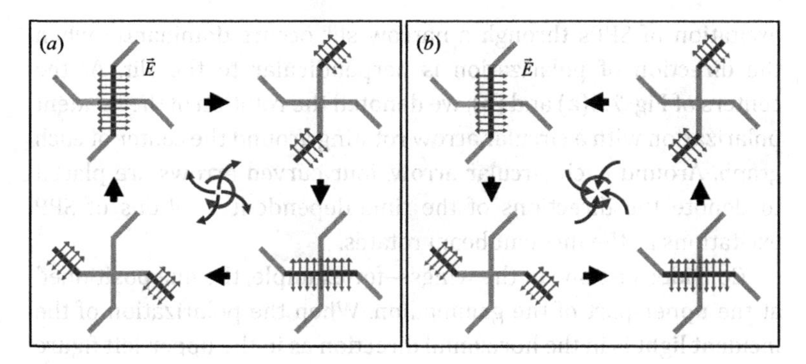

Figure 7.6 Time-dependent evolution of excited segments of a gammadion with an obtuse bent angle. The polarizations of the incident lights are (a) LCP and (b) RCP.

The bent angle dependent characteristics of the gammadion aperture can be understood by examining a simplified model that explains the polarization-dependent excitation of surface plasmons in a metallic slit. We present such an analogy with some schematic diagrams of obtuse and acute gammadions in Figs. 7.6 and 7.7.

Let us start with an example of a gammadion with obtuse wings, as in Fig. 7.6. In this figure, the polarization states of the incident fields are denoted with small double-sided arrows near the segments of the wings that are dominantly excited. Note that the

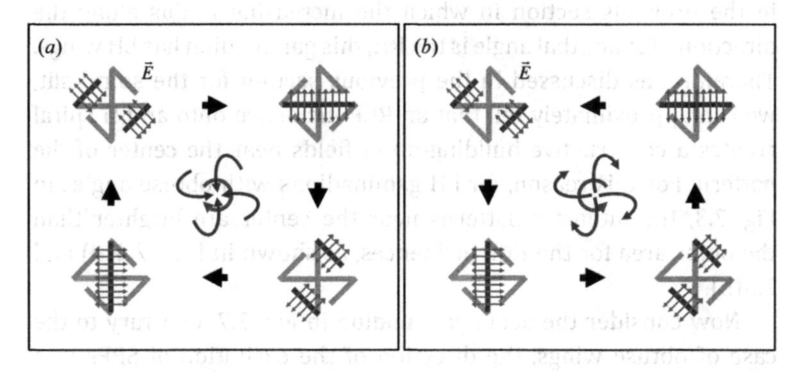

Figure 7.7 Time-dependent evolution of excited segments of a gammadion with an acute bent angle. The polarizations of the incident lights are (a) LCP and (b) RCP.

excitation of SPPs through a narrow slit occurs dominantly when the direction of polarization is perpendicular to the slit. At the centers of Fig. 7.6(a) and (b), we denoted the rotation of the incident polarization with a circular arrow rotating around the center of each graph. Around each circular arrow, four curved arrows are placed to denote the directions of the time-dependent locations of SPP excitations as the incident beam rotates.

Consider just one of the wings—for example, the one positioned at the upper-part of the gammadion. When the polarization of the incident light is in the horizontal direction as in the upper-left figure in Fig. 7.6(a), the excitation of the SPPs mainly occurs at the inner segment of that ring. As the direction of the polarization rotates in CW direction (LCP), the dominantly excited segment of that wing will subsequently be changed into the outer one as in the upper-right figure of Fig. 7.6(a). In a similar manner, the direction of the excitation of SPPs in a wing for an obtuse gammadion under an LCP incidence will move from the inner segments to the outer segments, as shown by the four curved arrows in the center of Fig. 7.6(a).

However, if the handedness of the incident beam is in the opposite direction (RCP), as in Fig. 7.6(b), the direction of the excitation of the SPs in a wing of the same obtuse gammadion under an RCP incidence will be inverted and directed inward, as depicted by four arrows in the center of Fig. 7.6(b). If we define the direction of this gammadion with the direction of helicity as in the previous section in which the increasing radius along the direction of azimuthal angle is the RH, this gammadion has LH wings. Therefore, as discussed in the previous section for the spiral slit, we can approximately tell that an RCP incidence onto an LH spiral creates a constructive building-up of fields near the center of the pattern. For this reason, for LH gammadions with obtuse angles in Fig. 7.3, the intensity patterns near the center are brighter than the outer area for the RCP incidences, as shown in Figs. 7.4(d) and 7.5(b).

Now consider the acute gammadion in Fig. 7.7. Contrary to the case of obtuse wings, the direction of the excitation of SPPs in a specific wing under an LCP incidence will move from the outer to the inner segments, as shown in four arrows in the center of Fig. 7.7(a). For the RCP incidence, the directions of these arrows are inverted as in Fig. 7.7(b). Therefore, although the handedness of the geometry can still be regarded to be an LH pattern, the dependence of the 'gathering' nature of the field to the center region on the polarization is inverted compared with the case of an obtuse angle.

This tendency of bent angle dependent concentration or distribution of the field intensity can be extended to patterns that are more complex. For example, if the wings of gammadion are not composed of segments of straight lines but in smooth curves, the curvature dependence of the 'focusing' nature of a gammadion-like plasmonic apertures can be analyzed in a similar manner.

7.4 Plasmonic Chirality in Complex Chiral Patterns: Vogel's Spiral Apertures

Let us consider more complex chiral patterns. For this objective, we choose quasiperiodic aperture patterns following the so-called Vogel's spiral [30]. The Vogel's spiral is well known as the phyllotaxis type geometry in nature—the pattern commonly found in the sunflower head, daisy, pinecone, and pineapple [30]. This pattern is the optimized arrangement for the seeds or leaves to be exposed to sunlight or rain. To construct such a pattern, Vogel [30] proposed placing the seeds on a Fermat's spiral with distances increasing with a ratio of consecutive Fibonacci numbers (with a golden ratio). Some studies on diffracted or scattered light by metallic particles or (non-metallic) apertures arranged by Vogel's spiral pattern have appeared recently [31–34].

Usually, the position of the n-th seed of a Vogel's spiral in polar coordinates $(r_{\rm n}, \theta_{\rm n})$ can be represented as,

$$r_{\rm n} = a\sqrt{n}, \tag{7.5}$$

$$\theta_{\rm n} = n\alpha, \tag{7.6}$$

where a is a constant scaling factor and α is the divergence angle usually given with an irrational number related with a golden ratio $\varphi = \left(1+\sqrt{5}\right)/2$:

$$\alpha = 2\pi/\varphi^2. \tag{7.7}$$

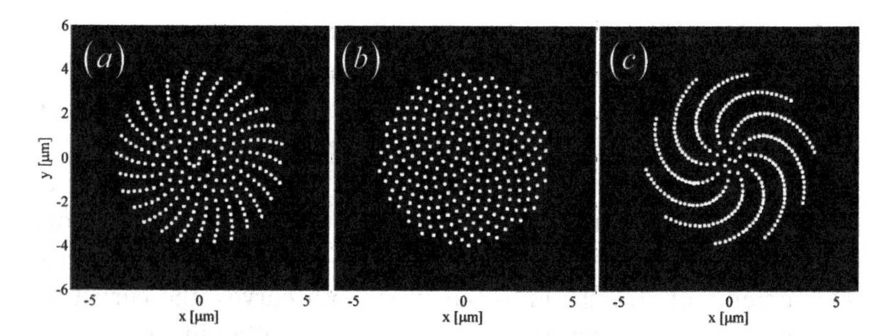

Figure 7.8 Several types of Vogel's spiral patterns obtained from three different divergence angles (α). The approximate values for the divergence angles are (a) $\alpha = 192.92$, (b) $\alpha = 137.51$, and (c)$\alpha = 108.34$.

By changing the value α, it becomes possible to obtain various spiral patterns. Some examples of Vogel's spiral patterns are shown in Fig. 7.8: The pattern in Fig. 7.8(b) is obtained by adopting the golden ratio to the divergence angle and is usually referred to as a golden spiral. The φ values for all the patterns in Fig. 7.8 are (a) $\varphi = \left(1 + \sqrt{3}\right)/2$, (b) $\varphi = \left(1 + \sqrt{5}\right)/2$, and (c) $\varphi = \left(1 + \sqrt{7}\right)/2$.

Among the patterns shown in Fig. 7.8, the golden spiral pattern in Fig. 7.8(b) has the most complex chirality. In this pattern, several winding paths are evident, in which the direction of rotation can be either CW or CCW. As a consequence, the nature of the response from the viewpoint of topological charge is very complex. Although slightly diminished, the pattern in Fig. 7.8(a) exhibits both RH and LH characteristics in the arrangement of the apertures. Although the pattern in Fig. 7.8(c) clearly shows an LH arrangement in the outer region, mixed LH and RH arrangements are observed in the small region near the center of the pattern. For the structures in Fig. 7.8(a) and (b), it would be expected that the handedness of the helicity in the transmitted fields for these patterns cannot be distinctively identified. However, for the structure in Fig. 7.8(c), a certain handedness can be observed.

For the structures in Fig. 7.8, let us examine the numerical results obtained from the RCWA simulations. The conditions used for the simulations are as follows: the incident wavelength, the thickness of silver metal plate, and diameter of each circular aperture are assumed to be 750 nm, 200 nm, and 100 nm, respectively. Due

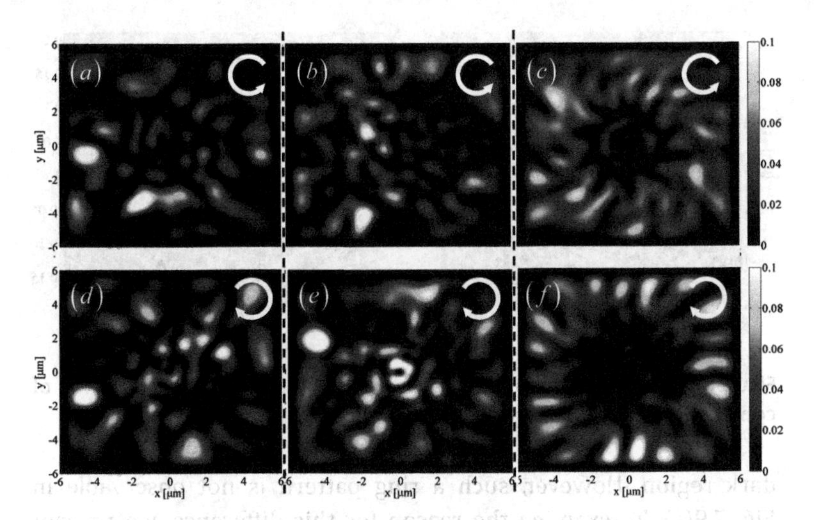

Figure 7.9 Polarization-dependent intensity distributions from Vogel's spiral patterns in Fig. 7.8. The incident field is RCP for (a), (b), and (c) and LCP for (d), (e), and (f). (a) and (d), (b) and (e), and (c) and (f) are the responses for the structure of Fig. 7.8(a), (b), and (c), respectively.

to computational restrictions, we limited the total number of the apertures to 200. In addition, as the RCWA method basically treats the periodic structures, we adopted perfectly matched layers (PMLs) [28] around the lateral boundaries of the domain of the calculation of which the lateral size is 12 μm × 12 μm in x- and y-directions. With this small boundary, it is difficult to observe the far-field diffraction patterns. Therefore, it is only possible to observe the central windowed region in the diffracted field. The calculated field intensity distributions observed at the plane of $z = 6$ μm are presented in Fig. 7.9.

In Fig. 7.9, certain differences between the images in the upper row (RCP) and in the lower row (LCP) are visible. However, in the responses for the structures in Fig. 7.8(a) and (b), which correspond to Fig. 7.9(a) and (d), and Fig. 7.9(b) and (e), respectively, the patterns are sufficiently complicated that not much of value can be found.

The interesting results are in Fig. 7.9(c) and (f), which are the responses from the structure in Fig. 7.8(c). In Fig. 7.9(c), a circular bright ring pattern can be seen near the outer edge of the central

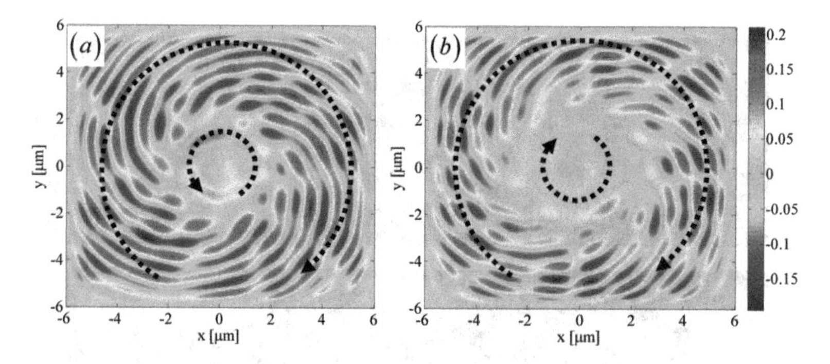

Figure 7.10 Field amplitudes obtained along circular paths. The radii of contours for (a) and (b) are 0.8 μm and 3 μm, respectively.

dark region. However, such a ring pattern is not observable in Fig. 7.9(f). To examine the reason for this difference, we present snapshots of the field distributions for a fixed time for these two responses in Fig. 7.10.

To represent the rotational properties that cannot be identified from snapshots of the field distributions, we added arrows in the figures. In Fig. 7.10(a), the field near the center region rotates in an opposite direction (CCW) to the field in the outer region (CW). However, in the field distribution in Fig. 7.10(b), the directions of rotation are approximately in the same direction (CW) for all regions. As the outer region of the pattern in Fig. 7.8(c) has 10 wings of which directions are LH, from Eq. (7.1), the topological charge numbers for Fig. 7.10(a) and (b) are found to be −9 and −11, respectively. As one can easily see from Fig. 7.10, these numbers coincide with the numbers identified by direct counting the peaks in the field distributions for the results shown in Fig. 7.10(a) and (b).

To show a more clear view to the topological charges in these cases, we present field amplitude plots in Fig. 7.11 that are measured along two different circular contours obtained from the field distributions in Fig. 7.10. The radii of the contours for Fig. 7.11(a) and (b) are 0.5 μm (central region) and 3 μm (outer region), respectively. Therefore, from Fig. 7.11(a) and (b), the topological charge numbers can be found more readily. From the geometry in Fig. 7.8(c), the helical directionality at the outer region corresponds to 10 wings in LH directions. However, at the central region, the

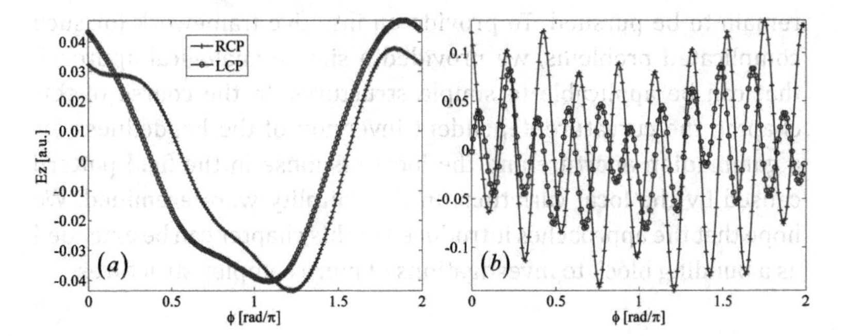

Figure 7.11 Polarization-dependent field distributions from Vogel's spiral patterns in Fig. 8(c). The incident field is RCP for (a) and LCP for (b). The arrows denote the rotational direction of the field in the center and outer regions.

helicity and approximate number of wings cannot be determined to a definite one. We can only find that at the proximity of the origin, the helicity disappears to be zero. Therefore, the topological charge numbers for the RCP and LCP incidences in the central regions follow the SAM of the incident field to be 1 and –1, respectively, as shown in Fig. 7.11(a). However, as we have examined in Fig. 7.10, the values in the outer regions are −9 and −11 [Fig. 7.11(b)].

It is interesting to note that the results for the structure in Fig. 7.8(c) show that the local properties of the handedness in the geometry affect the local nature of the diffracted field in the mid-field diffraction patterns.

7.5 Summary

In this chapter, we investigated the geometrical effect of field chirality caused by the excitation of SPPs. To accomplish this, we investigated patterns of a gammadion with several bent angles and quasi-periodic apertures of Vogel's spiral patterns. In this chapter, our main concern was chiral geometries in which the handedness or the geometrical contributions to topological charge are not simply determined by the handedness of the pattern but by considering the entire pattern. Performing an analysis and developing an understanding of such complex structures are not easy tasks and remain to be pursued. To provide an intuitive framework for such complicated problems, we provided a simple analogical approach that can be applicable to simple structures. In the course of this chapter, the curvature-dependent inversion of the handedness for a gammadion aperture and the local response in the field pattern caused by the local variations in the chirality were examined. We hope that the approaches introduced in this chapter can be extended as a building block to investigations of more complex structures.

Acknowledgment

This work was supported by the National Research Foundation and the Ministry of Education, Science and Technology of Korea through the Creative Research Initiatives Program (Active Plasmonics Application Systems).

References

1. Tamburini, F., Thidé, B., Molina-Terriza, G., and Anzolin, G. (2011). Twisting of light around rotating black holes, *Nat. Phys.*, **7**, pp. 195–197.
2. Andersen, M. F., Ryu, C., Cladé, P., Natarajan, V., Vaziri, A. Helmerson, K., and Phillips, W. D. (2006). Quantized rotation of atoms from photons with orbital angular momentum, *Phys. Rev. Lett.*, **97**, 170406.
3. Torres, J. P., and Torner, L., (2011). *Twisted Photons: Applications of Light with Orbital Angular Momentum* (Wiley-VCH, Berlin, Germany).
4. Franke-Arnold, S. Allen, L. and Padgett, M., (2008). Advances in optical angular momentum, *Laser Photonics Rev.*, **2**, pp. 299–313.
5. Fernandez-Corbaton, I., Zambrana-Puyalto, X, and Molina-Terriza, G. (2012). Helicity and angular momentum: A symmetry-based framework for the study of light-matter interactions, *Phys. Rev. A*, **86**, 042103.
6. Allen, L., Padgett, M., and Babiker, M. (1999). The orbital angular momentum of light, *Prog. Opt.*, **39**, pp. 291–372.
7. Zhao, Y., Edgar, J. S., Jeffries, G. D. M. McGloin, D, and Chiu, D. T. (2007). Spin-to-orbital angular momentum conversion in a strongly focused optical beam, *Phys. Rev. Lett.*, **99**, 033901.

8. Gorodetski, Y., Shitrit, N., Bretner, I., Kleiner, V., and Hasman, E. (2009). Observation of optical spin symmetry breaking in nanoapertures, *Nano Lett.*, **9**, pp. 3016–3019.
9. Vuong, L. T., L. Adam, A. J. Brok, J. M. Planken, P. C. M. and Urbach, H. P. (2010). Electromagnetic spin-orbit interactions via scattering of subwavelength apertures, *Phys. Rev. Lett.*, **104**, 083903.
10. Drezet, A., Genet, C. Laluet, J.-Y. and Ebbesen, T. W. (2008). Optical chirality without optical activity: How surface plasmons give a twist to light *Opt. Express*, **16**, pp. 12559–12570.
11. Barron, L. (1982). *Molecular Light Scattering and Optical Activity* (Cambridge University Press, Cambridge, England).
12. Fedotov, V. A. Mladyonov, P. L. Prosvirnin, S. L. Rogacheva, A. V., Chen, Y. and Zheludev, N. I. (2006). Asymmetric propagation of electromagnetic waves through a planar chiral structure *Phys. Rev. Lett.*, **97** 167401.
13. Raether, H. (1988). *Surface Plasmons on Smooth and Rough Surfaces and on Gratings* (Springer-Verlag, Berlin).
14. Kim, H., Park, J. Cho, S.-W., Lee, S.-Y., Kang, M., and Lee, B. (2010). Synthesis and dynamic switching of surface plasmon vortices with plasmonic vortex lens, *Nano Lett.*, **10**, pp. 529–536.
15. Drezet, A., Genet, C. Laluet, J.-Y. and Ebbesen, T. W. (2008). Optical chirality without optical activity: How surface plasmons give a twist to light, *Opt. Express*, **16**, pp. 12559–12570.
16. Ziegler J. I. and Haglund, Jr., R. F. (2010). Plasmonic response of nanoscale spirals, *Nano Lett.*, **10**, pp. 3013–3018.
17. Lee, S.-Y., Lee, I.-M., Park, J., Hwang, C.-Y., and Lee, B. (2011). Dynamic switching of the chiral beam on the spiral plasmonic bull's eye structure, *Appl. Opt.*, **50**, pp. G104–G112.
18. Konishi, K., Sugimoto, T., Bai, B., Svirko, Y., and Kuwata-Gonokami, M. (2007). Effect of surface plasmon resonance on the optical activity of chiral metal nanogratings, *Opt. Express*, **15**, pp. 9575–9583.
19. Krasavin, A. V. Schwanecke, A. S. Zheludev, N. I. Reichelt, M., Stroucken, T., Koch, S. W., and Wright, E. M. (2005). Polarization conversion and "focusing" of light propagating through a small chiral hole in a metallic screen, *Appl. Phys. Lett.*, **86**, 201105.
20. Krasavin, A. V. Schwanecke, A. S. and Zheludev, N. I. (2006). Extraordinary properties of light transmission through a small chiral hole in a metallic screen, *J. Opt. A-Pure Appl. Opt.*, **8**, pp. S98–S105.

21. Chen, W., Abeysinghe, D. C., Nelson, R. L., and Zhan, Q. (2010). Experimental confirmation of miniature spiral plasmonic lens as a circular polarization analyzer, *Nano Lett.*, **10**, pp. 2075–2079.
22. Ohno T., and Miyanishi, S. (2006). Study of surface plasmon chirality induced by Archimedes' spiral grooves, *Opt. Express*, **14**, pp. 6285–6290.
23. Liu, M., Zentgraf, T., Liu, Y., Bartal, G., and Zhang, X. (2010). Light-driven nanoscale plasmonic motors, *Nat. Nanotechnol.*, **5**, pp. 570–573.
24. Decker, M., Ruther, M., Kriegler, C. E., Zhou, J., Soukoulis, C. M., Linden, S., and Wegener, M. (2009). Strong optical activity from twisted-cross photonic metamaterials, *Opt. Lett.*, **34**, pp. 2501–2503.
25. Bozhevolnyi, S. I. (2009). *Plasmonic Nanoguides and Circuits* (Pan Stanford Publishing, Singapore).
26. Lee, S.-Y., Lee, I.-M. Park, J., Oh, S., Lee, W., Kim, K.-Y., and Lee, B. (2012). Role of magnetic induction currents in nanoslit excitation of surface plasmon polaritons, *Phys. Rev. Lett.*, **108**, 213907.
27. Kim, H., Lee, I.-M. and Lee, B. (2007). Extended scattering-matrix method for efficient full parallel implementation of rigorous coupled-wave analysis, *J. Opt. Soc. Am. A*, **24**, pp. 2313–2327.
28. Kim, H., Park, J., and Lee, B. (2012). *Fourier Modal Method and Its Applications in Computational Nanophotonics* (CRC Press, FL, USA).
29. Palik, D. (1998). *Handbook of Optical Constants of Solids* 2nd Ed. (Academic Press, USA).
30. Vogel, H. (1979). A better way to construct the sunflower head, *Math. Biosci.*, **44**, pp. 179–182.
31. Mihailescu, M. (2010). Natural quasy-periodic binary structure with focusing property in near field diffraction pattern, *Opt. Express*, **18**, pp. 12526–12536.
32. Lawrence, N., Trevino, J., and Negro, L. D. (2012). Control of optical orbital angular momentum by Vogel spiral arrays of metallic nanoparticles, *Opt. Lett.*, **37**, pp. 5076–5078.
33. Trevino, J., Cao, H., and Negro, L. D. (2012). Circularly symmetric light scattering from nanoplasmonic spirals, *Nano Lett.*, **11**, pp. 2008–2016.
34. Negro, L. D., Lawrence, N., and Trevino, J. (2012). Analytical light scattering and orbital angular momentum spectra of arbitrary Vogel spirals, *Opt. Express*, **20**, pp. 18209–18223.

Chapter 8

Local Field Topology behind Light Localization and Metamaterial Topological Transitions

Jonathan Tong, Alvin Mercedes, Gang Chen, and Svetlana V. Boriskina

Department of Mechanical Engineering, Massachusetts Institute of Technology, Cambridge, Massachusetts, USA
sborisk@mit.edu

We revisit the mechanisms governing the subwavelength spatial localization of light in surface plasmon polariton (SPP) modes by investigating both local and global features in optical powerflow at SPP frequencies. Close inspection of the instantaneous Poynting vector reveals formation of optical vortices—localized areas of cyclic powerflow—at the metal–dielectric interface. As a result, optical energy circulates through a subwavelength-thick 'conveyor belt' between the metal and dielectric where it creates a high density of optical states (DOS), tight optical energy localization, and low group velocity associated with SPP waves. The formation of bonding and anti-bonding SPP modes in metal–dielectric–metal waveguides can also be conveniently explained in terms of different spatial arrangements of localized powerflow vortices between two

Singular and Chiral Nanoplasmonics
Edited by Svetlana V. Boriskina and Nikolay I. Zheludev

ISBN 978-981-4613-17-0 (Hardcover), 978-981-4613-18-7 (eBook)
www.panstanford.com

metal interfaces. Finally, we investigate the underlying mechanisms of global topological transitions in metamaterials composed of multiple metal and dielectric films, that is, transitions of their iso-frequency surfaces from ellipsoids to hyperboloids, which are not accompanied by the breaking of lattice symmetry. Our analysis reveals that such global topological transitions are governed by the dynamic local re-arrangement of local topological features of the optical interference field, such as vortices and saddle points, which reconfigures global optical powerflow within the metamaterial. These new insights into plasmonic light localization and DOS manipulation not only help to explain the well-known properties of SPP waves but also provide useful guidelines for the design of plasmonic components and materials for a variety of practical applications.

8.1 Introduction

Tailored light interactions with metal surfaces and nanostructures can generate coherent collective oscillations of photons and free electrons, known as surface plasmons [1–3]. These hybrid collective states can be supported both by planar metal-dielectric interfaces in the form of surface plasmon polariton (SPP) waves and by metal particles and nanostructures in the form of localized surface plasmon (LSP) modes. Unique physical characteristics of plasmonic modes include extreme spatial field localization, high density of optical states (DOS), and low group velocity. These features open up new opportunities for nanoscale trapping and manipulation of light with applications in sensing [4–7], spectroscopy [8–13], on-chip communications [14, 15], and solar energy harvesting [16–24].

It has recently been revealed that some of the unique features associated with plasmonic effects on nanoparticles and particle clusters can be explained by the unusual pathways of nanoscale optical powerflow in the immediate vicinity of the metal nanostructures [25–29]. The local optical powerflow at each point in space is uniquely defined by the presence of local topological features in the phase of the optical interference field close to this point. Local topological features include phase singularities, points or

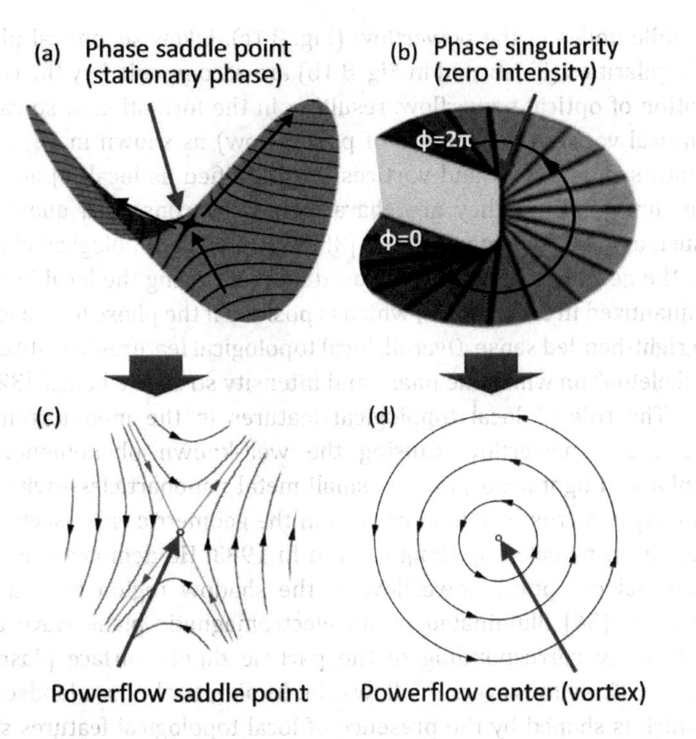

Figure 8.1 Local topological features in the optical phase field and powerflow. (a) Phase saddle node—a stationary point (or line) where the phase gradient vanishes. (b) Optical phase singularity—a point (or line) of destructive interference where the field intensity vanishes and the phase is undefined (i.e., all values of phase from 0 to 2π coexist). (c) The optical powerflow saddle point corresponding to the phase saddle point in (a). (d) The optical powerflow vortex corresponding to the phase singularity in (b).

lines in space where the field intensity is zero due to destructive interference. At these phase singularities, the phase of the field is undefined. Other types of local topological features are stationary phase nodes–points or lines in space where the phase gradient vanishes [27, 30–33]. The stationary phase nodes include local extrema (i.e., phase maxima and minima) as well as phase saddle nodes as shown in Fig. 8.1a.

Because the optical power always flows in the direction of the phase change, the phase maxima (minima) give rise to the powerflow sinks (sources) while phase saddle points give rise to saddle nodes in the powerflow (Fig. 8.1c). Likewise, optical phase singularities (illustrated in Fig. 8.1b) are accompanied by the circulation of optical powerflow, resulting in the formation of so-called optical vortices (or centers of power flow) as shown in Fig. 8.1d. Both saddle points and vortices are classified as local topological features because they are characterized by conserved quantities such as the topological charge [30, 34, 35]. The topological charge is the net phase change in a closed loop enclosing the local feature (quantized in units of 2π), which is positive if the phase increases in a right-handed sense. Overall, local topological features 'constitute a "skeleton" on which the phase and intensity structure hangs' [32].

The role of local topological features in the modification of nanoscale powerflow causing the well-known phenomenon of enhanced light absorption by small metal nanoparticles—with the absorption cross-section larger than the geometric cross-section—has been revealed by Craig Bohren in 1983. He demonstrated the reversal of optical powerflow in the shadow region behind the particle [36] illuminated by an electromagnetic plane wave at a frequency corresponding to the particle dipole surface plasmon mode. This flow reversal follows the local optical phase landscape, which is shaped by the presence of local topological features such as powerflow saddle points (Fig. 8.2a) above and inside the particle [27, 37–39]. In all the panels of Fig. 8.2, the large orange arrows indicate the direction of the incident wave, while the little arrows or streamlines illustrate the direction of the powerflow. The powerflow intensity at each point is defined by either the arrow length or the streamline density. The corresponding intensity distribution of the electric field is plotted in the background.

Two decades later, it was found that local topological features in the near-field region of metal nanoparticles can form not only at the frequency of their local surface plasmon resonance (LSP) but also under off-resonance illumination by light as shown in Fig. 8.2b [39]. These features include powerflow saddle points as well as nanoscale optical vortices (Fig. 8.2b) [39]. The complex near-field phase landscape around isolated nanoparticles can be manipulated by tuning their sizes, shapes, and materials characteristics. Continuous change of these parameters results in nucleation, spatial drift, and annihilation of local topological features [37, 40].

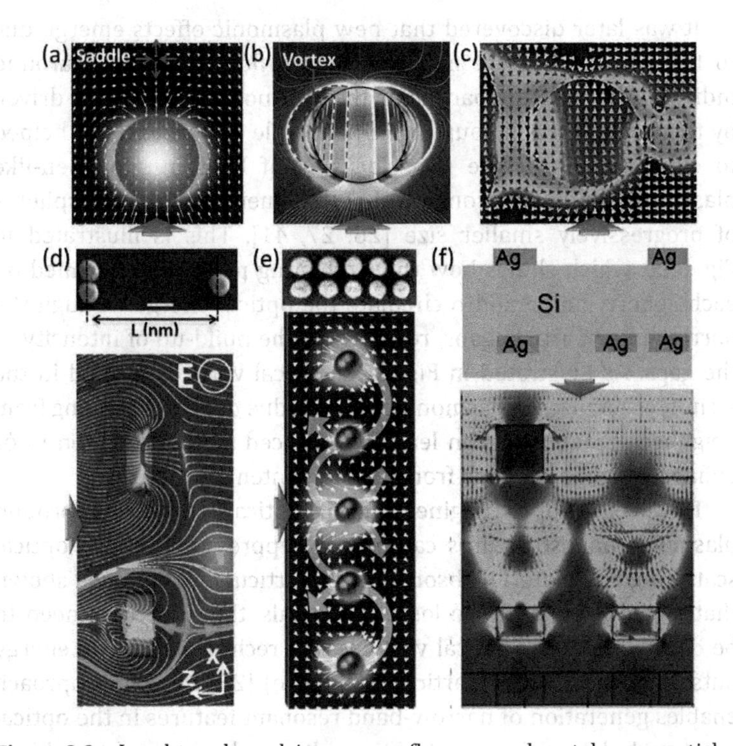

Figure 8.2 Local topology-driven powerflow around metal nanoparticles. (a) Enhanced light absorption by a nanoparticle owing to the powerflow saddle point in its shadow. Adapted with permission from Ref. [27] © The Royal Society of Chemistry. (b) Counter-rotating optical vortices on a nanosphere. Adapted with permission from Ref. [33] © The Optical Society. (c) Coupled optical vortices in the plasmonic nanolens. Adapted with permission from Ref. [27] © The Royal Society of Chemistry. (d) Optical vortices induced on plasmonic antennas by light scattered from another antenna. Reproduced with permission from Ref. [36] © 2013 Wiley-VCH Verlag GmbH & Co. KGaA, Weinheim. (e) Coupled optical vortices 'pinned' to a linear chain of nanoparticle dimers form a linear transmission-like sequence. Adapted with permission from Ref. [25] © American Chemical Society. (f) Vortices formed in the silicon slab due to light scattering from embedded metal nanoparticles. Adapted with permission from Ref. [38] © The Optical Society.

It was later discovered that new plasmonic effects emerge due to tailored coupling of nanoscale powerflows generated around individual particles. In particular, the local powerflow picture driven by the formation and coupling of nanoscale optical vortices helped to explain the extreme nano-focusing of light in snowmen-like plasmonic nanolenses composed of three neighboring nanospheres of progressively smaller size [26, 27, 41]. This is illustrated in Fig. 8.2c, which shows how the circulating powerflows formed on each sphere merge and recirculate the optical energy through the narrow inter-particle gaps, resulting in the build-up of intensity in the gaps. As illustrated in Fig. 8.2d, optical vortices formed in the vicinity of nanoscale plasmonic antennas due to light scattering from neighboring antennas can lead to enhanced light absorption in or enhanced light scattering from the nanoantennas [42].

However, proper engineering of optical powerflow around plasmonic nanostructures can enable suppression of both optical scattering and material absorption. In particular, it has been shown that to reduce dissipative losses in metals, the structures need to be designed to 'pin' optical vortices that recirculate optical energy outside of the metallic particles (Fig. 8.2e) [25–29]. This approach enables generation of narrow-band resonant features in the optical spectra of plasmonic nanostructures. It can also be used to achieve enhanced absorption in the host material (e.g., semiconductor) rather than in metal, as shown in Fig. 8.2f. These results therefore pave the way for vortex-pinning nanostructures to be used to enhance absorption in thin-film photovoltaic cells [25–27, 29, 43, 44].

The results presented in Fig. 8.2 clearly demonstrate that formation of local topological features and the resulting recirculation of the optical power through metal nanostructures are behind many interesting plasmonic effects such as extreme nano-focusing of light and electric field intensity enhancement. In this chapter, we demonstrate that localized topological features play a much larger role in various plasmonic effects than it has been recognized to date. We start with the simplest SPP wave on a metal–dielectric interface and then extend the analysis to metal-insulator-metal (MIM) waveguides and to so-called hyperbolic metamaterials [45]. Close inspection of the local field phase profiles and optical powerflow in the vicinity of the metal surfaces reveals formation of localized areas of circulating powerflow that recycles optical energy between metal and dielectric volumes. Such recirculation translates into the tight light localization of the SPP mode field on the surface, the large in-plane wavevector of SPP waves, and the resulting slowing of the wave propagation along the metal–dielectric interfaces.

In the following sections, we will discuss how the insights into the spatial structure of the localized topological features in the near field of plasmonic components and materials help to better understand their properties, to predict and exploit new plasmonic effects, and to design next-generation plasmonic devices with improved performance.

8.2 Back to Basics: Surface Plasmon Polariton

The most studied and written about plasmonic effect is the excitation of SPP waves propagating along metal–dielectric interfaces. SPP modes are TM polarized surface waves, which are characterized by large wavevectors parallel to the interface and low group velocities. As a result, they create strong electric field and high local DOS (LDOS) within the sub-wavelength thick layer adjacent to the interface. These waves can only exist on interfaces between materials having dielectric permittivity values of opposite signs. Dielectric permittivities of materials with a high density of free charge carriers—such as metals and highly doped semiconductors—are defined by the Drude model, $\varepsilon_{\mathrm{m}}(\omega) = \varepsilon_{\infty} - \omega_{\mathrm{p}}^2 \big/ (\omega^2 + i\gamma\omega)$, where ω_p^2 is the plasma frequency, ε_{∞} is the high-frequency permittivity limit, and γ is the electron collision frequency. The real part of the Drude permittivity becomes negative in the frequency range below the plasma frequency of the material. This makes possible SPP propagation along their interfaces with other materials.

An eigensolution of the electromagnetic boundary problem for the Maxwell equations on such an interface (shown in Fig. 8.3a) that describes the SPP mode has the following form [15]:

$$z > 0: \quad \begin{aligned} H_y(z) &= Ae^{i\beta x}e^{-k_2 z} \\ E_x(z) &= iA\frac{k_2}{\omega\varepsilon_0\varepsilon_2}e^{i\beta x}e^{-k_2 z} \\ E_z(z) &= -A\frac{\beta}{\omega\varepsilon_0\varepsilon_2}e^{i\beta x}e^{-k_2 z} \end{aligned} \tag{8.1}$$

and

$$z < 0: \quad \begin{aligned} H_y(z) &= Ae^{i\beta x}e^{k_1 z} \\ E_x(z) &= iA\frac{k_1}{\omega\varepsilon_0\varepsilon_1}e^{i\beta x}e^{k_1 z} \\ E_z(z) &= -A\frac{\beta}{\omega\varepsilon_0\varepsilon_1}e^{i\beta x}e^{k_1 z} \end{aligned} \tag{8.2}$$

Here, A is an arbitrary amplitude of the magnetic field, ω is the angular frequency, ε_0 is the vacuum permittivity, ε_i is the relative permittivity of the i-th medium, β is the component of the wave vector parallel to the interface, and k_i is the component of the wavevector in the i-th medium normal to the interface. Upon applying continuity boundary conditions to field expressions (8.2) and (8.3), a dispersion relation can be obtained as follows:

$$\beta = k_0\sqrt{\frac{\varepsilon_1\varepsilon_2}{\varepsilon_1 + \varepsilon_2}}, \tag{8.3}$$

where k_0 is the vacuum wave vector. The solution to Eq. (8.3) is plotted in Fig. 8.3b, and the branch corresponding to the SPP mode has a familiar flat dispersion form, reaching to infinite momentum values in the absence of dissipative losses in metal. Of course, some level of losses is inevitable in real materials; however, to make the following argument simple and straightforward without the loss of generality, we are going to consider an idealized case of lossless materials (i.e., $\gamma = 0$). In this case, the longitudinal component of the SPP wave vector that is a solution to dispersion equation (8.3) is purely real, whereas the normal component is purely imaginary. For all subsequent calculations in this section, the permittivities of the dielectric and metal are chosen to be silicon dioxide and silver, respectively [46, 47]. For Ag, we only used the data for the real part of the permittivity and neglected the dissipative term. However, as will be shown in the next section, the analysis and conclusions are valid for the general case of materials with dissipative losses.

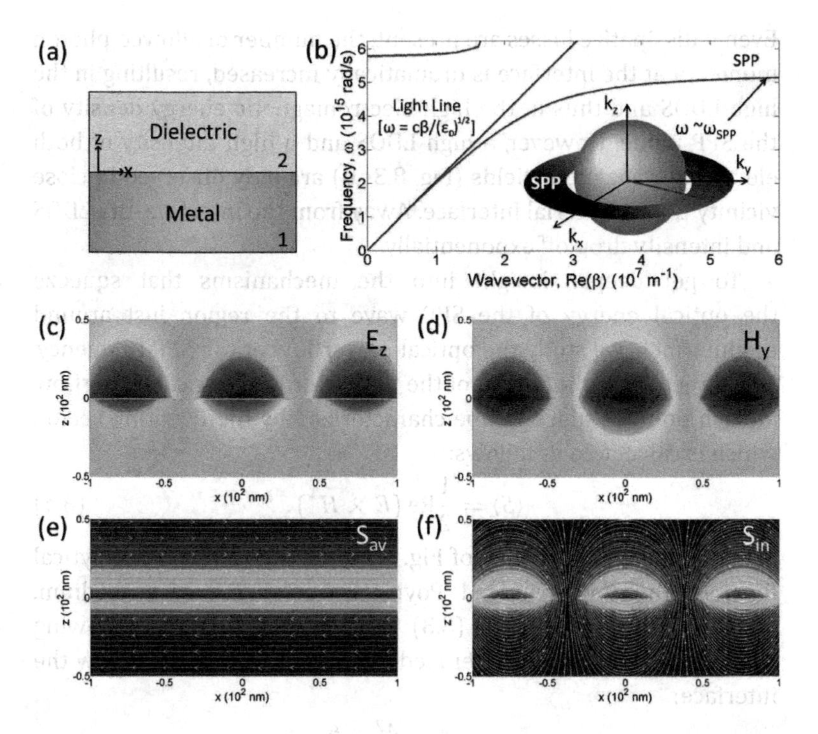

Figure 8.3 Circulating powerflow behind tight light localization and low group velocity in SPP waves. (a) Schematic of the planar metal–dielectric interface and the coordinate system used in the analysis. (b) The dispersion characteristics of the waves supported by the structure in (a). The inset shows an iso-frequency surface of the allowed photon momenta above the interface at the frequency close to the plasma frequency in the plasmonic material. (c,d) The electric (c) and magnetic field components of the SPP wave. (e) The time-averaged optical powerflow around the material interface. (f) The corresponding instantaneous powerflow at $t = 0$. All the field distributions were calculated at $w = 4.974 \times 10^{15}$ rad/s.

The momentum-space iso-frequency surface for the dielectric material just above the interface around the SPP frequency is shown in the inset to Fig. 8.3b [29]. This surface contains all the allowed k-vectors, and is a combination of a sphere corresponding to the propagating waves and a ring corresponding to the high-momentum SPP branch of the dispersion Eq. (8.3). In the absence of dissipative losses, the ring extends to infinity, which is in stark contrast with the finite-size spherical iso-frequency surface of regular materials.

Even if dissipative losses are present, the number of allowed photon momenta at the interface is dramatically increased, resulting in the high LDOS and thus in the high electromagnetic energy density of the SPP mode. However, a high LDOS and a high intensity of both electric and magnetic fields (Fig. 8.3c,d) are only observed in close vicinity to the material interface. Away from the interface, the LDOS and intensity drop off exponentially.

To get deeper insight into the mechanisms that squeeze the optical energy of the SPP wave to the region just around the interface, we study the optical powerflow at the SPP frequency. The direction and intensity of the time-averaged optical powerflow at each point of space can be characterized by the Poynting vector, which is calculated as follows:

$$\langle S \rangle = \frac{1}{2}\mathrm{Re}\left(E \times H^*\right). \tag{8.4}$$

In the simple geometry of Fig. 8.3a, we can derive an analytical form of the time-averaged Poynting vector for each medium. Substitution of (8.2) and (8.3) into (8.4) yields the following expressions for the time-averaged powerflow above and below the interface:

$$z > 0 : \langle S \rangle = \hat{x}\frac{A^2}{2}\frac{\beta}{\omega\varepsilon_0\varepsilon_2}e^{-2k_2 z} \tag{8.5}$$

$$z < 0 : \langle S \rangle = \hat{x}\frac{A^2}{2}\frac{\beta}{\omega\varepsilon_0\varepsilon_1}e^{2k_1 z} \tag{8.6}$$

Comparison of (8.5) and (8.6) immediately shows that the power flow parallel to the interface reverses direction abruptly when crossing from one medium to the other. The plot of the time-averaged powerflow shown in Fig. 8.3e visualizes the above observation. The time-averaged Poynting vector component in the vertical direction has a purely imaginary value, which is a signature of a reactive powerflow. In electromagnetic circuits, reactive power is the portion of power associated with the stored energy that returns to the source in each cycle and transfers no net energy to the load [48]. Reactive power is always a factor in alternating current circuits such as electrical grids, in which energy recycling through storage elements (i.e., inductors and capacitors) causes periodic reversals in the direction of energy flow. Although the reactive powerflow does not deliver any useful energy to the load, it assists in maintaining proper voltages across the power system. Reactive powerflow is manifested in measurable dissipative losses due to periodic energy recycling through the grid and sudden disruptions in the reactive powerflow pattern can cause a voltage drop along the line [49]. Likewise, the reactive powerflow away from the metal–dielectric interface associated with the SPP propagation along the interface does not contribute to the net energy transfer. The power is temporarily stored in the form of magnetic and electric fields.

As the SPP mode is an eigensolution of the Maxwell equations rather than a wave generated by either a localized source or a plane wave, an insight into its energy storage mechanism can be gained by plotting the instantaneous Poynting vector distribution [50]. The instantaneous optical powerflow at any given moment in time can be calculated as follows:

$$S = \frac{1}{2}\mathrm{Re}\left(E \times H^{*}\right) + \frac{1}{2}\mathrm{Re}\left(E \times He^{i\omega t}\right). \tag{8.7}$$

It can be seen that the time-averaged powerflow expression (8.4) is represented by the first term in (8.7), and the second term drops out due to its sinusoidal behavior when performing the averaging procedure. The instantaneous Poynting vector of the SPP wave is plotted in Fig. 8.3e and features repeating areas of circulating optical powerflow centered on the metal–dielectric interface. These data lead us to the conclusion that electromagnetic energy recycling through local optical vortices on the interface is behind tight field localization, high energy density, and reduced group velocity of the SPP waves. The new look at the old problem provided by revealing the circulating powerflow also offers a new take on the meaning of the large in-plane k-vector photon states that only exist very close to the interface supporting SPP waves. The circulating instantaneous SPP optical powerflow is characterized by the photon angular momentum, which has a conserved value at every point in space [51]. In turn, the value of the linear momentum that is tangential to the circulating powerflow scales inversely with the distance to the powerflow center (i.e., with the distance to the metal–dielectric interface). The tangential momentum in the direction perpendicular to the interface is canceled out by the time averaging, leaving only the in-plane tangential momentum, which, in turn, drops off away from the interface.

8.3 SPP Coupling via Shared Circulating Powerflow

Another interesting class of plasmonic waveguiding platforms is a MIM structure, which guides optical energy along a narrow slot between two metal interfaces. In such plasmon slot waveguides, optical mode volumes can be reduced to sub-wavelength scales while suffering low decay even for frequencies far from the plasmon resonance [52]. A schematic of the MIM plasmon slot waveguide is shown in Figs. 8.4a and 8.5a, and the dispersion characteristics

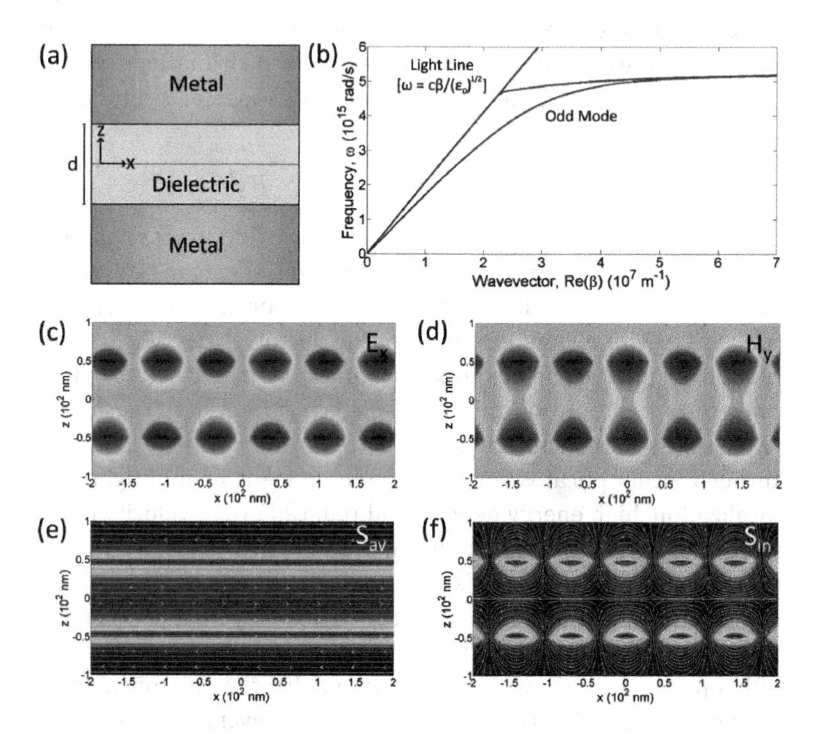

Figure 8.4 Shared circulating powerflow of the odd SPP mode in a MIM plasmon slot waveguide. (a) Schematic of the MIM waveguide and the coordinate system. (b) The dispersion characteristics of the SPP modes supported by the structure in (a). The green curve is the dispersion branch corresponding to the odd mode. (c,d) The electric (c) and magnetic field components of the odd SPP mode. (e) The time-averaged optical powerflow in the MIM waveguide. (f) The corresponding instantaneous powerflow at $t = 0$. All field distributions were calculated at $w = 4.974 \times 10^{15}$ rad/s.

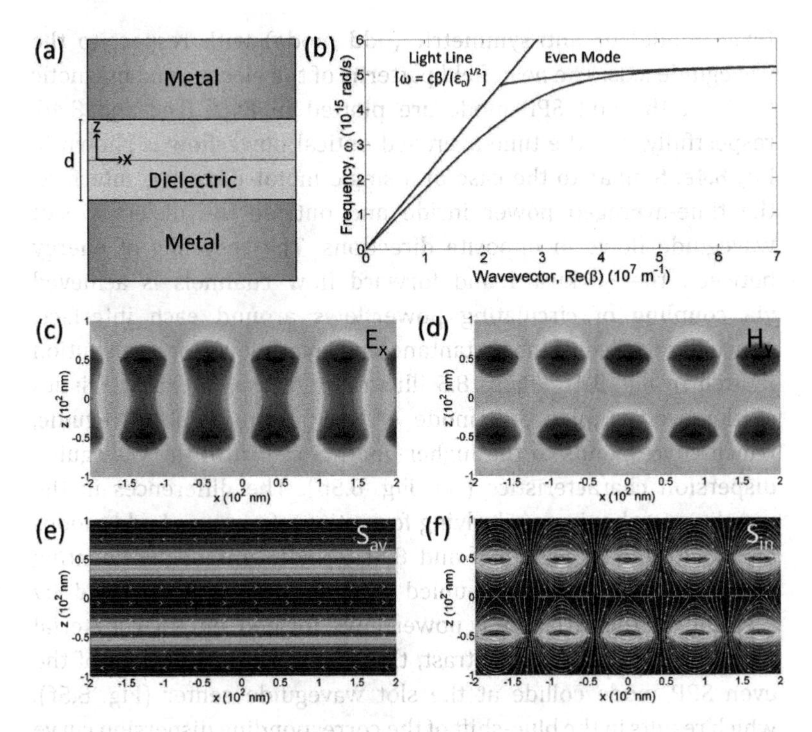

Figure 8.5 Shared circulating powerflow of the even SPP mode in a MIM plasmon slot waveguide. (a) Schematic of the MIM waveguide and the coordinate system. (b) The dispersion characteristics of the SPP modes supported by the structure in (a). The red curve is the dispersion branch corresponding to the even mode. (c,d) The electric (c) and magnetic field components of the even SPP mode. (e) The time-averaged optical powerflow in the MIM waveguide. (f) The corresponding instantaneous powerflow at $t = 0$. All field distributions were calculated at $w = 4.974 \times 10^{15}$ rads.

are plotted in Figs. 8.4b and 8.5b. The dispersion characteristics are obtained by solving the matrix equation with the appropriate boundary conditions on both metal–dielectric interfaces:

$$\text{Odd mode}: \tanh(k_1 a) = -\frac{k_2 \varepsilon_1}{k_1 \varepsilon_2} \tag{8.8}$$

$$\text{Even mode}: \tanh(k_1 a) = -\frac{k_1 \varepsilon_2}{k_2 \varepsilon_1} \tag{8.9}$$

Dispersion relations (8.8) and (8.9) correspond to the modes with tangential electric field distributions that are either symmetric (even mode) or anti-symmetric (odd mode) with respect to the waveguide axis. The near-field patterns of the electric and magnetic fields of the odd SPP mode are plotted in Figs. 8.4c and 8.4d, respectfully, and the time-averaged optical powerflow is shown in Fig. 8.4e. Similar to the case of a single metal–dielectric interface, the time-averaged power inside and outside the dielectric slot waveguide flows in opposite directions. The recycling of energy between the backward and forward flow channels is achieved via coupling of circulating powerflows around each interface, which is revealed in the instantaneous Poynting vector distribution plotted in Fig. 8.4f. Figure 8.5 illustrates the same characteristics of the even-coupled SPP mode of the plasmon slot waveguide, which corresponds to the higher-energy branch in the waveguide dispersion characteristics (see Fig. 8.5b). The differences in the coupling mechanism underlying formation of even and odd modes are illustrated in Figs. 8.4f and 8.5f. The instantaneous Poynting vector field of the odd-coupled SPP mode is characterized by merging of the circulating powerflows formed on each material interface (Fig. 8.4f). In contrast, the circulating powerflows of the even SPP mode collide at the slot waveguide center (Fig. 8.5f), which results in the blue-shift of the corresponding dispersion curve (Fig. 8.5b).

8.4 Hyperbolic Metamaterials: Global Field Topology Defined by Local Topological Features

Even more interesting optical effects can be engineered via near-field optical coupling between SPP waves formed on multiple stacked M-I interfaces. The resulting anisotropic nanostructured metal-dielectric material is schematically shown in Fig. 8.6a. Mutual electromagnetic coupling of SPP modes across multiple M-I interfaces results in the formation of several SPP branches in the dispersion characteristics of such anisotropic structures (shown in Fig. 8.6b). Here and in the following figures, the thicknesses of the Ag and TiO_2 layers are 9 nm and 22 nm, respectively, and the dissipative losses in Ag are fully accounted for. The presence of multiple high-k branches seen in Fig. 8.6b increases the bandwidth of the plasmonic-

enhanced high-DOS spectral region over that of the SPP on a single interface [29].

Accurate solutions for the dispersion of the multilayered metamaterial plotted in Fig. 8.6b were obtained by using the analytical transfer matrix method. However, in the limit of an infinite number of ultra-thin layers, the multilayered metamaterials shown in Fig. 8.6a can be described within an effective index model by a uniaxial effective dielectric tensor $\hat{\varepsilon} = \text{diag}\left[\varepsilon_{xx}, \varepsilon_{yy}, \varepsilon_{zz}\right]$ [45, 53–58]. If $\varepsilon_{xx} = \varepsilon_{yy} = \varepsilon_{||}$ and $\varepsilon_{||} \cdot \varepsilon_{zz} < 0$, the anisotropic metamaterial dispersion relation $\omega^2/c^2 = \left(k_x^2 + k_y^2\right)/\varepsilon_{zz} + k_z^2/\varepsilon_{||}$ transforms from having a closed shape, that is, ellipsoid, to an

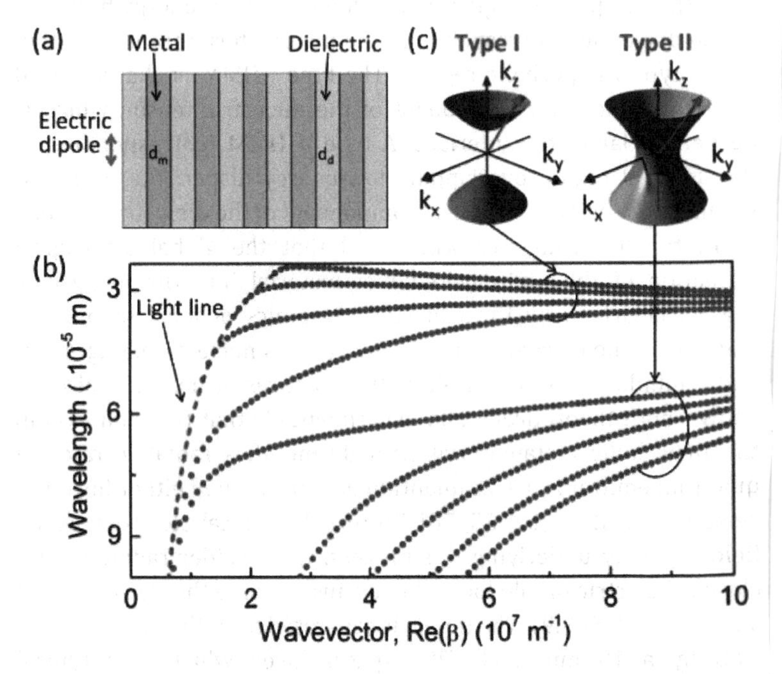

Figure 8.6 Metal-dielectric hyperbolic metamaterials: SPP modes dispersion and iso-frequency surfaces. (a) Schematic of a multilayered metamaterial (metal layers are grey colored). (b) Modal dispersion characteristics of the metamaterial. (c) Momentum-space representation of the dispersion relations (iso-frequency surfaces) in two frequency ranges where the metamaterial undergoes a topological transition from an ordinary material to HMM, with allowed wavevectors shown as red arrows.

open one, that is, hyperboloid as shown in Fig. 8.6c [45, 53–58]. This topological transition from an ordinary to a so-called hyperbolic photonic metamaterial (HMM) is an analog of the Lifshitz topological transition in superconductors [59] when their Fermi surfaces undergo transformation under the influence of external factors such as pressure or magnetic fields. In the same manner that the Lifshitz transition leads to dramatic changes in the electron transport in metals [60, 61], the topological transition of the photonic metamaterial has a dramatic effect on its DOS and photon transport characteristics. This enables development of novel devices with enhanced optical properties including super-resolution imaging, optical cloaking, and enhanced radiative heat transfer [45, 53–58]. The spectral region that includes the high-energy branches of the dispersion corresponds to the metamaterial transformation into a type I hyperbolic regime. The type I HMM is characterized by a single negative component of the dielectric tensor, which is perpendicular to the interface. A type II HMM (corresponding to the spectral region overlapping low-energy dispersion branches) features two negative in-plane components of the dielectric tensor.

In the following, we will reveal that the global topological transition of the multilayered metamaterial into the hyperbolic regime is driven by the collective dynamics of local topological features in the electromagnetic field such as nucleation, migration, and annihilation of nanoscale optical vortices. It has been already shown both theoretically and experimentally that the high DOS in the hyperbolic metamaterial strongly modifies radiative rates of quantum emitters such as quantum dots positioned either inside or close to the HMM [53, 55, 56]. To reveal the local electromagnetic field topology underlying this process, we consider radiation of a classical electric dipole located in air just outside the HMM slab (as shown Figs. 8.6–8.9). The dipole is separated from the metamaterial slab by a 10 nm thick TiO_2 spacer layer, which is a typical configuration in the experiment that helps to avoid quenching of quantum emitters such as quantum dots via non-radiative energy transfer to the metamaterial slab [53, 55, 56].

Figures 8.7 and 8.8 demonstrate how the metamaterial in the type I and type II hyperbolic regimes modifies radiation and transport of photons emitted by a dipole source. The availability

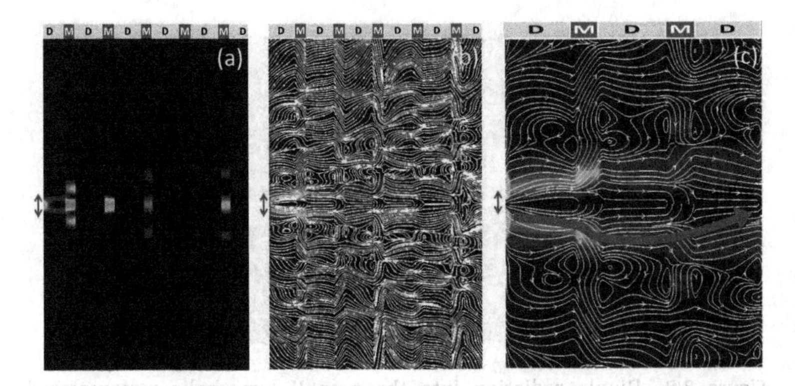

Figure 8.7 Dipole radiation into the type I hyperbolic metamaterial. (a) Electric field intensity distribution inside a 10 period thick Ag/TiO_2 multilayer structure generated by a classical electric dipole shown as the orange arrow. (b) Optical powerflow from the dipole source through the metamaterial. (c) A magnified view of the region in (b) inside the orange rectangle. In (b) and (c), the Poynting vector streamlines (white) indicate the local direction of the powerflow, and the background color map shows the time-averaged Poynting vector intensity distribution. The orange arrows in (c) highlight the global powerflow direction. In all the panels, the excitation wavelength is 350 nm.

of high-k states in the HMM provides additional channels for the dipole to radiatively decay and increases the electromagnetic energy density within the metamaterial. As a result, the dipole source emission pattern becomes strongly asymmetrical with most of the energy being channeled through the metamaterial. Furthermore, as revealed by Fig. 8.7, for the metamaterial in the type I hyperbolic regime, energy transport through the material is very directional. Most of the optical power flows through a narrow channel across the metamaterial without experiencing significant lateral spread within the metamaterial slab. A detailed picture of the local topological features that drive the directional powerflow across the metamaterial is shown in Fig. 8.7b and 8.7c. It should be noted that in Figs. 8.7–8.9, we plot the time-averaged powerflow defined by (8.4) rather than the instantaneous powerflow. Multiple areas of coupled vortex powerflow are clearly visible and their spatial arrangement favors an overall directional power flow by preventing lateral energy spread. The local topology-driven directional energy

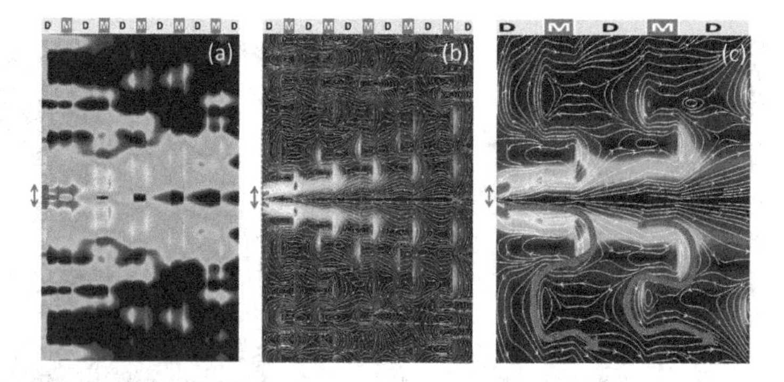

Figure 8.8 Dipole radiation into the type II hyperbolic metamaterial. (a) Electric field intensity distribution inside a 10 period thick Ag/TiO_2 multilayer structure generated by a classical electric dipole shown as the orange arrow. (b) Optical powerflow from the dipole source through the metamaterial. (c) A magnified view of the region in (b) inside the orange rectangle. In (b) and (c), the Poynting vector streamlines (white) indicate the local direction of the powerflow, and the background color map shows the time-averaged Poynting vector intensity distribution. The orange arrows in (c) highlight the global powerflow direction. In all the panels, the excitation wavelength is 650 nm.

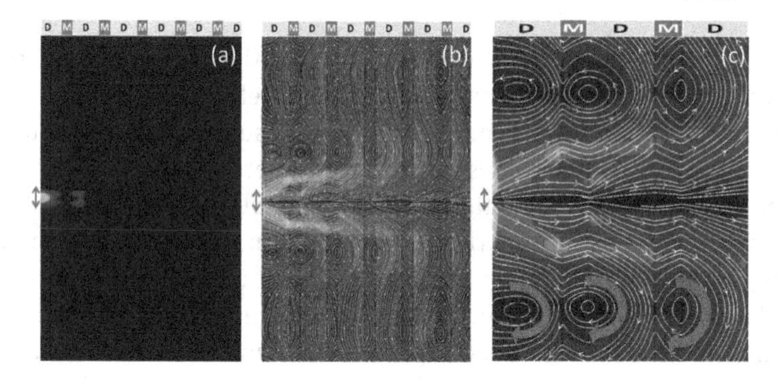

Figure 8.9 Dipole radiation into the multilayer stack outside of the hyperbolic regime. (a) Electric field intensity distribution inside a 10 period thick Ag/TiO_2 multilayer structure generated by a classical electric dipole shown as the orange arrow. (b) Optical powerflow from the dipole source through the multilayer. (c) A magnified view of the region in (b) inside the orange rectangle. In (b) and (c), the time-averaged Poynting vector streamlines (white) indicate the local direction of the powerflow, and the background color map shows the Poynting vector intensity distribution. The orange arrows in (c) highlight the global powerflow direction. In all the panels, the excitation wavelength is 500 nm.

transport across the HMM slab also results in the directional emission of the energy by the HMM surface, which can be used for the design of directional light sources [62].

Likewise, the high DOS within the metamaterial in the type II hyperbolic regime results in emission from the dipole source that is predominantly into the material. However, the powerflow driven by the emitted photons is markedly different from the type I HMM. In contrast, the optical powerflow through the type II HMM features significant lateral spread, as shown in Fig. 8.8. The local topology of the electromagnetic interference field in the metamaterial slab plotted in Fig. 8.8b and 8.8c reveals the mechanism driving the lateral energy spread. Once again, the formation of multiple coupled counter-rotating nanoscale optical vortices drives the global powerflow by recirculating the energy around local circulation points. The arrangement of the vortices in this case is different from that in the type I HMM, resulting in the markedly different overall powerflow pattern.

Finally, Fig. 8.9 illustrates the dramatic changes in the powerflow through the metamaterial in the photon energy range where the material in not in the hyperbolic regime. In this case, the metamaterial DOS is significantly lower than that in either type of HMM. This reduces the dipole radiative rate resulting in the low-energy density within the material as shown in Fig. 8.9a. The time-averaged Poynting vector field within the material still features areas of circulating powerflow as shown in Fig. 8.9b and 8.9c. However, in this case, the vortices once again are spatially rearranged and form a global energy recirculation network that inhibits powerflow across the metamaterial slab. Overall, we can conclude that the re-arrangement of the local field topology features is behind the global topological transitions in hyperbolic metamaterials.

8.5 Conclusion and Outlook

We have demonstrated that the formation of optical vortices—localized areas of circulating optical powerflow—is a hidden mechanism behind many unique characteristics of SPP modes on metal-dielectric interfaces. These include tight energy localization in the vicinity of the M-I interface and 'structural slow light' characterized by the reduced group velocity of SPP waves. Furthermore, a detailed understanding of the SPP powerflow characteristics helps to explain the existence of photon states with high linear lateral momentum only in close proximity to the interface. This is a manifestation of the angular momentum of photons recycled through coupled optical vortices formed on the interface. Furthermore, our analysis reveals that the formation and dynamical reconfiguring of connected networks of coupled optical vortices underlies the global topological transitions of artificial M-I materials into the hyperbolic regime.

The above results provide further support for a rational strategy we recently developed for the design of photonic components with novel functionalities [26, 27]. This bottom-up design strategy is based on the accurate positioning of local phase singularities and connecting them into coupled networks with the aim of tailoring the global spatial structure of the interference field. Understanding the origins and exploiting wave effects associated with phase singularities have proven to be of high importance in many branches of physics, including hydrodynamics, acoustics, quantum physics, and singular optics [63, 64]. To date, the most significant advances in optics driven by controllable formation of optical vortices have been related to the generation of propagating light beams and fiber modes carrying orbital angular momentum. This research has far-reaching applications in optical trapping and manipulation and offers a way to increase the data transmission rates via angular momentum multiplexing mechanisms [51, 65, 66]. By revealing the role of optical vortices in the unique characteristics of surface plasmon waves, we not only explain well-known plasmonic effects but also offer a new bottom-up approach to design plasmonic nanostructures and metamaterials with tailored energy transport characteristics. This can pave the way to new applications of plasmonic materials in optical communications [27, 67, 68], energy harvesting from solar and terrestrial heat sources [29], radiative heat transfer [29], imaging [69], and sensing [25, 26, 28].

Acknowledgments

This work has been supported by the US Department of Energy DOE-BES Grant No. DE-FG02-02ER45977 (for near-field energy transport) and by the US Department of Energy under the SunShot grant No. 6924527 (for alternative ways of manipulating photon trapping and recycling).

References

1. Stockman, M. I., Nanoplasmonics: Past, present, and glimpse into future. *Opt. Express*, 2011, **19**(22), pp. 22029–22106.
2. Halas, N. J., et al., Plasmons in strongly coupled metallic nanostructures. *Chem. Rev.*, 2011, **111**(6), pp. 3913–3961.
3. Schuller, J.A., et al., Plasmonics for extreme light concentration and manipulation. *Nat. Mater.*, 2010, **9**(3), pp. 193–204.
4. Sherry, L.J., et al., Localized surface plasmon resonance spectroscopy of single silver nanocubes. *Nano Lett.*, 2005, **5**(10), pp. 2034–2038.
5. Lal, S., S. Link, and N.J. Halas, Nano-optics from sensing to waveguiding. *Nat. Photon.*, 2007, **1**(11), pp. 641–648.
6. Lee, S.Y., et al., Spatial and spectral detection of protein monolayers with deterministic aperiodic arrays of metal nanoparticles. *Proc. Natl. Acad. Sci. U S A*, 2010, **107**(27), pp. 12086–12090.
7. Wang, J., et al., Illuminating epidermal growth factor receptor densities on filopodia through plasmon coupling. *ACS Nano*, 2011, **5**(8), pp. 6619–6628.
8. Gopinath, A., et al., Deterministic aperiodic arrays of metal nanoparticles for surface-enhanced Raman scattering (SERS). *Opt. Express*, 2009, **17**(5), pp. 3741–3753.
9. Wang, J., et al., Spectroscopic ultra-trace detection of nitro-aromatic gas vapor on rationally designed nanoparticle cluster arrays. *Anal. Chem.*, 2011, **83**(6), pp. 2243–2249.
10. Lee, S.J., et al., Surface-enhanced Raman spectroscopy and nanogeometry: The plasmonic origin of SERS. *J. Phys. Chem. C*, 2007, **111**(49), pp. 17985–17988.
11. Moskovits, M., et al., SERS and the single molecule, in *Optical Properties of Nanostructured Random Media,* 2002, Berlin, Springer-Verlag, pp. 215–226.

12. Kneipp, K., et al., Detection and identification of a single DNA base molecule using surface-enhanced Raman scattering (SERS). *Phys. Rev. E*, 1998, **57**(6), p. R6281.

13. Lakowicz, J.R., J. Malicka, and I. Gryczynski, Silver particles enhance emission of fluorescent DNA oligomers. *BioTechniques*, 2003, **34**(1), p. 62.

14. Pacifici, D., H.J. Lezec, and H.A. Atwater, All-optical modulation by plasmonic excitation of CdSe quantum dots. *Nature Photon.*, 2007, **1**(7), pp. 402–406.

15. Maier, S.A., et al., Local detection of electromagnetic energy transport below the diffraction limit in metal nanoparticle plasmon waveguides. *Nature Mater.*, 2003, **2**(4), pp. 229–232.

16. Atwater, H.A. and A. Polman, Plasmonics for improved photovoltaic devices. *Nature Mater.*, 2010, **9**(3), pp. 205–213.

17. Aydin, K., et al., Broadband polarization-independent resonant light absorption using ultrathin plasmonic super absorbers. *Nature Communicat.*, 2010, **2**, p. 517.

18. Atre, A., C., et al., Toward high-efficiency solar upconversion with plasmonic nanostructures. *J Optics*, 2012, **14**(2), p. 024008.

19. Green, M.A. and S. Pillai, Harnessing plasmonics for solar cells. *Nature Photon.*, 2012, **6**(3), pp. 130–132.

20. Chen, X., et al., Broadband enhancement in thin-film amorphous silicon solar cells enabled by nucleated silver nanoparticles. *Nano Lett.*, 2012, **12**(5), pp. 2187–2192.

21. Hu, L., X. Chen, and G. Chen, Surface-plasmon enhanced near-bandgap light absorption in silicon photovoltaics. *J. Comput. Theor. Nanosci.*, 2008, **5**(11), pp. 2096–2101.

22. Lee, J., et al., Plasmonic photoanodes for solar water splitting with visible light. *Nano Lett.*, 2012, **12**(9), pp. 5014–5019.

23. Linic, S., P. Christopher, and D.B. Ingram, Plasmonic-metal nanostructures for efficient conversion of solar to chemical energy. *Nature Mater.*, 2011, **10**(12), pp. 911–921.

24. Neumann, O., et al., Solar vapor generation enabled by nanoparticles, *ACS Nano*, 2013, **7**(1), pp. 42–49.

25. Ahn, W., et al., Electromagnetic field enhancement and spectrum shaping through plasmonically integrated optical vortices. *Nano Lett.*, 2012, **12**(1), pp. 219–227.

26. Boriskina, S. V., Plasmonics with a twist: taming optical tornadoes on the nanoscale, in *Plasmonics: Theory and Applications*, T. V Shahbazyan & M. I. Stockman (Eds.), pp. 431–461, vol. 5 of *Challenges and Advances in Computational Chemistry and Physics*, 2013, Springer.
27. Boriskina, S.V. and B.M. Reinhard, Molding the flow of light on the nanoscale: from vortex nanogears to phase-operated plasmonic machinery. *Nanoscale*, 2012, **4**(1), pp. 76–90.
28. Hong, Y., et al., Enhanced light focusing in self-assembled optoplasmonic clusters with subwavelength dimensions. *Adv. Mat.*, 2013, **25**(1), pp. 115–119.
29. Boriskina, S.V., H. Ghasemi, and G. Chen, Plasmonic materials for energy, From physics to applications. *Materials Today*, 2013, **16**(10), pp. 375–386.
30. Schouten, H., F., T.D. Visser, and D. Lenstra, Optical vortices near sub-wavelength structures. *J. Opt. B*, 2004, **6**(5), p. S404.
31. Dennis, M.R., et al., Singular optics: More ado about nothing. *J. Opt. A*, 2009, **11**, p. 090201.
32. Dennis, M.R., K. O'Holleran, and M.J. Padgett, Singular optics: optical vortices and polarization singularities. *Prog. Opt.*, 2009, **53**, pp. 293–363.
33. Soskin, M.S. and M.V. Vasnetsov, Singular optics. *Prog. Opt.*, 2001, **42**, pp. 219–276.
34. Allen, L., et al., Orbital angular momentum of light and the transformation of Laguerre-Gaussian laser modes. *Phys. Rev. A*, 1992, **45**(11), p. 8185.
35. Thomas, J.-L. and R. Marchiano, Pseudo angular momentum and topological charge conservation for nonlinear acoustical vortices. *Phys. Rev. Lett.*, 2003, **91**(24), p. 244302.
36. Bohren, C.F., How can a particle absorb more than the light incident on it? *Am. J. Phys.*, 1983, **51**(4), pp. 323–327.
37. Tribelsky, M.I. and B.S. Luk'yanchuk, Anomalous light scattering by small particles. *Phys. Rev. Lett.*, 2006, **97**(26), p. 263902.
38. Alù, A. and N. Engheta, Higher-order resonant power flow inside and around superdirective plasmonic nanoparticles. *J. Opt. Soc. Am. B*, 2007, **24**(10), pp. A89–A97.
39. Bashevoy, M., V. Fedotov, and N. Zheludev, Optical whirlpool on an absorbing metallic nanoparticle. *Opt. Express*, 2005, **13**(21), pp. 8372–8379.

40. Lukyanchuk, B.S. and et al., Peculiarities of light scattering by nanoparticles and nanowires near plasmon resonance frequencies. *J. Phys.: Conf. Series*, 2007, **59**(1), p. 234.
41. Li, K., M.I. Stockman, and D.J. Bergman, Self-similar chain of metal nanospheres as an efficient nanolens. *Phys. Rev. Lett.*, 2003, **91**(22), p. 227402.
42. Rahmani, M., et al., Beyond the hybridization effects in plasmonic nanoclusters: diffraction-induced enhanced absorption and scattering. *Small*, 2013, **10**(3), pp. 417–616.
43. Sundararajan, S.P., et al., Nanoparticle-induced enhancement and suppression of photocurrent in a silicon photodiode. *Nano Lett.*, 2008, **8**(2), pp. 624–630.
44. Lin, A., et al., An optimized surface plasmon photovoltaic structure using energy transfer between discrete nano-particles. *Opt. Express*, 2013, **21**(S1), pp. A131–A145.
45. Jacob, Z., et al., Engineering photonic density of states using metamaterials. *Appl. Phys. B*, 2010, **100**(1), pp. 215–218.
46. Philipp, H.R., *Silicon Dioxide (SiO2) (Glass)*, in *Handbook of Optical Constants of Solids*, E.D. Palik (Ed.), 1997, Academic Press, Burlington, pp. 749–763.
47. Johnson, P.B. and R.W. Christy, Optical constants of the noble metals. *Phys. Rev. B*, 1972, **6**(12), pp. 4370–4379.
48. IEEE, *IEEE 100 The Authoritative Dictionary of IEEE Standards Terms Seventh Edition*, in *IEEE Std 100-2000*, 2000.
49. Federal Energy Regulatory Commission and U.S./Canada Power Outage Task Force, *August 14, 2003 Outage sequence of events*, 2003.
50. Rosenblatt, G., E. Feigenbaum, and M. Orenstein, Circular motion of electromagnetic power shaping the dispersion of surface plasmon polaritons. *Optics Express*, 2010, **18**(25), pp. 25861–25872.
51. Allen, L., S.M. Barnett, and M.J. Padgett, *Optical Angular Momentum*, 2003, Bristol and Philadelphia, IOP Publishing.
52. Dionne, J.A., et al., Plasmon slot waveguides: Towards chip-scale propagation with subwavelength-scale localization. *Phys. Rev. B*, 2006, **73**(3), p. 035407.
53. Noginov, M.A., et al., Controlling spontaneous emission with metamaterials. *Opt. Lett.*, 2010, **35**(11), pp. 1863–1865.
54. Guo, Y. and Z. Jacob, Thermal hyperbolic metamaterials. *Opt. Express*, 2013, **21**(12), pp. 15014–15019.

55. Jacob, Z., I.I. Smolyaninov, and E.E. Narimanov, Broadband Purcell effect: Radiative decay engineering with metamaterials. *Appl. Phys. Lett.*, 2012, **100**(18), pp. 181105–181114.
56. Krishnamoorthy, H.N.S., et al., Topological transitions in metamaterials. *Science*, 2012, **336**(6078), pp. 205–209.
57. Smolyaninov, I.I., Y.-J. Hung, and C.C. Davis, Magnifying superlens in the visible frequency range. *Science*, 2007, **315**(5819), pp. 1699–1701.
58. Fang, N., et al., Sub-diffraction-limited optical imaging with a silver superlens. *Science*, 2005, **308**(5721), pp. 534–537.
59. Lifshitz, I.M., Anomalies of electron characteristics of a metal in the high-pressure region. *Sov. Phys. JETP*, 1960, **11**, pp. 1130–1135.
60. Liu, C., et al., Evidence for a Lifshitz transition in electron-doped iron arsenic superconductors at the onset of superconductivity. *Nature Phys.*, 2010, **6**(6), pp. 419–423.
61. Schlottmann, P., Calculation of electric transport close to a Lifshitz transition in a high magnetic field. *Euro. Phys. J. B*, 2013, **86**(3), pp. 1–7.
62. Molesky, S., C.J. Dewalt, and Z. Jacob, High temperature epsilon-near-zero and epsilon-near-pole metamaterial emitters for thermophotovoltaics. *Optics Express*, 2013, **21**(S1), pp. A96–A110.
63. Nye, J.F. and M.V. Berry, Dislocations in wave trains. *Proc. Royal Soc. London A. Math. Phys. Sci.*, 1974, **336**(1605), pp. 165–190.
64. Berry, M., Making waves in physics. *Nature*, 2000, **403**(6765), p. 21.
65. Allen, L., et al., *IV The Orbital Angular Momentum of Light*, in *Progress in Optics*. 1999, Elsevier, pp. 291–372.
66. Bozinovic, N., et al., Terabit-scale orbital angular momentum mode division multiplexing in fibers. *Science*, 2013, **340**(6140), pp. 1545–1548.
67. Sukhorukov, A.A., et al., Slow-light vortices in periodic waveguides. *J. Opt. A*, 2009, **11**, p. 094016.
68. Boriskina, S.V. and B.M. Reinhard, Adaptive on-chip control of nano-optical fields with optoplasmonic vortex nanogates. *Opt. Express*, 2011, **19**(22), pp. 22305–22315.
69. D'Aguanno, G., et al., Optical vortices during a superresolution process in a metamaterial. *Phys. Rev. A*, 2008, **77**(4), p. 043825.

Chapter 9

Nano-Fano Resonances and Topological Optics

Boris Luk'Yanchuk,[a] Zengbo Wang,[b] Andrey E. Miroshnichenko,[c] Yuri S. Kivshar,[c] Arseniy I. Kuznetsov,[a] Dongliang Gao,[d] Lei Gao,[d] and Cheng-Wei Qiu[e]

[a] *Data Storage Institute, Agency for Science, Technology and Research, 117608 Singapore, Singapore*
[b] *School of Electronic Engineering, Bangor University, Bangor LL57 1UT, UK*
[c] *Nonlinear Physics Centre, Research School of Physics and Engineering Australian National University, Canberra ACT 0200, Australia*
[d] *School of Physical Science and Technology, Soochow University, Suzhou 215006, China*
[e] *Department of Electrical and Computer Engineering, National University of Singapore, 119260 Singapore, Singapore*

Fano resonances and optical vortices are two well-known interference phenomena associated with the scattering of light. Usually, these two phenomena are considered to be completely independent, and in many cases, Fano resonances are observed without vortices and the vortices with the singular phase structure are not accompanied by Fano resonances. However, this situation changes dramatically when we move to the nanoscale. In this chapter, we demonstrate that Fano resonances observed for the light scattering by nanoparticles are accompanied by the singular phase effects (usually associated with topological optics) and the

Singular and Chiral Nanoplasmonics
Edited by Svetlana V. Boriskina and Nikolay I. Zheludev

ISBN 978-981-4613-17-0 (Hardcover), 978-981-4613-18-7 (eBook)
www.panstanford.com

generation of optical vortices with the characteristic core size well beyond the diffraction limit. Such effects are found for weakly dissipative metallic nanoparticles within the Mie theory. Important peculiarities of the far-field scattering and near-field Poynting flux are manifested in the so-called "nano-Fano resonances" introduced and discussed here. Control of the orbital momentum of photons with the help of nanostructures is a novel research direction, which is very attractive for many applications in quantum optics and information technologies, and it opens an unprecedented way for manipulating optical vortices at the nanoscale.

9.1 Introduction

The fascinating physics of light scattering offers two classes of important interference phenomena: optical vortices and Fano resonances. The optical vortices are observed in the structure of the wave phase that carries topological singularities, and therefore, they are associated with the so-called topological optics (also called singular optics and dislocations of the wave front). The Fano resonances are associated with a sharp asymmetric variation of the wave transmission. In general, these two phenomena are seemingly unrelated, and Fano resonances are observed without topological effects as well as singular optics is not necessary accompanied by Fano resonance effect. However, the situations change dramatically as soon as we reduce the size of the scattering structure and approach the nanoscale.

For a majority of plasmonic nanostructures with weak dissipation, the pronounced Fano resonance appears when the typical size parameter becomes of the order of unity, $q = 2\pi R/\lambda \approx 1$, where R is the characteristic scale of the structure and λ is the radiation wavelength. The term "pronounced" used here means that the corresponding Fano resonance is distinguished in comparison with the characteristic scale of the dipole resonance (Rayleigh scattering). This condition implies that a majority of plasmonic nanostructures with Fano resonances suffer from scaling.

Meanwhile, it is possible to generate Fano resonances within the nanostructures with a very small size parameter $q \ll 1$

associated with very weak scattering. The fascinating property of such structures is related to the coexistence of the Fano resonance and topological optics effects, where the characteristic size of vortices is well beyond the diffraction limit. This property can be found, for example, for weakly dissipative plasmonic nanoparticles in the framework of the Mie theory. In this chapter, we demonstrate that the limitation $q \approx 1$ for the pronounced Fano resonances can be overtaken for the cylindrical plasmonic structures, which exhibit a Fano resonance at the nanoscale, $q << 1$. Important peculiarities of the far-field scattering and near-field Poynting flux are presented for this novel type of "nano-Fano resonance." We believe that our results provide an insightful mechanism for the manipulation of Fano resonances at the extreme nanoscale, and also open an unprecedented way for controlling vortices in topological optics.

9.2 Fano Resonance and its Mechanical Analogue

The fascinating phenomenon of Fano resonance refers to the interference of a broad and narrow spectral radiation [1, 2]. In the case of light scattering by small plasmonic particles, the dipole Rayleigh scattering plays a role of a broad spectral radiation and the surface plasmon resonance (e.g., quadrupole or higher-order resonance) plays a role of a narrow spectral line interacting with the broad radiation. In the framework of the well-known Mie theory [3, 4], such a Fano resonance manifests itself in the differential scattering efficiency cross-sections. In terms of the interference of electrical or magnetic scattering amplitudes (namely, a_ℓ or b_ℓ within the Mie theory), it looks like an overlap of broad and narrow spectral lines, for example, broad dipole and narrow quadrupole lines (see Fig. 9.1). Within the Mie theory, the scattering amplitudes are defined by the well-known formulas:

$$a_\ell = \frac{\Re_\ell^{(a)}}{\Re_\ell^{(a)} + i\,\Im_\ell^{(a)}}, \quad b_\ell = \frac{\Re_\ell^{(b)}}{\Re_\ell^{(b)} + i\,\Im_\ell^{(b)}}, \tag{9.1}$$

where the functions $\Re$ and $\Im$ are combinations of the spherical Bessel and Neumann functions, see details in Ref. [5]. The

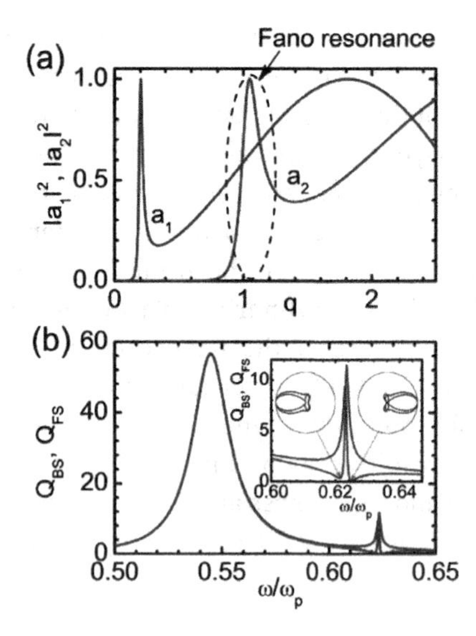

Figure 9.1 (a) Dipole a_1 (blue) and quadrupole a_2 (red) electrical scattering amplitudes versus the size parameter q at $\varepsilon \approx -2.1$. (b) Back scattering (BS; red) and forward scattering (FS; blue) cross-sections versus normalized frequency ω/ω_p. Dielectric permittivity ε is described by the Drude formula with $\gamma/\omega_\text{p} = 10^{-3}$ (weak dissipation), where ω_p is the plasma frequency, and γ is the collision frequency. Parameter $q = \omega_\text{p} R/c = 0.7$. Inset shows polar scattering diagrams in the x–z plane (azimuthal angle $\varphi = 0$ in the Mie theory) near the quadrupole resonance of a plasmonic particle. Red curve corresponds to linearly polarized light; blue lines represent non-polarized light.

resonances correspond to the condition $\Im_\ell^{(a,b)} = 0$, and for nondissipative materials, these conditions correspond to the conditions $a_\ell, b_\ell = 1$.

Figure 9.1a shows the dependences of the scattering amplitudes on the size parameter q. These dependences show two distinct peaks associated with the resonances at $q = 0.2$ (for the dipole mode) and $q = 1$ (for the quadrupole mode). For nondissipative materials, these resonances correspond to the condition $a_\ell = 1$. Periodically repeated "broad" maxima in the case of dielectric particles correspond to volume dipole resonances.

In the case of metallic particles, these maxima do not correspond to resonant excitation of plasmons. Usually, for small particles, the dipole scattering is dominating, $|a_1| >> |a_2|$. Amplitudes of $|a_1|$ and $|a_2|$ become comparable when the size parameter becomes of the order of unity (see Fig. 9.1a). At the same time, for the Fano resonance, we need these two amplitudes being comparable. Thus, with the small size parameter, one of the scattering amplitudes with necessity is very small. As a result, the Fano resonance is "not pronounced" for $q << 1$. However, two scattering modes can interfere either constructively or destructively.

The basic features of the Fano resonance can be understood from a simple classical problem of two coupled oscillators. The existence of interference effects in such a system has been known for a long time, and it has been extensively employed, for example, in mechanical systems for dynamic damping [6]. The dynamics of two coupled oscillators can be described in terms of their displacements x_1 and x_2 from the equilibrium positions.

$$\ddot{x}_1 + \gamma_1 \dot{x}_1 + \omega_1^2 x_1 = \Omega^2 (x_2 - x_1) + f_1 e^{-i\omega t},$$

$$\ddot{x}_2 + \gamma_2 \dot{x}_2 + \omega_2^2 x_2 = \Omega^2 (x_1 - x_2) + f_2 e^{-i\omega t}. \quad (9.2)$$

Here, ω is the frequency of an external force, ω_1 and ω_2 are the eigenfrequencies, Ω describes the coupling between the oscillators, and γ_1 and γ_2 are the dissipation coefficients. The steady-state solutions for the displacement of the oscillators are periodic, $x_1 = x_{10}e^{-i\omega t}$, $x_2 = x_{20}e^{-i\omega t}$, where the amplitudes are given by the expressions:

$$x_{10} = \frac{(f_1 + f_2)\,\Omega^2 - f_1\left(i\gamma_2\omega + \omega^2 - \omega_2^2\right)}{\left(\omega_1^2 - \omega^2 - i\gamma_1\omega\right)\left(\omega_2^2 - \omega^2 - i\gamma_2\omega\right) - \Omega^4},$$

$$x_{20} = \frac{(f_1 + f_2)\,\Omega^2 - f_2\left(i\gamma_1\omega + \omega^2 - \omega_1^2\right)}{\left(\omega_1^2 - \omega^2 - i\gamma_1\omega\right)\left(\omega_2^2 - \omega^2 - i\gamma_2\omega\right) - \Omega^4}. \quad (9.3)$$

The paradigm of the classical analog of the Fano resonance is that light excites only the broad mode, for example x_1, while the narrow resonance mode x_2 (dark mode) is excited just only due to the coupling [7, 8]. In this sense, only one oscillator is driven by a harmonic force, so that we may put $f_2 = 0$. A particular example of this situation is shown in Fig. 9.2. In the resonance region, we identify both constructive (at $\omega = \omega_{10}$) and destructive (at $\omega = \omega_{20}$) interferences.

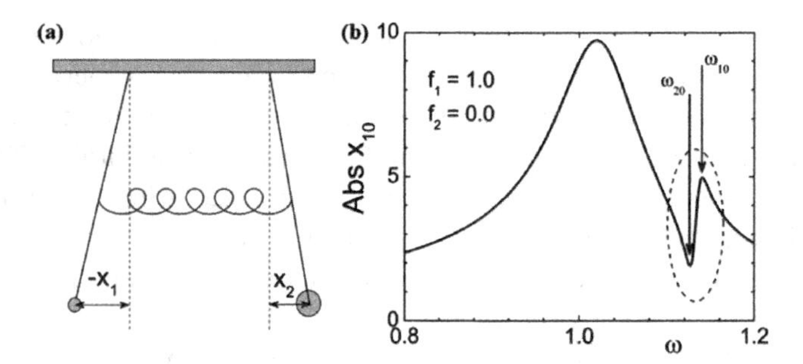

Figure 9.2 (a) Schematic of two dissimilar coupled oscillators. (b) Oscillator amplitude x_{10} calculated from Eq. (9.3) for $\omega_1 = 1$, $\omega_2 = 1.1$, $\Omega = 0.25$, $\gamma_1 = 0.1$, and $\gamma_2 = 0.01$. Constructive and destructive interference occur correspondingly at the resonant frequencies $\omega = \omega_{10}$ and $\omega = \omega_{20}$, shown by arrows. These frequencies are in the vicinity of the hybrid eigenfrequency [8].

We notice that in the vicinity of the Fano resonance, there appears a π-jump in the phase of the second oscillator (see Fig. 9.3). Below that resonant frequency, both oscillators are in phase, whereas above the resonant frequency, they oscillate out of phase.

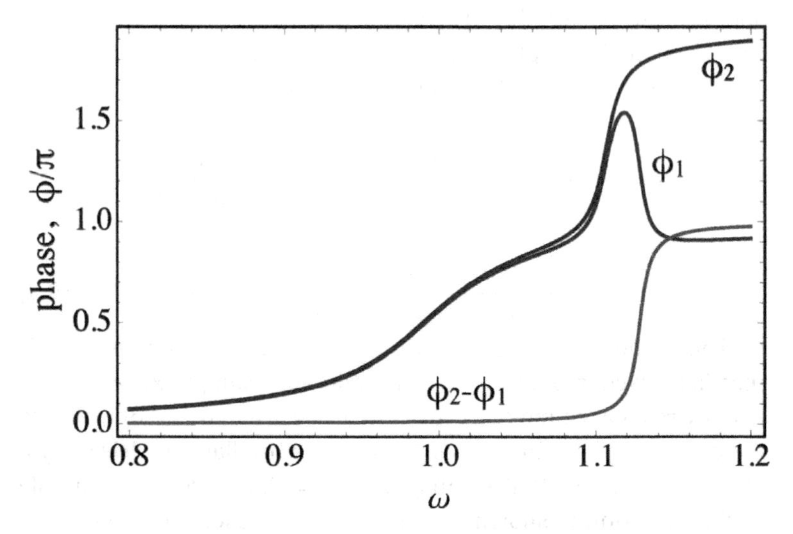

Figure 9.3 Phases of the first (ϕ_1) and second (ϕ_2) oscillators, and their difference as a function of the driving frequency.

The models of coupled oscillators can be employed for the qualitative explanation of many experimental results on the Fano resonances in plasmonic structures (see, e.g., Refs. [9, 10]). However, for more adequate comparison of theoretical and experimental results, the numerical solutions of Maxwell's equations are required. Although the basic physics of Fano resonances in plasmonic materials is well understood, some important issues are not resolved yet. We believe that the scaling of the Fano resonances for small structures is among those important unsolved problems.

9.3 Scaling of the Fano Resonances within the Mie Theory

The basic problem with scaling of Fano resonances can be observed for nondissipative plasmonic nanoparticles within the Mie theory. As mentioned above, the problem of the Mie scattering can be solved analytically for a metallic sphere in which we can employ the exact Mie solutions [3, 4]. We find the Fano resonance in the directional scattering efficiencies for the forward scattering (FS) and backward scattering (BS). The scattering efficiencies are presented by [1]

$$Q_{\mathrm{BS}} = \frac{1}{q^2}\left|\sum_{\ell=1}^{\infty}(2\ell+1)\,(-1)^{\ell}\,[a_\ell - b_\ell]\right|^2,$$

$$Q_{\mathrm{FS}} = \frac{1}{q^2}\left|\sum_{\ell=1}^{\infty}(2\ell+1)\,[a_\ell + b_\ell]\right|^2. \tag{9.4}$$

For $q << 1$ we expand the scattering amplitudes up to q^5

$$a_1 = -\frac{2i}{3}\frac{\varepsilon-1}{\varepsilon+2}q^3 - \frac{2i}{5}\frac{(\varepsilon-1)(\varepsilon-2)}{(\varepsilon+2)^2}q^5, \quad a_2 = -\frac{i}{15}\frac{\varepsilon-1}{2\varepsilon+3}q^5, \tag{9.5}$$

$$b_1 = -\frac{i}{45}q^5(\varepsilon-1), \quad b_2 = 0. \tag{9.6}$$

Importantly, the expansions contain singularities in the electric amplitudes, for example, in the dipole amplitude a_1 at $\varepsilon = -2$ (this singularity is seen in the formula for the Rayleigh scattering), in the quadrupole amplitude a_2 at $\varepsilon = -3/2$, and so on. At the same time, the magnetic amplitudes b_ℓ have no singularities.

However, there are no real singularities at exact plasmon resonances where $\Im_\ell^{(a)} = 0$. Electrical amplitudes a_ℓ tend to unity at the corresponding resonant frequencies, as can be seen clearly from the formula (9.1). This means that the Rayleigh approximation is not applicable at the points of the plasmon resonances where the Rayleigh scattering is replaced by the so-called anomalous light scattering [5]. To escape singularities at the scattering amplitudes, it is sufficient to expand numerator and denominator in (9.1) independently. However, it is not so important for the determination of anisotropy in scattering where the positions of the forward scattering and back scattering resonances can be found from Eqs. (9.5) and (9.6) with sufficient accuracy. Applying expansions (9.5) and (9.6) to Eq. (9.4), one can find that the solution of the equations $Q_{\mathrm{FS}} = 0$ and $Q_{\mathrm{BS}} = 0$ yields the size parameters

$$q^2 = q_{\mathrm{FS}}^2 = 15\,\frac{(\varepsilon+2)\,(2\varepsilon+3)}{(1-\varepsilon)\left(38+27\varepsilon+\varepsilon^2\right)},$$

$$q^2 = q_{\mathrm{BS}}^2 = 15\,\frac{(\varepsilon+2)\,(2\varepsilon+3)}{70+29\,\varepsilon-10\,\varepsilon^2+\varepsilon^3}. \tag{9.7}$$

Both quantities $q_{\mathrm{FS}}(\varepsilon)$ and $q_{\mathrm{BS}}(\varepsilon)$ vanish in the vicinity of the dipole ($\varepsilon \to -2$) and quadrupole ($\varepsilon \to -1.5$) resonances, see Fig. 9.4. In Fig. 9.4, we do not show nonphysical brunches of Eq. (9.7) corresponding to negative values of q^2.

At the corresponding frequencies, one can observe the transformation of the far-field scattering diagrams typical for the Fano resonance [1]. These polar diagrams are calculated by a standard way, see Ref. [4]. Corresponding scattering intensities are

$$I_{\mathrm{II}}^{(s)} = C\left|\sum_{\ell=1}^{\infty}(-1)^\ell\left[{}^{e}B_\ell P_\ell^{(1)'}(\cos\theta)\sin\theta - {}^{m}B_\ell\frac{P_\ell^{(1)}(\cos\theta)}{\sin\theta}\right]\right|^2,$$

$$I_{\perp}^{(s)} = C\left|\sum_{\ell=1}^{\infty}(-1)^\ell\left[{}^{e}B_\ell\frac{P_\ell^{(1)}(\cos\theta)}{\sin\theta} - {}^{m}B_\ell P_\ell^{(1)'}(\cos\theta)\sin\theta\right]\right|^2. \tag{9.8}$$

In Eq. (9.8), C is the normalization coefficient; other values are the same as in Ref. [4]. For small values of the size parameter $q << 1$, we observe the Fano resonance in weakly dissipated plasmonic

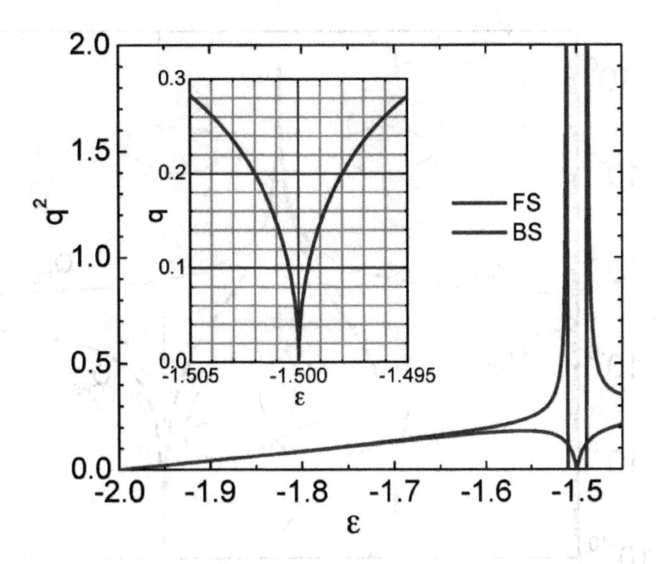

Figure 9.4 Scattering efficiencies versus permittivity. Forward-scattering efficiency vanishes along the blue curve given by the first equation of Eq. (9.7). Red curve corresponds to the second equation of Eq. (9.7), and it presents the back scattering efficiency. Inset shows the region of small q.

nanoparticles. The asymmetrical shapes in differential scattering efficiencies are shown in Fig. 9.5 in the vicinity of the quadrupole resonance. Variation of the scattering diagrams from the forward scattering to backward scattering is shown in the circular insets to Fig. 9.5.

Now, we can illustrate the basic problem that arises with scaling of the Fano resonance. Any real metal has a finite dissipation that strongly influences the behavior of the differential scattering, see Fig. 9.6. When the size parameter q becomes of the order of unity, it is possible to observe the Fano resonance in the media with weak dissipation. However, when $q << 1$, it is impossible to observe the Fano resonance even with weakly dissipating plasmonic media. We notice that for weakly dissipating metals such as K, Na, and Al, the minimum value of the dissipation parameter is about $\text{Im}\varepsilon \approx 0.14 - 0.18$ near the plasmon resonance frequencies [11].

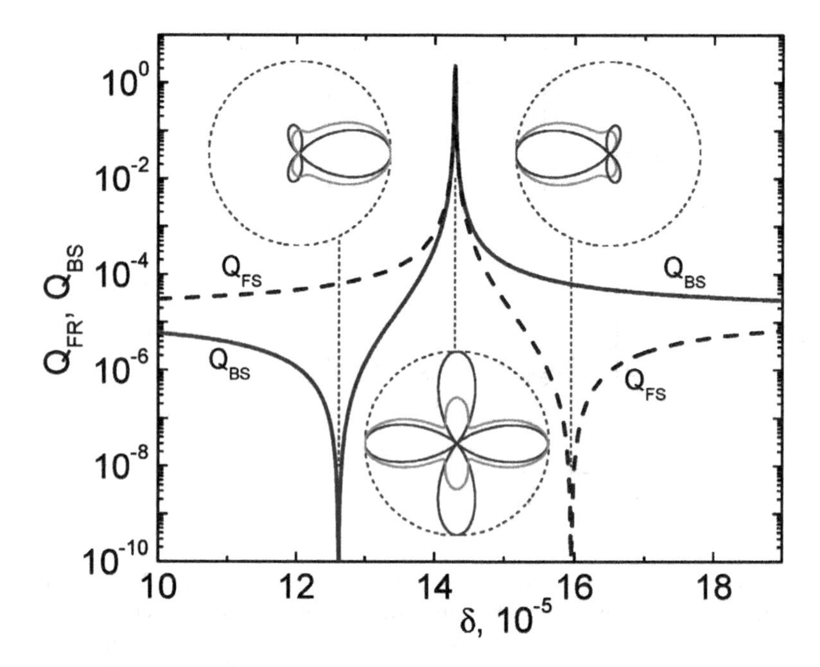

Figure 9.5 Scattering efficiencies Q_{FS} and Q_{BS} for the particle with $q = 0.02$ versus ε. Insets show the far-field scattering diagrams at the corresponding frequencies, calculated with Eq. (9.8) at the corresponding frequencies: $\varepsilon_1 = -1.5 - \delta_1$, where $\delta_1 = 1.67 \cdot 10^{-5}$ corresponds to zero forward scattering, $\varepsilon_2 = -1.5 + \delta_2$ where $\delta_2 = 1.67 \cdot 10^{-5}$ corresponds to zero backward scattering.

9.4 Formation of Vortices within the Mie Theory

Another type of interference phenomena is related to topological optics, and it is associated with terms such as singular optics, vortices, and dislocations of the wave front [12–15]. The basic point of the topological optics is an undefined phase of electromagnetic field at the points where the intensity of the wave vanishes. Any $2\pi n$ phase values can be continuously conjugated at this point. The integer n presents the so-called topological charge, which shows the number of twists that light does in one wavelength. There are many methods to create optical vortices by diffraction, computer-generated holograms, and a special light modulator. Recently, it was shown that the vortices can be created during

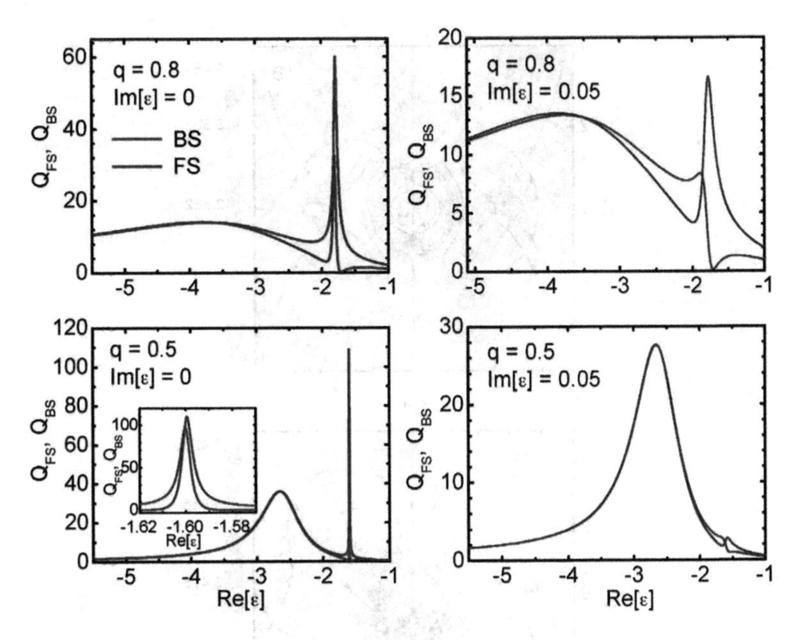

Figure 9.6 Scattering efficiencies Q_{FS} and Q_{BS} for Im $\varepsilon = 0$ and nanoparticles with weak dissipation (Im $\varepsilon = 0.05$). For $q < 0.5$, the Fano resonance disappears for Im $\varepsilon > 0.05$.

light scattering on plasmonic nanostructures [16–18]. For weakly dissipating plasmonic spheres, vortices can be found in the vicinity of dipole and quadrupole plasmon resonances, see the examples in [16–18].

Distribution of the Poynting vector for nondissipative plasmonic material and $q = 0.02$ at the point corresponding to the symmetrical quadrupole resonance and forward and back scattering are presented in Fig. 9.7. One can see great modifications of the vortices structures in the near field. The vortices are created during light scattering the plane wave in contrast to Ref. [19], in which the special emitter of whispering gallery modes was used.

The lowest-order Fano resonance within plasmonic nanoparticles is observed due to interference of the electric dipole and electric quadrupole [1]. Pronounced interference occurs when both interfering amplitudes are comparable at the same frequency. As we have mentioned above, the pronounced interference occurs

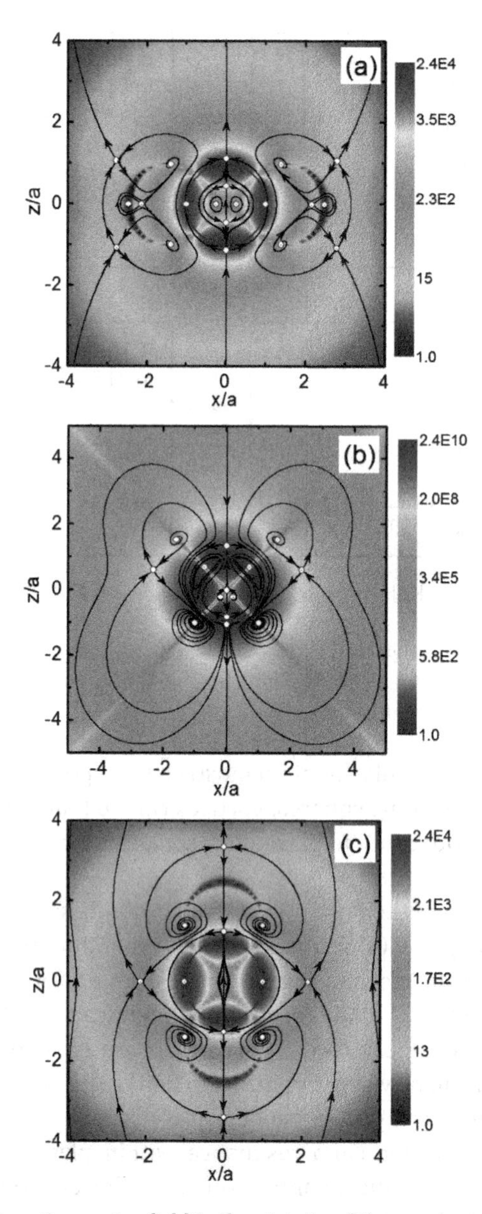

Figure 9.7 Poynting vector field in the vicinity of the quadrupole resonance at $q = 0.02$. (a–c) correspond to different values of ε with the far-field scattering diagrams shown in Fig. 9.5.

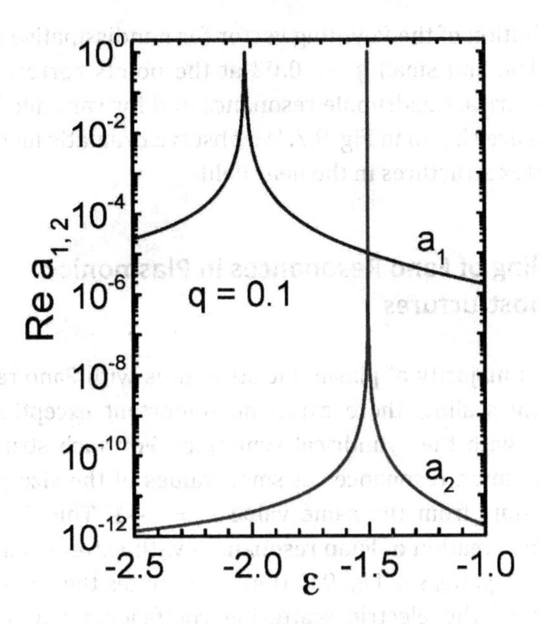

Figure 9.8 The dipole and quadrupole amplitudes versus dielectric permittivity for $q = 0.1$. Interference and Fano resonance shown in Fig. 9.4 occurs in the vicinity of $\varepsilon \approx -1.5$ where the dipole amplitude is five orders of magnitude smaller than the resonance quadrupole amplitude.

when the size parameter q is of the order of unity. With small values of the size parameter $q \to 0$, these dipole and quadrupole resonances appear at different frequencies, which correspond to $\varepsilon = -2$ for dipole and $\varepsilon = -1.5$ for quadrupole. Thus, the interference occurs in the range of frequencies where one of the interference amplitude is small. For example, when $q = 0.1$, the dipole amplitude is five orders of magnitude smaller than the quadrupole amplitude in the range of interference, see Fig. 9.8. Thus, it is very questionable to observe Fano resonances during the scattering light on small spheres. Unfortunately, a similar effect takes place on the hybridization diagrams for a majority of plasmonic nanostructures, that is, dipole and quadrupole resonances require different values of ε as $q \to 0$. This result suggests that Fano resonance is not observable in nanostructures with a size much smaller than the wavelength.

Distribution of the Poynting vector for nondissipative plasmonic nanoparticle and small $q = 0.02$ at the points corresponding to the symmetrical quadrupole resonance and forward and backward scattering are shown in Fig. 9.7. We observe dramatic modifications of the vortex structures in the near field.

9.5 Scaling of Fano Resonances in Plasmonic Nanostructures

Although a majority of plasmonic structures with Fano resonances suffer from scaling, there exists an important exception for the structures with the cylindrical symmetry. For such structures, all surface plasmon resonances at small values of the size parameter $q \to 0$ start from the same value $\varepsilon = -1$. This, in principle, permits the creation of Fano resonances with extreme nano-size in the visible regime, see Fig. 9.9. Here, $\bar{a}_l$ denotes the corresponding amplitude of the electric scattering coefficients within the Mie theory for the cylinder [20–22]. Here, a bar is used to distinguish these scattering amplitudes for spheres and cylinders. In the cylindrical nanowire, the surface plasmons are excited in the case of the perpendicular polarization, $\mathbf{E} \perp \boldsymbol{z}$ (TE-mode), and, thus are not excited with $\mathbf{E} \parallel \boldsymbol{z}$ (TM-mode) [20]. For the case of TE-mode and normal incidence, the scattering efficiency Q_{sca} is given by the expression

$$Q_{\text{sca}} = \frac{2}{q} \sum_{\ell = -\infty}^{\infty} |\, \bar{a}_\ell \,|^2. \tag{9.9}$$

In this definition of $q = kR$, R is the physical radius of the nanowire. The scattering amplitudes $\bar{a}_\ell$ (electric) are defined by the well-known formulas [3, 20]. However, it is convenient to write these formulas by separating the real and imaginary parts

$$\bar{a}_\ell = \frac{\Re_\ell}{\Re_\ell + i\,\Im_\ell}, \tag{9.10}$$

where the functions $\Re_\ell$ and $\Im_\ell$ are given by:

$$\begin{aligned} \Re_\ell &= n\, J_\ell\,(nq)\; J_\ell'\,(q) - J_\ell'\,(nq)\; J_\ell\,(q), \\ \Im_\ell &= n\, J_\ell\,(nq)\; N_\ell'\,(q) - J_\ell'\,(nq)\; N_\ell\,(q), \end{aligned} \tag{9.11}$$

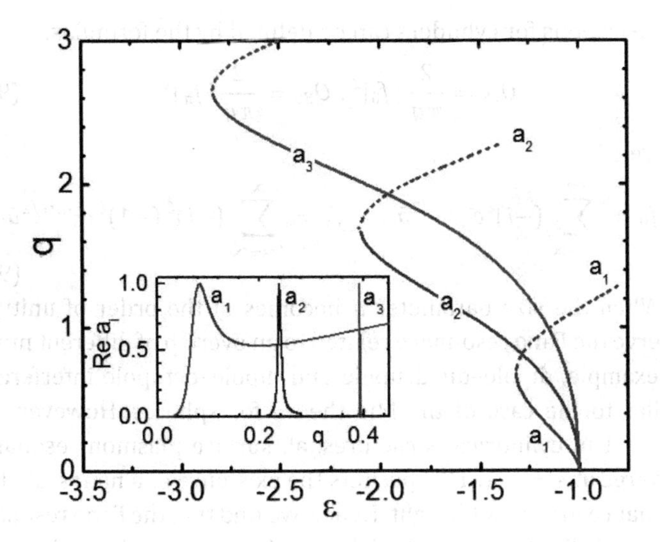

Figure 9.9 Trajectories of the first three optical electric resonances: a_1 (dipole), a_2 (quadrupole), and a_3 (octupole) for a cylinder. Insets show plots of dipole (red) and quadrupole (blue) and octupole (green) resonances versus the size parameter q at $\varepsilon = -1.02$. Monopole mode a_0 for a cylinder is not shown because it appears at higher values of q values, see in Ref. [20].

where $J_\ell(z)$ and $N_\ell(z)$ are the Bessel and the Neumann functions. The strokes in (11) indicate differentiation with respect to the argument of the function, that is, $J'_\ell(z) \equiv dJ_\ell(z)/dz$, etc. The refractive index $n = \sqrt{\varepsilon}$, ε is the relative dielectric permittivity, $\varepsilon = \varepsilon_{\text{p}}/\varepsilon_{\text{m}}$. Indices "p" and "m" indicate wire and ambient media. The coefficients $\bar{a}_\ell$ are symmetric: $\bar{a}_{-\ell} = \bar{a}_\ell$. The scattering cross-section is defined as $\sigma_{\text{sca}} = 2\,R\,L\,Q_{\text{sca}}$, where $2\,R\,L$ is a geometrical cross section (the length of the cylinder $L >> R$). In the case of nondissipative material, $Im\,\varepsilon = 0$, amplitudes $\bar{a}_\ell$ reach the maximal value $\bar{a}_\ell = 1$ along the trajectories on the $\{\varepsilon, q\}$ plane defined by the equation $\Im_\ell(\varepsilon, q) = 0$. Three trajectories for $\ell = 1$ (dipole), $\ell = 2$ (quadrupole), and $\ell = 3$ (octupole) are shown in Fig. 9.9.

The lowest Fano resonance appears due to the interference of the dipole and quadrupole modes. Similar to the spherical counterpart [1], this resonance can be predicted and interpreted with differential forward and back scattering cross-sections, Q_{FS} and Q_{BS}. These cross-sections for cylinders can be defined by the formulas:

$$Q_{\mathrm{FS}} = \frac{2}{\pi q} \left| f_0 \right|^2, \; Q_{\mathrm{BS}} = \frac{2}{\pi q} \left| f_\pi \right|^2, \tag{9.12}$$

where

$$f_0 = \sum_{\ell=-\infty}^{\infty} (-i)^\ell e^{-i\pi\ell/2} \bar{a}_\ell, \quad f_\pi = \sum_{\ell=-\infty}^{\infty} (-i)^\ell (-1)^\ell e^{-i\pi\ell/2} \bar{a}_\ell. \tag{9.13}$$

When the size parameter q becomes of the order of unity, we observe the Fano resonance related to an overlap of different modes, for example, dipole–quadrupole and dipole–octupole interference, similar to the case of the Mie theory for spheres. However, with $q << 1$ in cylindrical structures, all surface plasmon resonances converge at $\varepsilon = -1$. This permits the existence of a nanoscale Fano resonance in the visible light. Hence, we find that the Fano resonance has no size limitation in cylindrical configuration and nondissipative media.

An attractive property of plasmonic structures with Fano resonances is that the spatial distribution of scattered light strongly depends on the frequency [20]. In the vicinity of the Fano resonance, the far-field radiation exhibits strong asymmetry in the radiation pattern, as depicted in Fig. 9.10. In Fig. 9.11, we show the scattering diagrams for the frequencies near the Fano resonance for a nondissipative plasmonic cylinder. The switching is very sensitive to the perturbation of the frequency of the incident light (consequently a perturbation in material permittivity), which is verified in Fig. 9.11. It is important that this variation with cylindrical structures can be reached with very small values of the size parameter $q << 1$, and it can be employed in applications such as optical storage, sensing, or optical switching.

As mentioned above, dissipation strongly suppresses the ability to observe Fano resonances for spherical particles. In fact, the problem of dissipation is critical for any structures with Fano resonance. Thus, the minimal size on the structure with Fano resonance is limited by dissipation. Fortunately, the effect of dissipation is not so strongly pronounced for the structures with the cylindrical symmetry. We consider the frequency dependence of the

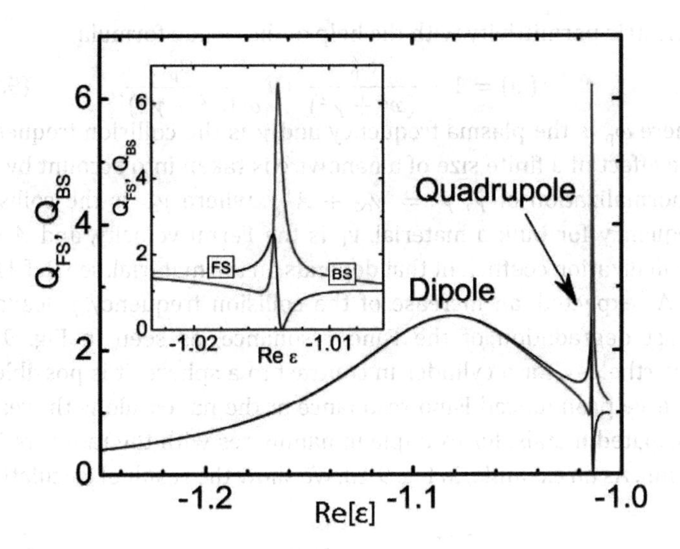

Figure 9.10 Forward and back scattering cross-sections for a nanowire with the size parameter $q = 0.2$. Inset is a zoomed-in view near the new Fano resonance for nanoparticles.

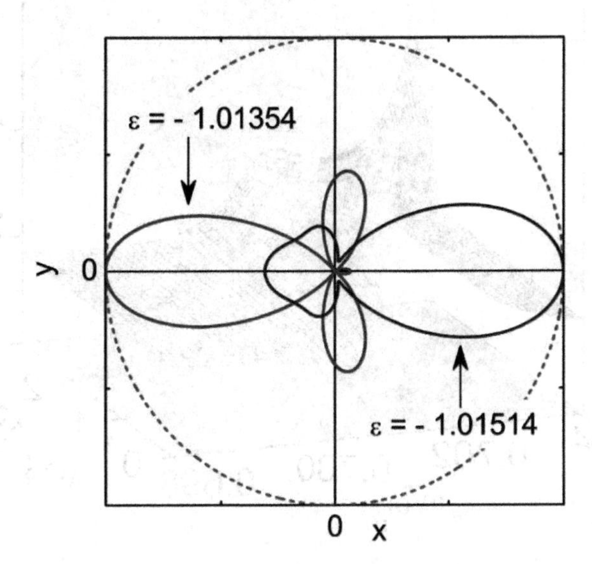

Figure 9.11 Scattering diagrams for the frequencies near the Fano resonance for a nanocylinder with the size parameter $q = 0.2$. A dramatic variation in scattering is observed with $\Delta\varepsilon/\varepsilon \approx 1.6 \cdot 10^{-3}$.

dielectric permittivity with the help of the Drude formula

$$\varepsilon(\omega) = 1 - \frac{\omega_{\mathrm{p}}^2}{(\omega^2 + \gamma^2)} + i\frac{\gamma}{\omega}\frac{\omega_{\mathrm{p}}^2}{(\omega^2 + \gamma^2)}, \tag{9.14}$$

where ω_{p} is the plasma frequency and γ is the collision frequency. The effect of a finite size of a nanowire is taken into account by the renormalization of γ, $\gamma = \gamma_{\infty} + A\frac{v_F}{R}$, where γ_{∞} is the collision frequency for bulk a material, v_{F} is the Fermi velocity, and A is a normalization coefficient that depends on the material, see Ref. [11].

As expected, an increase of the collision frequency γ leads to a fast degradation of the Fano resonance, as seen in Fig. 9.12. Nevertheless, for a cylinder in contrast to a sphere, it is possible to observe pronounced Fano resonance at the nanoscale with weakly dissipated metals, for example in nanowires with the radius of 30–40 nm. As an example, in Fig. 9.13, we show the result of calculations

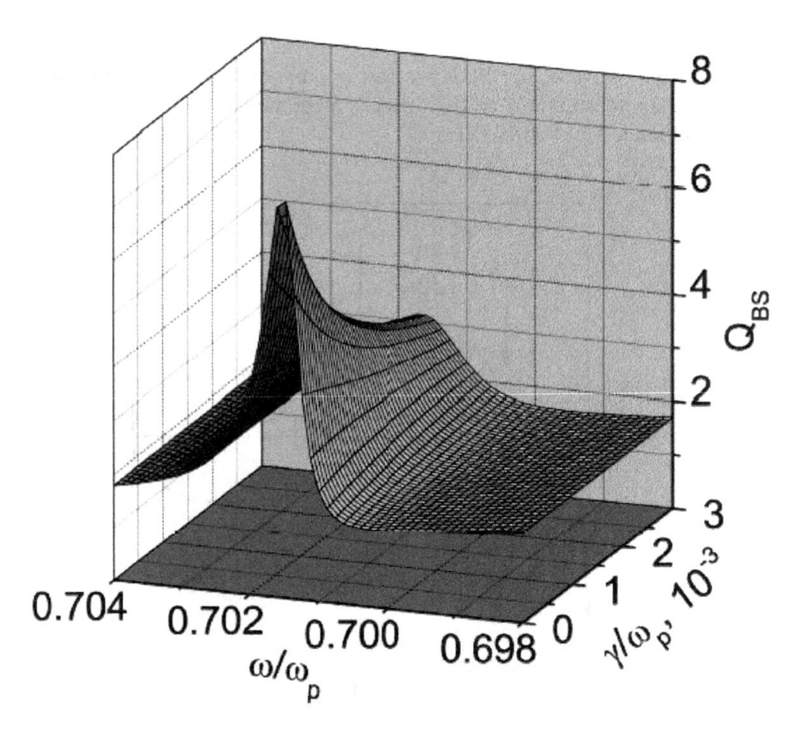

Figure 9.12 Fano resonance in the backscattering cross-section caused by the dipole-quadrupole interference for a nanowire with the size parameter $q = 0.3$.

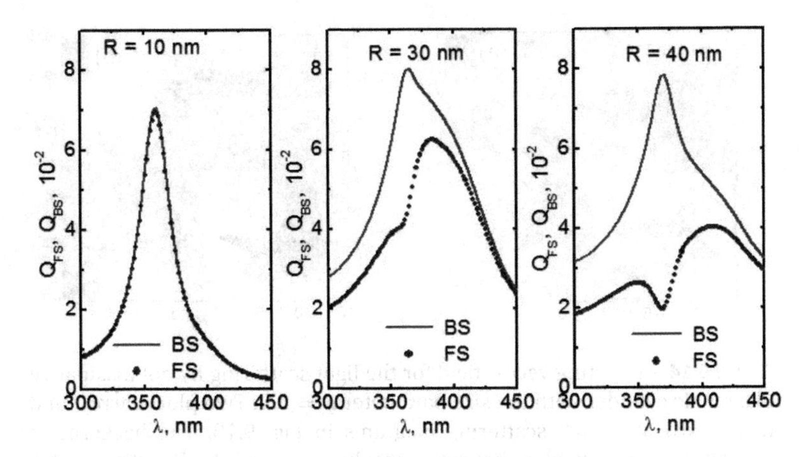

Figure 9.13 Efficiencies of the FS and BS for Na nanowires of a different size embedded in a NaCl matrix.

for a Na nanowire in the NaCl matrix. In this case, Fano resonance can be clearly seen at $R > 30$ nm. Approximately the same results for minimal R follow for a K nanowire in the KCl matrix and an Al nanowire in vacuum.

9.6 Vortices near Fano Resonances in Plasmonic Nanowires

Light scattering by a thin nanowire with a surface plasmon resonance is accompanied by bifurcations of the Poynting vector field exhibiting singular points and optical vortices [20]. In contrast to vortices shown in Fig. 9.7, optical vortices in cylindrical structures appear near the well pronounced Fano resonances similar to that shown in Fig. 9.10. An example of vortices in the near-field scattering of cylindrical structures with Fano resonance can be seen within the distributions of the Poynting vector shown in Fig. 9.14.

A majority of practically realized optical vortices have the characteristic size larger than the diffraction limit, for example, the core of an optical vortex is of the order of 10 μm [19]. In contrast, here we demonstrate optical vortices localized on the scale smaller than 100 nm, that is, two orders of magnitude smaller.

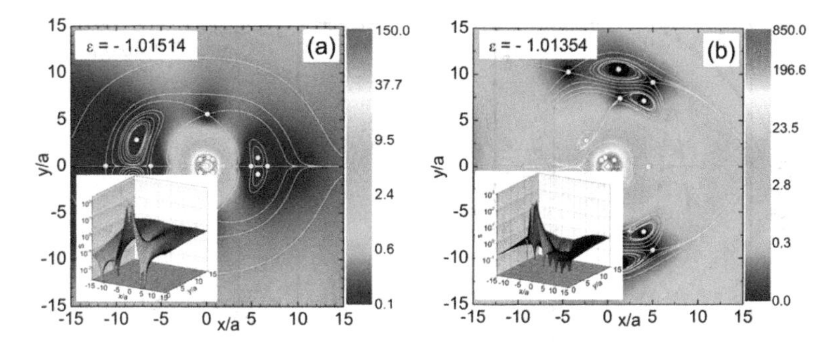

Figure 9.14 Poynting vector field for the light scattering by nondissipative plasmonic cylinder with the size parameter $q = 0.2$. Two plots correspond to forward and back scattering diagrams in Fig. 9.10. The background color in the contour plot shows the absolute value of the Poynting vector $S = \sqrt{S_r^2 + S_\phi^2}$ (logarithmic scale, see 3D plots in the insets). Lines show separatrices of the Poynting vector. Red circles mark the singular points. In the case of forward scattering (a), one can see 14 singular points out of the particle, and 4 vortices; (b) there are 14 singular points and 6 optical vortices. The insets show the magnitude of energy flux S in the x-y plane.

Thus, the "nano-Fano" resonances provide a route toward nanoscale optical vortices substantially different from the singularities of conventional optical fields. The unique ability to control topology of the electromagnetic wave distribution by small variation of the light frequency in the vicinity of the Fano resonance looks very promising for applications of topological optics, for example, in quantum optics.

The further enhancement of Fano resonance and vortices can be achieve by introducing anisotropy effects [22]. As an example, we consider a nanorod with radial anisotropy in the electrical and magnetic properties, which are different in normal and tangential directions. In the cylindrical coordinates, both tensors ε_{ij} and μ_{ij} are diagonal with the components $\varepsilon_{rr} = \varepsilon_r, \quad \varepsilon_{\theta\theta} = \varepsilon_{zz} = \varepsilon_t$, and $\mu_{rr} = \varepsilon_r$, $\mu_{\theta\theta} = \mu_{zz} = \mu_t$. Assuming for the fields the harmonic time dependence, $\propto \exp[-i\omega t]$ we can convert Maxwell's equations to the wave equation, see details in [22]. Solution of this equation for the scattered field $H_z^{(s)}$ is given by

$$H_z^{(s)} = \sum_{\ell=-\infty}^{\ell=\infty} i^\ell \left[J_\ell (k_0 r) + \bar{b}_\ell H_\ell^{(1)} (k_0 r) \right] e^{i\ell\theta}, \tag{9.15}$$

where k_0 is the wavevector in vacuum, $J_\ell(q)$ and $H_\ell^{(1)}(q)$ are the Bessel and Hankel functions, and electric scattering amplitude $\bar{b}_\ell$ is presented by the expressions

$$\bar{b}_\ell = -\frac{\Re_\ell}{\Re_\ell + i\,\Im_\ell}, \tag{9.16}$$

$$\Re_\ell = \sqrt{\varepsilon_t\mu_t}\,J_{\ell\sqrt{\frac{\varepsilon_r}{\varepsilon_t}}}\left(\sqrt{\varepsilon_t\mu_t}q\right)J_\ell'(q) - \mu_t J_\ell(q)\,J'_{\ell\sqrt{\frac{\varepsilon_r}{\varepsilon_t}}}\left(\sqrt{\varepsilon_t\mu_t}q\right), \tag{9.17}$$

$$\Im_\ell = \sqrt{\varepsilon_t\mu_t}\,J_{\ell\sqrt{\frac{\varepsilon_r}{\varepsilon_t}}}\left(\sqrt{\varepsilon_t\mu_t}q\right)\,Y_\ell'(q) - \mu_t Y_\ell(q)\,J'_{\ell\sqrt{\frac{\varepsilon_r}{\varepsilon_t}}}\left(\sqrt{\varepsilon_t\mu_t}q\right), \tag{9.18}$$

where $Y_\ell(q)$ is the Neumann function. Note that the Bessel function index $\ell\sqrt{\frac{\varepsilon_r}{\varepsilon_t}}$ is no longer a conventional integer. The exact optical resonance corresponds to the situation when $\Im_\ell = 0$, which leads to the condition $\left|\bar{b}_\ell\right| = 1$. A small variation of the ratio $\varepsilon_r/\varepsilon_t$ leads to a shift of the surface plasmon resonances and to the Fano resonance, see Fig. 9.15. It demostrates directional scattering versus asymmetry, similar to Figs. 9.1b and 9.5.

Finally, in Fig. 9.16, we show the distribution of the Poynting vector field with vortices on the scale of the cylinder cross-section. Importantly, the positions of some boundary singularities and corresponding vortices depend on the degree of anisotropy.

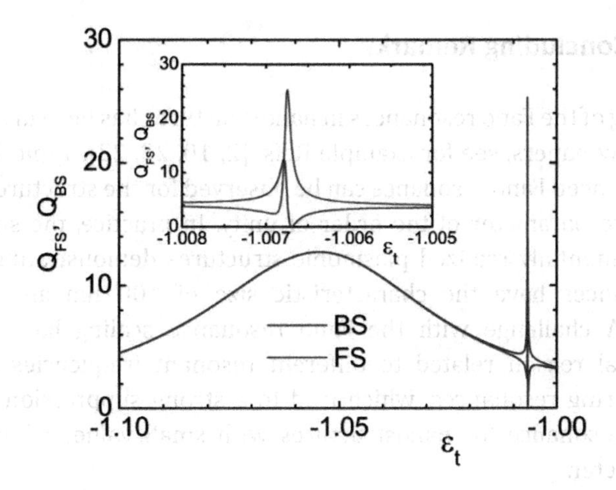

Figure 9.15 FS and BS for a nanowire with size parameter $q = 0.1$ and $\varepsilon_r = -1$ versus ε_t.

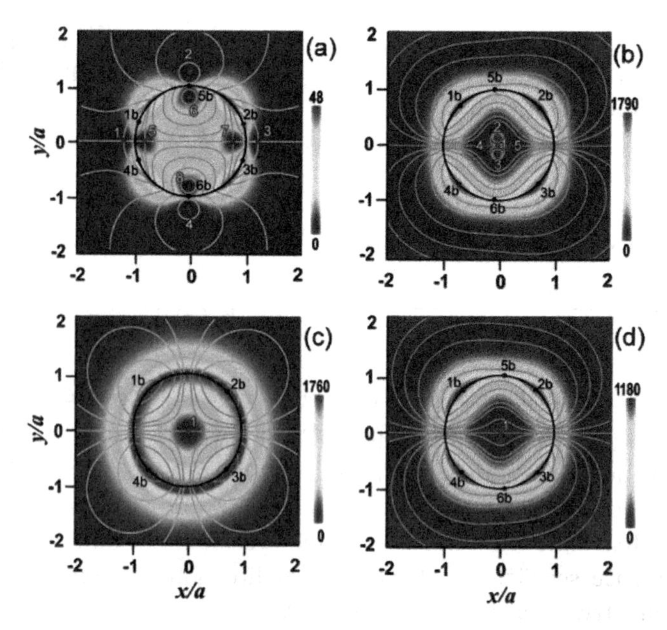

Figure 9.16 Distribution of the Poynting vector |S| and singular points for a nanowire with size parameter q = 0.1 and $\varepsilon_r = -1$: (a) isotropic cylinder $\varepsilon_t = -1$; (b) and anisotropic ones $\varepsilon_t = -1.0065$, (c)$\varepsilon_t = -1.0067835$, and (d) $\varepsilon_t = -1.0071$, respectively.

9.7 Concluding Remarks

Scaling of the Fano resonances in nanostructures has been discussed in a few papers, see for example Refs. [2, 10, 22, 23]. Typically, the pronounced Fano resonance can be observed for the structures with the size parameter of the order of unity. In practice, the smallest experimentally realized plasmonic structures demonstrating Fano resonances have the characteristic size of 100 nm and above [10]. A challenge with the Fano resonance scaling has a deep physical reason related to different resonant frequencies of the interfering resonances, which lead to a strong suppression of the Fano resonance for nanostructures with small values of the size parameter.

Here, we have studied the scaling of the Fano resonance for the nanostructures with spherical and cylindrical symmetries. We have found that the cylindrical geometry is less sensitive to the material dissipation in comparison with the spherical geometry. In contrast to the spherical nanostructures, all surface plasmon resonances in small cylindrical metallic structures appear at the same frequency corresponding to the value $\varepsilon = -1$. This permits the scaling and allows finding the intricate interconnection between near-field distribution and far-field scattering. The attractive property of small plasmonic structures is the coexistence of the Fano resonance and singular optics effects that allow to control optical vortices at the nanoscale, with the size of vortices being two orders of magnitude smaller than the conventional vortices discussed previously [14, 24].

As we mentioned above, the realization of the "nano-Fano" plasmonic structures requires materials with weak dissipation. In addition to natural low loss materials such as K, Na, and Al, a promising way is to synthesize new weakly dissipative materials such as alloys [25]. Another approach is to compensate losses by employing the concepts of active plasmonics [26] or to use anisotropy effects [22, 27]. Realization of extremely small structures with a pronounced Fano resonance is very attractive for various applications in the data storage technology, nanosensors, and topological optics.

Acknowledgments

We thank Drs. G. Vienne, R. Bakker, and A. Desyatnikov for useful discussions. This work was supported by the Agency for Science, Technology and Research (A*STAR) of Singapore: SERC Metamaterials Program on Superlens, grant no. 092 154 0099; SERC grant no. TSRP-1021520018; the grant no. JCOAG03-FG04-2009 from the Joint Council of A*STAR. The work of AEM and YSK was supported by the Australian Research Council.

References

1. B. Luk'yanchuk, N. I. Zheludev, S. A. Maier, N. J. Halas, P. Nordlander, H. Giessen, and T. C. Chong, The Fano resonance in plasmonic nanostructures and metamaterials, *Nat. Mater.* **9**, 707 (2010).

2. A. E. Miroshnichenko, S. Flach, and Yu. S. Kivshar, Fano resonances in nanoscale structures, *Rev. Mod. Phys.* **82**, 2257 (2010).
3. C. F. Bohren, and D. R. Huffman, *Absorption and Scattering of Light by Small Particles*, New York: Wiley (1998).
4. M. Born, and E. Wolf, *Principles of Optics*, 7th Ed., Cambridge: Cambridge University Press (1999).
5. M. I. Tribelsky, and B. S. Luk'yanchuk, Anomalous light scattering by small particles, *Phys. Rev. Lett.* **97**, 263902 (2006).
6. M. I. Rabinovitch, and D. I. Trubetskov, *Oscillations and Waves in Linear and Nonlinear Systems*, Kluwer Academic Publishers, The Netherlands (1989).
7. C. L. Garrido Alzar, M. A. G. Martinez, and P. Nussenzveig, Classical analog of electromagnetically induced transparency, *Am. J. Phys.* **70**, 37 (2002).
8. Y. S. Joe, A. M. Satanin, and C. S. Kim, Classical analogy of Fano resonances, *Phys. Scr.* **74**, 259 (2006).
9. N. Papasimakis and N. I. Zheludev, Metamaterial-induced transparency: Sharp fano resonances and slow light, *Opt. Photonics News*, **20**, 22 (2009).
10. M. Rahmani, B. Lukiyanchuk, and M. H. Hong, Fano resonance in novel plasmonic nanostructures, *Laser Photonics Rev.*, doi: 10.1002/lpor.201200021 (2012).
11. B. Luk'yanchuk, A. E. Miroshnichenko, M. I. Tribelsky, Y. S. Kivshar, and A. R. Khokhlov, Paradoxes in laser heating of plasmonic nanoparticles, *N. J. Phys.* **14**, 093022 (2012).
12. J. F. Nye and M. V. Berry, Dislocations in wave trains, *Proc. R. Soc. A* **336**, 165 (1974).
13. M. S. Soskin and M. V. Vasnetsov, Singular optics, *Prog. Optics* **42**, 219 (2001).
14. M. R. Dennis, K. O'Holleran, and M. J. Padgett, Singular optics: optical vortices and polarization singularities, *Prog. Optics* **53**, 293 (2009).
15. X. Yi, A.E. Miroshnichenko, and A.S. Desyatnikov, Optical vortices at Fano resonances, *Opt. Lett.* **37**, 4985 (2012).
16. Z. B. Wang, B. S. Luk'yanchuk, M. H. Hong, Y. Lin, and T. C. Chong, Energy flows around a small particle investigated by classical Mie theory, *Phys. Rev. B* **70**, 035418 (2004).
17. M. V. Bashevoy, V. A. Fedotov, and N. I. Zheludev, Optical whirlpool on an absorbing metallic nanoparticle, *Opt. Express* **13**, 8372 (2005).

18. B. S. Luk'yanchuk, Z. B. Wang, M. Tribelsky, V. Ternovsky, M. H. Hong and T. C. Chong, Peculiarities of light scattering by nanoparticles and nanowires near plasmon resonance frequencies, *JPCS* **59**, 234 (2007).
19. X. Cai, J. Wang, M. J. Strain, B. Johnson-Morris, J. B. Zhu, M. Sorel, J. L. O'Brien, M. G. Thompson, S. Yu, Integrated compact optical vortex beam emitters, *Science* **338**, 363 (2012).
20. B. S. Luk'yanchuk, and V. Ternovsky, Light scattering by thin wire with surface plasmon resonance: bifurcations of the Poynting vector field, *Phys. Rev. B* **73**, 235432 (2006).
21. B. S. Luk'yanchuk, M. I. Tribelsky, V. Ternovsky, Z. B. Wang, M. H. Hong, L. P. Shi, and T. C. Chong, Peculiarities of light scattering by nanoparticles and nanowires near plasmon resonance frequencies in weakly dissipating materials, *J. Opt. A: Pure Appl. Opt.* **9**, S294 (2007).
22. C. W. Qiu, A. Novitsky, L. Gao, J. W. Dong, and B. Luk'yanchuk, Anisotropy-induced Fano resonance, ArXiv:1202.5613v1 [physics.optics], 25 Feb 2012.
23. M. I. Tribelsky, A. E. Miroshnichenko, and Y. S. Kivshar, Unconventional Fano resonances in light scattering by small particles, *Europhys. Lett.* **97**, 44005 (2012).
24. M. Vasnetsov and K. Staliunas, *Optical Vortices*, Commack: Nova Science (1999).
25. A. Boltasseva and H. A. Atwater, Low-loss plasmonic metamaterials, *Science* **331**, 290 (2011).
26. M. A. Noginov et al., Demonstration of a spaser-based nanolaser, *Nature* **460**, 1110 (2009).
27. B. S. Luk'yanchuk and C.-W. Qiu, Enhancement scattering efficiencies in spherical particles with weakly dissipating anisotropic materials, *Appl. Phys. A* **92**, 773 (2008).

Chapter 10

Chiral Nanostructures Fabricated by Twisted Light with Spin

Takashige Omatsu[a] and Ryuji Morita[b]

[a]*Graduate School of Advanced Integration Science, Chiba University, 1-33, Yayoi-cho, Inage-ku, Chiba 263-8522, Japan*
[b]*Department of Applied Physics, Hokkaido University, Kita-13, Nishi-8, Kita-ku, Sapporo 060-8628, Japan*
omatsu@faculty.chiba-u.jp

Optical vortices (twisted light), widely used in optical tweezers and superresolution microscopes, exhibit wavefront helicity known as orbital angular momentum caused by a phase singularity.

We propose a new approach to next-generation materials processing by employing vortex lasers. We demonstrated for the first time that optical vortices can twist material to fabricate chiral nanostructures. The constituent elements (melted or vaporized material) of the irradiated material receive the helicity of optical vortices, thereby forming chiral nanostructures.

Such chiral nanostructures with a tip curvature of less than 40 nm (less than 1/25th of the laser wavelength) will enable us to provide new physical insight into laser-oriented materials science including structured materials and metamaterials as well as novel nanoscale imaging technologies for selective identification of the chirality and optical activity of molecules and chemical composites.

Singular and Chiral Nanoplasmonics
Edited by Svetlana V. Boriskina and Nikolay I. Zheludev

ISBN 978-981-4613-17-0 (Hardcover), 978-981-4613-18-7 (eBook)
www.panstanford.com

10.1 Introduction

Optical vortices [1–5], for instance, the Laguerre–Gaussian mode (an eigenmode of the electromagnetic equation in a cylindrical coordinate system), have been widely investigated in many areas, such as optical tweezers [6, 7], superresolution microscopes, [8–10], and quantum information [11, 12], because they exhibit unique features including annular intensity profiles and helical (twisted) wavefronts due to a phase singularity shown by $l\varphi$ (where l is an integer known as the topological charge and φ is the azimuthal angle) in the transverse plane and they carry orbital angular momentum (l).

Lights with circular polarization also possess a helical electric field and a spin angular momentum (s) associated with their circular polarization. Consequently, circularly polarized optical vortices, known as "twisted light with spin," carry a helicity referred to as the total angular momentum ($j = l + s$) (see Fig. 10.1), defined by the vector sum of the orbital (l) and spin (s) angular momenta [13, 14]. The total angular momentum has been evidenced by causing acceleration (or deceleration) of orbital motion of submicron particles in optical tweezers.

Recently, we discovered, for the first time, that the helicity (wavefront, polarization helicities, or both helicities characterized

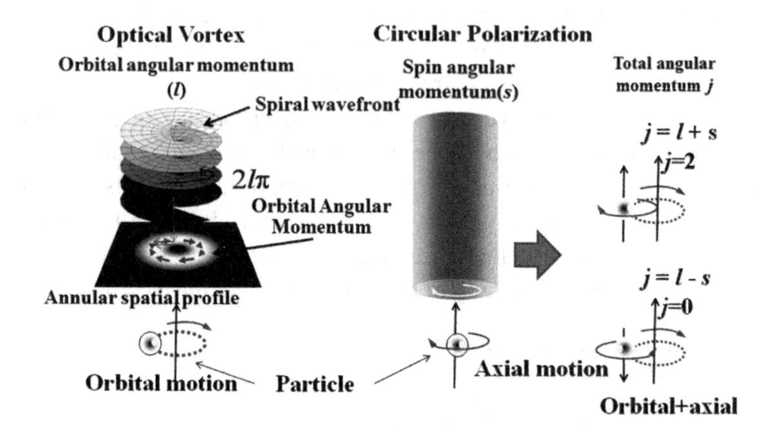

Figure 10.1 Angular momentum of light.

by the total angular momentum) of the optical vortex (OV) can be directly transferred to an irradiated metal sample through an ablation process, forming twisted nanoneedles (which we term chiral nanoneedles) [15–17], and the twisted direction (chilarity) of these nanoneedles can be controlled merely by changing the sign of the OV helicity. We further found that the helicities of the OV can also provide smooth processed surfaces with little debris. The ablation process based on helicity transfer of the optical vortices is termed OV laser ablation.

Chiral nanostructures fabricated by OV laser ablation will enable us to produce many new material structures, such as planar chiral metamaterials [18, 19] and plasmonic nanostructures [20, 21]. They might also have potential to distinguish the chirality and optical activity of molecules and chemical composites on a nanoscale [22].

In this chapter, we review OV laser ablation, including the formation of chiral metal nanoneedles. The chiral metal nanoneedles have a twisted conical surface, and their minimum tip curvature is measured to be less than 40 nm, which is less than 1/25th of the laser wavelength (1064 nm). We also briefly address a recent progress concerning OV laser technologies based on fiber laser architecture. Over 10 W picosecond (or nanosecond) vortex-pulse generation is possible using a stressed fiber amplifier with a large mode-area active core.

10.2 Radiation Force of the Light

When light is incident on particles, they become trapped by optical radiation forces including gradient and scattering forces [23, 24]. If the particles are sufficiently smaller than the wavelength of light, they can be treated as a dipole. Therefore, these optical radiation forces can be considered as interactions of the electric field, $\boldsymbol{E}(\boldsymbol{r},t)$, of light with the dipole (Rayleigh approximation). The gradient and scattering forces depend in a large part on the polarizability, α, of the particles. The α of nano-scale dielectric spheres (particles) is expressed by

$$\alpha = 4\pi n_2^2 a^3 \varepsilon_0 \frac{m^2 - 1}{m^2 + 2}, \tag{10.1}$$

where n_2 is the refractive index of the surrounding medium, m ($= n_1/n_2$) is the relative index of the particle, and a is its radius.

The gradient force, $\boldsymbol{F}_{\text{grad}}$, acting on the particles is represented by the following equation:

$$\mathbf{F}_{\text{grad}} = (\mathbf{p} \cdot \nabla)\,\mathbf{E} \propto \nabla I, \tag{10.2}$$

where $\boldsymbol{p}(= \alpha \boldsymbol{E})$ is the dipole moment, and I is the intensity profile of light.

The scattering force $\boldsymbol{F}_{\text{scatter}}$ is also given as

$$\mathbf{F}_{\text{scatter}} \propto \frac{\varepsilon_0}{2}\left(\mathbf{E}^* \times \mathbf{B} + \mathbf{E} \times \mathbf{B}^*\right), \tag{10.3}$$

where ε is the dielectric constant of a vacuum and $\boldsymbol{B}$ is the magnetic field.

10.2.1 *Angular Momentum Density*

The linear momentum density of the light $\boldsymbol{P}$, given by the time average real part of the Poynting vector, is written as

$$\begin{aligned} P &= \frac{\varepsilon_0}{2}\left(\mathbf{E}^* \times \mathbf{B} + \mathbf{E} \times \mathbf{B}^*\right) = i\omega_0 \frac{\varepsilon_0}{2}\left(u^* \nabla u - u \nabla u^*\right) \\ &\quad + \omega_0 k \varepsilon_0 \left|u\right|^2 \mathbf{z} - \omega s \frac{\varepsilon_0}{2}\frac{\partial \left|u\right|^2}{\partial r}\varphi, \end{aligned} \tag{10.4}$$

where u is the amplitude of the light field, ω_0 is the frequency of the light, and k is the wavenumber; s is the spin angular momentum (1 or -1) for clockwise or anti-clockwise circularly polarized light, and $\boldsymbol{z}$ and ϕ are unit vectors along the z- and azimuthal directions, respectively [14, 15]. Thus, the angular momentum $\boldsymbol{M}$, of the light is given by

$$\mathbf{M} = \mathbf{r} \times \mathbf{P}, \tag{10.5}$$

where $\boldsymbol{r}$ is a unit vector along the radial direction.

Laguerre–Gaussian modes (see Fig. 10.2), eigenmodes of the paraxial propagation electromagnetic equation in a cylindrical coordinate system are typical OV, and they have a doughnut-shaped spatial profile in the far-field and orbital angular momentum due to a phase singularity.

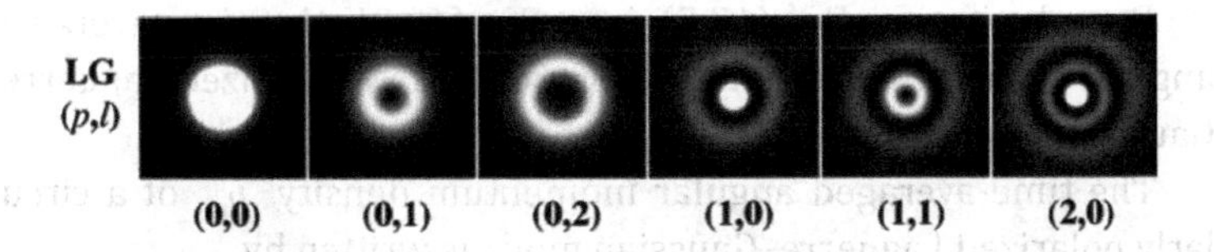

Figure 10.2 Intensity profiles of the Laguerre–Gaussian modes.

The amplitude $u_{pl}(r\varphi z)$ of the Laguerre–Gaussian mode can be written as

$$u_{p,l}(r,\varphi,z) = C_{p,l}\frac{1}{\omega(z)}\left(\frac{\sqrt{2}r}{\omega(z)}\right)^{|l|} L_p^{|l|}\left(\frac{2r^2}{\omega^2(z)}\right)\exp(-il\varphi)$$
$$\times \exp\left[-\frac{r^2}{\omega^2(z)}\right]\exp\left[-\frac{ikr^2}{2R(z)}\right]\exp[i(2p+|l|+1)\phi(z)], \tag{10.6}$$

where $L_p^{|l|}$ are generalized Laguerre polynomials, p is the radial index, l is the azimuthal index (known as the topological charge) $\omega(z)$ is the beam spot size , $R(z)$ is the radius of curvature, $\phi(z)$ is the phase parameter, and $C_{p,l}$ is the normalization constant. The spiral wavefront of OV is visualized by fringes formed by interference between the OV and a spherical (or plane) reference wavefront. The spherical referenced fringes as shown in Fig. 10.3 had a single spiral indicating the existence of a phase singularity [25, 26].

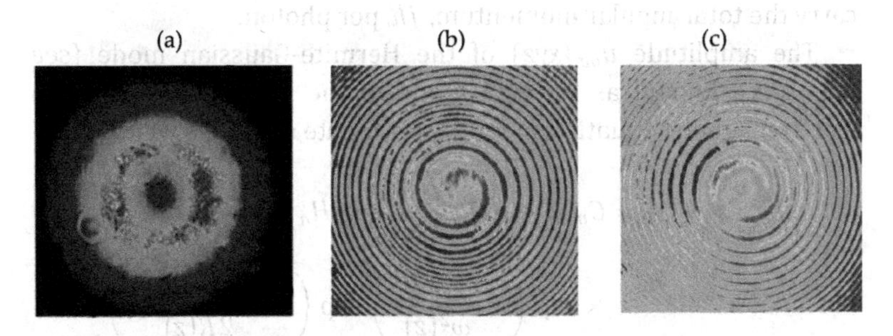

Figure 10.3 Spatial forms of OV with $|l| = 1$. (a) Intensity profile of the OV. Spherical referenced fringes with (b) clockwise and (c) anti-clockwise directions. Reprinted from Ref. [27] with permission of the Optical Society, Copyright 2012.

By substituting Eq. (10.5) into Eq. (10.3), the time-averaged angular momentum density, $j_{l,s}$, of a circularly polarized Laguerre-Gaussian mode is obtained.

The time-averaged angular momentum density, $j_{l,s}$, of a circularly polarized Laguerre–Gaussian mode is written by

$$j_{l,s} = \varepsilon_0 \omega_0 \left\{ l \left| u_{l,p} \right|^2 - \frac{1}{2} s r \frac{\partial \left| u_{l,p} \right|^2}{\partial r} \right\}, \tag{10.7}$$

where ω_0 is the frequency of the OV.

By substituting Eq. (10.6), $p = 0$, and $z =$ into Eq. (10.7), the spatial distribution of the angular momentum density, $j_{l,s}$, can be expressed by

$$j_{l,s} \propto \frac{\omega_0}{|l|!} \left(l - |l|\, s + s \left(\frac{\sqrt{2}r}{\omega(0)} \right)^2 \right) \left(\frac{\sqrt{2}r}{\omega(0)} \right)^{2|l|} \exp\left(-\frac{2r^2}{\omega(0)^2} \right), \tag{10.8}$$

where $\omega(0)$ is the beam waist. By reversing the sign of l and s in Eq. (10.8), we can find that the spatial distribution of the angular momentum density, $j_{l,s}$, is inverted.

The total angular momentum $j\hbar$, obtained by integrating the angular momentum density, $j_{l,s}$, over the whole beam aperture, is given by the sum of the orbital $l\hbar$ and spin $s\hbar$ angular momenta per photon, resulting in $j = l + s$. Consequently, the circularly polarized Laguerre-Gaussian modes having a spiral wavefront characterized by the topological charge l, and possessing a helical electric field, carry the total angular momentum, $j\hbar$, per photon.

The amplitude $u_{mn}(xyz)$ of the Hermite-Gaussian mode (see Fig. 10.4), known as an eigenmode of the paraxial propagation electromagnetic equation in a x–y coordinate system is written as:

$$\begin{aligned} u_{m,n}(x, y, z) = {} & C_{m,n} \frac{1}{\omega(z)} H_m \left(\frac{\sqrt{2}x}{\omega(z)} \right) H_n \left(\frac{\sqrt{2}y}{\omega(z)} \right) \\ & \times \exp\left(-\frac{x_2 + y^2}{\omega^2(z)} \right) \exp\left(-\frac{ik(x^2 + y^2)}{2R(z)} \right) \\ & \times \exp\left(i(m + n + 1)\phi(z) \right), \end{aligned} \tag{10.9}$$

The Hermite–Gaussian mode can be easily transformed into the Laguerre–Gaussian modes. For instance, the Hermite–Gaussian

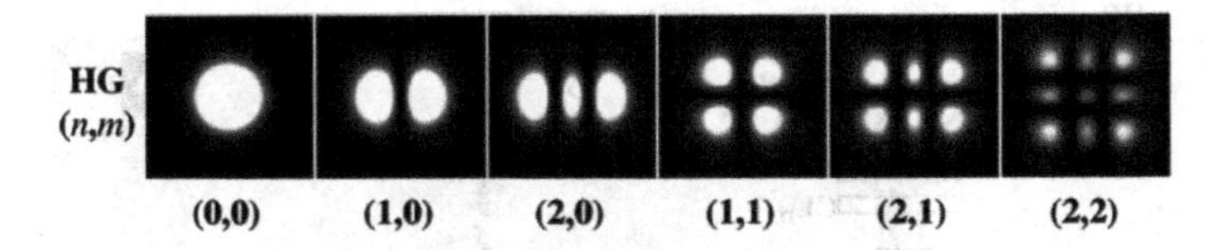

Figure 10.4 Intensity profiles of Hermite–Gaussian modes.

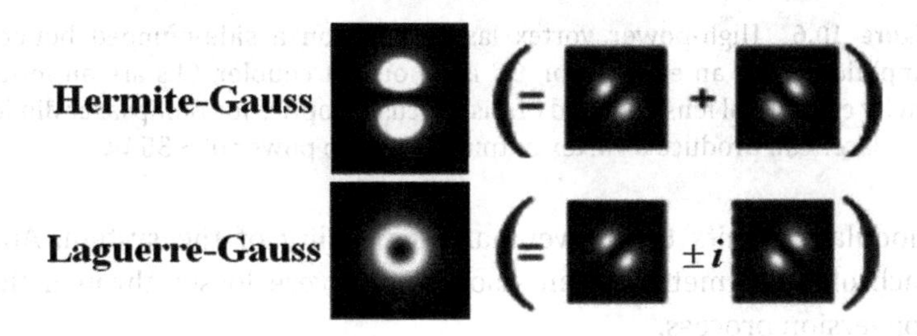

Figure 10.5 Mode conversion from the Hermite–Gaussian mode to the Laguerre–Gaussian mode.

TEM_{01} mode is decomposed at 45° into a pair (in-phase) of TEM_{01} modes. Laguerre-Gaussian LP^1 mode is also decomposed at 45° into a pair of TEM_{01} modes with $\pi/2$ or $-\pi/2$ out of phase with each other (see Fig. 10.5). Several methods to convert the Hermite-Gaussian TEM_{01} mode to the Laguerre-Gaussian LP^1 mode have been proposed. A conventional "mode-convertor" involving an astigmatic optics such as a cylindrical lens pair enables us to provide the necessary $\pi/2$ phase change to a pair of TEM_{01} modes. By using a "mode-convertor," the Hermite–Gaussian TEM_{mn} mode is converted to the Laguerre–Gaussian $LP^{|m,n|}{}_{\min(mn)}$ mode, or vice versa [28].

10.2.2 *How Can One Produce Optical Vortices?*

Mode conversion techniques from a Gaussian beam to an OV beam by using additional phase elements such as spiral phase plates (SPPs) and static (or dynamic) computer-generated holograms have been proposed. In particular, dynamic computer-generated holograms produced by utilizing a spatial light modulator enable us to create an arbitrary wavefront, such as a two-dimentional OV array [29, 30]. However, the damage threshold of the spatial light

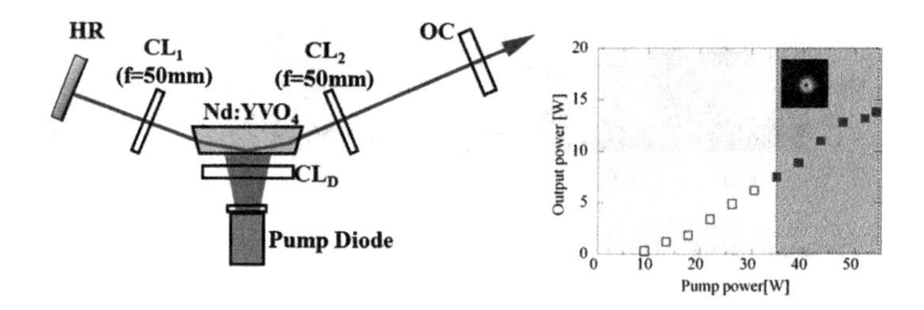

Figure 10.6 High-power vortex laser based on a side-pumped bounce amplifier. HR is an end mirror, OC is an output coupler, CLs are an intra-cavity cylindrical lens pair, and CL_D is a focusing optics for pump laser diode. The laser can produce a vortex output at a pump power of >35 W.

modulator limits the power handling ability of the system. And each of these methods can also suffer severe losses through the conversion process.

An alternative method to effectively generate high-quality vortex output is to directly force the laser to oscillate on an LG mode instead of a Hermite–Gaussian mode. To date, several methods have been demonstrated using an annular shaped beam pumping [31], intra-cavity SPP [32], and a damaged resonator mirror [33]. However, the fatal drawback to them is the requirement of additional phase elements and extremely precise alignment. The achieved output power has been limited to 1 W.

In recent years, we have proposed a thermal lens aperture technique for transversely diode-pumped solid-state lasers [34–36]. An ultrahigh-power (17.8 W) vortex output at a high efficiency of more than 30% has been achieved from a transversely diode-pumped $Nd:GdVO_4$ laser, in which the thermal lens in the laser material forms a limiting aperture for the desired LG mode in the gain medium without any additional phase elements (see Fig. 10.6).

We have also produced high-power, pulsed OV outputs from a diode-pumped solid-state master laser oscillator in combination with a stressed, ytterbium (Yb)-doped, large-mode-area fiber power amplifier [27, 37, 38]. A collimated master laser output with a Gaussian profile is off-axially injected into a large-mode area fiber, thereby yielding in-phase coupling of two orthogonal LP_{11} modes at a high efficiency. These modes are 90° or −90° out of phase with each other at the end of the fiber by providing appropriate stress to

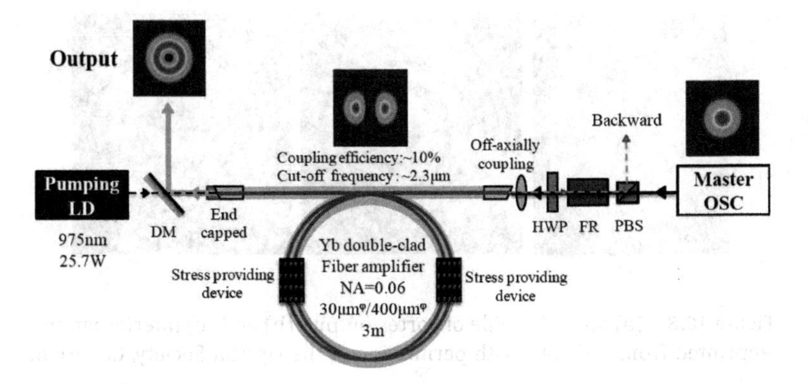

Figure 10.7 Schematic diagram of the OV fiber laser. PBS polarizing beam splitter; FR Faraday rotator; HWP half-wave plate, and DM dichroic mirror. The master OSC used is a CW-mode-locked (or Q-switched) $Nd:YVO_4$ laser. Reprinted from Ref. [38] with permission of the Optical Society, Copyright 2011.

the fiber. The system will then generate a vortex output. The sign of the spiral wavefront of the output is also controlled merely by varying the stress onto the fiber.

A schematic diagram of the experimental setup is shown in Fig. 10.7. A fiber amplifier used was a polarization-maintaining large-mode area Yb^{3+}-doped double-clad fiber (with a length of 4 m, a core diameter of 30 μm, a core NA of 0.06, a cladding diameter of 400 μm, and a cladding NA of 0.46). The cutoff value of the fiber amplifier was estimated to be 5.3. The fiber amplifier was pumped by a 975 nm fiber-coupled laser diode, and its exit facet was end-capped and mounted on a metal block cooled by a water chiller to prevent any optical damage to the fiber facet at a high pump level. The fiber was also bent into a hoop with a radius of nearly 13 cm to suppress undesired higher-order modes.

Over 25 W of picosecond vortex output from a stressed Yb-doped fiber amplifier system in combination with a continuous-wave mode-locked $Nd:YVO_4$ master laser has been achieved, corresponding to optical-optical efficiency of 47.9.%

Millijoule (0.83 mJ) nanosecond vortex pulses have been also generated from the fiber amplifier in combination with an actively

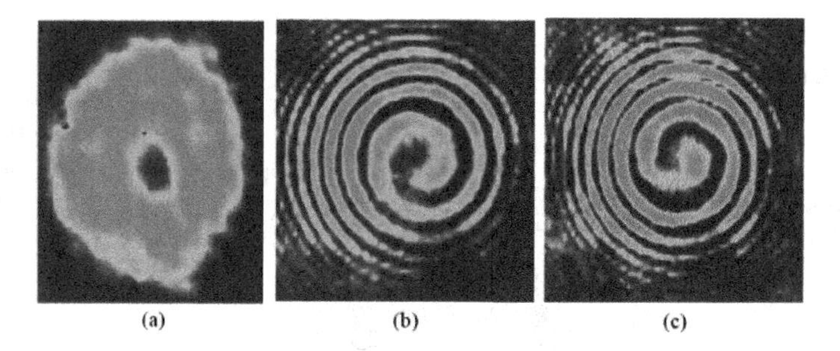

Figure 10.8 (a) Spatial profile of vortex output. (b) and (c) Interferograms. Reprinted from Ref. [38] with permission of the Optical Society, Copyright 2011.

Q-switched Nd:YVO_4 master laser. Spatial forms of the vortex output are summarized in Fig. 10.8.

To generate femtosecond OV pulses, several techniques have also been proposed using an achromatic spiral plate formed of two materials [39], and computer-generated holograms in combination with 4 f configuration [4]. We and our coworkers have successfully produced ultra-broadband OV pulses with a wavelength range of 500–800 nm by utilizing an achromatic optics method including an axially symmetric polarizer [41].

10.3 Optical Vortex Laser Ablation

The laser ablation process is a promising way to perform material processing, and it has always been developed together with evolution of laser technologies. For instance, ultrafast lasers, such as femtosecond or picosecond lasers, having extreme short pulse-width, enable us to create a very high peak intensity with low pulse energies. Such ultrashort pulses fundamentally provide new physical insight into the laser–matter interaction, for instance, three-dimentional micro-fabrication in transparent materials [42]. Further, they may also reduce the fluid dynamics and heat-affected zone, resulting in less debris at high energy and cost efficiencies [43–45].

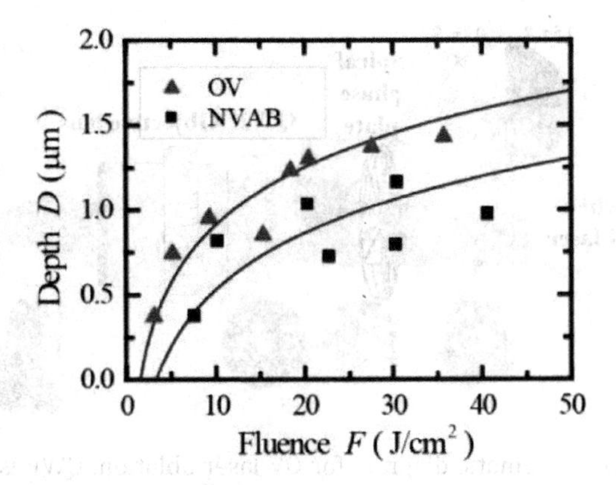

Figure 10.9 Experimental ablation threshold by OV and annular beam (NVAB) pumping. Reprinted from Ref. [15] with permission of the Optical Society, Copyright 2010.

The angular momentum of the OV, causing orbital motion of submicron particles in optical tweezers, can control the dynamics of compositional elements (a plasma of ions and electrons and melted matter) produced by the high intense laser pulses. In this way the OV can potentially improve the material processing with high accuracy and without needing expensive equipment such as ultrafast lasers. However, optical vortices have never been applied to laser ablation and their orbital angular momentum effects on the ablation processes have never been investigated, either.

We and our coworkers found that laser ablation by using optical vortices (termed as OV laser ablation) provided a proceeded surface with less debris in comparison with using an annular beam without a phase singularity. Further, OV laser ablation also reduced the ablation threshold (see Fig. 10.9).

10.3.1 *Nanoneedle Fabrication*

As stated above, the circularly polarized OV (twisted light with spin) has a total angular momentum (j) equal to the vector sum of the orbital (l) and spin (s) angular momenta. When the sign of the orbital angular momentum, l (for instance, 1), is the same as (or opposite to) that of the spin angular momentum, s (1 or −1), the

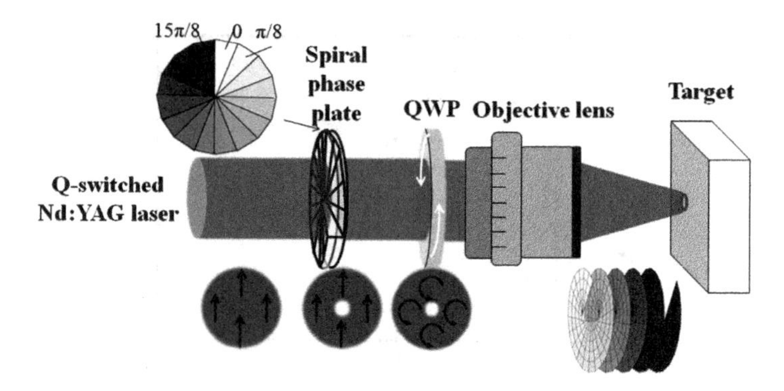

Figure 10.10 Schematic diagram for OV laser ablation. QWP is a quarter-wave plate. Reprinted from Ref. [16] with permission of the Optical Society, Copyright 2010.

total angular momentum of the OV is defined to be 2 (or 0), as shown in Fig. 10.1.

Figure 10.10 shows a schematic diagram of an experimental setup for OV laser ablation. A commercial Q-switched Nd:YAG laser (Quanta-Ray GCR190) with a wavelength of 1064 nm and a pulse duration of 30 ns was used for a pump laser, and its output with a Gaussian spatial form was converted into an OV with $l = 1$ by a SPP, fabricated by electron beam etching, azimuthally divided into 16 parts using a $n\pi/8$ phase shifter (where n is an integer between 0 and 15). To control the polarization (spin angular momentum) of the OV, a quarter-wave plate placed in the optical path between the SPP and an objective lens (NA = 0.08) was also used. The circularly polarized OV was then loosely focused by the objective lens onto a target, Ta plate (complex dielectric constant is approximately $-2.54 + 10i$ at 1064 nm [46]) to a ϕ 130 μm spot on the sample surface. In this system, using low NA focusing optics, the spatial form [in the near (or far) field] of the OV with $j = 2$ is annular, and it is fully identical to that of an OV with $j = 0$. The OV pulse energy was fixed at 2 mJ.

All experiments were performed at atmospheric pressure and room temperature. The morphology of the ablated target was observed using a confocal laser-scanning microscope (Keyence VK-

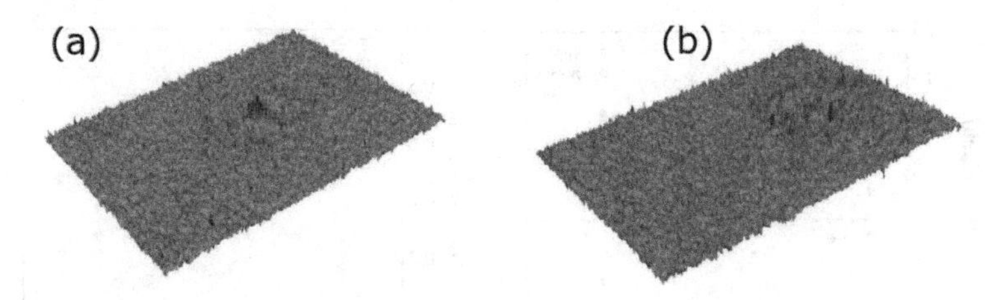

Figure 10.11 Ablated surfaces by a single OV pulse with (a) $j = 2$ and (b) $j = 0$. Reprinted from Ref. [16] with permission of the Optical Society, Copyright 2010.

9700/VK9710GS) with a spatial resolution of 0.02 μm in both depth and transverse displacements.

Laser-scanning microscopic images of target surfaces ablated by optical vortices with (a) $j = 2$ and (b) $j = 0$ are shown in Fig. 10.11.

After the single-shot irradiation of OV pulse with $j = 2$, a protuberance appears at the center of the surface. The height (the length between a top and a bottom end of the protuberance) and the thickness (FWHM) of the protuberance were approximately measured to be 4.4 ± 0.23 and 9.2 ± 1.7 μm, respectively. After three $j = 2$ pulses were overlaid, the protuberance was shaped to be a needle with a height (from the target surface) of nearly 10 μm (see Fig. 10.12).

The needle was narrowed by the deposition of a few more OV pulses, and its tip diameter of less than 0.3 μm, defined as a diameter measured 5% below the end of the needle, was achieved. The aspect ratio of the needle (the ratio of the length of the needle

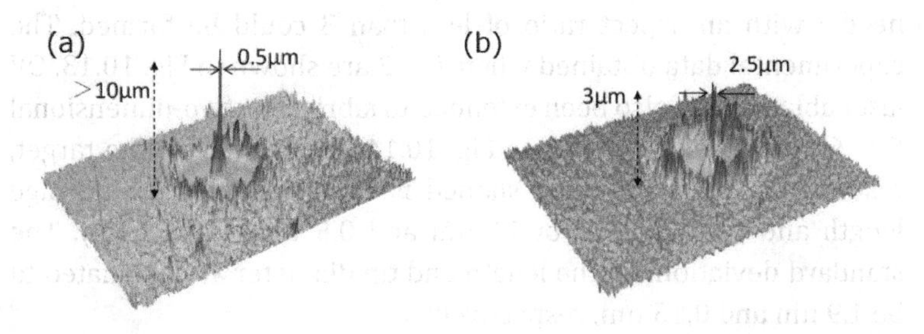

Figure 10.12 Surfaces processed by the deposition of four OV pulses having (a) $j = 2$ and (b) $j = 0$ [16].

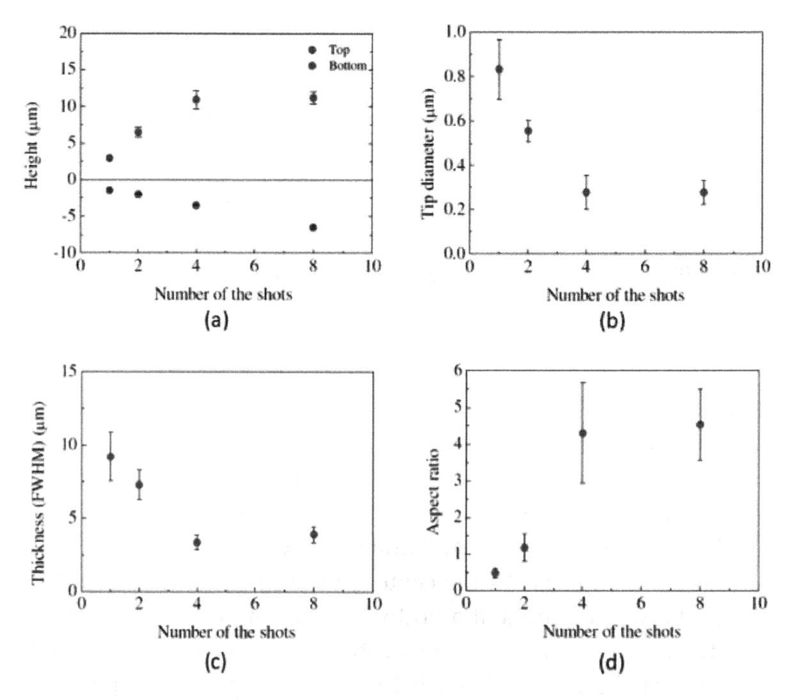

Figure 10.13 Experimental measurements of the (a) height, (b) tip diameter, (c) thickness and (d) aspect ratio of the nanostructures. Error bars show the standard deviations of the measured values [16].

to its diameter) was estimated to be 4.5. In contrast, when $j = 0$, the ablated surface had no structure at the center, and it also collected a lot of debris along the azimuthal direction around the outline. Even after several vortex pulses were overlaid, only a fat needle with an aspect ratio of less than 3 could be formed. The experimental data obtained when $j = 2$ are shown in Fig. 10.13. OV laser ablation has also been extended to fabricate a two-dimensional 5×6 microneedle array (see Fig. 10.14) by translating the target, resulting in uniformly well shaped nanoneedles with an average length and tip diameter of 11 μm and 0.5 μm, respectively. The standard deviations in the length and tip diameter are estimated to be 1.9 μm and 0.15 μm, respectively.

The ablated surface pumped by a Gaussian beam, even with clockwise (or anti-clockwise) circular polarization, had no nano-

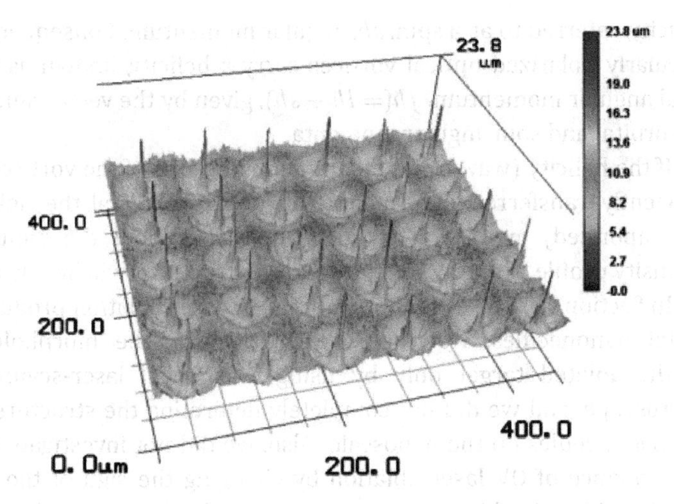

Figure 10.14 Two-dimensional 5 × 6 microneedle array fabricated by using OV pulses. Reprinted from Ref. [16] with permission of the Optical Society, Copyright 2010.

structures, such as needles. The formation of the nanoneedle requires the collection and confinement of the melted metal to an on-axis hole of the OV by the photon pressure. In Gaussian beams that do not have an on-axis hole, the collection of the melted metal does not occur efficiently.

10.3.2 *Chiral Nanoneedle Fabrication*

There is no known technique to twist material on a nanoscale. In particular, it is impossible to twist metal on a nanoscale to create spiral (chiral) metal nanoneedles, even by utilizing advanced chemical techniques. Laser materials processing breaks down a metal into its constituent elements (melted or vaporized metal) by high-intensity laser pulses, and it is unsuitable for recombining the constituent elements and structuring spiral materials. Therefore, laser materials processing has not been performed to produce chiral metal nanoneedles.

As mentioned above, optical vortices have a wavefront helicity due to a phase singularity, and they carry orbital angular momentum, $l\hbar$. Circularly polarized light also has a polarization helicity, referred to as a spin, $s\hbar$, angular momentum. Consequently, circularly polarized optical vortices carry a helicity, known as the total angular momentum, $j\hbar(= l\hbar + s\hbar)$, given by the vector sum of the orbital and spin angular momenta.

If the helicity (wavefront, polarization or both) of the vortices is efficiently transferred to the melted (or vaporized) metal, the melted (or vaporized) metal will rotate azimuthally about the annular intensity profile of the OV. Resulting spiral structures will be formed.

In Section 10.2.1, we mentioned that OV laser ablation produces metal nanoneedles. However, we investigated the morphology of the ablated target only by using a confocal laser-scanning microscope, and we did not completely determine the structure of the nanoneedles on the nanoscale. Also, we did not investigate the performance of OV laser ablation by changing the sign of the OV helicity, either. In this section, we report on how to determine the chirality of nanoscale metal structures by the OV helicity. We also mention that the laser energy used in the present experiments was controlled in the range of 0.075–0.3 mJ, which is less than one-sixth that (2 mJ) used in the experiments mentioned in Section 10.2.2. At an energy level above 1 mJ, dense plasma (vaporized material) produced by the leading edge of the OV pulse shielded the rest of the optical pulse, resulting in an insufficient transfer of the angular momentum to the melted material. And thus, only low-energy pulse deposition permitted chiral nanoneedles to form.

To reverse the helicity of the OV, the SPP and the quarter wave plate were inverted. Four vortex pulses with a ϕ ~65 μm annular spot onto the target were overlaid. The ablated target was observed by a scanning electron microscope, SEM (JEOL, JSM-6010LA), with a spatial resolution of 8 nm at 3 kV.

A target surface ablated by an OV pulse with $j = -2$ is shown in Fig. 10.15a. A needle was formed at the center of the ablated zone with a smooth outline, and it typically had a tip curvature of nearly 70 nm and a height of nearly 10 μm. The conical surface of the magnified needle (see Fig. 10.15c,d) was twisted azimuthally in the clockwise direction. In contrast, a needle fabricated by the irradiation of an OV with $j = 2$ was twisted azimuthally in the counter-clockwise direction. We call these twisted needles chiral nanoneedles (see Fig. 10.15e,f).

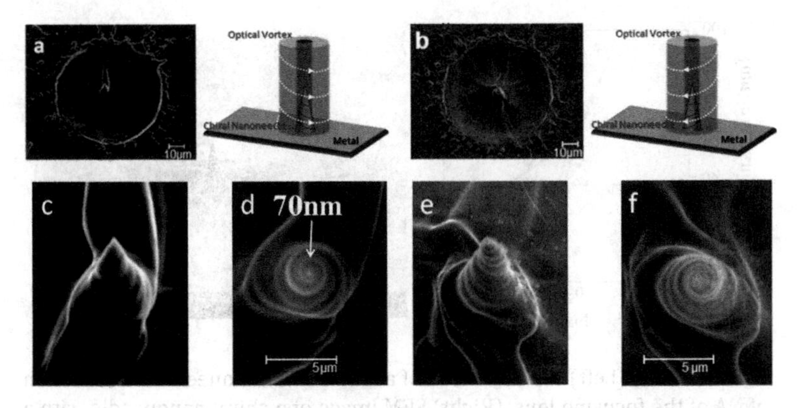

Figure 10.15 (a,c,d) SEM images of a processed surface and a twisted nanoneedle fabricated by an OV with $j = -2$ (clockwise). (b,d,f) SEM images of a processed surface and a twisted nanoneedle fabricated by an OV with $j = 2$ (counter-clockwise). (c,d) The 25° views and (e,f) top views of the nanoneedle. The focusing lens had an NA of 0.08. Reprinted with permission from Ref. [17]. Copyright 2012 American Chemical Society.

These results indicate that the twisting direction (chirality) of the chiral nanoneedle is selectively determined by the helicity of the OV pulse. We noticed that as the magnitude of j increased, the spiral frequency of the nanoneedle (defined as the winding number per 1 μm) increased. We must also mention that ultrashort (i.e., femtosecond) OV pulses cannot provide any chiral nanoneedles. The tip curvature of the chiral nanoneedle was also inversely proportional to the NA of the objective lens, and it was minimized to nearly 36 nm (Fig. 10.16), which was less than 1/25th of the laser wavelength (1064 nm) at NA = 0.18. The nanoneedle still had a twisted conical surface, and its height was measured to 7.5 μm.

The chiral nanoneedle may have been oxided, because all of the experiments were performed at atmospheric pressure and room temperature. We investigated the electrical properties of the nanoneedle by using two 50 μm diameter tungsten probes with an internal resistance of nearly 1.0 Ω. One probe contacted with the top of the nanoneedle, while the other was tightly pressed against the substrate surface so as to minimize the contact resistance. As shown in Fig. 10.17, the current between the nanoneedle and the substrate was found to be directly proportional to the supplied voltage. The

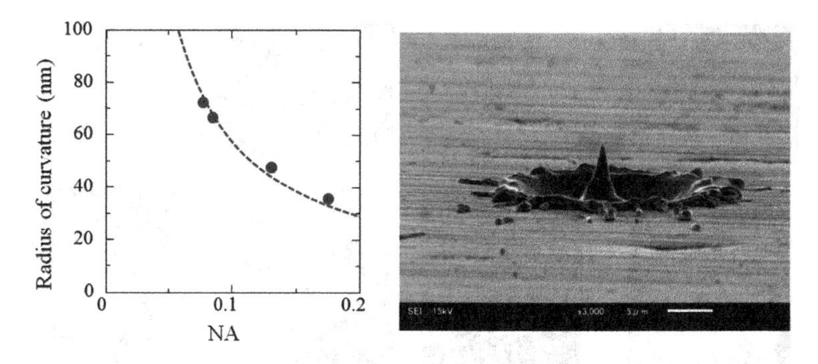

Figure 10.16 (Left) Tip curvature of a fabricated nanoneedle as a function of NA of the focusing lens. (Right) SEM image of a chiral nanoneedle with a tip curvature of 36 nm. Reprinted with permission from Ref. [17]. Copyright 2012 American Chemical Society.

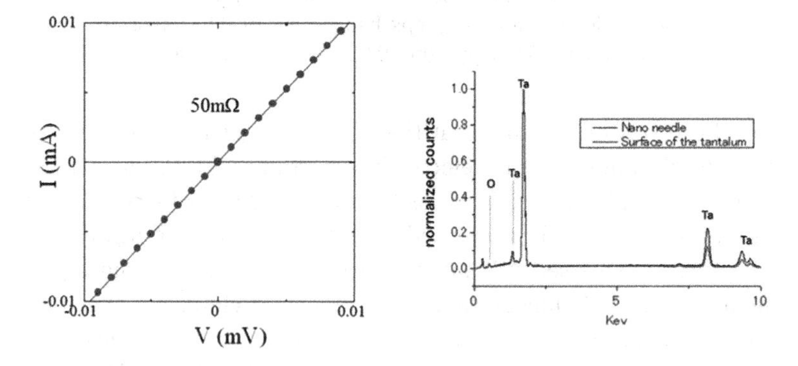

Figure 10.17 (Left) Measured current against supplied voltage between the nanoneedle and the substrate by using a tungsten probe. (Right) EDX spectra of the nanoneedle and the substrate. Reprinted with permission from Ref. [17]. Copyright 2012 American Chemical Society.

resistance of the needle excluding the internal resistance of the probe was measured to be nearly 50 mΩ. The measured EDX spectrum of the fabricated needle (shown by the red line) was also identical to that of the substrate surface (shown by the black line), indicating that the fabricated nanoneedle and the substrate had the same chemical compositions. And thus, the fabricated needle was perfectly metallic.

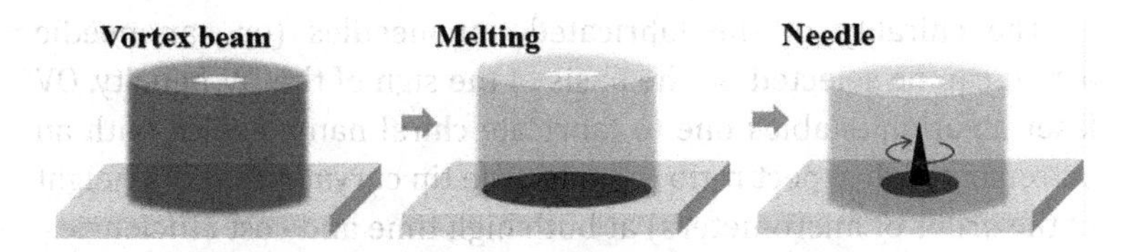

Figure 10.18 Model for chiral nanoneedle fabrication by OV laser ablation.

We can understand the above related results by adopting the following model, as shown in Fig. 10.18. The target metal is melted (or vaporized partially) by the irradiation of a focused vortex pulse. The melted metal then receives angular momentum from the OV to rotate azimuthally about the annular intensity profile of the OV. This is evidenced by the formation of a smooth processed surface with little debris. The melted metal revolves about the optical axis of the OV, and it is directed and confined in the hole (stable equilibrium position) by the photon pressure. Also, the angular momentum of the melted metal then is also damped by friction to form a chiral metal nanoneedle at the center of the processed metal surface. This model, in which the OV helicity, defined as the sign of the angular momentum, is transferred to the melted metal, is evidenced by the observation that the conical surface of the needle is twisted azimuthally with the sign of the OV helicity.

10.4 Conclusion

We discovered, for the first time, that OV helicity is transferred to a melted metal, twisting metal to form chiral nanostructures, such as needles with a tip curvature of ~40 nm and a height of 7.6 μm. Chiral nanoneedles fabrication based on OV laser ablation can also be extended to provide a two-dimensional chiral nanoneedles array merely by translating the target or irradiating two-dimensional vortex array on the target. In fact, by utilizing a highly repetitive Q-switched Nd:YAG (or Nd:YVO_4) laser with PRF more than 100 kHz [47, 48] in combination with a high-speed scanning stage, ultra-rapid fabrication with a speed of >25,000 needles/s should be attainable.

The chirality of the fabricated nanoneedles (or nanoneedle array) can be selected on the basis of the sign of the OV helicity. OV laser ablation enables one to fabricate chiral nanoneedles with an extremely high aspect ratio (a nanoscale tip curvature with a height of the order of micrometers) at both high time and cost efficiencies. This will open up potentially new advanced technologies, such as nanoimaging systems, energy-saving displays, and biomedical nanoelectromechanical systems.

Chiral twisted nanostructures also have the potential to selectively distinguish the chirality and optical activity of molecules and chemical composites on a nanoscale, thereby yielding nanoscale imaging systems with chiral selectivity, chemical reactions on plasmonic nanostructures, and planar metamaterials.

Acknowledgments

The authors acknowledge financial support of a Research Grant (CREST) from Japan Science and Technology Agency, and a Grant-in-Aid for Scientific Research (No. 21360026) from the Japan Society for the Promotion of Science.

References

1. Allen, L., Beijersbergen, M. W., Spreeuw, R. J. C., Woerdman, J. P., (1992). Orbital angular momentum of light and the transformation of Laguerre-Gaussian laser modes, *Phys. Rev. A*, **45**, pp. 8185–8189.
2. Indebetouw, G., (1993). Optical vortices and their propagation. *J. Mod. Opt.*, **40**, pp. 73–87.
3. Padgett, M., Courtial, J., Allen, L., (2004). Light's orbital angular momentum, *Phys. Today*, **57**, pp. 35–40.
4. Molina-Terrizagabriel, G., Torres, J. P., and Torner, L., (2007). Twisted photons, *Nat. Phys.* **3**, pp. 305–310.
5. Leach, J., Padgett, M.J., Barnett, S.M., Franke-Arnold, S., Courtial, J., (2002). Measuring the orbital angular momentum of a single photon, *Phys. Rev. Lett.*, **88**, pp. 2579011–2579014.

6. Gahagan, K. T., and Swartzlander, Jr., G. A., (1996). Optical vortex trapping of particles, *Opt. Lett.* **21**, pp. 827–829.
7. Paterson, L., MacDonald, M. P., Arlt, J., Sibbett, W., Bryant, P. E., and Dholakia, K., (2001). Controlled rotation of optically trapped microscopic particles, *Science*, **292**, pp. 912–914.
8. Klar, T. A., and Hell, S. W., (1999). Subdiffraction limit in far-field fluorescence dip microscopy, *Opt. Lett.* **24**, pp. 954–956.
9. Vicidomini, G., Moneron, G., Eggeling, C., Rittweger, E., and Hell, S. W., (2012). STED with wavelengths closer to the emission maximum, *Opt. Express*, **20**, pp. 5225–5236.
10. Watanabe, T., Iketaki, Y., Omatsu, T., Yamamoto, K., Sakai, M., and Fujii, M., (2003). Two-point-separation in super-resolution fluorescence microscope based on up-conversion fluorescence depletion technique, *Opt. Express*, **11**, pp. 3271–3276.
11. Kawase, D., Miyamoto, Y., Takeda, M., Sasaki, K., and Takeuchi, S., (2009). Effect of high-dimensional entanglement of Laguerre-Gaussian modes in parametric downconversion, *J. Opt. Soc. Am. B*, **26**, pp. 797–804.
12. Gibson, G., Courtial, J., Padgett, M. J., Vasnetsov, M., Pas'ko, V., Barnett, S. M., and Franke-Arnold, S., (2004). Free-space information transfer using light beams carrying orbital angular momentum, *Opt. Express*, **12**, pp. 5448–5456.
13. O'Neil, A. T., MacVicar, I., Allen, L., and Padgett, M. J., (2002). Intrinsic and Extrinsic Nature of the Orbital Angular Momentum of a Light Beam, *Phys. Rev. Lett.*, **88**, 053601-1-4.
14. Simpson, N. B., Dholakia, K., Allen, L., and Padgett, M. J. (1997). Mechanical equivalence of spin and orbital angular momentum of light: an optical spanner, *Opt. Lett.*, **22**, pp. 52–54.
15. Hamazaki, J., Morita, R., Chujo, K., Kobayashi, Y., Tanda, S., and Omatsu, T., (2010). Optical-vortex laser ablation, *Opt. Express*, **18**, pp. 2144–2151.
16. Omatsu, T., Chujo, K., Miyamoto, K., Okida, M., Nakamura, K., Aoki, N., Morita, R., (2010). Metal microneedle fabrication using twisted light with spin, *Opt. Express*, **18**, pp. 17967–17973.
17. Toyoda, K., Miyamoto, K., Aoki, N., Morita, R., and Omatsu, T., (2012). Using optical vortex to control the chirality of twisted metal nanostructures, *Nano Lett.*, **12**, pp. 3645–3649.
18. Konishi, K., Bai, B., Meng, X., Karvinen, P., Turunen, J., Svirko, Y. P., Kuwata-Gonokami, M., (2008). Observation of extraordinary optical activity in planar chiral photonic crystals, *Opt. Express*, **16**, pp. 7189–7196.

19. Ozbay, E., (2006). Plasmonics: merging photonics and electronics at nanoscale dimensions, *Science*, **311**, pp. 189–193.
20. Maier, S. A., Brongersma, M. L., Kik, P. G., Meltzer, S., Requicha, A. A. G., and Atwater, H. A., (2001). Plasmonics: a route to nanoscale optical devices, *Adv. Mater.*, **13**, pp. 1501–1505.
21. Bradshaw, D., Claridge, J. B., Cussen, E. J., Prior, T. J., Rosseinsky, and M. J., (2005). Design, chirality, and flexibility in nanoporous molecule-based materials, *Acc. Chem. Res.*, **38**, pp. 273–282.
22. Yao, Z., Ch., H. W., Postma, L., Balents, and Dekker, C., (1999). Carbon nanotube intramolecular junctions, *Nature*, **402**, pp. 273–276.
23. Harada, Y., and Asakura, T., (1996). Radiation forces on a dielectric sphere in the Rayleigh scattering regime, *Opt. Commun.*, **124**, pp. 529–541.
24. Neuman, K. C., ans Block. S. M., (2004). Optical trapping, *Rev. Science Instruments*, **75**, pp. 2787–2809.
25. Sueda, K., Miyaji, G., Miyanaga, N., Nakatsuka, M., (2004). Laguerre-Gaussian beam generated with a multilevel spiral phase plate for high intensity laser pulses, *Opt. Express*, **12**, pp. 3548–3553.
26. Vickers, J., Burch, M., Vyas, R., Singh, S., (2008). Phase and interference properties of optical vortex beams, *J. Opt. Soc. Am. A*, **25**, pp. 823–827.
27. Koyama, M., Hirose, T., Okida, M., Miyamoto, K., and Omatsu, T., (2011). Power scaling of a picosecond vortex laser based on a stressed Yb-doped fiber amplifier, *Opt. Express*, **19**, pp. 2994–2999.
28. Beijersbergen, M. W., Allem, L., van der Veen, H. E. L. O., and Woerdman, J. P., (1993). Astigmatic laser mode converters and transfer of orbital angular momentum, *Opt. Commun.* **96**, pp. 123–132.
29. Heckenberg, N. R., McDuff, R., Smith, C. P., and White, A. G., (1992). Generation of optical phase singularities by computer generated holograms, *Opt. Lett.*, **17**, pp. 221–223.
30. Guo, Z., Qu, S., Liu, S., (2007). Generating optical vortex with computer-generated hologram fabricated inside glass by femtosecond laser pulses, *Opt. Commun.*, **273**, pp. 286–289.
31. Bisson, J.-F., Senatsky, Y., and Ueda, K.-I., (2005). Generation of Laguerre-Gaussian modes in Nd:YAG laser using diffractive optical pumping, *Las. Phys. Lett.*, **2**, pp. 327–333.
32. Kotlyar, V. V., Almazov, A. A., Khonina, S. N., Soifer, V. A., Elfstrom, H., and Turnen, J., Generation of phase singularity through diffracting a plane or Gaussian beam by a spiral phase plate, *J. Opt. Soc. A*, **22**, pp. 849–861.

33. Ito, A., Kozawa, Y., and Sato, S., (2010). Generation of hollow scalar and vector beams using a spot-defect mirror, *J. Opt. Soc. A*, **27**, pp. 2072–2077.

34. Okida, M., Omatsu, T., Itoh, M., and Yatagai, T., (2007). Direct generation of high power Laguerre–Gaussian output from a diode-pumped Nd:YVO_4 1.3-μm bounce laser, *Opt. Express*, **15**, pp. 7616–7622.

35. Okida, M., Hayashi, Y., Omatsu, T., Hamazaki, J., and Morita, R., (2009). Characterization of 1.06 μm optical vortex laser based on a side-pumped Nd:$GdVO_4$ bounce oscillator, *Appl. Phys. B*, **95**, pp. 69–73.

36. Chard, S.P., Shardlow, P.C., Damzen, M.J., (2009). High-power non-astigmatic TEM00 and vortex mode generation in a compact bounce laser design, *Appl. Phys. B*, **97**, pp. 275–280.

37. Tanaka, Y., Okida, M., Miyamoto, K., and Omatsu, T., (2009). High power picosecond vortex laser based on a large-mode-area fiber amplifier, *Opt. Express*, **17**, pp. 14362–14366.

38. Koyama, M., Hirose, T., Okida, M., Miyamoto, K., and Omatsu, T., (2011), Nanosecond vortex laser pulses with millijoule pulse energies from an Yb-doped double-clad fiber power amplifier, *Opt. Express*, **19**, pp. 14420–14425.

39. Moh, K. J., Yuan, X.-C., Tang, D. Y., Cheong, W. C., Zhang, L. S., Low, D. K. Y., Peng, X., Niu, H. B., and Lin, Z. Y., (2006). Generation of femtosecond optical vortices using a single refractive optical element, *Appl. Phys. Lett.*, **88**, 091103-1–3.

40. Bezuhanov, K., Dreischuh, A., Paulus, G. G., Schätzel, M. G., and Walther, H., (2004). Vortices in femtosecond laser fields, *Opt. Lett.*, **29**, pp. 1942–1944.

41. Tokizane, Y., Oka, K., and Morita, R., (2009). Supercontinuum optical vortex pulse generation without spatial or topological-charge dispersion, *Opt. Express*, **17**, pp. 14517–14525.

42. Kawata, S, Sun, H., Tanaka, T., Takada, K., (2001). Finer features for functional microdevices, *Nature*, **412**, pp. 697–698.

43. Liu, X., Du, D., Mourou, G., (1997). Laser ablation and micromachining with ultrashort laser pulses, *IEEE J. Quan. Electron.* **33**, pp. 1706–1716.

44. Willmott, P. R., and Huber, J. R., (2000). Pulsed laser vaporization and deposition, *Rev. Mod. Phys.*, **72**, pp. 315–328.

45. Weck, A., Crawford, T. H. R., Wilkinson, D. S., Haugen, H. K., and Preston, J. S., (2008). Laser drilling of high aspect ratio holes in copper with femtosecond, picosecond and nanosecond pulses, *Appl. Phys. A*, **90**, pp. 537–543.

46. Weaver, J. H., Lynch, D. W., and Olson, C. G., (1974). Optical properties of V, Ta, and Mo from 0.1 to 35 eV, *Phys. Rev. B*, **10**, pp. 501–516.

47. Omatsu, T., Isogami, T., Minassian, A., and Damzen, M. J., (2005). >100 kHz Q-switched operation intransversely diode pumped ceramic Nd:YAG laser in bounce geometry, *Opt. Commun.*, **249**, pp. 531–537.

48. Omatsu, T., Okida, M., Minassian, A., and Damzen, M. J, (2006). High repetition rate Q-switching performance in transversely diode-pumped Nd doped mixed gadolinium yttrium vanadate bounce laser, *Opt. Express*, **14**, pp. 2727–2734.

Chapter 11

Engineering the Orbital Angular Momentum of Light with Plasmonic Vogel Spiral Arrays

Luca Dal Negro, Nate Lawrence, and Jacob Trevino
Electrical and Computer Engineering, Boston University, 8 Saint Mary's Street, Boston, MA 02139, USA
dalnegro@bu.edu

In this chapter, we present our work on the engineering of structured light with orbital angular momentum (OAM) using aperiodic arrays of metallic nanoparticles arranged with Vogel spiral geometry. In particular, we introduce and discuss the distinctive structural properties of different types of Vogel spirals and we demonstrate the ability to encode specific numerical sequences, determined by the aperiodic geometry, onto the azimuthal OAM values of diffracted optical beams. Finally, using Fourier–Hankel mode decomposition and interferometric reconstruction of the complex scattered far-fields, we prove experimentally the controlled generation of large OAM azimuthal values, in excellent agreement with analytical scattering theory. The engineering of diffracted beams with controllable OAM spectra using flat aperiodic plasmon

Singular and Chiral Nanoplasmonics
Edited by Svetlana V. Boriskina and Nikolay I. Zheludev

ISBN 978-981-4613-17-0 (Hardcover), 978-981-4613-18-7 (eBook)
www.panstanford.com

arrays is relevant to a number of applications in singular optics, classical, and quantum cryptography.

11.1 Introduction to Aperiodic Optical Structures

Periodic optical media support extended Bloch eigenmodes and feature continuous energy bands [1]. On the contrary, in the absence of inelastic interactions, random media can feature exponentially localized eigenmodes and energy spectra characterized by isolated δ-peaks [2]. A substantial amount of work has been devoted in the past few years to understand transport, localization, and wave scattering phenomena in disordered random media [2–6]. These activities unveiled fascinating analogies between the behavior of electronic and optical waves, such as disorder-induced Anderson light localization [6, 7], the photonic Hall effect [8], optical magnetoresistance [9], universal conductance fluctuations of light waves [10], and optical negative temperature coefficient resistance [11]. However, the engineering of light scattering and transport phenomena in complex optical systems, such as random media, is still very limited. In fact, random structures, while providing a convenient path to field localization, are not reproducible and lack simple engineering design rules for deterministic optimization. These difficulties have strongly limited our ability to conceive, explore, and manipulate optical resonances and light scattering phenomena in systems devoid of spatial periodicity. On the contrary, aperiodic optical media generated by deterministic mathematical rules, known as deterministic aperiodic structures, have recently attracted significant attention in the optics and electronics communities due to their simplicity of design, fabrication, and compatibility with current materials deposition and device fabrication technologies [12–16]. Initial work, mostly confined to theoretical investigations of one-dimensional (1D) aperiodic systems [17–24], have fully succeeded in stimulating broader experimental/theoretical studies of optical nanostructures that leverage deterministic aperiodic order as a comprehensive strategy for the design and manipulation of optical modes and frequency spectra to achieve new device functionalities. In addition to the numerous applications to optical sensing, light emission, and photon dispersion management [13, 16], the study of deterministic aperiodic optical structures is a highly interdisciplinary research field, conceptually rooted in discrete geometry (i.e., tiling theory, point patterns theory, and mathematical crystallography [25–27]), dynamical systems theory (i.e., specifically, discrete dynamical systems and automatic sequences) [28–30], and number theory [31–33]. As a result, the optics of aperiodic media has the potential to generate many exciting and cross-disciplinary opportunities that only recently began to be actively pursued [24].

Understanding the optical response of aperiodic deterministic systems offers an almost unexplored potential for the manipulation of the complex diffractive phenomena of interest to singular optics applications, such as the generation of structured light beams with compound optical vortices.

The scope of this chapter is to provide a background and to discuss the manipulation of OAM light states using plasmonic deterministic aperiodic nano structures (DANS) with Vogel spiral geometry [16, 34]. DANS are optical structures in which the refractive index varies over length scales comparable or smaller than the wavelength of light. They include dielectric and metallic structures, metallo-dielectric plasmonic nanostructures, and metamaterials. DANS can be conveniently fabricated using conventional nanolithographic techniques while displaying transport and localization properties akin to random systems. Most importantly, the Fourier space of DANS can be simply designed to range from a pure-point discrete spectrum, such as for periodic and quasi-periodic crystals, to a diffused spectrum with short-range correlations, as for disordered amorphous systems. In addition, the aperiodic reciprocal space of DANS results in distinctively complex diffraction patterns encoding non-crystallographic point symmetries of arbitrary order, as well as more abstract mathematical symmetries [13].

As we will specifically discuss in the rest of this book chapter, DANS with isotropic Fourier space can be designed in the absence of Bragg peaks based on the concept of aperiodic Vogel spiral geometry [35]. Vogel spiral arrays of metallic nanoparticles define a novel engineering approach, based on the geometric control of polarization-insensitive planar diffraction phenomena, to enhance light-matter coupling in photonic-plasmonic nanostructures [34, 36, 37]. Deterministic aperiodic media with increasing degree of rotational symmetry in their reciprocal space will be briefly reviewed in the next section providing the necessary background to introduce the fascinating class of Vogel spiral structures.

11.2 Rotational Symmetry: From Tilings to Vogel Spirals

One of the deepest results of classical crystallography states that the combination of translations with rotations restricts the total number of available rotational symmetries to the ones compatible with the periodicity of the lattice [27]. This important result is known as the crystallographic restriction. We say that a structure possesses an n-fold rotational symmetry if it is left unchanged when rotated by an angle $2\pi/n$, and the integer n is called the order of the rotational symmetry (or the order of its symmetry axis). It can be shown that only rotational symmetries of order $n = 2, 3, 4, 6$ fulfill the translational symmetry requirements of 2D and 3D periodic lattices in Euclidean space [25–27], therefore excluding $n = 5$ and $n > 6$.

As a result, the pentagonal symmetry, very often encountered in nature as in the pentamerism of viruses, micro-organisms such as radiolarians, plants, and a number of marine animals (i.e., sea stars, urchins, crinoids, and so on) has been traditionally excluded from the mineral kingdom until the non-crystallographic symmetries were discovered in quasicrystals. Moreover, it was recently shown that aperiodic tilings displaying an arbitrary degree of rotational symmetry can be deterministically constructed using a purely algebraic approach [38]. In Fig. 11.1, we display four remarkable aperiodic point patterns featuring increasing rotational symmetry, along with their corresponding reciprocal space, obtained by Fourier transformation. We notice that aperiodic structures possess a non-periodic reciprocal space. As a result, the diffraction k vectors of aperiodic Fourier space lose their global meaning and should be regarded merely as locally defined spatial frequency components.

In Fig. 11.1a, we show a point pattern obtained by positioning particles at the vertices of the celebrated Penrose tiling, named after the mathematician and physicist Roger Penrose who investigated

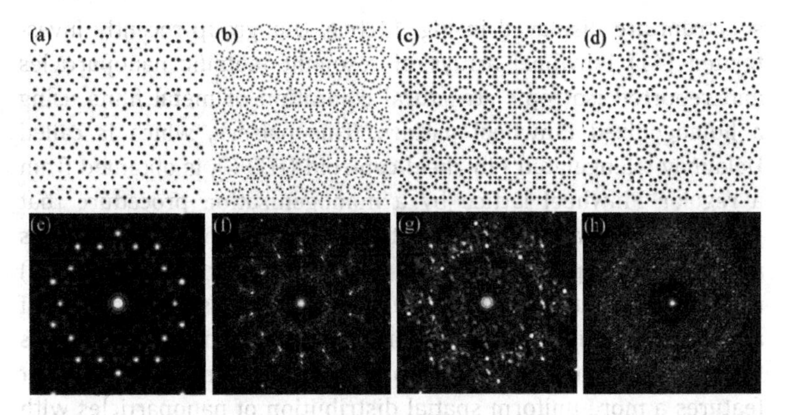

Figure 11.1 (a) Penrose array, generation 12; (b) Danzer array, generation 4; (c) Pinwheel array, generation 5; (d) Delaunay-triangulated Pinwheel centroid (DTPC) array, generation 5; (d) Penrose reciprocal space; (e) Danzer reciprocal space; (f) Pinwheel reciprocal space (h) DTPC reciprocal space.

such aperiodic tiling of the plane in the 1970s. He discovered that only two planar shapes, or prototiles known as the kite and dart tiles, are sufficient to aperiodically cover the entire plane without leaving any gaps. Moreover, the Fourier spectrum of the Penrose tiling, shown in Fig. 11.1e, features δ Bragg peaks arranged in a pattern with 10-fold rotational symmetry. In Fig. 11.1b, we show a deterministic point pattern obtained using the Danzer inflation rule [39], which results in a 14-fold rotational symmetry. The corresponding reciprocal space is shown in Fig. 11.1f.

Deterministic point patterns with increasing degree of rotational symmetry up to an infinite order (i.e., continuous circular symmetry) have also been demonstrated [40] using a simple iterative procedure that decomposes a triangle into congruent copies.

The resulting structure, called the Pinwheel tiling, has triangular elements (i.e., tiles) that appear in infinitely many orientations and, in the limit of infinite-size, the diffraction pattern displays continuous ("infinity-fold") rotational symmetry. Radin has shown that there are no discrete components in the Pinwheel diffraction spectrum [40]. However, it is currently unknown whether the spectrum is continuous or singular continuous. A point pattern obtained from the Pinwheel tiling and the corresponding Fourier spectrum are displayed in Fig. 11.1c,g. Our group recently investigated [41] Pinwheel arrays of resonant metallic nanoparticles and reported isotropic structural coloration of metal films using a homogenized Pinwheel pattern. This pattern, which is called Delaunay triangulated Pinwheel centroid (DTPC), is obtained from a regular Pinwheel lattice by a homogenization procedure that performs a Delaunay triangulation of the array and positions additional nanoparticles in the center of mass (i.e., baricenter) of the triangular elements. The DTPC, which is an example of a deterministic isotropic and homogeneous particle array, shares the same rotational symmetry of the regular Pinwheel array but features a more uniform spatial distribution of nanoparticles with strongly reduced clustering, as evident in Fig. 11.1d. The calculated reciprocal space of the DTPC pattern is displayed in Fig. 11.1h. We notice the higher degree of spatial uniformity of the DTPC when compared with the regular Pinwheel (Fig. 11.1c) pattern. Moreover, local structural correlations, which are due to the finite size of the regular Pinwheel array, result in well-defined scattering peaks in the Pinwheel reciprocal space, shown in Fig. 11.1g. These correlations are absent in the reciprocal space of the DTPC array (Fig. 11.1h) due to its higher degree of spatial uniformity that better approximates uncorrelated disorder.

Using DTPC plasmonic arrays, it was possible to obtain bright green coloration of Au films with greatly reduced angular sensitivity and enhanced color uniformity compared with both periodic and random arrays [41]. Finally, we can appreciate from Fig. 11.1 that the Fourier spectra become more diffuse when increasing the degree of rotational symmetry toward isotropic structures. This general behavior is manifested by amorphous/liquid random media and it is displayed by the broad family of deterministic aperiodic structures known as Vogel's spirals. Interestingly, the diffraction spectra of Vogel spiral particle arrays do not possess any discrete Bragg component and feature almost continuous circular symmetry, as shown in Fig. 11.2 for representative types of structures. The structural properties of Vogel spiral point patterns will be discussed in the next section.

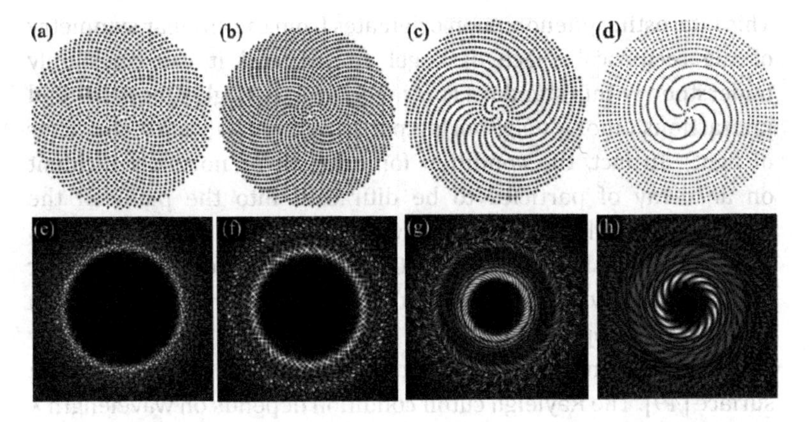

Figure 11.2 (a) GA spiral array, (b) τ spiral array, (c) μ spiral array, (d) π spiral array. (e) GA spiral reciprocal; (f) τ spiral reciprocal space; (g) μ spiral reciprocal space; (h) π spiral reciprocal space.

11.2.1 *Structural Properties of Vogel Spiral Arrays*

Vogel's spiral structures have been investigated by mathematicians, botanists, and theoretical biologists [42] in relation to the geometrical problems of phyllotaxis [43–45], which concerns the understanding of the spatial arrangements of leaves, bracts, and florets on plant stems.

Aperiodic Vogel spiral arrays of nanoparticles are rapidly emerging as a powerful nanophotonics platform with distinctive optical properties of interest to a number of engineering applications [36, 46–48]. This fascinating category of deterministic aperiodic media features circularly symmetric scattering rings in Fourier space entirely controlled by simple generation rules that induce a very rich structural complexity. Trevino et al. [37] have recently described the structure of Vogel spiral arrays of nanoparticles using multi-fractal geometry and discovered that such structures feature a degree of local order in between short-range correlated amorphous/liquid systems and uncorrelated random systems. Moreover, it has been recently demonstrated that Vogel spiral arrays of metallic nanoparticles feature distinctive structural resonances and produce polarization-insensitive, planar light diffraction across a broad spectral range, referred to as circular light scattering [34].

This interesting phenomenon originates from the circular symmetry of the reciprocal space of Vogel spirals, and it can be simply understood within standard Fourier optics (i.e., neglecting near-field interactions among neighboring particles and the vector character of light). In fact, the condition for light waves normally incident on an array of particles to be diffracted into the plane of the array requires the longitudinal wavevector component of light to identically vanish, that is, $k_z = 0$. This requirement is equivalent to the well-known Rayleigh condition that determines the transition between the propagation and the cutoff of the first diffractive order in periodic gratings, so that diffracted waves travel along the grating surface [49]. The Rayleigh cutoff condition depends on wavelength λ and on the transverse spatial frequencies ν_x and ν_y of the diffracting element, according to:

$$k_z = 2\pi\sqrt{(1/\lambda)^2 - \nu_x^2 - \nu_y^2} = 0 \tag{11.1}$$

Equation (11.1) is satisfied on a circle of radius $1/\lambda$ in the reciprocal space of the array. As a result, diffractive elements possessing circularly symmetric Fourier space naturally satisfy the Rayleigh condition irrespectively of the incident polarization and diffract normal incident radiation into evanescent surface modes.

We say that the resonant condition expressed by Eq. (11.1) gives rise to planar diffraction. It is important to notice that, differently from periodic crystals and quasicrystals with rotational symmetries of finite order, Vogel spirals satisfy the resonant condition for planar diffraction over a broader range of wavelengths because they display "thick isotropic rings" in reciprocal space (see Fig. 11.2), uniquely determined by the generation parameters of the structures. The planar diffraction property of Vogel spirals is ideally suited to enhance light-matter interactions on planar substrates [34], and recently led to the demonstration of thin-film solar cell absorption enhancement [36], light emission [46, 50], and second harmonic generation enhancement [48] using metal-dielectric arrays.

Another fascinating feature of Vogel spiral diffracting elements is their ability to support distinctive scattering resonances carrying well-defined numerical sequences of orbital angular momentum (OAM) of light, potentially leading to novel applications in singular optics and optical cryptography [51–53]. In the rest of this chapter, after reviewing the main structural characteristics of Vogel spirals, we will specifically focus on the manipulation of OAM spectra.

Vogel spiral point patterns are defined in polar coordinates (r, θ) by the following equations [13, 54, 55]:

$$r_n = a_0\sqrt{n} \qquad \theta_n = n\alpha \tag{11.2}$$

where $n = 0, 1, 2, \ldots$ is an integer index, a_0 is a constant scaling factor, and α is an irrational number known as the divergence angle. This angle specifies the constant aperture between successive point particles in the array. Irrational numbers (ξ) can be used to generate irrational angles (α°, in degrees) by the relationship $\alpha^\circ = 360^\circ - [\xi - floor(\xi)] \times 360^\circ$. The value $\alpha \approx 137.508^\circ$ approximates the irrational number known as the golden angle, which generates the so-called Fibonacci Golden Angle spiral (GA), shown in Fig. 11.2a. Rational approximations to the golden angle can be obtained by the formula $\alpha = 360 \times (1 + p/q) - 1$ where p and $q < p$ are consecutive Fibonacci numbers.

The spatial structure of a GA spiral can be decomposed into clockwise and counterclockwise families of out-spiraling particles lines, known as parastichies, which stretch out from the center of the structure. Interestingly, the number of spiral arms in each family of parastichies is given by consecutive Fibonacci numbers [54]. We notice that as the golden angle is an irrational number, the GA spiral lacks both translational and rotational symmetry. Accordingly, its spatial Fourier spectrum does not exhibit well-defined Bragg peaks, as for standard photonic crystals and quasicrystals, but rather it features a broad and diffuse circular ring whose spectral position is determined by the particles geometry (Fig. 11.2e).

More generally, different types of aperiodic Vogel spirals can be generated by choosing other irrational values of the divergence angle giving rise to vastly different spiral geometries all characterized by isotropic Fourier spectra. We show three additional examples of aperiodic spirals generated by irrational divergence angles in Fig. 11.2, known as τ, μ , and π spiral arrays, obtained using the approximated values listed in Table 11.1. We notice that Vogel spirals with remarkably different structural properties can be obtained by choosing only slightly different values for the divergence angle, thus providing the opportunity to control and explore distinctively

Table 11.1 Divergence angles utilized for the irrational and perturbed GA spirals

Label	ξ	Divergence angle (°)	Labeling	ξ	Divergence angle (°)
1.1 τ	$(2+\sqrt{8})/2$	210.883118	GA	$(1+\sqrt{5})/2$	137.507764
1.2 μ	$(5+\sqrt{29})/2$	290.670335			
1.3 π	3.14159...	309.026645			
α_1		137.300000			
α_2		137.369255	β_1		137.523137
α_3		137.403882	β_2		137.553882
α_4		137.473137	β_3		137.569255
GA	$(1+\sqrt{5})/2$	137.507764	β_4		137.600000

different degrees of aperiodic structural order. Previous studies have focused on the three most investigated types of aperiodic spirals, including the GA spiral and two other Vogel spirals obtained by the following choice of divergence angles: 137.3° (i.e., α_1 spiral) and 137.6° (i.e., β_4 spiral) [34, 43, 54]. The α_1 and β_4 spirals are called "nearly golden spirals" because their divergence angles are numerically very close to the GA value, but their families of diverging arms, known as parastichies, are considerably fewer. In a recent study, Trevino et al. [37] extended the analysis of aperiodic Vogel spirals to structures generated with divergence angles that are equi-spaced between the α_1-spiral and the GA spiral, and between the golden angle and β_4, as summarized in Table 11.1.

These spiral structures can be considered as one-parameter (i.e., the divergence angle α) structural perturbation of the GA spiral and possess fascinating geometrical features, which are responsible for unique mode localization properties and optical spectra [37].

It is well known that in the GA spiral, there are many parastichies in both clockwise (CW) and counter-clockwise (CCW) directions [56] and that the numbers of parastichies are consecutive numbers in the Fibonacci series, the ratio of which approximates the golden ratio. However, as the divergence angle is varied either above (supra-GA or β-series) or below the golden angle (sub-GA or α-series), the center region of the spiral where both sets of parastichies (CW and CCW) exist shrinks to a point. The outer regions are left with parastichies that rotate only CW for divergence

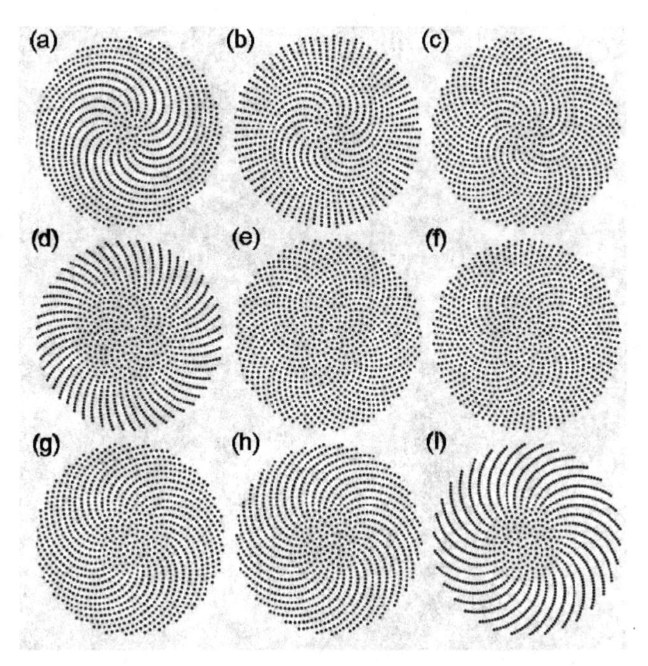

Figure 11.3 Vogel spiral array consisting of 1000 particles, created with a divergence angle of (a) 137.3° (α_1), (b) 137.3692546° (α_2), (c) 137.4038819° (α_3), (d) 137.4731367° (α_4), (e) 137.5077641° (GA), (f) 137.5231367° (β_1), (g) 137.553882° (β_2), (h) 137.5692547° (β_3), (i) 137.6° (β_1). Reproduced from Ref. [37] with permission of the Optical Society, Copyright 2012.

angles greater than the golden angle and CCW for those below, thus providing deterministic aperiodic structures with distinctively chiral geometry. For the spirals with larger deviation from the golden angle (α_1 and β_4 in Fig. 11.3), gaps appear in the center head of the spirals and the resulting point patterns mostly consist of either CW or CCW spiraling arms. We also notice that stronger structural perturbations (i.e., further increase in the diverge angle) lead to less interesting spiral structures containing only radially diverging parastichies (not investigated here).

To better understand the consequences of the divergence angle perturbation on the optical properties of Vogel spiral arrays, we first investigate their spatial Fourier spectra. Figure 11.4 displays the 2D spatial Fourier spectra obtained by calculating the amplitude of

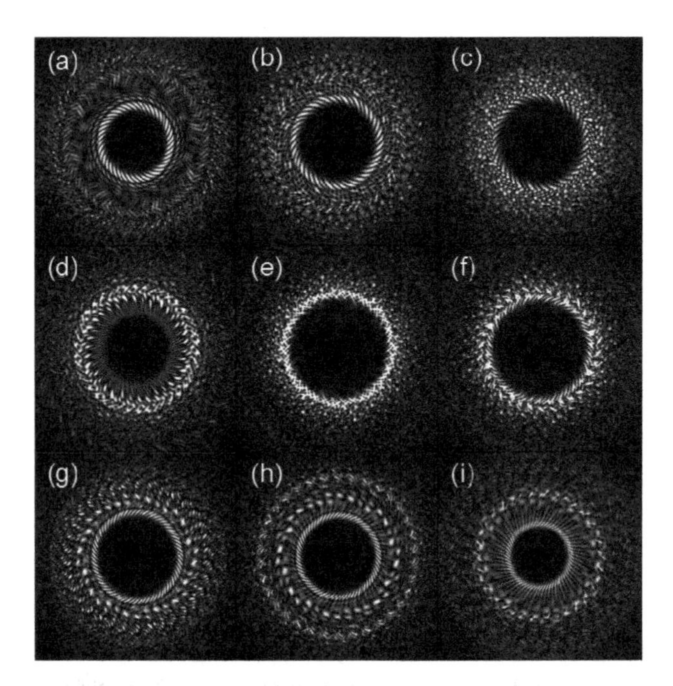

Figure 11.4 Calculated spatial Fourier spectrum of the spiral structures shown in Fig. 11.1. The reciprocal space structure of a (a) α_1 spiral, (b) α_2 spiral, (c) α_3 spiral, (d) α_4 spiral, (e) GA spiral, (f) β_1 spiral (g) β_2 spiral, (h) β_3 spiral, and (i) β_4 spiral are plotted. Reproduced from Ref. [37] with permission of the Optical Society, Copyright 2012.

the discrete Fourier transform (DFT) of the spiral arrays shown in Fig. 11.3.

We can appreciate that all the spectra in Fig. 11.4 lack Bragg peaks and display diffuse circular rings (Fig. 11.4a–i). The many spatial frequency components in Vogel's spirals give rise to a diffuse background, as for amorphous and random systems. Interestingly, despite the lack of rotational symmetry of Vogel spirals, their Fourier spectra are highly isotropic (approaching circular symmetry), as a consequence of a high degree of statistical isotropy [34, 52].

As previously reported [34, 52, 57], the GA spiral features a well-defined and broad scattering ring in the center of the reciprocal space (Fig. 11.4e), which corresponds to the dominant spatial frequency of the structure [52]. Perturbing the GA spiral by varying the divergence angle from the golden angle creates more "disordered" Vogel spirals and results in the formation of multiple scattering rings, associated with additional characteristic length scales, embedded in a diffuse background of fluctuating spots with weaker intensity. In the perturbed Vogel spirals (i.e., Fig. 11.4a,b,d,g–i), different patterns of spatial organization at finer scales are clearly discernable in the diffuse background. The onset of these substructures in Fourier space reflects the gradual removal of statistical isotropy of the GA spiral, which "transitions" to less homogeneous sub-structures with variable degrees of local order.

In order to better characterize the degree of local order in Vogel spirals we resort to, analytical tools are more suitable for the detection of local spatial variations. Our group has recently studied the local geometrical structure of Vogel spirals by the powerful methods of spatial correlation functions [37]. In addition, we applied wavelet-based multifractal analysis to describe the distinctive fluctuations (i.e., the singularity spectrum) of the Vogel spiral geometries and their local density of optical states (LDOS) in the frequency domain, unveiling their multi-fractal nature for the first time [37].

We will now adopt the correlation functions approach to discuss the geometrical structure of Vogel spirals and the main consequences on the resulting optical properties. A comprehensive discussion of all these aspects can be found in Ref. [37]. The pair correlation function, $g(\mathrm{r})$, also known as the radial density distribution function, is employed to evaluate the probability of finding two particles separated by a distance r, thus measuring the local (correlation) order in the structure. Figure 11.5a displays the calculated $g(\mathrm{r})$ for spiral arrays with divergence angles between α_1 and the golden angle (α series), while Fig. 11.5b shows the results of the analysis for arrays generated with divergence angles between the golden angle and β_4 (β series). In order to better capture the geometrical features associated with the structure (i.e., array pattern) of Vogel spirals, the $g(\mathrm{r})$ was calculated directly from the array point patterns (i.e., no element pattern associated with finite-size particles) using the library spatstat [58] within the R statistical

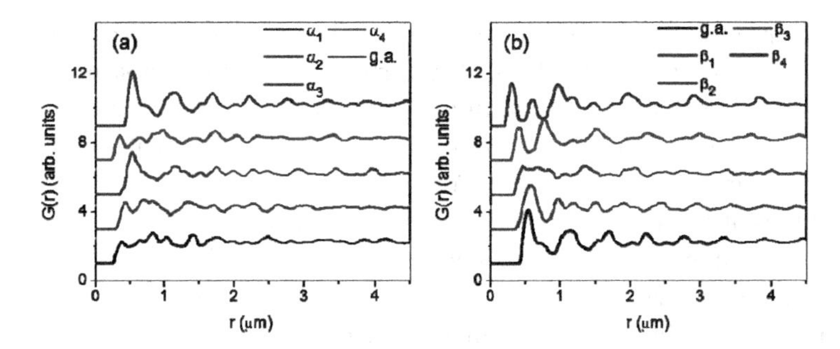

Figure 11.5 Pair correlation function $g(r)$ for spiral arrays with divergence angles between (a) α_1 and the golden angle and (b) between the golden angle and β_4. Reproduced from Ref. [37] with permission of the Optical Society, Copyright 2012.

analysis package. The pair correlation function is calculated as:

$$g(r) = \frac{K'(r)}{2\pi r} \tag{11.3}$$

where r is the radius of the observation window and K'(r) is the first derivative of the reduced second moment function ("Ripley's K function") [59].

The results of the pair correlation analysis shown in Fig. 11.5 reveal a fascinating aspect of the geometry of Vogel spirals, namely their structural similarity to monoatomic gases and liquids. We can clearly appreciate from Fig. 11.5 that the GA spiral exhibits several oscillating peaks, indicating that for certain radial separations, corresponding to local coordination shells, it is more likely to find particles in the array. A similar oscillating behavior for $g(r)$ can be observed when studying the structure of liquids by X-ray scattering [60]. We also notice that the $g(r)$ of the most perturbed (i.e., more disordered) Vogel spiral (Fig. 11.5a, α_1-spiral) features strongly damped oscillations against a constant background, similarly to the $g(r)$ measured for a gas of random particles. Between these two extremes (α_2 to α_4 and β_1 to β_3), a varying degree of local order can be observed for the other spirals in the series. These results demonstrate that the degree of local order in Vogel spiral structures can be deterministically controlled between the correlation properties of photonic amorphous structures [61, 62] and uncorrelated random systems by continuously varying the divergence angle α, which acts as an order parameter.

To deepen our understanding of the complex particle arrangement in Vogel spirals, we have also investigated the spatial distribution of the distance d between first neighboring particles by performing a Delaunay triangulation of the spiral array [52, 57]. This technique provides information on the statistical distribution of the first neighbor distance d and provides a measure of the spatial uniformity of point patterns [63].

In Fig. 11.6, we show the calculated statistical distribution, obtained by the Delaunay triangulation, of the parameter d normalized by d_0, which corresponds to the most probable value (where the distribution is peaked). In all the investigated structures, the most probable value d_0 is generally found to be close to the average

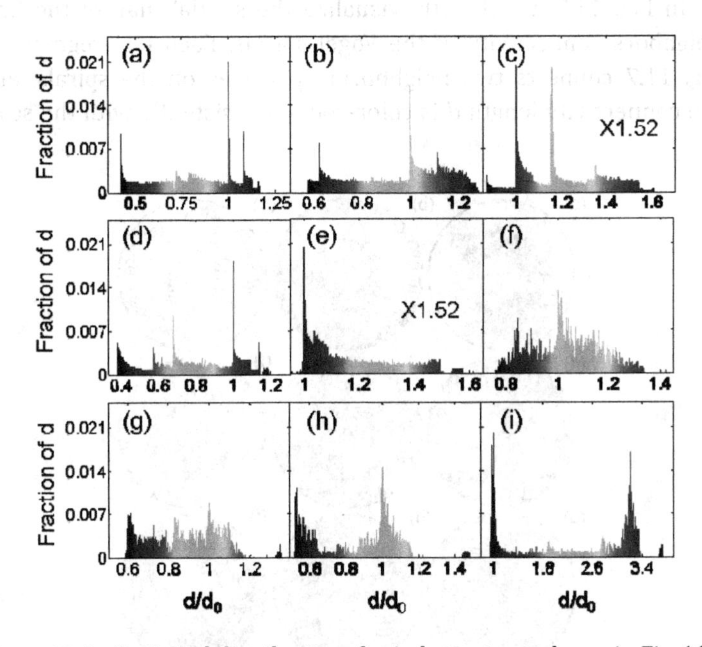

Figure 11.6 Statistical distribution of spiral structures shown in Fig. 11.1. Values represent the distance between neighboring particles d normalized to the most probable value d_0, obtained by Delaunay triangulation (increasing numerical values from blue to red colors). The y-axis displays the fraction of d in the total distribution. Reproduced from Ref. [37] with permission of the Optical Society, Copyright 2012.

inter-particle separation. However, the distributions of neighboring particles shown in Fig. 11.6 are distinctively non-Gaussian in nature and display slowly decaying tails, similar to the "heavy tails" often encountered in mathematical finance (i.e., extreme value theory), suddenly interrupted by large fluctuations or "spikes" in the particle arrangement. These characteristic fluctuations are very pronounced for the two series of perturbed GA spirals, consistently with their reduced degree of spatial homogeneity. All the distributions in Fig. 11.6 are broad with varying numbers of sharp peaks corresponding to different correlation lengths, consistent with the presence of the fine sub-structures already captured in Fourier space (Fig. 11.4).

Next, we perform spatial Delaunay triangulation analysis in order to visualize the spatial locations on the spirals where the different correlation lengths (i.e., distribution spikes) appear more frequently.

In Fig. 11.7, we directly visualize the spatial map of the first neighbors connectivity of the Vogel spirals. Each line segment in Fig. 11.7 connects two neighboring particles on the spirals, and the connectivity length d is color-coded consistently with the scale

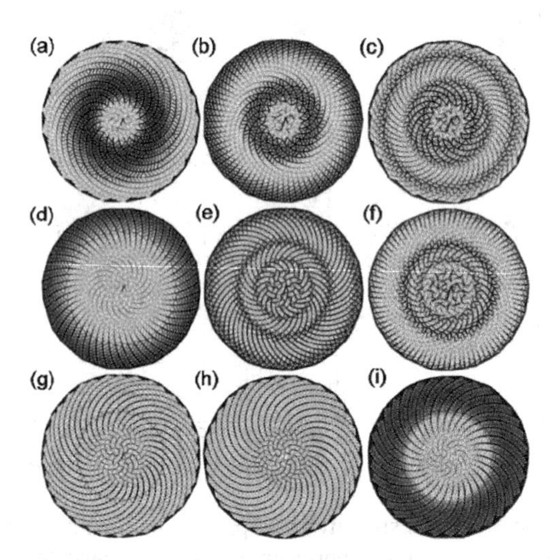

Figure 11.7 Delaunay triangulation of spiral structures shown in Fig. 11.3. The line segments that connect neighboring circles are color-coded by their lengths d. The colors are consistent to those in Fig. 11.12. Reproduced from Ref. [37] with permission of the Optical Society, Copyright 2012.

of Fig. 11.6 (i.e., increasing numerical values from blue to red colors). The non-uniform color distributions shown in Figs. 11.7 graphically represent the distinctive spatial order of Vogel's spirals. In particular, we clearly notice that circular symmetry is found in the distribution of particles for all the spirals, including the strongly inhomogeneous α and β-series. As recently demonstrated by Trevino et al. [37], regions of markedly different values of d define "radial heterostructures" that can efficiently trap radiation in regions with different lattice constants, similarly to the case of the concentric rings of omniguide Bragg fibers. The sharp contrast between adjacent rings radially traps radiation by Bragg scattering along different circular loops. The circular regions evidenced in the spatial map of local particle coordination in Fig. 11.7 well correspond to the scattering rings observed in the Fourier spectra (Fig. 11.4), and are also at the origin of the recently discovered circular scattering resonances carrying OAM in Vogel spirals [34, 52]. Moreover, as discussed in Refs. [37, 52], the characteristic "circular Bragg scattering" occurring between dielectric rods in Vogel spirals gives rise to localized resonant modes with well-defined radial and azimuthal numbers similar to the whispering gallery modes of micro-disk resonators.

We observe that characteristic radial heterostructures are present in the Vogel spiral geometry regardless of the particular choice of the irrational divergence angle. In Fig. 11.8, we make this point by displaying structural results obtained on the τ, μ, and π-irrational spiral arrays, which vastly differ in their real-space structures (Fig. 11.2b-d). This analysis unveils very general structural features of aperiodic Vogel spirals generated by irrational divergence angles and can guide the design of photonic-plasmonic structures that support a large density of distinctively localized optical resonances with a large degree of azimuthal symmetry. In particular, the radially localized azimuthal modes of perturbed and irrational-angle aperiodic spirals are extremely attractive for the engineering of novel light sources, laser devices, and optical sensors that combine a broad spectrum of localized eigenmodes with open dielectric pillar structures for increased refractive index sensitivity to environmental perturbations.

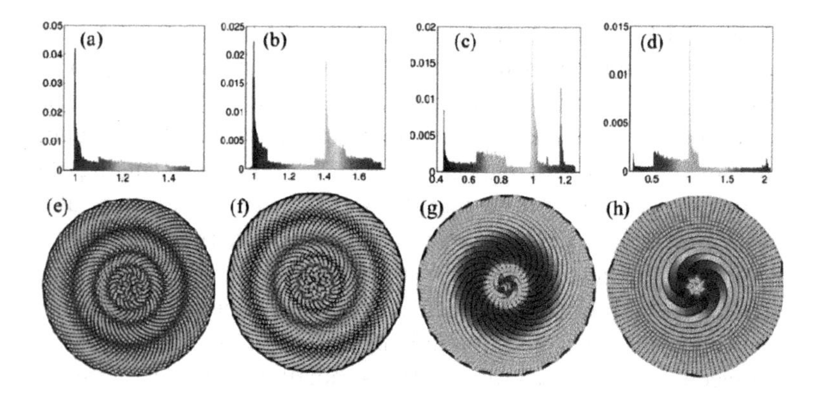

Figure 11.8 Statistical distribution first neighbor distances in irrational-angle spiral structures and corresponding spatial Delaunay triangulation maps for (a,e) GA spiral array, (b,f) τ spiral array, (c,g) μ spiral array, and (d,h) π spiral array. Values represent the distance between neighboring particles d normalized to the most probable value d_0, obtained by (increasing numerical values from blue to red colors). The vertical axis displays the fraction of d in the total distribution.

In the next section, we will focus on the application of aperiodic spiral plasmonic arrays for the generation of structured light that encodes numerical sequences in the OAM azimuthal spectrum.

11.3 Engineering Orbital Angular Momentum of Light

It was recently discovered that Vogel spiral arrays of metallic nanoparticles, when illuminated by optical beams, give rise to diffracted radiation carrying very complex OAM patterns [34, 51, 52]. Recently, Dal Negro et al. [51] developed an analytical model that captured the Fourier–Hankel spectral properties of arbitrary Vogel spiral arrays in closed form solution, thus providing the toolsets necessary to engineer complex OAM states in the far-field diffraction regions. Within the framework of scalar Fourier optics, it was shown that the Fourier spectrum (i.e., Fraunhofer diffraction pattern) of Vogel spirals is described by the complex sum [51]:

$$E_{\infty}(\nu_r, \nu_\theta) = E_0 \sum_{n=1}^{N} e^{j2\pi\sqrt{n}a_0\nu_r\cos(\nu_\theta - n\alpha)} \tag{11.4}$$

where the variables (ν_r, ν_θ) are Fourier conjugate of the direct-space cylindrical coordinates (r, θ) used to represent the Vogel spiral array, α is the irrational divergence angle, a_0 is a constant scaling factor, and N is the number of particles in the array [51].

Fourier–Hankel modal decomposition can be used to analyze a superposition state of OAM states carrying modes in the far-field pattern and determine their relative contribution to the overall diffracted beam. Decomposition of $\rho(r, \theta)$ into a basis [52, 64] set with helical phase fronts is accomplished through Fourier–Hankel decomposition (FHD) according to:

$$f(m, k_r) = \frac{1}{2\pi} \int_0^\infty \int_0^{2\pi} r dr d\theta \rho(r, \theta) J_m(k_r r) e^{im\theta} \quad (11.5)$$

where J_m is the m-th order Bessel function. In this decomposition, the m-th order function identifies OAM states with azimuthal number m, accommodating both positive and negative integer m values.

By analytically performing FHD analysis, Dal Negro et al. [51] demonstrated that diffracted optical beams by Vogel spirals carry OAM values arranged in aperiodic numerical sequences determined by the number-theoretic properties of the irrational divergence angle α. More precisely, the OAM values transmitted in the far-field region are directly determined by the rational approximations of the continued fraction expansion of the irrational divergence angles of Vogel spirals [51]. In particular, wave diffraction by GA arrays generates a Fibonacci sequence of OAM values in the Fraunhofer far-field region. This fascinating property of Vogel spirals can be understood clearly by considering the analytical solution of the FHD of the far-field radiation pattern, given by [51]:

$$f(m, k_r) = \sum_{n=1}^{N} A(k_r) e^{imn\alpha} \quad (11.6)$$

where $A(k_r)$ is a k_r-dependent coefficient, which can be ignored as we are concerned with the azimuthal dependence contained in $f(m)$.

We see from the result in Eq. (11.6) that when the product $m\alpha$ is an integer, the N contributing waves will be exactly in phase to produce an OAM peak with azimuthal number m. However, for an irrational value of the angle α, this condition will never be exactly met. Nevertheless, using the theory of continued expansion of rational fractions, we can design structures that approximately match the integer condition for the $m\alpha$ product, thus producing well-defined OAM peaks. This is so because an arbitrary divergence angle α is directly determined by an irrational number ζ that admits precisely one infinite continued fraction representation (and vice versa) in the form [32, 33]:

$$\zeta = [a_0; a_1, a_2, a_3,] = a_0 + \cfrac{1}{a_1 + \cfrac{1}{a_2 + \cfrac{1}{a_3 + \cfrac{1}{a_4 +} \cdots}}} \tag{11.7}$$

Such infinite continued fraction representation is very useful because its initial segments provide excellent rational approximations to the irrational numbers. The rational approximations (i.e., fractions) are called the convergents of the continued fraction, and it can be shown that even-numbered convergents are smaller than the original number ζ while odd-numbered ones are bigger. Moreover, once the continued fraction expansion of ζ has been obtained, well-defined recursion rules exist to quickly generate the successive convergents. In fact, each convergent can be expressed explicitly in terms of the continued fraction as the ratio of certain multivariate polynomials called continuants. If two convergents are found, with numerators $p_1, p_2, \ldots$ and denominators $q_1, q_2, \ldots$ then the successive convergents are given by the formula:

$$\frac{p_n}{q_n} = \frac{a_n p_{n-1} + p_{n-2}}{a_n q_{n-1} + q_{n-2}} \tag{11.8}$$

Thus, to generate new terms into a rational approximation, only the two previous convergents are necessary. The initial or seed values required for the evaluation of the first two terms are (0,1) and (1,0) for (p_{-2}, p_{-1}) and (q_{-2}, q_{-1}), respectively. It is clear from the discussion above that for spirals generated from an arbitrary irrational number ζ, azimuthal peaks of order m (i.e., Bessel order m) will appear in its FHD due to the denominators q_n of the rational approximations (i.e., the convergents) of $\zeta \approx p_n/q_n$. In fact, for all integer Bessel orders $m = q_n$, the exponential sum in Eq. (11.6) will give in-phase contributions to the FHD and produce strong azimuthal peaks. Therefore, once the rational approximations of the irrational number ζ have been identified on the basis of the well-established continued fraction theory, Vogel spirals that "encode" in their OAM spectra well-defined numeric sequence of azimuthal orders, specified by the denominators q_{n} in Eq. (11.8), can be readily designed. This general strategy enables to generalize the OAM Fibonacci coding recently discovered for GA spirals. It is important to realize that using the approach just introduced, it becomes possible to generate structured light with controllable OAM spectra by plasmon-enhanced diffraction in aperiodic arrays of nanoparticles. The use of very compact Vogel spiral arrays additionally enables the generation of large OAM azimuthal orders that are intrinsic to their geometric structures.

In Fig. 11.9, we show the calculated Fraunhofer far-fields and the corresponding azimuthal OAM spectra of GA and π spirals. As we are primarily concerned with the azimuthal component $f(m)$ of OAM, we can sum $f(m, k_r)$ over radial the wavenumbers k_r.

Figure 11.9c,d demonstrate the very rich structure of OAM peaks of the scattered radiation enabled by the aperiodic geometry control of Vogel spirals. These peaks exactly occur at azimuthal numbers (labeled in the figures) corresponding to the different denominators of the rational approximations of the irrational divergence angles used to generate the spirals, as previously discussed. In particular, the Fibonacci sequence of OAM values is coded in the far-field region of the radiation scattered by the GA spiral (Fig. 11.9c).

Using phase-delayed interferometric measurements, Lawrence et al. [47] have recently measured experimentally the complex electric field distribution of scattered radiation by Vogel spirals and demonstrated for the first time structured light carrying multiple values of OAM in the far-field region of arrays of metallic nanoparticles, in excellent agreement with the analytical theory [51]. In addition, direct excitation of optical modes carrying OAM values has recently been demonstrated in light-emitting dielectric GA spiral structures consisting of Er-doped SiN nanopillars [46].

Before discussing more in depth the experimental demonstration of coded OAM spectra with large azimuthal values, we will briefly

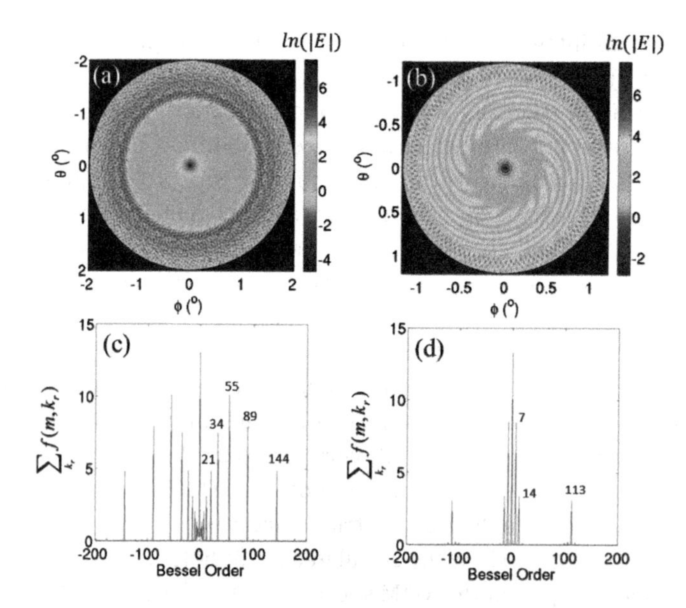

Figure 11.9 (a,b) Analytically calculated far-field radiation patterns of GA and π spiral with 2000 particles at a wavelength of 633 nm for structures with $a_0 = 14.5$ μm. The far-field radiation patterns have been truncated with an angular aperture of 4° and 3°, respectively. (c,d) Fourier–Hankel transforms of far-field scattered radiation by GA and π spiral, respectively, summed over the radial wavenumber k_r. The numbers in the figures indicate the azimuthal Bessel order of the corresponding FHD peaks.

review several unique aspects related to the Fresnel diffraction of optical fields by Vogel spirals with well-defined chirality in the next section.

11.3.1 *Diffracted Beam Propagation from Aperiodic Chiral Spirals*

In this section, we will show that the interplay between the structural disorder and the well-defined CCW chirality of the α_1 spiral gives rise to fascinating wave diffraction and self-imaging properties that can be of interest to a number of applications in singular optics. In particular, we will discuss the evolution of a scattered plane wave as it propagates away from the plane of a diffracting array of apertures with α_1 chiral geometry toward the far-field region.

This analysis will also provide physical insights on the formation of the structured far-fields with multiple OAM states, which are a distinctive feature of diffracting arrays. Field propagation and diffraction from Vogel spiral arrays of dielectric nanopillars has recently been experimentally investigated in Ref. [65].

In order to efficiently investigate the paraxial free propagation of the scattered field intensity over a larger range of distances from the spiral (i.e., object) plane, we have resorted to the method of fractional Fourier transformation (FRFT). This approach provides an equivalent formulation of the paraxial wave propagation and Fresnel scalar diffraction theory [66] and considers light propagation as a process of continual fractional transformation of increasing order. The FRFT is a well-known generalization in fractional calculus of the familiar Fourier transform operation [67], and it has successfully been applied to the study of quadratic phase systems, imaging systems, and diffraction problems in general [66, 68, 69].

Given a function $f(u)$, under the same conditions in which the standard Fourier transform exists, we can define the a-th order FRFT $f_a(u)$ with a being a real number in several equivalent ways [66]. The most direct definition of the FRFT is given in terms of the linear integral transform:

$$f_a(u) \equiv \int_{-\infty}^{\infty} K_a(u, u') f(u') du' \tag{11.9}$$

with a kernel defined by [66, 70]:

$$\begin{aligned} K_a(u, u') &\equiv A_\alpha \exp\left[i\pi(\cot\alpha u^2 - 2\csc\alpha u u' + \cot\alpha u'^2)\right] \\ A_\alpha &\equiv \sqrt{1 - i\cot\alpha} \\ \alpha &\equiv \frac{a\pi}{2} \end{aligned} \tag{11.10}$$

for $a \neq 2n$ and $K_a(u, u') = \delta(u - u')$ when $a = 4n$ and $K_a(u, u') = \delta(u + u')$ when $a = 4n \pm 2$, with n being an integer. The a-th order fractional transform defined above is sometimes called the α-th order transform and it coincides with the standard (i.e., integer) Fourier transform for $a = 1$ or $\alpha = \pi/2$. More information on alternative definitions, generalizations, and the many fascinating properties of FRFTs can be found in Refs. [66, 71]. The FRFT of a function can be roughly thought of as the Fourier transform to the n-th power, where n need not be an integer. Moreover, being a particular type of linear canonical transformation, the FRFT maps a function to any intermediate domain between time and frequency and can be interpreted as a rotation in the time-frequency domain [66].

In optics, the main advantage of the FRFT method compared with numerical simulation relies on its superior computational efficiency, which enables a more extended investigation of the qualitative behavior of the intensity propagation over a large propagation range. In what follows, we computed the two-dimensional FRFT according to Refs. [70, 72] and studied the propagation of the diffracted field up to 50 μm above the spiral plane.

In Fig. 11.10, we show few representative simulations that display the calculated diffracted intensity at different distances from the spiral plane, as specified in the caption. We can clearly appreciate situations in which the diffracted field patterns are clearly rotated with respect to the geometry of the parastichies arms in Fig. 11.10(i,j). The emergence of this characteristic rotation phenomenon from the FRFT field simulations, which do not consider material dispersion properties or the detailed particle shape, demonstrates its very robust nature. In fact, we have recently confirmed by full-vector 3D finite difference time domain of Au nanoparticle arrays that this intriguing effect originates from the coherent interactions of (singly) diffracted wavelets. In particular, we can appreciate from the field evolution shown in Fig. 11.10 that the central area of the spiral couples radiation into directions that are orthogonal to the surrounding parastichies arms. These secondary lines of scattered radiation spatially define a complementary set of parastichies arms that are responsible for the inversion of the intensity pattern at short distances from the object plane. As the intensity propagation unfolds, the diffracted wavelets coherently reinforce each other along distinctively rotating ring-like structures observed in Fig. 11.10(h–j), which gradually transition at larger distances into the characteristic circularly symmetric far-field patterns of Vogel spirals (Fig. 11.10l) [34, 37]. This distinct rotation of the diffracted intensity pattern observed in chiral spiral arrays provides a clear evidence of net OAM transfer within few micrometers from the object, which could be exploited

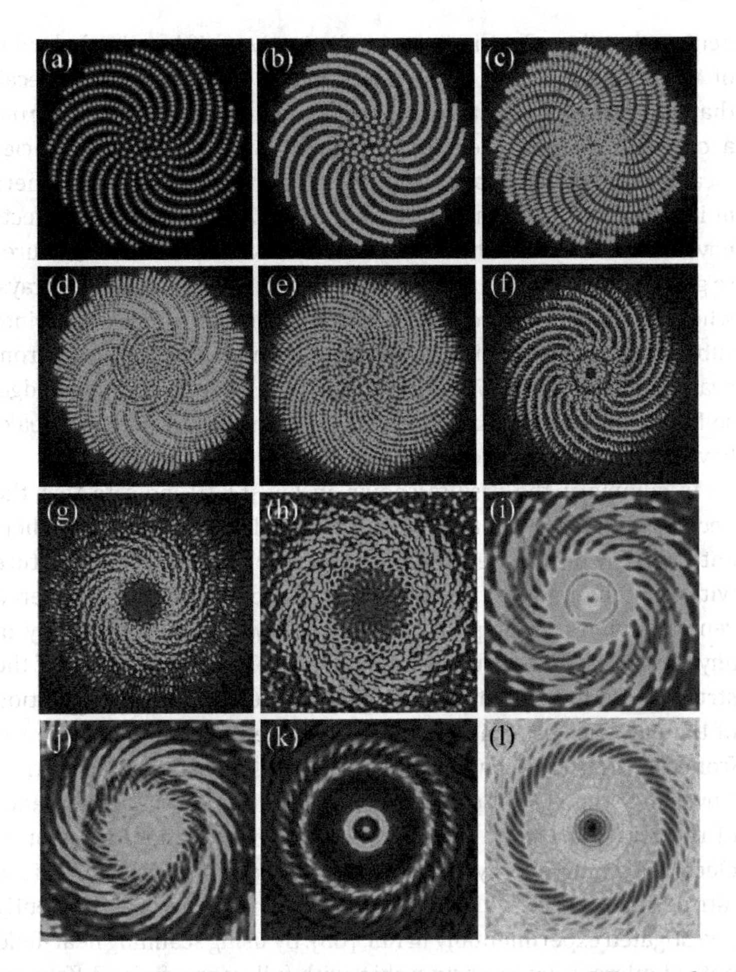

Figure 11.10 Magnitude of electric field propagated to different planes using fractional Fourier transform. Propagation distance for panels (a–k) are 10 nm, 43 nm, 78 nm, 107 nm, 188 nm, 0.457 μm, 0.606 μm, 0.836 μm, 1.95 μm, 2.68 μm and 6.5 μm, respectively. Panel (l) shows the analytical far-field in $\log_{10}$ scale. The structure is excited by a normally incident plane wave at 1550 nm.

to realize optical torques and novel tweezers on the basis of planar plasmonic nanostructures. Moreover, depending on the propagation distance, we have observed a characteristic oscillation between the formation of ring-like structures and the inversion of the field intensity patterns with respect to the spiral geometry. We believe that this effect indicates a distinctive self-imaging behavior of Vogel spirals, which is currently under investigation. We recall that full image reconstruction will happen at a distance L from a coherently illuminated array with discrete spatial frequencies located in reciprocal space at rings of radii $\rho^2 = 1/\lambda^2 - (m/L)^2$ where m is an integer such that $0 \leq m \leq L/\lambda$ [73]. Self-imaging effects have been vastly investigated in the context of periodic structures (e.g., the Talbot effect) as well as in quasi-periodic Penrose arrays, where it has been recently shown that they can focus light into sub-wavelength spots in the far-field without contributions from evanescent fields [74–76]. However, to the best of our knowledge, self-imaging effects in aperiodic media that lack diffraction peaks have never been reported.

In summary, the results shown in Fig. 11.10 indicate that the peculiar diffractive behavior of the α_1 spiral results from the coherent interplay of two well-separated spatial regions of the structure with very dissimilar structural order: on one side the disordered central region of the spiral that diffracts wavelets isotropically in any direction, and on the other side the surrounding region of the structure with well-defined chirality defined by the CCW rotation of the parastichies arms. In this region, the radiated optical power from each scatterer couples into orthogonal parastichies arms, thus "inverting" the spatial pattern at intermediate self-imaging planes of the array. The near-field coupling and propagation behavior of electromagnetic energy scattered at 1.56 μm by arrays of silicon nitride nanopillars with Vogel spiral geometry has been recently investigated experimentally in Ref. [65]. By using scanning near-field optical microscopy in partnership with full-vector finite difference time domain numerical simulations, Caselli et al. [65] demonstrate a characteristic rotation of the scattered field pattern by a Vogel spiral consistent with net transfer of OAM in the Fresnel zone, within few micrometres from the plane of the array. We believe that the unique interplay between aperiodic order and chiral structures such as the investigated α_1 spiral can provide novel opportunities for the manipulation of sub-wavelength optical fields and disclose richer scenarios for the engineering of focusing and self-imaging phenomena in nanophotonics.

11.3.2 *Experimental Demonstration of OAM Generation and Control*

In this final section, we will discuss the recent demonstration of broad OAM azimuthal spectra produced by light scattering in plasmonic Vogel spirals.

It has long been established that electromagnetic fields carry linear momentum and angular spin. However, the possibility for optical fields to additionally carry OAM has only recently been realized [77, 78].

In general, OAM arises through the azimuthal phase dependence of the complex optical field. Optical OAM has recently found uses in rotating optical traps [79], secure optical communication [80], and increasing data transfer rates through OAM multiplexing for fiber-based systems [81]. Moreover, recent advances in the science and technology of OAM have provided the possibility to detect light waves carrying simultaneously multiple OAM values [82]. However, the controlled generation of optical waves that can simultaneously carry large values of OAM still remains very challenging.

Currently, the generation of OAM is achieved by converting Gaussian laser modes to Laguerre-Gaussian (LG) modes with explicit azimuthal phase dependence. This can be accomplished using a system of cylindrical lenses [83] or spatial light modulators (SLMs) [84]. However, SLMs are expensive and their pixel size limits the complexity of patterns that can be created. More recently, the generation of OAM from planar plasmonic devices has also been demonstrated [34, 85]. Unfortunately though, most of the current generation methods are limited to creating OAM states with only a few azimuthal values.

The controlled generation and manipulation of OAM states with large values of azimuthal numbers has very recently been demonstrated experimentally by Lawrence et al. [46] using various types of Vogel arrays of metal nanoparticles. The OAM content of a diffracted optical wave at 633 nm was analyzed using phase stepped interferometric measurements to recover the complex optical field [47]. Fourier–Hankel modal decomposition of scattered radiation was performed to demonstrate the generation of OAM sequences from Vogel spirals.

The nanoparticle arrays were fabricated using electron beam lithography (EBL) on quartz substrates. The fabrication process flow begins with 180 nm of PMMA 950 (Poly-Methyl MethAcrylate) spin coated on top of the substrate, followed by a soft bake at 180°C for 20 min. The nanopatterns were written using a Zeiss SUPRA 40VP SEM equipped with Raith beam blanker and NPGS for nanopatterning. After developing the resist in a 3:1 solution of MIBK (methylisobutyle ketone) and isopropanol, a thin metal stack was deposited on the patterned surface by electron-beam evaporation. The stack consisted of a 2 nm Cr adhesion layer followed by a 28 nm Au layer. A lift-off process was then performed using heated acetone.

The SEMs of fabricated arrays are shown in Figs. 11.11 and 11.12. In particular, GA, τ, μ, and π spirals were fabricated and the corresponding far-field diffraction patterns experimentally measured (Fig. 11.11c,d and Fig. 11.12c,d).

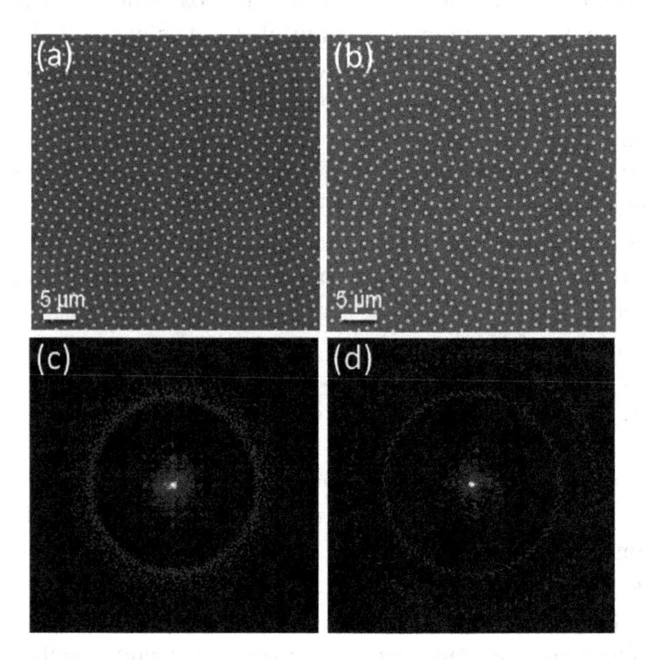

Figure 11.11 SEM micrographs of (a) GA spiral and (b) τ spiral gold nanoparticle arrays on a fused silica substrate. Measured far-field diffraction patterns for (c) φ spiral and (d) τ spiral, taken from respective arrays shown in Fig. 11.a and b.

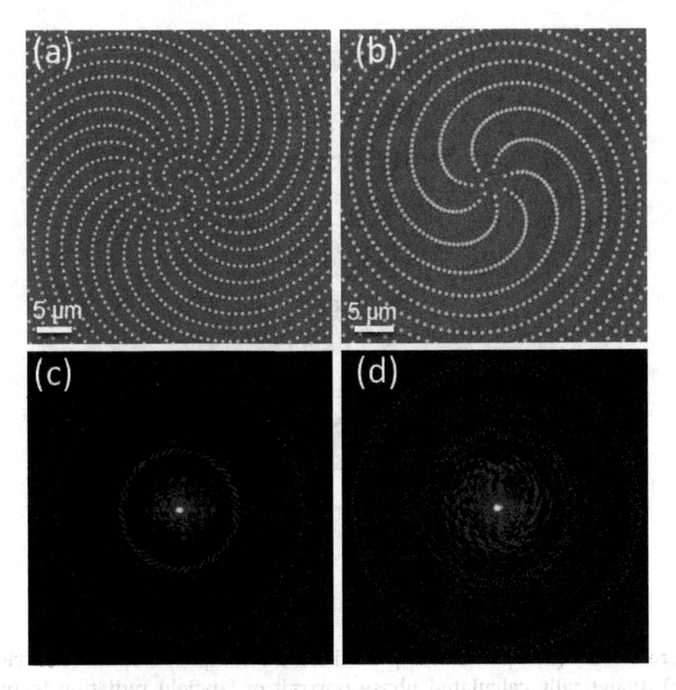

Figure 11.12 SEM micrographs of (a) μ spiral and (b) π spiral gold nanoparticle arrays on a fused silica substrate. Measured far-field diffraction patterns for (c) μ spiral and (d) π spiral, taken from respective arrays shown in Fig. 11.12a and b.

In Fig. 11.13a, we show the optical set-up used to measure the complex far-field of scattered radiation from the fabricated structures. A HeNe laser was utilized in these experiments and weakly focused to a spot size of 50 μm, from the rear of the sample and the scattered light from the array was collected by a 50X (NA = 0.75) objective. Two additional lenses were added to image the far-field at the plane of the CCD [47].

In order to measure the phase of the far-field radiation, a reference beam was reflected from the piezo stage mounted mirrors, expanded and directed to the CCD. Multiple interference patterns were collected by increasing the piezo bias voltage, scanning the phase of the reference beam. A phase retrieval algorithm was finally used to recover the phase of the scattered light relative to the one of the reference beam. The analytically calculated phase of the far-field

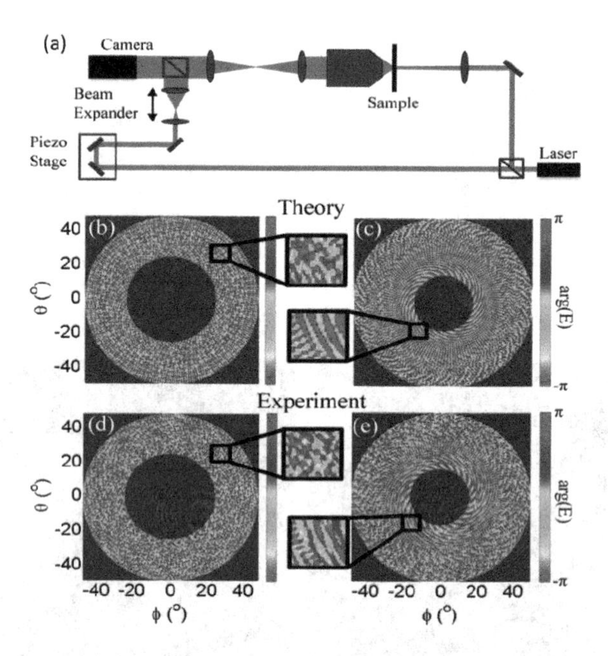

Figure 11.13 (a) Optical set up used for the complex amplitude retrieval. (b,c) Analytically calculated phase portrait of far-field radiation from GA and μ spirals, respectively. (d,e) Experimentally measured phase portrait from GA and μ spirals, respectively. Shaded areas indicate regions where the field intensity is too low to measure the phase accurately. These regions are ignored in the FHD. Reproduced from Ref. [51], with permission of the Optical Society.

scattered radiation from GA and μ spirals is shown in Fig. 11.13b,c, respectively. The measured phase from the GA spiral is shown in Fig. 11.13d and the phase from a μ spiral is shown in Fig. 11.13e. The shaded areas in Fig. 11.13 indicate the limits (angular range) where the measurement of the scattered light is inaccurate due to the limitations of the experimental setup [47]. Very similar results were obtained for the μ and π spirals (not shown here). The insets in the reconstructed phase portraits in Fig. 11.13 highlight the formation of a distribution of optical vortices at positions that well match the analytical calculations.

Modal decomposition was used to analyze a superposition of OAM carrying modes in the far-field pattern and determine their relative contribution [64]. Decomposition into a basis set with

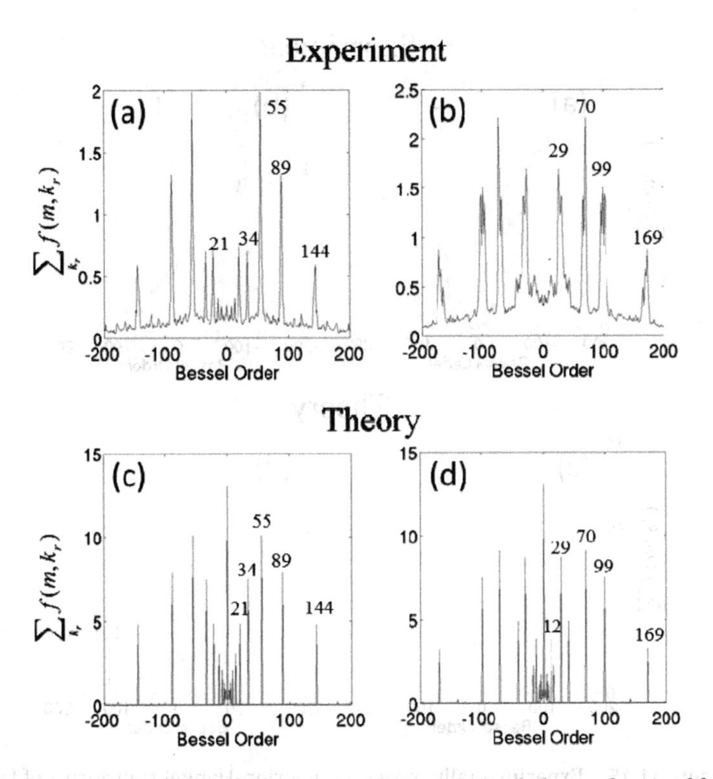

Figure 11.14 Experimentally measured Fourier–Hankel transforms of far-field scattered radiation by (a) GA spiral and (b) τ spiral, respectively, summed over the radial wavenumber k_r. The numbers in the figures indicate the azimuthal Bessel order of the corresponding FHD peaks. Analytically calculated Fourier–Hankel transforms of far-field scattered radiation by (c) GA spiral and (d) τ spiral.

azimuthal dependence is accomplished through FHD according to the Eq. (11.5). We recall that in this decomposition, the m-th order function carries OAM with azimuthal number m, accommodating both positive and negative integer values for m.

In Figs. 11.14 and 11.15, the summed FHD of the calculated and experimentally measured far-field radiation from the fabricated spirals are shown. Here, we have summed the $f(m, k_r)$ over the radial wavevectors k_r as we are primarily concerned with the azimuthal component of the field. We should notice in Figs. 11.14 and 11.15 that the azimuthal spectra exhibits a number of peaks

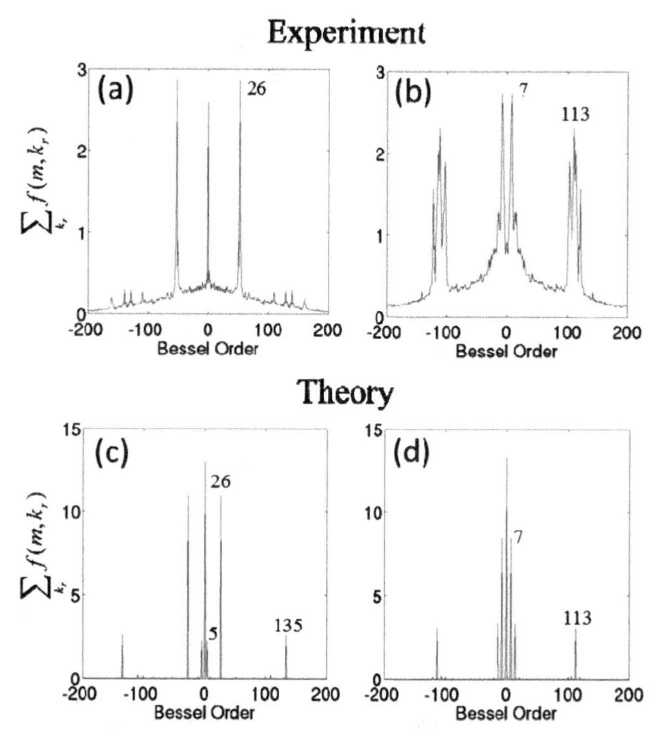

Figure 11.15 Experimentally measured Fourier–Hankel transforms of far-field scattered radiation by (a) μ spiral and (b) π spiral, respectively, summed over the radial wavenumber k_r. The numbers in the figures indicate the azimuthal Bessel order of the corresponding FHD peaks. Analytically calculated Fourier–Hankel transforms of far-field scattered radiation by (c) μ spiral and (d) π spiral.

at numbers corresponding to the denominators of the rational approximations (i.e., convergents) of the irrational divergence angles used to generate the spirals, as previously discussed [51].

For the GA spiral, the encoding of the Fibonacci sequence up to azimuthal number 144 was experimentally demonstrated in the measured OAM spectrum. The τ spiral has a similar number of convergents in our region of interest and allowed us to demonstrate however OAM peaks up to azimuthal number 169. However, these peaks are significantly broadened by the aperturing of the beam [51]. On the contrary, we can see in Fig. 11.15 that the μ and

π spirals have significantly fewer convergents in our region of interest. However, OAM peaks are still clearly observed at linear combinations of the convergents despite the significant broadening due to the imperfect experimental alignment.

It is worth mentioning that the measured values of OAM peaks are fully explained on the basis of the analytical far-field diffraction theory [51] reviewed in Section 11.3. In particular, peaks in the measured OAM spectra are observed at numbers corresponding to the Fibonacci sequence for the GA spiral and more generally, as demonstrated by Figs. 11.14 and 11.15, peaks are observed matching the theoretically predicted values based on the fractional expansion theory of irrational spirals. A full list of analytically calculated and experimentally measured FHD peak positions can be found in Ref. [47].

These results demonstrate the ability to design and generate structured light with extremely complex orbital patterns simultaneously carrying a large number of azimuthal values in the far-field diffraction zone. Moreover, the reconstructed complex field images (Fig. 11.13) indicate the presence of a large number of optical vortices that result from wave diffraction in aperiodic Vogel spirals.

It is very interesting to speculate on the engineering of novel light sources that could radiate such complex optical beams or couple them into conventional fiber-based communication systems to achieve greatly expanded spatial multiplexing capabilities. Moreover, information coding in higher dimensional azimuthal bases could drastically expand the present capabilities of both classical and quantum information systems [53].

The successful demonstration of the superposition of many OAM modes in engineered Vogel spiral arrays of Au nanoparticles provides novel exciting opportunities for singular optics that are of interest also to secure optical communication, optical trapping, and plasmonic sensing technology.

11.4 Outlook and Conclusion

In this chapter, we presented a comprehensive overview of our recent work on the engineering of structured light carrying OAM using aperiodic plasmon arrays with Vogel spiral geometry. Specifically, we discussed the relevant background, structure-property relations, and the light-scattering properties of Au nanoparticle arrays with different types of Vogel spiral order. These structures define a fascinating novel platform for the manipulation of OAM of light potentially leading to novel applications in secure communications and singular optics. Our discussion has particularly emphasized the importance of photonic diffraction phenomena in isotropic Fourier space, which could provide novel opportunities to manipulate light-matter interactions on the nanoscale using plasmonic resonance.

The computational and experimental results presented in this chapter demonstrate the significance of aperiodic structures in the context of singular optics and their potential for innovation in basic science and technological applications. However, the connubium of aperiodic deterministic media with singular optics is still in its infancy and our results represent only a starting point for future work in this fascinating and highly interdisciplinary research area. The ability to create novel complex optical beams with programmable OAM spectra using compact plasmonic nanoparticle arrays suitable for planar device integration promises to impact several fields in optical trapping and sensing technologies. Moreover, light scattering by aperiodic Vogel spirals makes possible the information encoding in a large set of azimuthal bases, which is currently of great relevance for both classical and quantum cryptography [53]. We believe that the engineering of aperiodic optical devices that leverage complex beams with structured phase properties will produce significant advances to both fundamental science and optical technology.

Acknowledgments

This work was partly supported by the AFOSR programs under Award numbers FA9550-10-1-0019 and FA9550-13-1-001 and by the NSF Career Award No. ECCS-0846651.

References

1. Joannopoulos, J. D., Johnson, S. G., Winn, J. N. and Meade, R. D. (2008). *Photonic Crystals: Molding the Flow of Light* (Princeton University Press, Singapore).
2. Sheng, P. (2006). *Introduction to Wave Scattering, Localization and Mesoscopic Phenomena* (Springer Verlag, Germany).
3. Wolf, P.-E. and Maret, G. (1985). Weak localization and coherent backscattering of photons in disordered media, *Phys. Rev. Lett.*, **55**, 2696.
4. Albada, M. P. V. and Lagendijk, A. (1985). Observation of weak localization of light in a random medium, *Phys. Rev. Lett.*, **55**, 2692.
5. Wiersma, D. S., van Albada, M. P. and Lagendijk, A. (1995). Coherent backscattering of light from amplifying random media, *Phys. Rev. Lett.*, **75**, 1739.
6. Lagendijk, A., van Tiggelen, B. and Wiersma, D. S. (2009). Fifty years of Anderson localization, *Phys. Today*, **62**, 24–29.
7. Anderson, P. W. (1985). The question of classical localization: A theory of white paint? *Phil. Mag. Part B*, **52**, 505–509.
8. van Tiggelen, B. A. (1995). Transverse diffusion of light in Faraday-active media, *Phys. Rev. Lett.*, **75**, 422.
9. Sparenberg, A., Rikken, G. L. J. A. and van Tiggelen, B. A. (1997). Observation of photonic magnetoresistance, *Phys. Rev. Lett.*, **79**, 757.
10. Scheffold, F. and Maret, G. (1998). Universal conductance fluctuations of light, *Phys. Rev. Lett.*, **81**, 5800.
11. Wiersma, D. S., Colocci, M., Righini, R. and Aliev, F. (2001). Temperature-controlled light diffusion in random media, *Phys. Rev. B*, **64**, 144208.
12. Maciá, E. (2006). The role of aperiodic order in science and technology, *Rep. Progress Phys.*, **69**, 397–441.
13. Maciá, E. (2009). *Aperiodic Structures in Condensed Matter: Fundamentals and Applications* (CRC Press Taylor & Francis, Boca Raton).
14. Steurer, W. and Sutter-Widmer, D. (2007). Photonic and phononic quasicrystals, *J. Phys. D: Appl. Phys.*, **40**, R229–R247.
15. Poddubny, A. N. and Ivchenko, E. L. Photonic quasicrystalline and aperiodic structures, *Physica E: Low Dimens. Syst. Nanostruct.*, **42**, 1871–1895.
16. Dal Negro, L., Boriskina, S.V. (2012). Deterministic aperiodic nanostructures for photonics and plasmonics applications, *Laser Photon. Rev.*, **6**, 178–218.

17. Kohmoto, M., Kadanoff, L. P. and Tang, C. (1983). Localization problem in one dimension: Mapping and escape, *Phys. Rev. Lett.*, **50**, 1870.
18. Kohmoto, M., Sutherland, B. and Iguchi, K. (1987). Localization of optics: Quasiperiodic media, *Phys. Rev. Lett.*, **58**, 2436.
19. Vasconcelos, M. S. and Albuquerque, E. L. (1999). Transmission fingerprints in quasiperiodic dielectric multilayers, *Phys. Rev. B*, **59**, 11128.
20. Dulea, M., Severin, M. and Riklund, R. (1990). Transmission of light through deterministic aperiodic non-Fibonaccian multilayers, *Phys. Rev. B*, **42**, 3680.
21. Gellermann, W., Kohmoto, M., Sutherland, B. and Taylor, P. C. (1994). Localization of light waves in Fibonacci dielectric multilayers, *Phys. Rev. Lett.*, **72**, 633.
22. Merlin, R., Bajema, K., Clarke, R., Juang, F. Y. and Bhattacharya, P. K. (1985). Quasiperiodic GaAs-AlAs heterostructures, *Phys. Rev. Lett.*, **55**, 1768
23. Maciá, E. (1998). Optical engineering with Fibonacci dielectric multilayers, *Appl. Phys. Lett.*, **73**, 3330–3332.
24. Maciá, E. (2001). Exploiting quasiperiodic order in the design of optical devices, *Phys. Rev. B*, **63**, 205421.
25. Senechal, M. (1996). *Quasicrystals and Geometry* (Cambridge University Press, New York).
26. Janssen, T., Chapuis, G. and De Boissieu, M. (2007). *Aperiodic Crystals: From Modulated Phases to Quasicrystals* (Oxford University Press, USA).
27. De Graef, M., McHenry, M. E. and Keppens, V. (2008). Structure of Materials: An Introduction to Crystallography, Diffraction, and Symmetry, *J. Acoustical Soc. Am.*, **124**, 1385–1386.
28. Allouche, J. P. and Shallit, J. O. (2003). *Automatic Sequences: Theory, Applications, Generalizations* (Cambridge University Press, New York).
29. Queffélec, M. (2010). *Substitution dynamical systems-spectral analysis* (Springer Verlag, Berlin Heidelberg).
30. Lind, D. A. and Marcus, B. (1995). *An Introduction to Symbolic Dynamics and Coding* (Cambridge University Press).
31. Schroeder, M. (2009). *Number Theory in Science and Communication: With Applications in Cryptography, Physics, Digital Information, Computing, and Self-Similarity* (Springer Verlag, Berlin Heidelberg).
32. Hardy, G. H. and Wright, E. M. (2008). An Introduction to the Theory of Numbers (Oxford University Press, Oxford).

33. Miller, S. J. and Takloo-Bighash, R. (2006). *An Invitation to Modern Number Theory* (Princeton University Press, New Jersey).
34. Trevino, J., Cao, H. and Dal Negro, L. (2011). Circularly symmetric light scattering from nanoplasmonic spirals, *Nano Lett.*, **11**, 2008–2016.
35. Vogel, H. (1979). A better way to construct the sunflower head, *Math. Bio.*, **44**, 179enc9.
36. Trevino, J., Forestiere, C., Di Martino, G., Yerci, S., Priolo, F. and Dal Negro, L. (2012). Plasmonic-photonic arrays with aperiodic spiral order for ultra-thin film solar cells., *Opt. Express*, **20**, A418–A430.
37. Trevino, J., Liew, S. F., Noh, H., Cao, H. and Dal Negro, L. (2012). Geometrical structure, multifractal spectra and localized optical modes of aperiodic Vogel spirals, *Opt. Express*, **20**, 3015–3033.
38. De Bruijn, N. G. (1981). Algebraic theory of Penrose's non-periodic tilings of the plane, *Kon. Nederl. Akad. Wetensch. Proc. Ser. A*, **43**, 84.
39. Danzer, L. (1989). Three-dimensional analogs of the planar Penrose tilings and quasicrystals, *Discrete Math.*, **76**, 1–7.
40. Radin, C. (1994). The pinwheel tilings of the plane, *Ann. Math.*, **139**, 661–702.
41. Lee, S. Y., Forestiere, C., Pasquale, A. J., Trevino, J., Walsh, G., Galli, P., Romagnoli, M. and Dal Negro, L. Plasmon-enhanced structural coloration of metal films with isotropic Pinwheel nanoparticle arrays, *Opt. Express*, **19**, 23818–23830.
42. Prusinkiewicz, P. and Lindenmayer, A. (1990). *The Algorithmic Beauty of Plants* (Springer-Verlag, New York).
43. Ball, P. (2009). *Shapes* (Oxford University Press, New York).
44. Jean, R. V. (1995). *Phyllotaxis* (Cambridge Unicersity Press, Cambridge).
45. Thompson, D. A. W. (1992). *On Growth and Form* (Dover, New York).
46. Lawrence, N., Trevino, J. and Dal Negro, L. (2012). Aperiodic arrays of active nanopillars for radiation engineering, *J. Appl. Phys.*, **111**, 113101.
47. Lawrence, N., Trevino, J. and Dal Negro, L. (2012). Control of optical orbital angular momentum by Vogel spiral arrays of metallic nanoparticles, *Opt. Lett.*, **37**, 5076–5078.
48. Capretti, A., Walsh, G. F., Minissale, S., Trevino, J., Forestiere, C., Miano, G. and Dal Negro, L. (2012). Multipolar second harmonic generation from planar arrays of Au nanoparticles., *Opt. Express*, **20**, 15797–15806.
49. Fano, U. (1941). The theory of anomalous diffraction gratings and of quasi-stationary waves on metallic surfaces (Sommerfeld's waves), *JOSA*, **31**, 213–222.

50. Pecora, E. F., Lawrence, N., Gregg, P., Trevino, J., Artoni, P., Irrera, A., Priolo, F. and Dal Negro, L. (2012). Nanopatterning of silicon nanowires for enhancing visible photoluminescence., *Nanoscale*, **4**, 2863–2866.
51. Dal Negro, L., Lawrence, N. and Trevino, J. (2012). Analytical light scattering and orbital angular mementum spectra of arbitrary Vogel spirals, *Opt. Express*, **20**, 18209–18223.
52. Liew, S. F., Noh, H., Trevino, J., Negro, L. D. and Cao, H. (2011). Localized photonic band edge modes and orbital angular momenta of light in a golden-angle spiral, *Optics Express*, **19**, 23631–23642.
53. Simon, D. S., Lawrence, N., Trevino, J., Dal Negro, L., Sergienko, A. V. (2013). high capacity quantum Fibonacci coding for key distribution, *Phys. Rev. A.*, **87**, 032312-1/032312-10.
54. Naylor, M. (2002). Golden, $\sqrt{2}$, and π flowers: A spiral story, *Mathematics Magazine*, **75**, 163–172.
55. Mitchison, G. J. (1977). Phyllotaxis and the Fibonacci series, *Science*, **196**, 270–275.
56. Adam, J. A. (2009). *A Mathematical Nature Walk* (Priceton University Press, New Jersey).
57. Pollard, M. E. and Parker, G. J. (2009). Low-contrast bandgaps of a planar parabolic spiral lattice, *Opt. Lett.*, **34**, 2805–2807.
58. Baddeley, A. and Turner, R. (2005). Spatstat: An R package for analyzing spatial point patterns, *J. Stat. Softw.*, **12**, 1–42.
59. Ripley, B. D. (1977). Modelling spatial patterns, *J. R. Stat. Soc. Series B (Stat. Methodol.)*, 172–212.
60. Janot, C. (1997). *Quasicrystals: A Primer* (Oxford University Press, USA).
61. Yang, J. K., Noh, H., Liew, S. F., Rooks, M. J., Solomon, G. S. and Cao, H. (2011). Lasing modes in polycrystalline and amorphous photonic structures, *Phys. Rev. A*, **84**, 033820.
62. Torquato, S. and Stillinger, F. H. (2003). Local density fluctuations, hyperuniformity, and order metrics, *Phys. Rev. E*, **68**, 041113.
63. Illian, J., Penttinen, A., Stoyan, H. and Stoyan, D. (2008). *Statistical Analysis and Modeling of Spatial Point Patterns* (John Wiley, West Sussex).
64. Chavez-Cerda, S., P. M. J., Allison, I., New, G. H. C., Gutierrez-Vega, J. C., O'Neil, A. T., MacVicar, I., Courtial, J. (2002). Holographic generation and orbital angular momentum of high-order Mathieu beams *J. Opt. B: Quantum Semiclass. Opt.*, **4**, S52–S57.

65. Intonti, F., Caselli, C., Lawrence, N., Trevino, J., Wiersma, D. S. and Dal Negro, L. (2013). Near-field distribution and propagation of scattering resonances in Vogel spiral arrays of dielectric nanopillars, *New J. Phys.*, **15**, 085023.

66. Ozaktas, H. M., Zalevsky, Z., Alper Kutay, M. (2001). *The Fractional Fourier Transform with Applications in Optics and Signal Processing* (John Wiley, New York).

67. West, B. J., Bologna, M., Grigolini, P. (2003). *Physics of Fractal Operators*, (Springer, New York).

68. Mendlovic, D., Ozaktas, H.M. (1993). Fractional Fourier transforms and their optical implementation I, *J. Opt. Soc. Am. A.*, **10**, 1875.

69. Mendlovic, D., Ozaktas, H.M. (1993). Fractional Fourier transforms and their optical implementation II, *J. Opt. Soc. Am. A.*, **10**, 2522.

70. Bultheel, A., Martínez-Sulbaran, H. (2004). Computation of the fractional Fourier transform, *Appl. Comput. Harmon. Anal.*, **16**, 182.

71. Narayanan, V. A., Prabhu, K.M.M. (2003). The fractional Fourier transform: theory, implementation and error analysis, *Microprocess. Microsyst.*, **27**, 511.

72. Numerical Approximation and Linear Algebra Group (NALAG), University of Leuven. http://nalag.cs.kuleuven.be/research/software/FRFT/)

73. Montgomery, W. D. (1967). Self-imaging objects of infinite apertures, *J. Opt. Soc. Am. A.*, **57**, 772.

74. Huang, F. M., Chen, Y., Javier Garcia de Abaco, F., Zheludev, N.I. (2007). Optical super-resolution through super-oscillations, *J. Opt. A: Pure Appl. Opt.*, **9**, 285.

75. Huang, F. M., Kao, T. S., Fedotov, V. A., Chen, Y., Zheludev, N. I. (2008). Nanohole array as a lens, *Nano Lett.*, **8**, 2469.

76. Huang, F. M., Zheludev, N. I., Chen, Y., Javier Garcia de Abaco, F. (2007). Focusing of light by a nanohole array, *Appl. Phys. Lett.*, **90**, 091119.

77. Allen L., P. M. J., Babiker M. (1999). The orbital angular momentum of light, *Prog. Opt.* , **39**, 291–372.

78. Allen, L., Beijersbergen, M. W., Spreeuw, R. J. C. and Woerdman, J. P. (1992). Orbital angular momentum of light and the transformation of Laguerre-Gaussian laser modes, *Phys. Rev. A*, **45**, 8185–8189.

79. Grier, D. G. (2003). A revolution in optical manipulation, *Nature*, **424**, 810–816.

80. Mair, A., Vaziri, A., Weihs, G. and Zeilinger, A. (2001). Entanglement of the orbital angular momentum states of photons, *Nature*, **412**, 313–316.

81. Wang, J., Yang, J.-Y., Fazal, I. M., Ahmed, N., Yan, Y., Huang, H., Ren, Y., Yue, Y., Dolinar, S., Tur, M. and Willner, A. E. Terabit free-space data transmission employing orbital angular momentum multiplexing, *Nat. Photon*, **6**, 488–496.

82. Berkhout, G. C. G., Lavery, Martin P. J., Courtial, Johannes, Beijersbergen, Marco W., Padgett, Miles J. (2010). Efficient sorting of orbital angular momentum states of light, *Phys. Rev. Lett.*, **105**, 153601.

83. Padgett, M., Courtial, J. and Allen, L. (2004). Light's orbital angular momentum, *Physics Today*, **57**, 35–40.

84. Heckenberg, N. R., McDuff, R., Smith, C. P., White, A. G. (1992). Generation of optical phase singularities by computer-generated holograms, *Opt. Lett.*, **17**, 221–223.

85. Yu, N., Genevet, P., Kats, M. A., Aieta, F., Tetienne, J., Capasso, F., Gaburro, Z. Light propagation with phase discontinuities: Generalized laws of reflection and refraction, *Science*, **334**, 333–337.

Chapter 12

Probing Magnetic Plasmons with Vortex Electron Beams

Reuven Gordon

Department of Electrical and Computer Engineering, University of Victoria, Victoria, BC V8P5C2, Canada

rgordon@uvic.ca

12.1 Introduction

Recent advances in nanotechnology have inspired researchers to search for artificial materials with any desired optical response. Of particular interest are materials with negative refraction, which allows for perfect lensing (Pendry, 2000). Negative refraction requires both negative permittivity and permeability. Such a response can be achieved by nanostructuring metals and dielectrics at a scale much smaller than the optical wavelength (Smith et al., 2004). Indeed, the approach of metamaterials is to create extreme subwavelength features that have a strong response, and usually this means a resonance, in the frequency range of interest.

The question arises: how can we achieve strong electric and magnetic resonances in the visible regime? To help answer this

Singular and Chiral Nanoplasmonics
Edited by Svetlana V. Boriskina and Nikolay I. Zheludev

ISBN 978-981-4613-17-0 (Hardcover), 978-981-4613-18-7 (eBook)
www.panstanford.com

question, it would be useful to be able to probe the electric and magnetic response of nanostructured materials at the extreme subwavelength scale. Electron energy loss spectroscopy (EELS) allows for spectral characterization at such a fine scale, yet it has typically been applied only to the electric response. Recent advances in vortex-EELS open up new possibilities for probing the magnetic response of nanostructures as well, as will be described in this chapter, focusing on metal nanostructures.

12.2 Nanoplasmonics on a Cylindrical Wire

Plasmonics refers to the optical response of metals, typically in the ultraviolet to near-infrared regime. The optical response is usually dominated by the free electron motion (i.e., the plasma), allowing for negative permittivity, and consequently, the appearance of surface waves. The quantum of the electromagnetic wave at a metal surface is called a plasmon. As plasmonics can refer to excitations on large metal structures or even entirely flat surfaces, the term nanoplasmonics is used to emphasize the tailoring of the plasmonic response by nanostructuring.

In the following subsections, the cylindrical wire will be used as a canonical example of nanoplasmonics, showing the features of surface wave propagation, localized surface plasmon resonance, and a magnetic plasmon response.

12.2.1 *Surface Waves*

In 1899, Sommerfeld proposed the solution of a surface wave supported by a cylindrical wire (Sommerfeld, 1899). This solution predates that of surface waves on a flat surface by almost 8 years (Zenneck, 1907). For the case of an infinite radius cylinder, the cylinder solution and the flat surface solution are identical. Here, we consider only the cylindrical solution, using cylindrical co-ordinates (ρ, ϕ, z), which shows the influence of reducing the size to the extreme subwavelength regime. The wave is transverse magnetic, and its azimuthal magnetic field is given by:

$$H_\phi(\rho) = \begin{cases} I_1(p_m\rho)\exp(i\beta z - i\omega t)/I_1(p_m a), & \text{if } \rho < a \\ K_1(p_d\rho)\exp(i\beta z - i\omega t)/K_1(p_d a), & \text{if } \rho > a \end{cases} \quad (12.1)$$

with I_n, K_n being the modified Bessel functions of the first and second kind of order n, $p_{m,d} = \sqrt{\beta^2 - k_0^2\epsilon_{m,d}}$ where β is the propagation constant, $k_0 = \omega/c$ is the free-space propagation constant (angular frequency divided by the speed of light), and $\epsilon_{m,d}$ is the permittivity of the metal, dielectric. The corresponding electric field components can be solved from the Maxwell-Ampère equation. Matching the longitudinal electric field at the boundary gives the relation:

$$\frac{K_0(p_d a)I_1(p_m a)}{K_1(p_d a)I_0(p_m a)} = -\frac{\epsilon_d p_m}{\epsilon_m p_d} \quad (12.2)$$

which is solved for β. In the limit of $a \to \infty$, it is easy to show that this equation gives:

$$\frac{\beta}{k_0} = \sqrt{\frac{\epsilon_m \epsilon_d}{\epsilon_m + \epsilon_d}} \quad (12.3)$$

the usual surface plasmon dispersion for a flat metal-dielectric interface.

For the opposite limit of an infinitesimally thin wire, it can be shown that the following scaling holds: $\beta \propto 1/a$. This means that both the group velocity and phase velocity slow down as the wire is reduced in radius and the electromagnetic energy becomes more tightly confined to the wire. It has been shown that for an adiabatic taper, this leads to strong field enhancement toward the end of the wire as the waves slow and the field becomes more confined (Stockman, 2004). Figure 12.1 shows the dependence of the propagation constant squared on the wire diameter, with the two limiting cases of thick and thin wires.

It is interesting to note that Sommerfeld and Zenneck considered how a radiating source would excite such surface modes. Scaled down to the visible regime, this is essentially the problem of how quantum light emitter will excite a nanowire. Indeed, the increased local density of optical states on a nanowire allows for efficient transfer of information along the wire, for example, from a quantum dot (Akimov et al., 2007). Due to the mode confinement and the

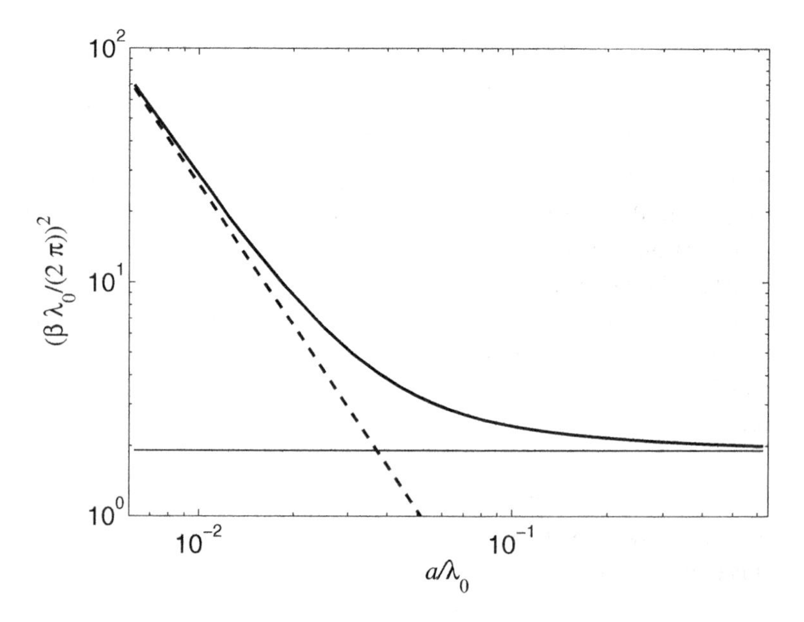

Figure 12.1 The dependence of the square of the normalized propagation constant squared (i.e., the effective index squared) on the wire radius. For gold at free-space wavelength of 800 nm, the value $\epsilon_m = -24$ was used (ignoring the imaginary part, which is around 1.6), and for water, the value $\epsilon_d = 1.7689$ was used. The two limiting cases are for a flat surface (thin line) and for an infinitessimal wire (dashed line) are shown.

group velocity decrease with wire radius, the emission rate into the plasmonic wire scales as a^{-3}. As the damping rate scales as the third power of the inverse distance between the emitter and the wire, it is possible to have greatly enhanced emission into the plasmon mode without quenching out the emitter (Chang et al., 2006).

12.2.2 *Localized Surface Plasmons*

We now consider a cylindrical wire of finite extent. The surface waves on such a wire will reflect at the ends, due to impedance and mode-shape mismatch. This allows for Fabry–Pérot resonances, or standing waves, to be set up along the wire. The Fabry–Pérot condition for the resonance along the wire is given by:

$$\beta l + \phi = m\pi \tag{12.4}$$

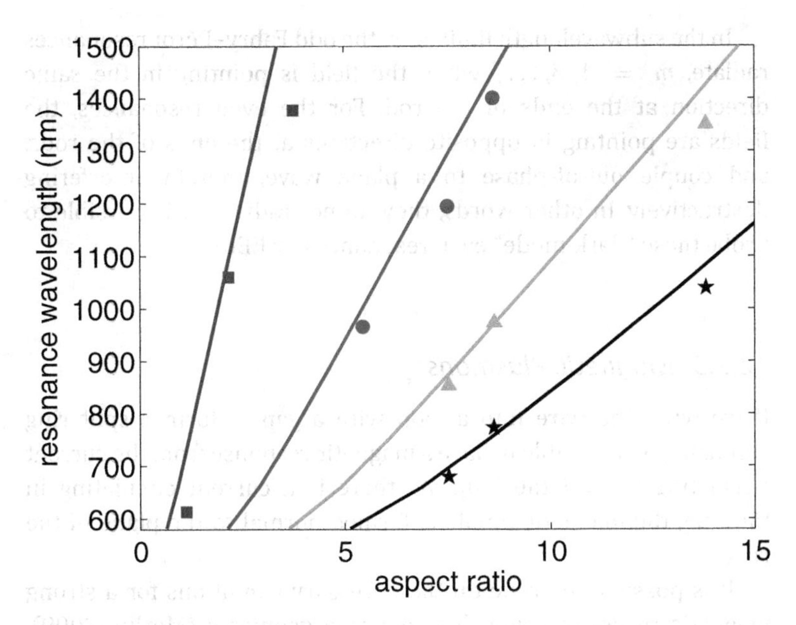

Figure 12.2 Fabry–Pérot resonance wavelengths of an 85 nm wide gold nanorod as compared with experimentally measured resonances. Adapted from Ref. (Gordon, 2009) with permission.

where β is given by Eq. 12.2 above, ϕ is the phase of reflection. Of course, this equation requires a way to solve for the phase of reflection.

Simple single mode matching can be used to derive the reflection amplitude and phase for a narrow wire, where it is assumed that there is only a single mode along the wire and that this is matched to a continuum of free-space modes with the same uniform azimuthal symmetry (Gordon, 2009). Figure 12.2 shows the Fabry–Pérot resonances calculated for a 85 nm wide gold wire with aspect ratios, and compared with the measured resonances from experiment (Payne et al., 2006).

For rounded ends, an approximate expression for the phase of reflection has been assumed (Novotny, 2007):

$$\phi = 2\beta a \tag{12.5}$$

to give an effective wavelength scaling in designing resonant nanoantennas.

In the subwavelength limit, only the odd Fabry–Pérot resonances radiate, $m = 1, 3, \ldots$, when the field is pointing in the same direction at the ends of the rod. For the even resonances, the fields are pointing in opposite directions at the ends of the rods, and couple out-of-phase to a plane wave, thereby interfering destructively. In other words, they do not radiate. It is possible to probe these "dark-mode" even resonances by EELS.

12.2.3 *Magnetic Plasmons*

If we wrap the wire into a loop with a gap to form a split ring resonator, it is possible to have a magnetic response from the current circulating around the loop. As there is a current circulating in the ring, the magnetic dipole is formed normal to the plane of the ring.

It is possible to argue on the necessary conditions for a strong magnetic response using this split ring geometry (Merlin, 2009). In particular, the field in Eq. 12.1 also determines the current in the wire, which can be found from the magnetic field. Further, assuming that the field in the gap is nearly constant, it can be shown that the resonance becomes strongly damped in the case in which the magnitude of the permittivity is small. For large permittivity, on the contrary, it is possible to have a strong LC (inductance-capacitance) resonance, as the field is effectively expelled from the metal region, and a correspondingly strong paramagnetic response. This is consistent with the interpretation of earlier works that found that the additional inductances (Zhou et al., 2005) and capacitances (Tretyakov, 2007) from the field penetration into the metal limit the resonance frequency.

A consequence of this result is that it is difficult to achieve a strong magnetic response in the visible regime from nanostructured metals because the noble metals (i.e., good conductors) have a smaller magnitude of permittivity there. At the same time, other metals may prove to be better candidates, such as aluminum (Lahiri et al., 2010), and with advances in nanofabrication, there is still promise for achieving a strong magnetic response in the visible regime (Shalaev, 2007).

12.2.4 *Summary*

In summary, we have seen that the key results in nanoplasmonics can be found by considering a cylindrical wire. In particular, these results are:

- Metal dielectric waveguides support guided modes;
- As their size is reduced, these modes become confined to the extreme subwavelength scale and both their group velocity and phase velocity slow down;
- Localized plasmons can be thought of as standing wave Fabry–Pérot resonances, where the end reflection is included in the analysis; and,
- By wrapping the wire in a loop, a magnetic response can be obtained, but it has been argued that this can only be strong for the large permittivity limit.

In the next section, we will describe how fast electrons can be used as a nanoplasmonic probe.

12.3 Electron Energy Loss Spectroscopy

Among the first predictions of observing plasmons considered the use of EELS (Blackstock, A.W. and Ritchie, R.H. and Birkhoff, R.D., 1955). Since that time, there have been significant advances in the use of EELS to map out plasmons on nanostructured metals. Here, we will focus on the theory of EELS and discuss applications of EELS to nanoplasmonics.

12.3.1 *Theory*

An electron traveling past a metal nanostructure is essentially a delta-like current source that induces an electric field. The electron then loses energy, ΔE through this self-induced field via the relation:

$$\Delta E = e \int_{-\infty}^{\infty} \vec{v} \cdot \vec{E}^{\text{ind}} \left(\vec{r}_e(t), t\right) dt = \int_0^{\infty} \hbar\omega\Gamma(\omega)\, d\omega \qquad (12.6)$$

where e is the electron charge, $\vec{v} = v_z\hat{z}$ is its velocity (along the z-direction), $\vec{E}^{\text{ind}}(\vec{r}_e(t), t)$ is the induced electric field along the path of the electron, $\vec{r}_e(t)$, as a function of time t. The last equality is the scattering probability per unit frequency, $\Gamma(\omega)$, where $\hbar\omega$ is the photon/plasmon energy.

Using the properties of Fourier transforms, asserting that the field is real, gives:

$$\Gamma(\omega) = \frac{e}{\pi\hbar\omega}\int_{-\infty}^{\infty} \Re\{\exp(i\omega t)\vec{E}^{\text{ind}}(\vec{r}_e(t), t)\}\,dt \tag{12.7}$$

where $\Re$ is the real part.

The electron may be considered as a moving point charge, so that the current density can be expressed:

$$\vec{j}(\vec{r}, \omega) = -e\vec{v}\int_{-\infty}^{\infty} \delta(\vec{r} - \vec{r}_e(t))\exp(-i\omega t)\,dt \tag{12.8}$$

where $\delta(\vec{r} - \vec{r}_e(t))$ is the Dirac delta function. Using this expression for the current density as a source, it is possible to calculate the induced field through the dyadic Green's function, and thereby find:

$$\Gamma(\omega) = \frac{e^2 v_z^2 \mu_0}{\pi\hbar}\iint \Im\{\exp(i\omega(t - t'))\,G_{zz}(\vec{r}_e(t), \vec{r}_e(t'); \omega)\}\,dt\,dt' \tag{12.9}$$

where both integrals are over all time, $\Im$ is the imaginary part, and the trajectory of the electron is chosen to be along the z direction. The z component of the dyadic Green's function is the particular solution to the equation:

$$\nabla \times \nabla \times \vec{G}_z(\vec{r}, \vec{r'}) - k^2\vec{G}_z(\vec{r}, \vec{r'}) = \delta(\vec{r} - \vec{r'})\hat{z} \tag{12.10}$$

where $k^2 = \epsilon_r\omega^2/c^2$, and ϵ_r is the relative permittivity (assuming a non-magnetic material). When integrating over the trajectory of the electron, the integral in Eq. 12.9 can be taken over the z, z' co-ordinates by noting that $t = z/v_z$ and $t' = z'/v_z$.

12.3.2 *EELS for Nanoplasmonics*

As is clear from the theory of the previous subsection, EELS allows for mapping out the induced electric response of a nanostructure

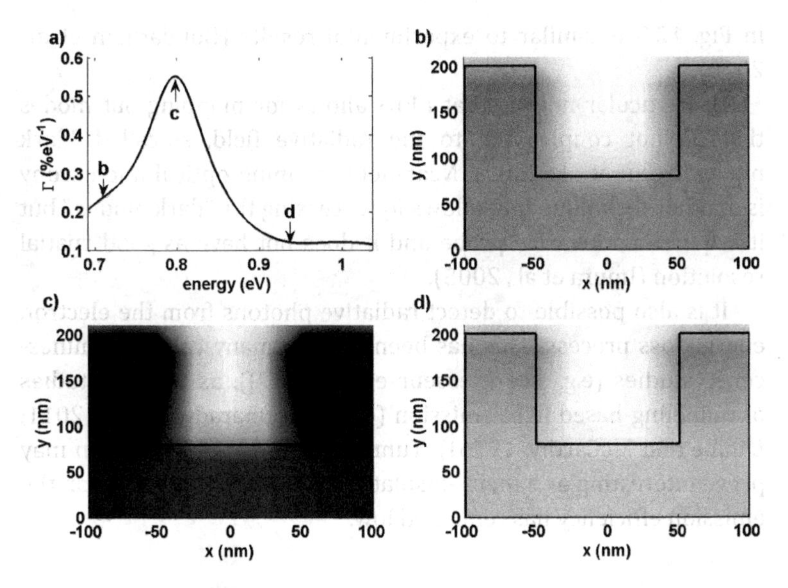

Figure 12.3 (a) EELS scattering spectrum for a split ring resonator structure in gold at the point of maximum scattering at the edge of the arms. (b–d) EELS map of the same split ring resonator at the energy values marked in part (a). The height of the gold is 30 nm.

from a nanoscopic current source, albeit integrated with the appropriate phase along a linear trajectory. This property allows for mapping out the plasmonic modes of metal nanonanostructures. It has been noticed that there is a relation between the local density of optical states and the EELS map (García de Abajo and Kociak, 2008); however, care should be taken in excercising this correspondence, as only a single component of the local density of optical states contributes and only through an integral (Hohenester et al., 2009).

Figure 12.3 shows the EELS response of a split ring resonator in gold as calculated numerically using the finite-difference time-domain (FDTD) technique. This calculation uses a series of z-oriented dipole sources along lines along z to calculate the Green's function response in Eq. 12.9. Care should be taken when using a dipole source in a lossy material because of the divergence of both the real and imaginary parts of the Green's function at the origin; however, this can be dealt with using standard approaches (Van Vlack and Hughes, 2012). The distribution shown in Fig. 12.3 is similar to experimental results (Boudarham et al., 2010).

Of particular note is that EELS allows for mapping out modes that do not couple well to the radiative field, so-called "dark modes" (Chu et al., 2009). Near-field scanning optical microscopy is another technique that allows for accessing the "dark modes," but it requires an invasive probe and it does not have as good spatial resolution (Imura et al., 2005).

It is also possible to detect radiative photons from the electron energy loss process. This has been used in many cathodoluminescence studies (e.g., see (Vesseur et al., 2007)), as well as studies of tunneling-based light emission (e.g., see (Bharadwaj et al., 2011; Lambe and McCarthy, 1976)). Tunneling-based light emission may prove interesting as a metal-insulator source of light, but so far, the emission efficiency has remained low.

12.4 Vortex-EELS and the Magnetic Plasmon

Recently, advances in electron beam optics have allowed for the creation of vortex beams with orbital angular momentum. As these beams act like an effective magnetic current density, they may be used to probe the magnetic response of nanostructures through energy-loss spectroscopy. In the following sections, the vortex electron beam will be introduced, as well as the effective magnetic charge that it allows. This will enable the calculation of vortex-EELS, for which it is shown that the scattering probability can be comparable to conventional EELS, opening up the door for a new probe for the local magnetic response.

12.4.1 *Vortex Electron Beams*

In 2010, it was demonstrated that point dislocations in crystal lattices allow for the production of diffracted electron beams with orbital angular momentum (Uchida and Tonomura, 2010). The point defect creates diffraction planes that are the superposition of a phase singularity with a linear grating. Later, man-made vortex diffraction structure was created in a thin platinum film to produce diffracted vortex beams (Verbeeck et al., 2010). Since that time, there have been advances in the phase plates to produce beams of large orbital angular momentum quantum number, by looking at higher order diffraction beams or creating phase plates with a larger winding number (McMorran et al., 2011).

12.4.2 *Effective Magnetic Charge for a Vortex Beam*

A key aspect of the theory of vortex-EELS is the formulation of the effective magnetic charge. This is found by noting that the spiraling electron possesses quanta of $\hbar$, the reduced Planck's constant, in orbital angular momentum.

The electron spiral has radius of a, while moving with a constant velocity in the z-direction of v_z. The total electron velocity is

$$\vec{v} = -v_\theta \frac{y}{a}\hat{x} + v_\theta \frac{x}{a}\hat{y} + v_z\hat{z}, \tag{12.11}$$

where $\hat{x}$ is the x-directed unit vector. The electron current density is given by:

$$\vec{j}(r,t) = -e\delta\left(\vec{r} - \vec{r}_e(t)\right)\vec{v} \tag{12.12}$$

From this, the magnetic current density in the frequency domain is given by

$$\vec{j}_m = \frac{i}{\omega\epsilon_0}\nabla \times \vec{j}/\epsilon_r \tag{12.13}$$

where ϵ_0, ϵ_r are the free-space and relative permittivity values.

The electric current has angular momentum along the z axis, so that

$$(\nabla \times \vec{v}) \cdot \hat{z} = \frac{2}{a}v_\theta. \tag{12.14}$$

To find v_θ, we use the angular momentum given by

$$L_z = mav_\theta, \tag{12.15}$$

where m is the electron mass (ignoring relativistic effects). In addition, we use the fact that the vortex beam has quantized orbital angular momentum given by

$$L_z = l\hbar, \tag{12.16}$$

where l is the orbital angular momentum quantum number. Combining these two equations for the angular momentum gives

$$v_\theta = \frac{l\hbar}{ma}. \tag{12.17}$$

We find the magnitude of the effective magnetic charge, e_m, by writing the magnetic charge density along z axis as follows:

$$e_m \delta\left(\mathbf{r} - \mathbf{r}_e(t)\right) = \frac{j_m}{v_z} = \frac{2e}{a\omega\epsilon_0\epsilon_r}\frac{v_\theta}{v_z}\delta\left(\mathbf{r} - \mathbf{r}_e(t)\right). \tag{12.18}$$

Subsequently, we obtain the effective magnetic charge:

$$e_m = \frac{2el\hbar}{\omega\epsilon_0\epsilon_r v_z a^2}. \tag{12.19}$$

12.4.3 *Vortex-EELS Scattering Loss Probability*

The analysis to find the vortex-EELS scattering probability follows the same approach as with conventional EELS with the duality replacements $\vec{E} \to Z\vec{H}$, $Z\vec{H} \to -\vec{E}$, $\vec{j} \to \vec{j}_m$, and $e \to e_m$, where $Z = \sqrt{\mu/\epsilon}$. With these transforms, the scattering probability can be found from Eq. 12.9 above:

$$\Gamma(\omega) = \frac{e_m^2 v_z^2 \epsilon_0}{\pi\hbar} \iint \Im\{\exp\left(i\omega(t-t')\right) G^H{}_{zz}\left(\vec{r}_e(t), \vec{r}_e(t'); \omega\right)\} \, dt \, dt' \tag{12.20}$$

The z component of the magnetic dyadic Green's function is the particular solution to the equation:

$$\nabla \times \frac{1}{\epsilon_r}\nabla \times \vec{G}_z^H(\vec{r}, \vec{r}') - k^2\vec{G}_z^H(\vec{r}, \vec{r}') = \delta(\vec{r} - \vec{r}')\hat{z}. \tag{12.21}$$

Equation 12.20 is the key result of this chapter, providing a relation that may be used to calculate the electron energy loss scattering expected from a vortex electron beam by means of coupling to the local magnetic response. As with Eq. 12.9, the integrals may take over the spatial co-ordinates z, z', by noting the electron velocity. A more detailed derivation and discussion of this equation is given elsewhere (Mohammadi et al., 2012).

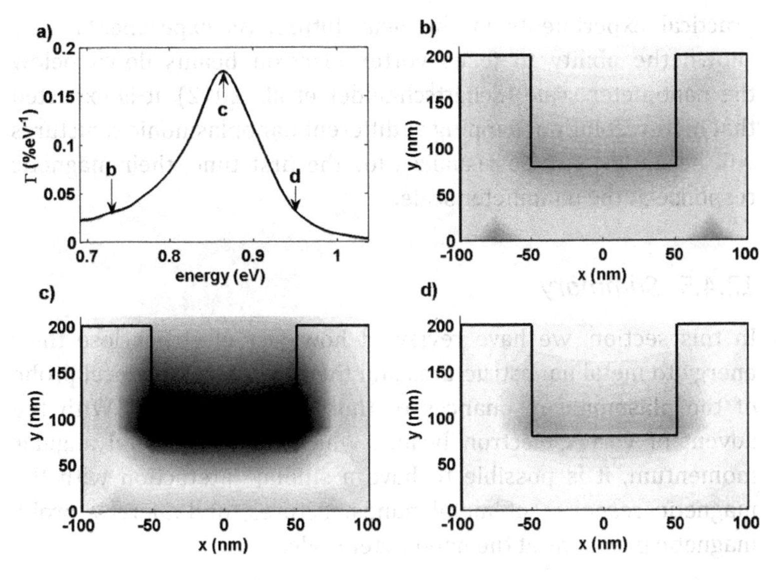

Figure 12.4 (a) Vortex-EELS scattering spectrum for a split ring resonator structure in gold at the point of maximum scattering in the middle of the ring. (b–d) Vortex-EELS map of the same split ring resonator at the energy values marked in part (a).

12.4.4 *Probing the Magnetic Plasmon*

The generation of vortex electron beams opens up exciting possibilities in probing the magnetic response of nanostructures, with the goal of improving our understanding of microscopic magnetic contributions to their optical response. Figure 12.4 shows an example calculation given for a split ring resonator structure in gold, with the same dimensions as Figure 12.3. As expected, the center of the ring, with the highest magnetic contribution to the local density of optical states, shows the greatest scattering at the magnetic plasmon resonance of 0.863 eV. This scattering profile was calculated in the same way as Figure 12.3, except for the use of a line of magnetic dipole sources instead of electric dipole sources.

From a practical point-of-view, it is interesting to note that the scattering probability seen here is comparable in magnitude to that of regular EELS featured in Figure 12.3. Therefore, it is reasonable to expect that vortex-EELS may be demonstrated in practical experiments in the near future. As experiments have shown the ability to focus vortex electron beams down below the nanometer scale (Schattschneider et al., 2012), it is expected that high-resolution mapping of different nanoplasmonic structures will be undertaken to visualize, for the first time, their magnetic response at the nanometer scale.

12.4.5 *Summary*

In this section, we have reviewed how fast electrons lose their energy to metal nanostructures and thereby allow for a local probe of the plasmonic resonances of those nanostructures. With the advent of vortex electron beams, which possess orbital angular momentum, it is possible to have a similar interaction with the magnetic response of metal nanostructures, and thereby probe magnetic plasmons at the nanometer scale.

12.5 Outlook

Nanoplasmonics uses the nanostructuring of metals to control the flow of light at the nanometer scale. It is possible to achieve a strong electric and magnetic response from nanostructured metals in visible to infrared regime, and this is highly desired for applications, especially those involving metamaterials, such as perfect lensing. Although EELS has long been used to probe the plasmon, the advent of vortex electron beams opens up the possibility, for the first time, of probing the magnetic plasmon. With this new capability, a greater understanding of the microscopic response of nanoplasmonic structures, including the magnetic response, may be achieved.

References

1. Akimov, A., Mukherjee, A., Yu, C., Chang, D., Zibrov, A., Hemmer, P., Park, H. and Lukin, M. (2007). Generation of single optical plasmons in metallic nanowires coupled to quantum dots, *Nature* **450**, 7168, pp. 402–406.

2. Bharadwaj, P., Bouhelier, A. and Novotny, L. (2011). Electrical excitation of surface plasmons, *Physical Review Letters* **106**, 22, p. 226802.
3. Blackstock, A. W. and Ritchie, R. H. and Birkhoff, R. D. (1955). Mean free path for discrete electron energy losses in metallic foils, *Physical Review* **100**, 4, p. 1078.
4. Boudarham, G., Feth, N., Myroshnychenko, V., Linden, S., García de Abajo, J., Wegener, M. and Kociak, M. (2010). Spectral imaging of individual split-ring resonators, *Phys. Rev. Lett.* **105**, p. 255501.
5. Chang, D., Sørensen, A., Hemmer, P. and Lukin, M. (2006). Quantum optics with surface plasmons, *Physical Review Letters* **97**, 5, p. 53002.
6. Chu, M. W., and Myroshnychenko, V., and Chen, CH., and Deng, J. P., and Mou, C. Y., and García de Abajo, F. J. (2009). Probing bright and dark surface-plasmon modes in individual and coupled nobel metal nanoparticles using an electron beam, *Nano Lett.* **1**, 9, pp. 399–404.
7. García de Abajo, F. J. and Kociak, M. (2008). Probing the photonic local density of states with electron energy loss spectroscopy, *Phys. Rev. Lett.* **100**, p. 106804.
8. Gordon, R. (2009). Reflection of cylindrical surface waves, *Optics Express* **17**, 21, pp. 18621–18629.
9. Hohenester, U., Ditlbacher, H. and Krenn, J. R. (2009). Electron-energy-loss spectra of plasmonic nanoparticles, *Phys. Rev. Lett.* **103**, p. 106801.
10. Imura, K., Nagahara, T. and Okamoto, H. (2005). Near-field optical imaging of plasmon modes in gold nanorods, *The Journal of Chemical Physics* **122**, p. 154701.
11. Lahiri, B., McMeekin, S., Khokhar, A., De La Rue, R. and Johnson, N. (2010). Magnetic response of split ring resonators (srrs) at visible frequencies, *Optics Express* **18**, 3, pp. 3210–3218.
12. Lambe, J. and McCarthy, S. (1976). Light emission from inelastic electron tunneling, *Physical Review Letters* **37**, 14, pp. 923–925.
13. McMorran, B. J., Agrawal, A., Anderson, I. M., Herzing, A. A., Lezec, H. J., McClelland, J. J. and Unguris, J. (2011). Electron vortex beams with high quanta of orbital angular momentum, *Science* **331**, 6014, pp. 192–195.
14. Merlin, R. (2009). Metamaterials and the landau-lifshitz permeability argument: Large permittivity begets high-frequency magnetism, *Proceedings of the National Academy of Sciences of the United States of America* **106**, 6, pp. 1693–1698.

15. Mohammadi, Z., Van Vlack, C., Hughes, S., Bornemann, J. and Gordon, R. (2012). Vortex electron energy loss spectroscopy for near-field mapping of magnetic plasmons, *Optics Express* **20**, 14, pp. 15024–15034.
16. Novotny, L. (2007). Effective wavelength scaling for optical antennas, *Physical Review Letters* **98**, 26, p. 266802.
17. Payne, E., Shuford, K., Park, S., Schatz, G. and Mirkin, C. (2006). Multipole plasmon resonances in gold nanorods, *Journal of Physical Chemistry B* **110**, 5, pp. 2150–2154.
18. Pendry, J. (2000). Negative refraction makes a perfect lens, *Physical Review Letters* **85**, 18, pp. 3966–3969.
19. Schattschneider, P., Stöger-Pollach, M., Löffler, S., Steiger-Thirsfeld, A., Hell, J. and Verbeeck, J. (2012). Sub-nanometer free electrons with topological charge, *Ultramicroscopy* **115**, 0, pp. 21–25.
20. Shalaev, V. M. (2007). Optical negative-index metamaterials, *Nature Photonics* **1**, pp. 41–48.
21. Smith, D. R., Pendry, J. B. and Wiltshire, M. C. K. (2004). Metamaterials and negative refractive index, *Science* **305**, 5685, pp. 788–792.
22. Sommerfeld, A. (1899). Über die fortpflanzung elektrodynamischer wellen an längs eines drahtes, *Annalen der Physik und Chemie* **67**, pp. 233–290.
23. Stockman, M. (2004). Nanofocusing of optical energy in tapered plasmonic waveguides, *Physical Review Letters* **93**, 13, p. 137404.
24. Tretyakov, S. (2007). On geometrical scaling of split-ring and double-bar resonators at optical frequencies, *Metamaterials* **1**, 1, pp. 40–43.
25. Uchida, M. and Tonomura, A. (2010). Generation of electron beams carrying orbital angular momentum, *Nature* **464**, pp. 737–739.
26. Van Vlack, C. and Hughes, S. (2012). Finite-difference time-domain technique as an efficient tool for calculating the regularized Green's function: applications to the local-field problem in quantum optics for inhomogeneous lossy materials, *Optics Letters* **37**, 14, pp. 2880–2882.
27. Verbeeck, J., Tian, H. and Schattschneider, P. (2010). Production and application of electron vortex beams, *Nature* **467**, 7313, pp. 301–304.
28. Vesseur, E., de Waele, R., Kuttge, M. and Polman, A. (2007). Direct observation of plasmonic modes in au nanowires using high-resolution cathodoluminescence spectroscopy, *Nano letters* **7**, 9, pp. 2843–2846.

29. Zenneck, J. (1907). Über die fortpflanzung elektrodynamischer wellen längs einer ebenen leiterfläche und ihre beziehung zur drahtlosen telegraphie, *Annalen der Physik* **328**, pp. 846–866.
30. Zhou, J., Koschny, T., Kafesaki, M., Economou, E., Pendry, J. and Soukoulis, C. (2005). Saturation of the magnetic response of split-ring resonators at optical frequencies, *Physical Review Letters* **95**, 22, p. 223902.

Chapter 13

Electromagnetic Optical Vortices of Plasmonic Taiji Marks

Wei Ting Chen, Pin Chieh Wu, and Din Ping Tsai
Department of Physics, National Taiwan University, No. 1, Sec. 4, Roosevelt Road, Taipei, 10617, Taiwan
dptsai@phys.ntu.edu.tw

Optical vortex, which is associated with optical singularities, was found a hundred years ago. The optical singularity, which is a point where the Poynting vector of a wave is zero, has been developed into a branch of optics (singular optics). The optical vortex can also be created in the near-field region of a special low symmetric Taiji pattern due to its plasmonic resonance. In this chapter, the optical vortex of Taiji mark will be carefully analyzed. Furthermore, the optical torque of the Taiji mark induced by optical vortex is studied for the opto-mechanical oscillator.

13.1 Introduction: An Explanation of the Process and Approach

In nature, vortices can be observed on both microscopic and macroscopic scales. For example, vortex-like structures within

Singular and Chiral Nanoplasmonics
Edited by Svetlana V. Boriskina and Nikolay I. Zheludev

ISBN 978-981-4613-17-0 (Hardcover), 978-981-4613-18-7 (eBook)
www.panstanford.com

Bose-Einstein condensates [1] and tornados are two common examples of vortex formation on micro-scales and macro-scales, respectively. Vortices could also be found in nano-optics, for example, in electromagnetically excited metallic nanoparticles [2–4], within sub-wavelength slits [5], in nonlinear Kerr media [6, 7], and in a super lens [8]. In the 1970s, Nye and Berry theoretically pointed out that the optical vertices can be created within phase singularities, that is, points at which the Poynting vector vanishes, although there exists a circulating component of the Poynting vector in the surrounding region [9]. This fascinating phenomenon has attracted widespread attentions [10–12] and has been developed into a new field of optics, called singular optics [13, 14].

The collective oscillation of electrons of noble metal, which is so-called surface plasmons resonance, has offered a new way to manipulate as well as to enhance optical energy in the near-field region of metallic particles [15–18]. It also provides a way to control the wavefront of light by acting on its phase and amplitude [19]. Using subwavelength metallic structures, plasmonics has been demonstrated to create phase singularity with the vortex-like pattern of the time-averaged Poynting vector at any arbitrary frequency [20–22]. The applications of optical energy vortex include optical trapping of bio-molecule and nano-particles [23, 24], optical force and momentum manipulation [25–27], synthesis and dynamic switching [28], and fluorescence microscopy with nanoscale resolution [29–31]. Intriguingly, the optical vortex can be created without phase singularity by a specific geometry of plasmonic Taiji structures [32]. The origination of Taiji pattern can be traced back to Zhou dynasty in China, which is more than two thousand years ago. It is a very old schematic representation of two opposing but complementary patterns in the oriental civilization [33]. In Chinese culture, Taiji is a combination of Yin and Yang, the Yin represents for the negative part and corresponds to Moon in nature, and the Yang represents for the positive part and corresponds to Sun in nature. The modern structure of Taiji diagram, which consists of two fish-shaped structures, had come out in the 12th century (see Fig. 13.1a). The geometry of Taiji mark is very similar to the optical vortex profile (see Fig. 13.1b) [34]. Due to the special structure of the Taiji mark, the vortex-like pattern

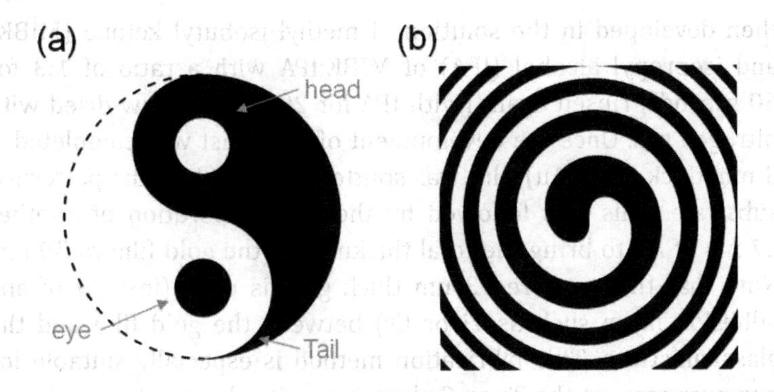

Figure 13.1 (a) Schematic diagram of Taiji mark consists of two complementary fish-shaped structures. The arrows mark the head, eye, and tail of a fish-shaped structure. (b) A calculated electromagnetic energy pattern of the vortex beam interferes with a co-propagating Gaussian beam [34]. The profile at the center part is very similar to the Taiji mark.

of the time-averaged Poynting vector exhibits one or more dips at plasmonic resonance(s). Furthermore, the plasmonic resonance could be adjusted by modifying the tail structure and the periodicity of the Taiji pattern.

13.2 Electromagnetic Energy Vortex without Optical Singularity

13.2.1 *Fabrication Method*

An array with 70 × 70 gold Taiji marks and with a 30 nm thickness covering a total area of 50 × 50 μm^2 was fabricated by electron beam lithography on a fused silica substrate. In order to eliminate charging effect, a conductive polymer called Espacer (from Kokusai Eisei Co., Showa Denso Group, Japan) was used. A 200 nm thick PMMA layer was spin-coated on the fused silica wafer, and then was put back on a hot plate for 3 min at 180°C. Espacer was spin-coated at 1500 rpm over the PMMA layer. The sample was written using an e-beam lithography system (Elionix ELS-7000) at an acceleration voltage of 100 keV with a current of 30 pico-ampere. After exposure, the sample was first rinsed with de-ionized water to remove Espacer, then developed in the solution of methyl isobutyl ketone (MIBK) and isopropyl alcohol (IPA) of MIBK:IPA with a ratio of 1:3 for 60 seconds, rinsed again (with IPA for 20 s), and blow-dried with nitrogen gas. Once the development of the resist was completed, a 3 nm–thick gold (Au) film was sputter-deposited on the patterned substrate. This was followed by thermal evaporation of another 27 nm of Au to bring the total thickness of the gold film to 30 nm. Note that the sputtered 3 nm thick gold is used (instead of any adhesion layer, such as Ti or Cr) between the gold film and the glass substrate. This fabrication method is especially suitable for our purposes, as the Ti or Cr layers usually change the resonance features of the Taiji mark. Finally, such fabricated sample was soaked in acetone for over 12 h, and then the un-patterned regions were lifted off in an ultrasonic cleaner.

13.2.2 *Taiji Mark Design*

Figure 13.2a shows a SEM micrograph of the fabricated Taiji pattern; a schematic drawing of the ideal Taiji pattern is shown in Fig. 13.2b.

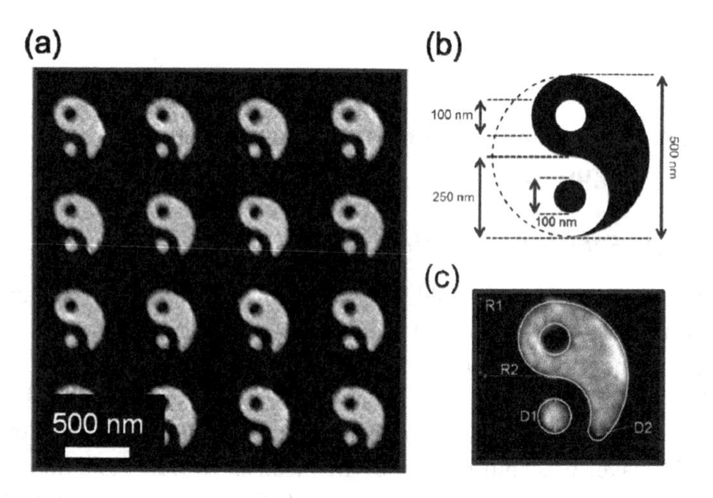

Figure 13.2 (a) SEM image of a small region from the fabricated sample, showing diameter of tails. (b) The ideal design of Taiji mark. (c) Magnified view of a single cell in the SEM image of frame (c) showing the fitting curves of a fabricated Taiji mark (R1 = 275 nm, R2 = 250 nm, D1 = 100 nm, diameter of the tangent circle D2 = 60 nm).

It is evident that the feature of sharp tail is lost due to resolution limitation, and therefore the tail of each Taiji mark deviates from the ideal Taiji pattern. According to the SEM observation results, the best fitting curve for the tail feature of the fabricated Taiji pattern is a tangent circle between an ellipse (long axis D1 = 275 nm, short axis D2 = 250 nm) and a circle (diameter = 250 nm). The geometrical size of more than 20 Taiji marks is carefully measured and the average radius of the tangent circle is determined to be around 30 nm with a standard deviation of nearly 2 nm. The numerical simulation model was subsequently set up according to these SEM observations.

13.2.3 *Measurement and Simulation Results*

Figure 13.3 shows both the experimental and simulated transmittance spectra of the Taiji array illuminated at normal-incidence with a linearly polarized light source along x-direction and y-direction, respectively. The transmission spectra from wavelength λ = 900–1700 nm were measured using a Fourier-transform infrared spectrometer with an infrared microscope (15$\times$ Cassegrain objective, numerical aperture NA = 0.4, near-infrared polarizer, and an InGaAs detector). An iris was used to collect the incident light to a square area of about 50 $\times$ 50 μm^2. The transmission spectra are normalized by those of the fused silica wafer.

The simulation spectra were obtained with the finite-element method (FEM). For simulating an array structure of a Taiji pattern, the boundary conditions are set as periodic boundary conditions with normal-incidence. The simulated Taiji structure has a 60 nm diameter tangent circle at the tail of the Taiji pattern, in accordance with our SEM observations. The refractive index of the fused silica glass substrate is set as 1.4584. The permittivity of gold was described by the Drude–Lorentz model using a damping constant of 0.14 eV with a plasma frequency of 8.997 eV.

There are two dips and one peak in the experimental as well as simulated spectra in Fig. 13.3; these features of the transmission spectra are similar under x- and y-polarized illumination in Fig. 13.3a and 13.3b. The second broader transmittance dip in both cases of x- and y-polarized illumination shows a particular interesting

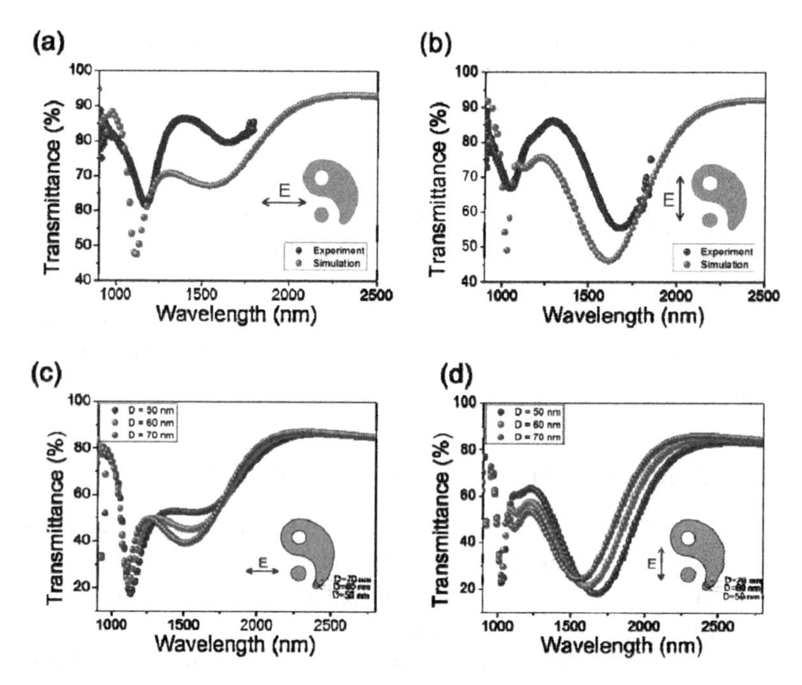

Figure 13.3 (Top row) Experimental and simulated transmittance spectra for (a) x-polarized illumination; (b) y-polarized illumination. The illumination wavelengths available in our experiments ranged from min λ = 900 nm to max. λ = 1700 nm. (Bottom row) Evolution of the simulated transmittance spectra upon varying the diameter of the tangent circle: (c) x-polarized illumination; (d) y-polarized illumination.

vortex-like time-averaged Poynting vector profile at the near-field region of Taiji pattern (see Fig. 13.4) and will be discussed in the following sessions.

In Fig. 13.3a and 13.3b, the intensity differences between the experimental and the simulated spectra are due to the presence of surface roughness in the actual sample (which is ignored in the simulations); the inaccuracies of the Drude–Lorentz model of the bulk dielectric constant of Au may also be responsible for some of the observed differences. The experimentally determined resonance wavelengths are coherent with simulations results; however, due to the variation of the tail structure of Taiji pattern, the results of experiment and simulation have shown slight differences.

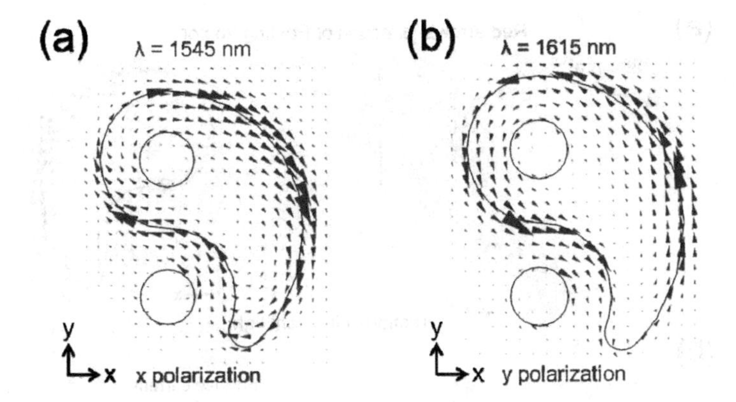

Figure 13.4 (a) Vortex-like profile of the time-averaged Poynting vector (red arrows) in the case of x-polarized illumination at 1545 nm (b) and in the case of y-polarized illumination at 1615 nm, respectively.

Figure 13.3c and 13.3d show the evolution of the simulated spectra as the tail diameter of the Taiji pattern is changed; clearly, the tail feature has a substantial influence over the resonant behavior of the structure, especially for the second dip showing electromagnetic energy vortex. As a result, the experimentally determined resonance wavelengths slightly shift away from the experimental measurement.

13.2.4 *Vortex-Like Poynting Vector Profile*

The electromagnetic energy vortex only exists in the near-field region of Taiji mark and corresponds to different swirling directions for x- and y-polarized illumination. In order to understand the vortex-like Poynting vector profile, a simple physical model, consisted by an electric dipole perpendicular to the direction of the surface currents in the near-field region of the Taiji marks, is given. The near-field electromagnetic field distribution for the dipole and also for the surface currents could be seen in the electrostatic approximation. In Fig. 13.5b, the Poynting vector, the electric field (induced by the electric dipole), and the magnetic field (induced by the surface current) are denoted as solid red lines, dashed blue lines, and orange symbols, respectively. It should be evident that, in the cross-product of the electric and magnetic fields, a vortex-like

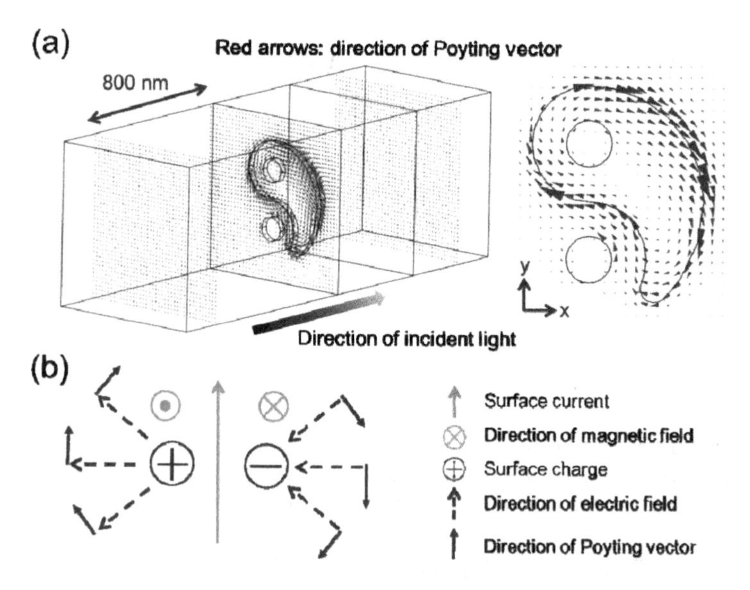

Figure 13.5 (a) Vortex-like profile of the time-averaged Poynting vector (red arrows) at different planes in the near-field region of the illuminated Taiji structure for the case of x-polarized illumination at wavelength = 1545 nm. (b) Schematic diagram of the physical model used to explain the origin of the vortex-like profile. The vortex in this diagram is produced by a horizontal electric dipole and a vertical surface current.

Poynting vector profile in the near-field region could be achieved if an electric dipole perpendicular to the direction of the surface current was present. However, the schematic physical model in Fig. 13.5b seems impossible because the oscillation direction of the electric dipole is typically the same as that of the current. That the Taiji pattern can form a vortex-like Poynting vector profile is rooted in its special structure. Figure 13.6a, corresponding to x-polarized illumination at wavelength $\lambda = 1545$ nm, shows that the sharp tail of the Taiji pattern is associated with a high concentration of negative electric charge, and therefore in a lowered electric potential. It is in the wake of this lowered potential that an electric current flows to the tail. As a result, the dipole in Fig. 13.6a is seen to oscillate horizontally, while the current has vertical component, in agreement with the proposed physical model of Fig. 13.5b. For the case in y-polarized illumination, shown in Fig. 13.6b, the dipole is seen to oscillate vertically with the horizontal oscillation current.

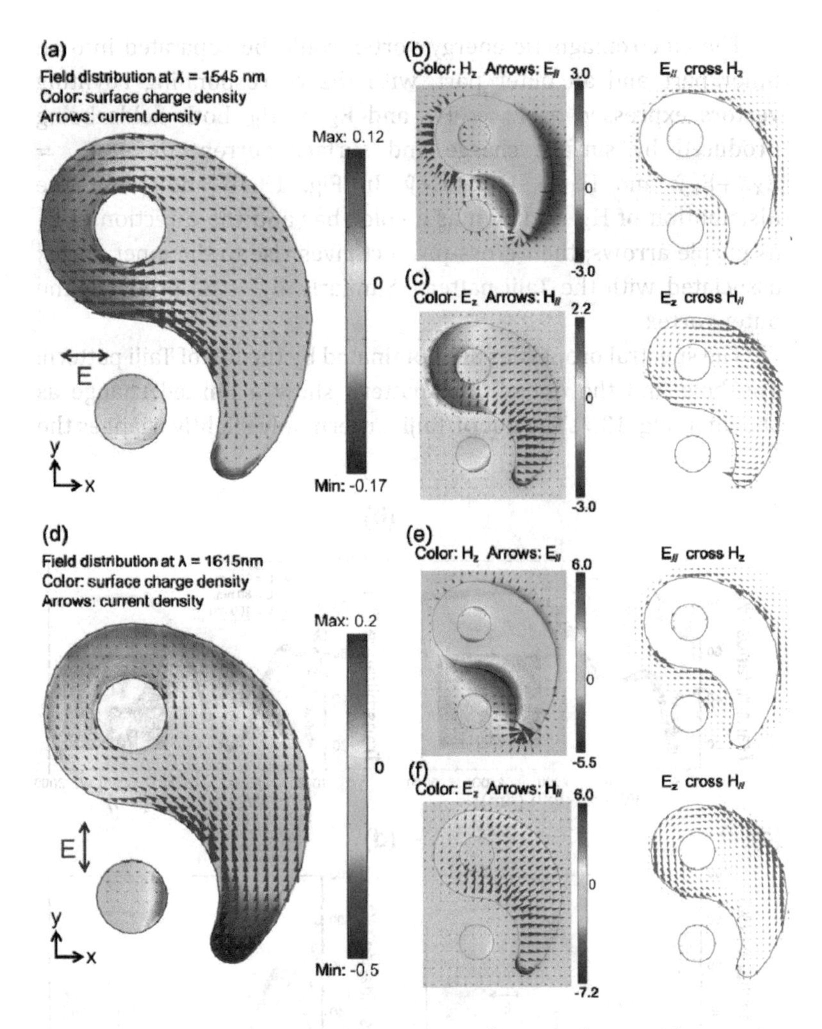

Figure 13.6 Simulation results corresponding to the interface region between air and gold (Taiji mark), obtained for the case of x-polarized and y-polarized illumination at wavelength $\lambda = 1545$ nm and 1615 nm, respectively. (a,d) Surface charge density in arbitrary units (color-coded) and surface current density (purple arrows). (b,e) Vertical magnetic field in arbitrary units (color-coded) and horizontal electric field (purple arrows). (c,f) Vertical electric field in arbitrary units (color-coded) and horizontal magnetic field (purple arrows); shown on the right-hand-side in each case are the resulting time-averaged Poynting vector profiles.

The electromagnetic energy vortex could be separated into an inner part and an outer part, with the corresponding Poynting vectors expressed as $E_{||} \times H_Z$ and $E_Z \times H_{||}$, both fields being produced by surface charge and surface current; here, $E_{||} = E_X\hat{x}+E_Y\hat{y}$ and $H_{||} = H_X\hat{x}+H_Y\hat{y}$. In Fig. 13.6b and 13.6e, the distribution of H_Z is shown as a color bar and the direction of $E_{||}$ as purple arrows; their cross-product gives rise to the inner vortex associated with the Taiji pattern. Similarly, $E_Z \times H_{||}$ produces the outer vortex.

The spectral properties are dominated by the tail of Taiji pattern. The hole and the dot of Taiji pattern show a limited change as shown in Fig. 13.7. The dot of Taiji pattern only slightly changes the

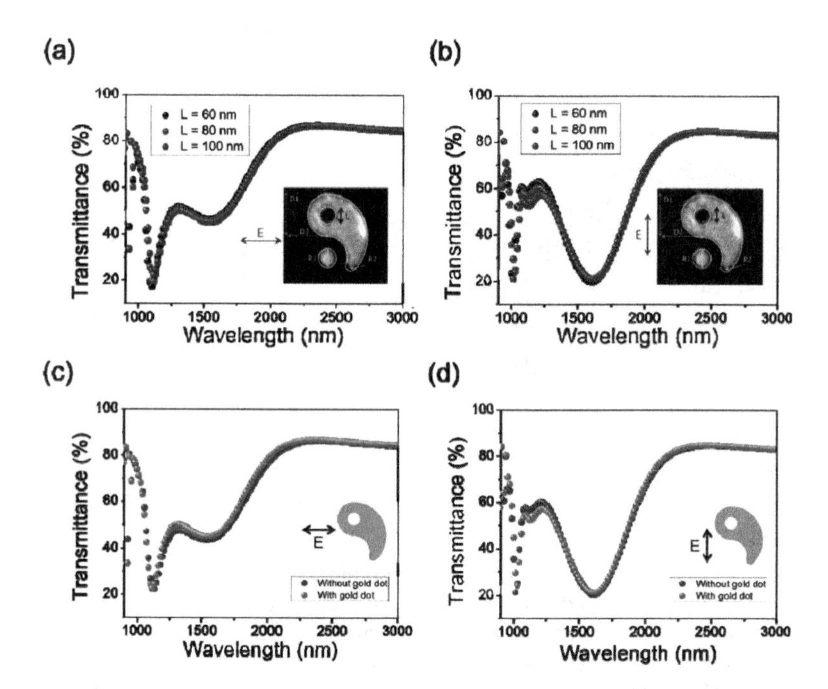

Figure 13.7 Simulation transmittance spectra for various diameter of nanohole under (a) x-polarized and (b) y-polarized illumination. Purpure arrow denotes the direction of electric field. (c) x-polarized and (d) y-polarized illumination for comparing the plasmonic Taiji mark with/without gold dot. The insets show the designed pattern of plasmonic Taiji mark without gold dot and polarization state.

intensity, as its plasmonic resonance is located at the wavelength below 900 nm.

13.2.5 *Taiji Mark for Opto-Mechanical Oscillator*

We now consider the application of electromagnetic energy vortex of Taiji mark. It is well known that the interaction between light and matter is usually accompanied with force or torque [35–38]. The optical force was first observed experimentally more than a century ago [39, 40], and the optical torque has been reported by Beth more than 70 years ago that the photons can induce a mechanical torque via scattering or absorption [41]. The study on optical force/torque grows dramatically after the invention of the laser in 1960s [42] and it can be applied to optomechanics for optical motors [43–45], optical wrenches [46, 47], and laser trapping [48, 49].

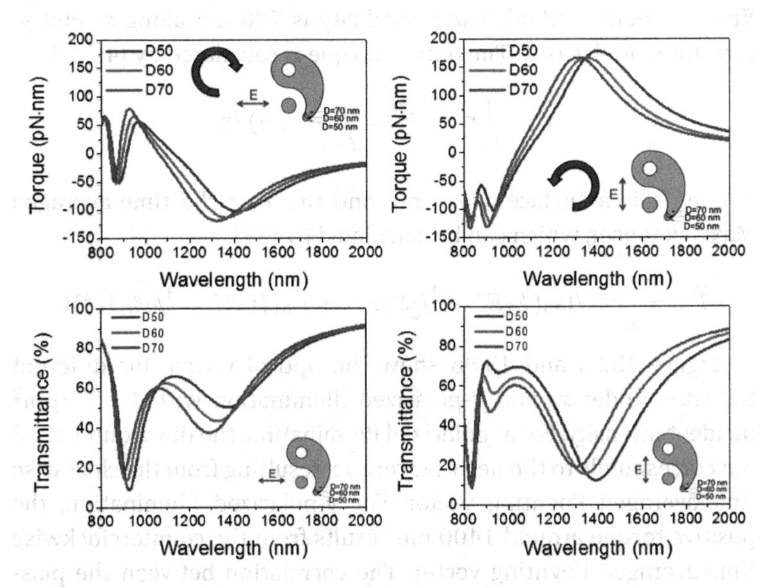

Figure 13.8 Evolution of the simulated optical torque and transmittance spectra upon varying the diameter of the tangent circle. (a) and (b) show the optical torque for 1 mW/μm^2 incident intensity. The positive and negative values of optical torque correspond to anticlockwise and clockwise rotation of Taiji pattern, denoted by red circular mark. (c) and (d) show transmittance spectra with the similar tendency of Fig. 13.3c and d. The red arrow shows the polarization of incident wave.

The electromagnetic energy vortex produces optical force on the Taiji mark and therefore the Taiji mark will experience optical torque due to its asymmetric structure and optical energy vortex. The rotation direction of the Taiji mark can be controlled by changing the polarization of incident light and is expected as clockwise and counterclockwise for x- and y-polarized illumination, in the same direction of optical energy vortex. The Taiji mark could be used as a nano-oscillator. For example, the Taiji mark rotates clockwise under x-polarized illumination; after 90° rotation, the Taiji mark is under y-polarized illumination and the Taiji mark rotates counterclockwise. As a result, the rotation of Taiji mark is sinusoidal as an oscillator. The simulated results of optical torque are shown below for this concept.

Three different tail sizes of Taiji mark array without substrate under x- and y-polarized illumination are calculated on the basis of finite element method. The periodicity is 700 nm along x- and y-direction for all cases. The optical torque is calculated by [45, 50]

$$\vec{\tau} = \oint_s \vec{r} \times \cdot \overrightarrow{ds} = \oint_s \vec{r} \times (\cdot \hat{n}) ds,$$

where ds is a surface area unit, and the $\overleftrightarrow{T}$ is the time-averaged Maxwell tensor, which can be calculated by [37]

$$\overleftrightarrow{T}_{ij} = \frac{1}{2} real[\varepsilon_0 (E_i E_i^* - {}^1\!/_2 \delta_{ij} E^2) + \mu_0 (H_i H_i^* - {}^1\!/_2 \delta_{ij} H^2)].$$

Figure 13.8a and 13.8b show the optical torque for different tail sizes under x- and y-polarized illumination with 1 mW/μm^2 incident intensity. For x-polarized illumination, the dip around 1300 nm corresponds to the negative torque, resulting from the clockwise time-averaged Poynting vector. For y-polarized illumination, the positive torque around 1400 nm results from the counterclockwise time-averaged Poynting vector. The correlation between the plasmon resonances and the optical torque is clearly observed in the relation between the transmittance spectra (Fig. 13.8c and 13.8d) and the optical torques (Fig. 13.8a and 13.8b) of Taiji pattern, where the plasmonic resonance wavelength that has the lowest transmittance is found to be near those exhibiting the maximum optical torque.

13.3 Summary

The special geometric features of the Taiji pattern produce a vortex-like Poynting vector profile in the near-field region of the Taiji mark when illuminated under the conditions of the plasmonic resonance. This vortex-like Poynting vector profile of Taiji pattern is rooted in an induced electric dipole oscillating perpendicularly to the induced surface current. The plasmonic resonance wavelength associated with the vortex can be fine-tuned by adjusting the tail structure of the Taiji mark and the period of the Taiji array, and the swirling direction of the Taiji pattern can be tuned by the polarization of the incident wave. The vortex-like Poynting vector of the Taiji pattern produces optical torques and can be applied for MEMs or NEMs, such as nano-gear and nano-oscillator.

References

1. Anderson, M. H. (1995). Observation of Bose-Einstein condensation in a dilute atomic vapor, *Science*, **269**, pp. 198–201.
2. Bashevoy, M. V. (2005). Optical whirlpool on an absorbing metallic nanoparticle, *Opt. Express*, **13**, pp. 8372–8379.
3. Lu, J. Y. (2009). Optical singularities associated with the energy flow of two closely spaced core-shell nanocylinders, *Opt. Express*, **17**, pp. 19451–19458.
4. Tribelsky, M. I. (2006). Anomalous light scattering by small particles, *Phys. Rev. Lett.*, **97**, p. 263902.
5. Schouten, H. F. (2003). Creation and annihilation of phase singularities near a sub-wavelength slit, *Opt. Express*, **11**, pp. 371–380.
6. Kivshar, Y. S. (2000). Self-focusing and transverse instabilities of solitary waves, *Phys. Lett.*, **331**, pp. 118–195.
7. Swartzlander, G. A. (1992). Optical vortex solitons observed in Kerr nonlinear media, *Phys. Rev. Lett.*, **69**, pp. 2503–2506.
8. Aguanno, G. D. (2008). Optical vortices during a superresolution process in a metamaterial, *Phys. Rev. A*, **77**, p. 043825.
9. Nye, J. F. and Berry, M. V. (1974). Dislocation in wave trains, *Proc. R. Soc. Lond. Series A: Math. Phys. Eng. Sci.*, **336**, pp. 165–190.

10. Gahagan, K. T. (1999). Simultaneous trapping of low-index and high-index microparticles observed with an optical-vortex trap, *J. Opt. Soc. Am. B*, **16**, pp. 533–537.
11. Ohta, A. (2007). Analyses of radiation force and torque on a spherical particle near a substrate illuminated by a focused Laguerre-Gaussian beam, *Opt. Commun.*, **274**, pp. 269–273.
12. Zhan, Q. W. (2006). Properties of circularly polarized vortex beams, *Opt. Lett.*, **31**, pp. 867–869.
13. Gbur, G. (2002). Singular optics, *Opt. Photonics News*, **13**, p. 55.
14. Soskin, M. S. (1998). Nonlinear singular optics, *Pure Appl. Opt.*, **7**, pp. 301–311.
15. Chen, H. M. (2012). Plasmon inducing effects for enhanced photoelectrochemical water splitting: X-ray absorption approach to electronic structures, *ACS Nano*, **6**, pp. 7362–7372.
16. Okamoto, K. (2004). Surface-plasmon-enhanced light emitters based on InGaN quantum wells, *Nat. Mater.*, **3**, pp. 601–605.
17. Chen, J. J. (2011). Plasmonic photocatalyst for H_2 evolution in photocatalytic water splitting, *J. Phys. Chem. C*, **115**, pp. 210–216.
18. Chen, W. T. (2011). Manipulation of multidimensional plasmonic spectra for information storage, *Appl. Phys. Lett.* **98**, p. 171106.
19. Tan, P. S. (2011). Phase singularity of surface plasmon polaritons generated by optical vortices, *Opt. Express*, **36**, pp. 3287–3289.
20. Genevet, P. (2012). Ultra-thin plasmonic optical vortex plate based on phase discontinuities, *Appl. Phys. Lett.*, **100**, p. 013101.
21. Kang, J.-H., (2011). Low-power nano-optical vortex trapping via plasmonic diabolo nanoantennas, *Nat. Commun.*, **2**, pp. 1–6.
22. Lembessis, V. E. (2011). Surface plasmon optical vortices and their influence on atoms, *J. Opt.*, **13**, p. 064002.
23. Ashkin, A. (1987). Optical trapping and manipulation of viruses and bacteria, *Science*, **235**, pp. 1517–1520.
24. Gahagan, K. T. (1999). Simultaneous trapping of low-index and high-index microparticles observed with an optical-vortex trap, *J. Opt. Soc. Am. B*, **16**, pp. 533–537.
25. Sato, S. (1991). Optical trapping and rotational manipulation of microscopic particles and biological cells using higher-order mode Nd:YAG laser beams, *Electron. Lett.*, **27**, pp. 1831–1832.
26. Galajda, P. (2001). Complex micromachines produced and driven by light, *Appl. Phys. Lett.*, **78**, pp. 249–251.

27. Knöner, G. (2007). Integrated optomechanical microelements, *Opt. Express*, **15**, pp. 5521–5530.

28. Kim, H. (2010). Synthesis and dynamic switching of surface plasmon vortices with plasmonic vortex lens, *Nano Lett.*, **10**, pp. 529–536.

29. Lee, J. H. (2006). Experimental verification of an optical vortex, *Phys. Rev. Lett.*, **97**, p. 053901.

30. Tamburini, F. (2006). Overcoming the Rayleigh criterion limit with optical vortices, *Phys. Rev. Lett.*, **97**, p. 163903.

31. Hell, S. H. (2007). Far-field optical nanoscopy, *Science*, **316**, pp. 1153–1158.

32. Chen, W. T. (2010). Electromagnetic energy vortex associated with sub-wavelength plasmonic Taiji marks, *Opt. Express*, **18**, pp. 19665–19671.

33. Browne, C. (2007). Taiji variations: Yin and Yang in multiple dimensions, *Comput. Graph.*, **31**, pp. 142–146.

34. Yu, N. (2011). Light propagation with phase discontinuities: Generalized laws of reflection and refraction, *Science*, **334**, pp. 333–337.

35. Wang, X. (1997). General expressions for dielectrophoretic force and electrorotational torque derived using the Maxwell stress tensor method, *J. Electrostatics*, **39**, pp. 277–295.

36. Zhao, R. (2010). Optical forces in nanowire pairs and metamaterials, *Opt. Express*, **18**, pp. 25665–25670.

37. Zhang, J. (2012). Optical gecko toe: Optically controlled attractive near-field forces between plasmonic metamaterials and dielectric or metal surfaces, *Phys. Rev. B*, **85**, p. 205123.

38. Volpe, G. (2006). Surface plasmon radiation forces, *Phys. Rev. Lett.*, **96**, p. 238101.

39. Lebedev, P. (1901). Untersuchungen über die Druckkräfte des Lichtes, *Annalen der Physik*, **6**, pp. 433–457.

40. Nichols, E. F. (1901). A preliminary communication on the pressure of light and heat radiation, *Science*, **14**, p. 588.

41. Beth, R. A. (1963). Mechanical detection and measurement of the angular momentum of light, *Phys. Rev.* **50**, pp. 115–125.

42. Maiman, T. H. (1960). Stimulated optical radiation in Ruby, *Nature*, **187**, pp. 493–494.

43. Galajda, P. (2001). Complex micromachines produced and driven by light, *Appl. Phys. Lett.*, **78**, pp. 249–251.

44. Kelemen, L. (2006). Integrated optical motor, *Appl. Opt.*, **45**, p. 2777.

45. Liu, M. (2010). Light-driven nanoscale plasmonic motors, *Nat. Nanotechnol.*, **5**, pp. 570–573.
46. Porta, A. L. (2004). Optical torque wrench: Angular trapping, rotation, and torque detection of quartz microparticles, *Phys. Rev. Lett.*, **92**, p. 190801.
47. Pedaci, F. (2010). Excitable particles in an optical torque wrench, *Nat. Phys.*, **7**, pp. 259–264.
48. Grier, D. G. (2003). A revolution in optical manipulation, *Nature*, **424**, pp. 810–816.
49. Moffitt, J. R. (2008). Recent advances in optical tweezers, *Ann. Rev. Biochem.*, **77**, pp. 205–228.
50. Rockstuhl, C. (2005). Calculation of the torque on dielectric elliptical cylinders, *J. Opt. Soc. Am. A: Opt. Image Sci. Vis.* **22**, pp. 109–116

Chapter 14

Passive and Active Nano-Antenna Systems

Samel Arslanagić,[a] Sawyer D. Campbell,[b] and Richard W. Ziolkowski[c]

[a]*Department of Electrical Engineering, Electromagnetic Systems, Technical University of Denmark (DTU), Ørsteds Plads, Building 348, DK-2800, Kgs. Lyngby, Denmark*
[b]*College of Optical Sciences, The University of Arizona, 1630 E. University Blvd., Tucson, AZ 85721-0094, USA*
[c]*Department of Electrical and Computer Engineering, The University of Arizona, 1230 E. Speedway Blvd., Tucson, AZ 85721-0104, USA*
ziolkowski@ece.arizona.edu

Interesting theoretical and practical radiation and scattering effects associated with passive and active core–shell nanoparticles (NPs) have been investigated and will be reviewed. Both plane wave and elementary Hertzian dipole excitations and a variety of gain media have been examined. Optimal configurations at visible wavelengths have been determined. Examples will illustrate how one can engineer these nano-scatterers to have cross-sections significantly larger than their geometrical size and how the corresponding nano-antenna structures can be matched to local quantum emitters to act as nano-amplifiers that significantly enhance the far-field responses. On the contrary, it will be shown how they can also be designed to actively jam these sources.

Singular and Chiral Nanoplasmonics
Edited by Svetlana V. Boriskina and Nikolay I. Zheludev

ISBN 978-981-4613-17-0 (Hardcover), 978-981-4613-18-7 (eBook)
www.panstanford.com

14.1 Introduction

One of the successful outcomes of metamaterial research has been the realization that the juxtaposition of two materials, one with positive material properties and the other with negative ones, can be used to create electrically small resonators [1]. This design paradigm has led to the investigation of ultra-thin planar [2] and electrically small spherical [3–18] cavities formed by combining a normal dielectric, that is, a double-positive (DPS) material, with an epsilon-negative (ENG) material. We have studied a variety of passive [3–9] and active [10–18] spherical core–shell designs, both at microwave and optical frequencies. Both plane wave and electric Hertzian dipole (EHD) excitations have been investigated. At microwave frequencies, it was demonstrated that the ENG shell acted as an impedance transformer that improved matching of a radiating element to its source and to its surrounding medium, hence, either (by design) enhancing or reducing the total radiated power and scattering and absorption cross-sections [3–9]. Tunability and multi-frequency systems were obtained by varying the geometry and materials [3–9] and by introducing multi-layered systems [6]. Gain was introduced to enhance the impedance bandwidth [10]. These canonical geometries have led to realistic electrically small, near-field resonant parasitic antennas [19]. On the contrary, at optical frequencies, metals such as gold and silver are naturally occurring ENG media. Moreover, they are rather lossy. As a consequence, gain was initially introduced to overcome those losses [11]. However, and more importantly, these electrically small resonator configurations were found to lase [11] and to act as nano-amplifiers [12–18]. Again, it was demonstrated that whatever these coated nanoparticle (CNP) geometries were, they could be engineered to be matched to the optical source [20, 21]. Furthermore, as nano-antennas, they act like their microwave counterparts and provide a means to impedance match quantum emitters to their surrounding medium while enhancing or reducing their radiated powers significantly [22].

In this chapter, we will highlight some of our results for optical spherical CNPs impregnated with gain media, both with plane wave and EHD excitations. First, we will illustrate with power flow streamlines how the very electrically small CNPs have cross-sections much larger than their physical size under plane wave excitations and how this behavior decreases the actual gain constants needed to achieve amplification. Next, we will discuss how their designs vary when ideal gain material and quantum dots are taken into account. Comparisons of resonant configurations with the gain medium being contained within the dielectric core surrounded by the metallic shell and its inverse with the gain medium covering a metallic core (inside-out or IO case) will be presented. We have identified the former as being the preferred one. Finally, we will illustrate how the active CNPs act as nano-amplifiers and nano-jammers of quantum emitters with EHD excitations.

14.2 Coated Nanoparticles Excited by a Plane Wave

Resonant, active CNPs, when properly designed, capture significantly more of the incident field energy than its physical size suggests is possible. The corresponding enhancement of its extinction cross-section is correlated with the concentration of the local field energy into its gain region. This energy localization can be visualized with the behavior of the flow lines of the Poynting vector field in the neighborhood of the CNP. Strong expulsion of the optical power generated from the interaction of the captured incident field energy with the gain medium creates an intense scattered field. As the interactions between the scattered field and the exciting plane wave increase, optical vortices form in the neighborhood of the active CNP. Gain depletion eventually occurs when the increase in the effective gain sufficiently detunes the resonance. A simple model for the gain enhancement effects observed in active CNPs relates the enhanced effective size of the CNP caused by the field localization to the required gain necessary to achieve its super-resonant state. A comparison of the metal-covered, gain core, active CNP studied previously to the experimentally realized gain-impregnated silica-covered metal "spaser" suggests that the active CNP design would require significantly less gain while offering a much larger enhancement of the incident field. Proposed modifications of both geometries that augment the field localization suggest further reductions in the gain values needed to achieve significant amplification of the output signal.

Experimentation of these concepts is possible because noble metals, such as silver or gold, are naturally occurring ENG materials at optical frequencies. One of the best known effects associated with optics at surfaces is the occurrence of plasmons on an interface between DPS and ENG materials [23]. Because plasmons are catalysts for wave coupling over a wide dynamic range of wave vectors, they have been intensely studied for numerous applications, including coherent light sources. It was recognized in the seminal work [24] that plasmon-based amplification is possible and desirable. This physics has stimulated, for example, a wide range of theoretical, simulation, and experimental efforts to compensate losses due to absorption in metals, which causes damping of localized surface plasmons (SPs) and propagating SP polaritons (SPPs), by introducing optical gain into the medium adjacent to a metallic surface (e.g., [25–30]). This issue has also been investigated vigorously to improve the loss figure of merit of optical metamaterials (e.g., [13, 31–34]) and to achieve lasing spasers [35]. The first experimental demonstration of compensating Joule losses in metallic photonic metamaterials using optically pumped PbS semiconductor quantum dots was reported in [36]. Lossless optical metamaterials have been demonstrated experimentally in a gain-impregnated fishnet structure [37]. Furthermore, the levels of miniaturization of lasers that have been reported over the past few years are truly impressive (see, e.g., the review [38]). A major approach is to use combinations of metals and dielectrics in conjunction with gain media to form nano-laser resonators.

Experimental verification of a NP version of these highly subwavelength lasers has been reported [39]. One very surprising aspect of this NP-laser demonstration was that the gain levels were sufficient to produce lasing despite initial theoretical predictions to the contrary. In fact, this "inside-out" (IO) CNP-laser configuration, which consisted of a layered spherical NP that had a gold core and a dye-impregnated silica coating, was predicted to have gain requirements that would be considerably more than the active CNP laser described in [11], which consisted of a gain-impregnated silica core with a metal coating. The experiments [37,

39] illustrated that there is a significant difference between material gain values and the effective gain values in the presence of resonant structures.

Additional insight into the effective gain behavior of the active CNPs was given in [17]. In contrast to the plasmon amplification concept promoted in [40], the idea was emphasized that despite the CNPs being electrically small, their ability to capture and localize the excitation fields leads to the enhanced effective gain values. The plasmons indeed play a crucial role, but it is to provide the feedback mechanism that creates the resonance. The ENG shell configuration is resonant and has a very large response even without the presence of the gain medium [3–10]. The resulting electrically small resonators foster the localization of the fields, which in turn promotes their stronger interaction with the gain material when it is present. Moreover, it has been recognized for some time that electrically small ENG metallic particles can have effective absorption cross-sections that are much larger than their geometrical cross-section values in the vicinity of their plasmon resonances, for example, at optical frequencies where $\varepsilon_{\text{rel}} \approx -2$ [41, 42]. In particular, visualizations of the power flow near resonant metallic (plasmonic) spheres have aided in understanding this effect. More recently, it has been recognized that these power flows have even more dramatic features, including vortices, and that they have a significant impact on how energy is transferred into and out of these ENG particles [43, 44]. We will illustrate that the flow of energy into the active CNPs has similar behaviors, that is, the effective cross-sections of the active CNPs are much larger than their geometrical value and, consequently, localize more excitation field in the gain region of the resonant cavity. Moreover, this captured and localized field energy in the gain region is then amplified by the plasmon-based electrically small resonant cavity. The flow of this amplified energy out of the gain region is further facilitated by the resonant geometry. This plasmon-engineered behavior leads to the smaller actual gain values that are sufficient to achieve lasing/amplification. Finally, we will contrast the active CNP geometry with the IO-CNP configuration within this model. We will then suggest enhanced versions of both systems, which further promote the localization of the fields in the gain region, thus increasing the degree of amplification while decreasing the actual gain values.

Let the incident, scattered, and total electromagnetic fields be denoted by ($\mathbf{E}_{inc}$, $\mathbf{H}_{inc}$), ($\mathbf{E}_{scat}$, $\mathbf{H}_{scat}$) and ($\mathbf{E}_{tot}$, $\mathbf{H}_{tot}$), respectively. The total flow of optical power is given by the time averaged Poynting vector

$$\mathbf{S}_{tot} = \frac{1}{2}\text{Re}\left\{\mathbf{E}_{tot} \times \mathbf{H}^*_{tot}\right\} = \mathbf{S}_{inc} + \mathbf{S}_{scat} + \mathbf{S}_{ext} \tag{14.1}$$

which has been broken up into the sum of three parts, respectively, the time-averaged incident, scattered, and extinction Poynting vectors: $\mathbf{S}_{inc} = \frac{1}{2}\text{Re}\left\{\mathbf{E}_{inc} \times \mathbf{H}^*_{inc}\right\}$, $\mathbf{S}_{scat} = \frac{1}{2}\text{Re}\left\{\mathbf{E}_{scat} \times \mathbf{H}^*_{scat}\right\}$, and $\mathbf{S}_{ext} = \frac{1}{2}\text{Re}\left\{\mathbf{E}_{inc} \times \mathbf{H}^*_{scat} + \mathbf{E}_{scat} \times \mathbf{H}^*_{inc}\right\}$.

If A is the closed surface surrounding the CNP with outward pointing unit normal vector $\mathbf{a}_n$, then one can introduce the total incident, scattered, extinction, and absorbed powers as the appropriate flux of these Poynting vectors through A:

$$P_{inc} = -\oiint_A \mathbf{S}_{inc} \bullet \mathbf{a}_n dA$$

$$P_{scat} = +\oiint_A \mathbf{S}_{scat} \bullet \mathbf{a}_n dA$$

$$P_{ext} = -\oiint_A \mathbf{S}_{ext} \bullet \mathbf{a}_n dA$$

$$P_{abs} = -\oiint_A \mathbf{S}_{tot} \bullet \mathbf{a}_n dA \tag{14.2}$$

where the absorbed power P_{abs} represents the amount of power captured by the CNP. Since the incident plane wave has its source outside of A and is defined independently of the presence of the CNP, $P_{inc} = 0$, yielding: $P_{abs} = P_{inc} - P_{scat} + P_{ext} = -P_{scat} + P_{ext}$, which means $P_{ext} = \text{P}_{scat} + P_{abs}$. The various cross-sections and corresponding efficiencies associated with a scatterer are defined in terms of these power terms and the power captured by its geometric cross-section: $P_0 = I_{inc}\,\sigma_{geom}$, where the incident intensity (power density) at the center of the scatterer: $\eta I_{inc} = |\mathbf{E}_{inc}|^2/2$, η being the wave impedance of the surrounding medium, and its geometric cross-section, and, for instance, if the scatterer is a spherical CNP of radius a, the geometrical cross-section is $\sigma_{geom} = \pi a^2$. In particular, the scattering, absorption and extinction cross-sections (σ's), and efficiencies (Q's) are given by the expressions:

$$\sigma_{scat} = P_{scat}/I_{inc} = Q_{scat} P_0$$
$$\sigma_{abs} = P_{abs}/I_{inc} = Q_{abs} P_0$$
$$\sigma_{ext} = P_{ext}/I_{inc} = Q_{ext} P_0 \quad (14.3)$$

The scattering, extinction, and absorption cross-sections for a spherical CNP are obtained with Mie theory multipole-based series expansions [17, 45]. The scattering (absorption) efficiency of the particle relates the amount of power scattered (absorbed) by the scatterer to the amount of power incident upon its surface. The extinction cross-section indicates how much power is removed from the incident field. The corresponding efficiencies are dimensionless quantities relating the effective cross-sections to the geometrical cross-section of the scatterer. Clearly, the extinction efficiency can be written as the sum of the absorption and scattering efficiencies: $Q_{ext} = Q_{scat} + Q_{abs}$. Note that for active scatterers, that is, scatterers containing a gain medium, it is possible for the absorbed power, P_{abs}, to become negative leading to a negative value for Q_{abs}. This behavior indicates that fields are being amplified inside the particle. If the negative absorption becomes greater than the scattering losses, the extinction cross-section becomes negative, that is, $Q_{ext} < 0$. This means the scattering losses have been overcome, and the incident field will then be amplified. To maximize the extinction efficiency, it is necessary to design a particle that strongly interacts electromagnetically with the incident field, that is, to design a particle with a large effective scattering cross-section.

A size and wavelength-dependent permittivity model has been used to determine the material properties of gold, Au, and silver, Ag [11], for all examples presented later. These Au and Ag models recover the measured results given, for example, in [46]. The original gain studies [11] considered a three-level rare earth and a canonical gain model. The Mie theory model that underlies our analysis starts conceptually with the assumption that the gain medium is describable at least by a three-level system. A pump signal is assumed that drives the gain medium into its excited state. The desired excitation frequency for either a plane wave or an EHD excitation is then the transition frequency of the gain medium to a lower energy state. This approach captures the steady-state physics of the gain medium. The CNP design itself must then overlap strongly with these transition frequencies, that is, the gain material's emission spectrum, in order for it to provide the mechanism for feedback. Also, it is assumed that all of the gain material participates in the absorption-scattering processes. In practice, an additional efficiency would have to be introduced since not all of the gain media will produce photons at the same time. Thus, the Mie results yield a best case answer to how large a response will occur. To establish the population inversion necessary to achieve the required gain in the active region, a higher energy (frequency) pump beam would be required.

Any gain model describes the active region's complex index of refraction, $N = n + i\kappa$, where κ is referred to as the extinction coefficient of the material. The gain constant and extinction coefficient (effective gain parameter) of a medium are related through the expression $g = -2k_0\kappa$, where $k_0 = 2\pi/\lambda_0 = 2\pi f_0/c$, f_0 and λ_0 being the excitation frequency and wavelength, respectively, and c being the speed of light in vacuum. Because the canonical model is easily integrated into multipole solutions and captures the essentials of the gain effects for design purposes, it has been used extensively in the literature.

To illustrate that an electrically small, active CNP design can capture much more power from the incident plane wave that anticipated from its geometrical size, consider a CNP that consists of a 30 nm outer radius silver (Ag) shell and a 24 nm radius gain-impregnated silica, SiO_2, core. This configuration has been shown to have extremely strong resonance behaviors [11]. The resonance effects of the active CNP are dependent on the effective gain parameter κ and the incident wavelength. The scattering and absorption efficiencies and the extinction efficiency are shown, respectively, in Fig. 14.1a and Fig. 14.1b. For each κ value, these efficiencies were obtained by first determining the maximum magnitude of the spectral profile of the scattering cross-section and the resonance frequency at which it occurs and then calculating the corresponding absorption and extinction efficiency values for that resonance frequency.

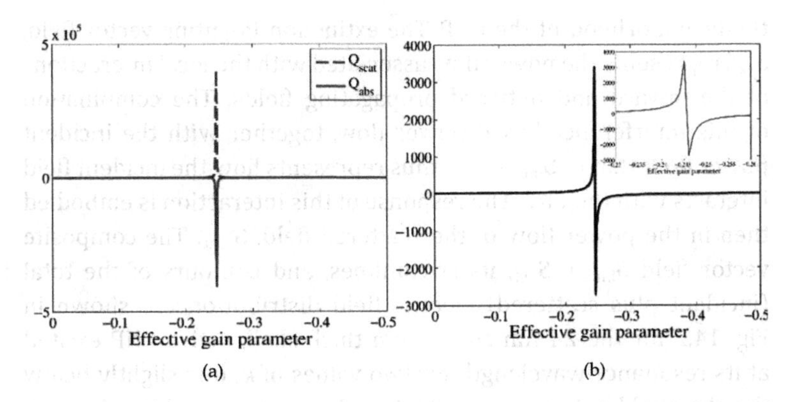

Figure 14.1 Scattering and absorption efficiencies (a) and extinction efficiency (b) versus the effective gain parameter κ for the active Ag/SiO_2 CNP [17].

Although the response of an active CNP depends on both the exciting wavelength and the value of κ, as well as its geometrical parameters, it is observed in Fig. 14.1 that the response of an active CNP is sensitive to the value of κ. When the value of κ is such that the response of the CNP achieves its largest negative value, that is, it produces its maximum amplification or lasing, we say that the CNP is in its super-resonant state. The threshold value (i.e., value of κ at which $Q_{\text{ext}} = 0$) occurs near $\kappa = -0.245$, as can be seen from the zoom subplot in Fig. 14.1b. One finds that slightly below this threshold value, the Q_{ext} value is extremely large positive, whereas slightly above threshold it is extremely large negative. The active CNP resonance is also sensitive to the incident wavelength. For κ values above and below the one defining the super-resonant state, the corresponding resonance wavelengths become slightly detuned from their super-resonant value. For the indicated active CNP geometry, the super-resonant state occurs at a wavelength of 501.9 nm when $\kappa = -0.25$, that is, Q_{scat} has its maximum value for those parameters, whereas Q_{abs} and Q_{ext} have their minimums, that is, their largest negative values.

There are strong interactions between the incident field and a passive or active CNP when either is responding in its super-resonant state. The strength and characteristics of these interactions are revealed by the behavior of the various Poynting vector fields in the neighborhood of the CNP. The extinction Poynting vector field, $\mathbf{S}_{ext}$, represents the power flow associated with the local interactions of the inward and outward propagating fields. The combination of this interference-based power flow, together with the incident power flow, that is, $\mathbf{S}_{inc} + \mathbf{S}_{ext}$, thus represents how the incident field interacts with the CNP. The response of this interaction is embodied then in the power flow of the scattered field, $\mathbf{S}_{scat}$. The composite vector field $\mathbf{S}_{inc} + \mathbf{S}_{ext}$, its streamlines, and contours of the total (incident plus scattered) electric field distribution are shown in Fig. 14.2 for the 24 nm core, 6 nm thick shell, active CNP excited at its resonance wavelength for two values of κ, one slightly below the threshold value, $\kappa = -0.23$, and the other slightly above it, $\kappa = -0.25$. The resonance frequency of the former is at 502.2 nm, while the latter is at 501.9 nm. In both cases, the gain is sufficiently large such that $Q_{abs} < 0$, $Q_{ext} > 0$ in the former case and $Q_{ext} < 0$ in the latter, as shown in Fig. 14.1.

For the below-threshold conditions in Fig. 14.2a, one observes that the power of the incident field is strongly converging onto the particle. This localization of the incident field corresponds to the fact that the absorption cross-section is resonantly large for the given conditions, that is, the particle acts as though it is much bigger than its physical size. This much stronger than expected interaction

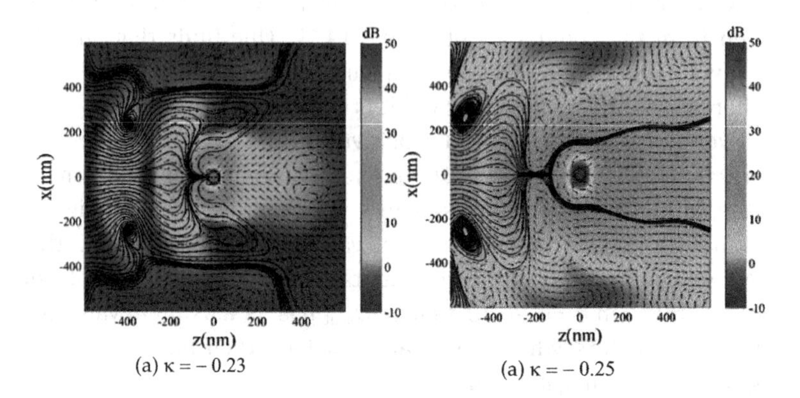

Figure 14.2 Contour plots of the electric field distribution along with the normalized Poynting vector field: $\mathbf{S}_{inc}+\mathbf{S}_{ext}$. The streamlines of this Poynting vector field show the extreme behavior of the flow of optical power in the presence of the active CNP [17].

between the incident field and the active CNP is also echoed in the dramatic increase in the scattering cross-section shown in Fig. 14.1. If the particle did not localize and capture a significant portion of the incident field, it would not have the opportunity to produce a large scattering response. One sees in Fig. 14.2b that above threshold, the flux of the outward flowing scattered field power generated from the induced sources in the gain region is large enough to overcome the losses. In fact, the strong expulsion of the power generated from within the CNP core and resonantly enhanced by the CNP structure appears to have pushed the incident field power away from the CNP. This means that the scattered field has become large enough that the interaction-based power flow is now dominating the incident power flow in the region around the CNP. As a result, one also observes the formation of optical vortices on the source side of the CNP and additional vector field zeros in its neighborhood, as the higher order modes associated with the scattered field become noticeably larger in the near field. Moreover, these vortices appear to be helping channel the incident power toward the center of the CNP.

The behavior of the composite normalized Poynting vector field: $\mathbf{S}_{\text{inc}} + \mathbf{S}_{\text{ext}}$, near the active CNP further highlights the behaviors of the combined incident and interaction-based power flows. Zoomed-in views of Fig. 14.2a and 14.2b are given, respectively, in Fig. 14.3a and 14.3b. The dominance of the resonant dipole fields is clearly

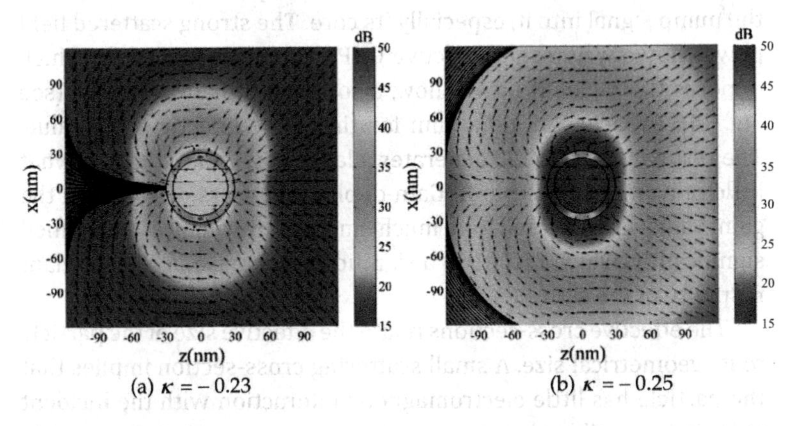

Figure 14.3 Zoomed-in view of the electric field distributions and the Poynting vector fields shown in Fig. 14.2 [17].

seen in the electric field distributions. Moreover, there is a clear localization and capture of a much larger portion of the incident field power than its geometrical size. Moreover, there is a channeling of this power into the core of the active CNP in Fig. 14.2a and 14.3a, both from its front and back sides. Note that this strong localization behavior physically occurs in time until the $\kappa = -0.23$ value is exceeded, as the gain medium is pumped into its super-resonance steady-state value $\kappa = -0.25$. Once that value is reached, Fig. 14.2b and 14.3b clearly show the interaction-based Poynting vector field dominates and indicates the presence of a strong outward power flow from the core of the active CNP. This behavior occurs and is strengthened once the threshold gain value has been exceeded and the losses are overcome. Because the enhanced scattering dominates the extinction cross-section rather than its absorption component, it is again recognized that the scattered field power flow generated by the resonant, active CNP dominates that of the incident field and begins to shield the CNP from it and leads to the presence of vortices and other types of zeros of the composite Poynting vector field, which appear in the neighborhood of the CNP in Fig. 14.2a and 14.2b.

Note that the below threshold behavior in Fig. 14.3a will also occur for any pump signal that excites the gain medium if the pump frequency is within the frequency bandwidth of the CNP's resonance, that is, the resonant CNP will help localize and channel the pump signal into it, especially its core. The strong scattered field power flow away from the active CNP evident in Fig. 14.3b, which opposes the incident power flow, also helps explain why Q_{abs} (see Fig. 14.1) reaches its minimum for this same effective gain value. The resonant active CNP generates a large power output from what little power is coupled to it. Gain-depletion effects occur when the gain value is increased too much and the resonance is detuned significantly, which results in a significant decrease in the resonant response of the active CNP.

The effective cross-sections relate the effective size of the particle to its geometrical size. A small scattering cross-section implies that the particle has little electromagnetic interaction with the incident field. It is possible that $Q_{scat} < 1$, which means that the particle behaves as one that is effectively smaller than its geometrical size.

On the contrary, in creating a nano-amplifier, the goal is to maximize the amplification of the scattered field and thus to maximize Q_{scat}. Consequently, the design of such an amplifier should result in a large localization of the incident field, which could then affect a large effective cross-section. In Fig. 14.2a and 14.2b, it is clear that the incident field interacts with the active CNP from distances much further than its physical extent. This behavior is emphasized in the localization of the incident power in those figures and the enhanced cross-section results shown in Fig. 14.1a.

As noted previously, the active CNP has a scattering cross-section that is effectively much larger than its geometrical size. This also means that the gain region is effectively larger as well. Since the excitation field interacts with an effectively larger particle, the total gain enhancement of the incident fields therefore will be effectively greater. To account for this behavior, the concept of an effective gain constant was introduced [17], that is, the active CNP acts as though its gain coefficient is $\kappa = \sqrt{Q_{scat}}\,\kappa_{bulk}$, which means the required theoretical gain coefficient is related to the actual physical bulk value as

$$g_{bulk} = g_{eff}/\sqrt{Q_{scat}}. \tag{14.4}$$

Consider again the super-resonant state of the active CNP and its cross-section results shown in Fig. 14.1. The effective and bulk gain coefficients versus the theoretical value κ are plotted in Fig. 14.4 [17]. Recall that the maximum of the scattering cross-section occurs at 501.9 nm when $\kappa = -0.25$ and gives $\sqrt{Q_{scat}} \approx 630$. To achieve this value of κ, the corresponding gain bulk constant would be $g_{eff} = 6.0\times10^4$ cm^{-1}. Although such high gain values have been obtained with quantum dots, they are not available with more common gain media such as dyes. This would imply that the realization of these active CNP designs could be quite difficult. Nevertheless, as shown in Fig. 14.4a, the reduction in the bulk values associated with the effects of the enlarged cross-section are significant. This means that the required physical bulk gain values are much smaller than their theoretical values, in agreement with the successful dye-based experiments [37, 39].

The first experimental NP spaser [39] employed a dye-doped silica shell surrounding a gold core. This IO configuration is

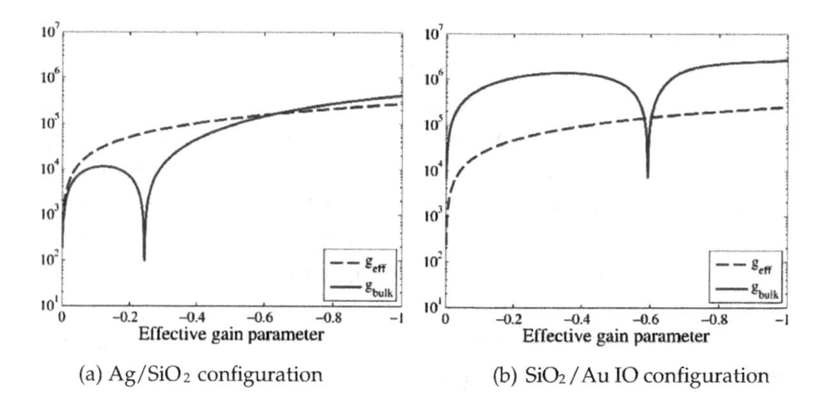

(a) Ag/SiO2 configuration (b) SiO2 / Au IO configuration

Figure 14.4 The theoretical effective and physical bulkgain coefficients in cm^{-1} versus the effective gain parameter κ for the Ag/SiO_2 and the corresponding SiO_2/Au IO active CNP designs [17].

advantageous from a fabrication point of view. The bulk and effective gain values for this IO design are shown in Fig. 14.4b. Again, it is clear that the scattering cross-section enhancement factor drastically reduces the bulk gain value required at resonance. On the contrary, a comparison of Fig. 14.4a and 14.4b shows that the required gain for the active CNP design is much less than that for the IO design. Furthermore, as discussed in [17], one can observe from Fig. 14.4 that the active CNP design would be less sensitive to variations in the experimental gain parameters.

The active CNP and IO systems are both two-layer designs. Although their configurations are complementary to one another, they exhibit similar behaviors. Since the active CNP design has a much stronger resonance than the one produced by the IO design, one could argue that adding an exterior metal shell coating to the IO design would be make it more like an active CNP. As shown in Fig. 14.5, this three-layer, alternate coated IO design does perform well and still radiates a dipolar mode. We note that given the experimentally realized IO particle was designed using OG-488 dye [39], which has a certain gain linewidth, any real design would have to have its geometry resonance matched to those lines. Consequently, only a thin 3.0 nm Ag shell was added to the IO design. It only shifts the resonance slightly down to the shorter wavelength

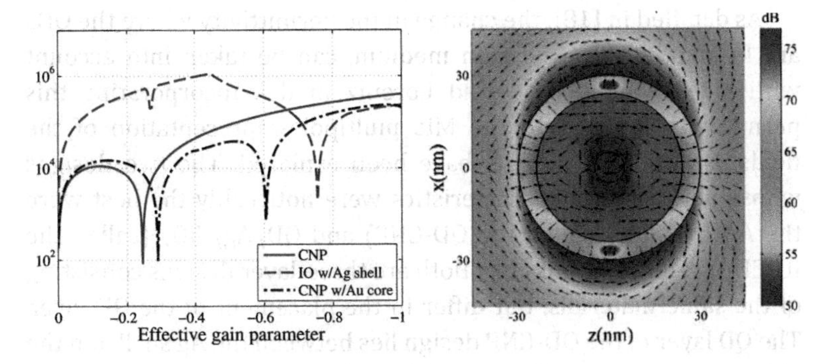

Figure 14.5 Three-layer CNP performance characteristics. (a) Bulk gain values, and (b) electric field distribution and the normalized Poynting vector field: $\mathbf{S}_{\mathrm{inc}} + \mathbf{S}_{\mathrm{ext}}$ [17].

503.3 nm from 513.3 nm. Any metal coated IO design would involve only a thin outer coating to not significantly change the resonance frequency and, as such, would be amenable to fabrication chemistries. As one observes in Fig. 14.5a, the addition of this thin layer significantly improves upon the original active IO CNP's performance characteristics. Similarly, the addition of a 5.0 nm Au core to the original two-layer Ag/SiO_2 design yields a three-layer design with two resonant frequencies. The first resonance occurs with $\kappa = -0.291$ at a wavelength of 504.4 nm, and the second resonance occurs at a wavelength of 507.7 nm with $\kappa = -0.605$. Its performance characteristics are quite competitive with the other two-layer and three-layer nano-amplifier designs.

These successful multilayer geometries encouraged our consideration [18] of realistic gain media in our active CNP designs. In particular, quantum-dots (QDs) provide an exciting option for the gain media because they possess large gain coefficients resulting from their extreme confinement effects. The optical properties of core/shell QDs can be tuned by changing the relative size of the core/shell, that is, by effectively changing their band gap structure. Similarly, as noted previously, the resonance of a CNP can be adjusted by changing the relative sizes of its layers. As shown in [18], locating the QDs inside a resonant CNP structure is again optimal; it greatly enhances the intrinsic amplifying behavior of the combined QD-CNP system.

As detailed in [18], the change in the permittivity where the QDs are located, that is, the gain medium, can be taken into account with a Maxwell-Garnet-based Lorentz model. Incorporating this permittivity model into the Mie multipole-representation of the fields, a variety of designs have been explored. The two designs whose performance characteristics were noticeably the best were the Ag/QD/SiO_2 (called the QD-CNP) and QD/Ag/SiO_2 (called the IO-CNP) configurations [18]. Both are three-layer designs consisting of the same materials, but differ in the placement of the QD layer. The QD layer of the QD-CNP design lies between the Ag shell and the SiO_2 core, while the QD layer of the IO-CNP is the outermost layer. The core and shell sizes were adjustable parameters; the QD size was fixed by the manufacturer's specifications. Both of the resulting optimized designs feature a 17.5 nm radius core and a 6.0 nm thick QD layer. For both cases, a single layer of QDs is sufficient to achieve a very large resonant response. The optimized thickness of the Ag shell in the QD-CNP design was 6.2 nm (29.7 nm total radius); it was 6.3 nm thick (29.8 nm total radius) for the IO-CNP design.

We have found that placing the active material near the shell boundary can greatly increase the performance of the active CNP. In fact, although we have studied designs with the active region near, but separated from the shell, we have seen that the optimal location of the active region is directly adjacent to the shell. Although it is known that placing the active region next to a metal will enhance the nonradiative decay rate of the active material, that is, it will quench the emission rate, recent theoretical and experimental studies have shown that plasmonic-related effects can actually significantly enhance the radiative decay rates [47–53]. In particular, by properly designing a multi-layer structure to retain the unique optical and electronic properties of the QD and CNP and to provide a much larger density of radiating states to which the emissions can couple, one can produce active CNPs with very efficient florescence behaviors, especially for materials with high internal quantum efficiencies that QDs possess. Even though the metal layer is thin and the QDs are not, which would lead to lower quenching effects in any event, designs exhibiting the large radiated power enhancements associated with a super-resonance state correspond to ones that foster enhanced radiative decays.

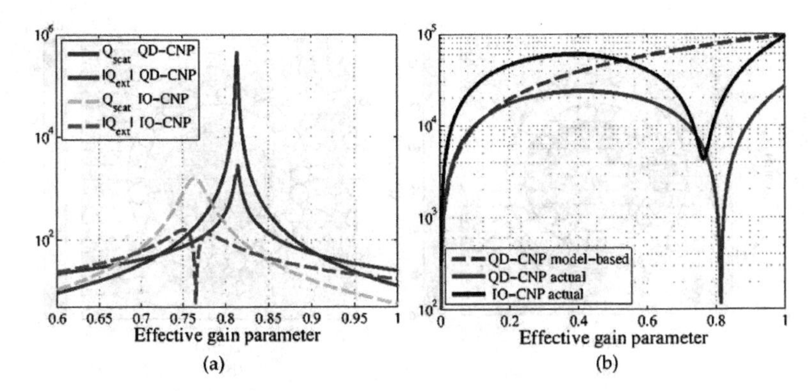

Figure 14.6 (a) Scattering and extinction efficiencies for the optimized QD-CNP and the IO-CNP configurations, and (b) The core–shell model required gain values (blue) and the actual required gain values of the QD-CNP (red) and IO-CNP (black) designs (in cm^{-1}) [18].

Thus, the proposed optimized QD-CNP and IO-CNP designs avoid quenching and significantly outperform the single QD, two-layer version.

The scattering efficiency and the absolute value of the extinction efficiency for these designs are plotted against the gain scale factor in Fig. 14.6a. We note that these optimized configurations require less than the peak gain value to achieve their resonant states, which would be quite advantageous in practice when fabrication and experimental tolerances tend to negatively impact the ideal component performance characteristics. Also note that the extinction efficiency achieves a negative value on the larger gain factor value (right) side of the resonant peaks of the scattering efficiency. (We note that this feature appears in the $|Q_{ext}|$ curves as the sharp nulls since the absolute value is being plotted.) Recall that when the extinction efficiency becomes negative, it means the losses are overcome and the incident field is amplified. One immediately observes that the scattering and extinction behaviors of the QD-CNP design are several orders of magnitude larger than those of the IO-CNP.

Figure 14.7 shows the power flow behavior of the total Poynting's vector field for both designs when the incident plane wave has a 560 nm wavelength. The QDs are drawn as circles representing their

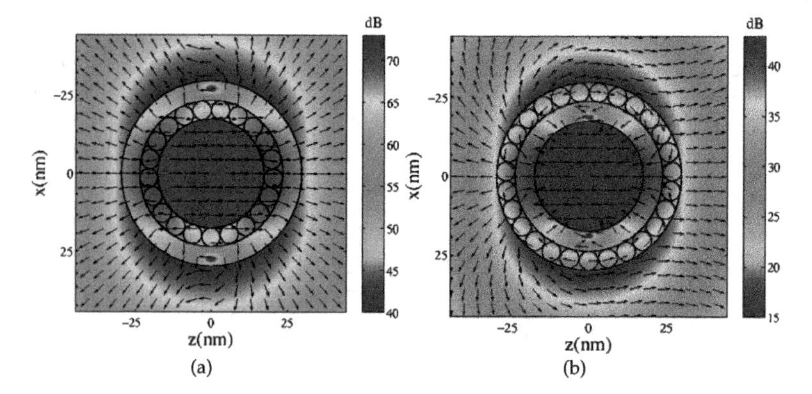

Figure 14.7 Contour plots of the electric field distribution along with the normalized total Poynting's vector field $\mathbf{S}_{\text{tot}}$ for the (a) QD-CNP and (b) IO-CNP designs [18].

actual size with respect to the size of the structure. In both the QD-CNP and IO-CNP cases, these structures produce a strong overall dipole resonance in their exteriors. However, it is clear that power is flowing more strongly outward from the QD layer in the QD-CNP structure than it is in the IO-CNP one (note that the minimum level in Fig. 14.7a is the maximum one in Fig. 14.7b), further emphasizing the large scattering cross-section differences shown in Fig. 14.6. Although the gain region in the three-layer QD-CNP and IO-CNP designs is only a single layer of QDs, it has a volume equivalent to 282 and 481 individual dots, respectively.

The field overlap in the gain region is strong and is caused, in part, by an increase in the incident power flow into the gain region. This strong coupling of the incident field to the gain region leads to large Q_{scat} values, which in turn reduces the actual gain values needed to achieve amplification. A comparison of the required gain values is given in Fig. 14.6b, in which the model-based gain value (blue) and the actual gain values of the QD-CNP (red) and IO-CNP (black) designs are plotted versus the QD gain scale factor. The largest gain enhancement factor for the QD-CNP and IO-CNP configurations are, respectively, 287 and 17.

We note that the IO-CNP design has a similar, but weaker behavior. It simply does not possess as large a gain enhancement factor. This is due to the QD layer being the outermost one. In the

QD-CNP case, the QD layer is within the cavity formed by the silica core and the metal shell. Thus, the coupling of the large fields in the core to the SPPs created in the Ag shell at the Ag-QD (i.e., an $\varepsilon < 0$, $\varepsilon > 0$) interface provides a much larger feedback mechanism through the QD gain layer. This feedback is much smaller in the IO-CNP case because it occurs between the outside vacuum region and the metal shell. These interpretations are supported by earlier three-layer observations that included an exterior metal coating to the basic IO-CNP configuration that substantially improved its performance.

14.3 Coated Nanoparticles Excited by Electric Hertizan Dipoles

The present section investigates the fundamental electromagnetic properties of spherical active CNPs with the aim of clarifying their suitability and potential as elements for nano-antennas, and as elements for the jamming applications of quantum emitters. The investigated CNPs again consist of a SiO_2 spherical nano-core covered with a plasmonic spherical concentric nano-shell. Attention is devoted to the near and far-field behavior of these particles in the presence of a single or multiple EHDs. In particular, the impacts of the plasmonic material, the dipole orientation, and locations on the resonant and transparent properties of the suggested CNPs have all been investigated thoroughly. In the analysis, a constant frequency canonical gain model has been used to account for the gain introduced in the dielectric part of the CNPs, whereas the size and frequency-dependent models for silver, gold, and copper have been used for the nano-shell layers of the CNPs.

14.3.1 *Problem Formulation*

The CNP consists of a nano-core (of radius r_1) covered concentrically with a nano-shell (of radius r_2), see Fig. 14.8. The nano-core and nano-shell consist of simple materials with permittivities, permeabilities, and wave numbers, respectively, denoted by ε_1, μ_1, and k_1, (for the nano-core) and ε_2, μ_2, and k_2 (for the nano-shell). The permittivity and permeability of the ambient free space are ε_0

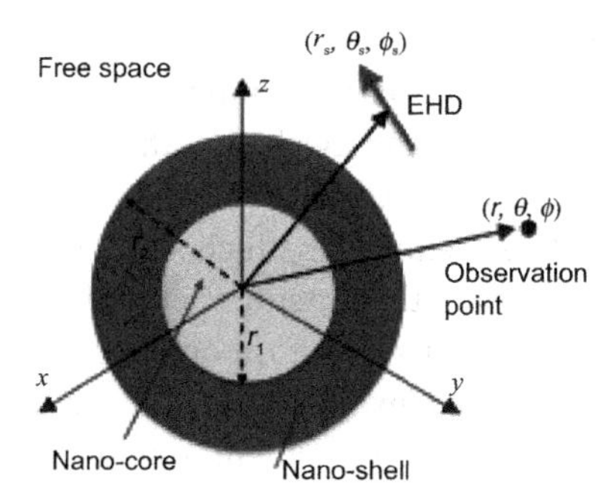

Figure 14.8 Single EHD illumination of a CNP.

and μ_0; thus, its wave number is $k_0 = \omega\sqrt{\varepsilon_0\mu_0}$, while its intrinsic impedance is $\eta_0 = \sqrt{\mu_0/\varepsilon_0}$.

The CNP is excited by an arbitrarily oriented and located EHDs with the dipole moment being equal to $\vec{p} = p\mathbf{a}_p$, where $\hat{p}$ is its orientation and p[Am] is its complex amplitude. This EHD amplitude is typically expressed as the product of the constant current I_e[A] applied to it and its length l[m], that is, $p = I_e l$. The EHDs are driven by a time harmonic source with the frequency f; the corresponding free-space wavelength is $\lambda = 2\pi/k_0$. We introduce a spherical coordinate system, with the coordinates (r, θ, ϕ) and unit vectors $(\mathbf{a}_r, \mathbf{a}_\theta, \mathbf{a}_\phi)$, and a Cartesian coordinate system, with the coordinates (x, y, z) and unit vectors $(\mathbf{a}_x, \mathbf{a}_y, \mathbf{a}_z)$, such that the origin coincides with the center of the CNP. The coordinates of the observation point and the EHD are (r, θ, ϕ) and (r_s, θ_s, ϕ_s), respectively. In the two coordinate systems, the dipole moment can be expressed, respectively, as $\vec{p} = \mathbf{a}_r p_r + \mathbf{a}_\theta p_\theta + \mathbf{a}_\phi p_\phi$ and $\vec{p} = \mathbf{a}_x p_x + \mathbf{a}_y p_y + \mathbf{a}_z p_z$.

14.3.2 *Theoretical Considerations*

The analytical solution for the single-EHD problem in Fig. 14.8 was derived in [7], and here we only summarize its main points. The known electromagnetic field due to the EHD is expanded in terms of transverse magnetic (TM) and transverse electric (TE) spherical waves with the known expansion coefficients $a_{nm}^{(c)}$ (TM coefficients), and $b_{nm}^{(c)}$, (TE coefficients). The index $c = 1$ applies for $r < r_s$ while

$c = 4$ applies for $r > r_s$. The unknown scattered fields due to the CNP, that is, the fields inside and outside the CNP, are also expanded in terms of TM and TE spherical waves. These expansions involve the unknown TM and TE expansion coefficients denoted by $A_{i,nm}$ and $B_{i,nm}$, respectively, where $i = 1$ for the fields inside the nano-core, $i = 2$ and 3 for the fields inside the nano-shell, and $i = 4$ for the fields outside the CNP. These expansion coefficients depend on the EHD location and orientation; they are easily obtained by enforcing the boundary conditions on the two spherical interfaces, $r = r_1$ and $r = r_2$. Once these coefficients are known, we have the complete knowledge of the fields in all regions. These fields will, in general, possess all three components. We have for the electric field: $\mathbf{E} = \mathbf{a}_r E_r + \mathbf{a}_\theta E_\theta + \mathbf{a}_\phi E_\phi$, where E_r, E_θ, and E_ϕ are, respectively, its r-, θ-, and ϕ-components. In the same manner, we have for the magnetic field: $\mathbf{H} = \mathbf{a}_r H_r + \mathbf{a}_\theta H_\theta + \mathbf{a}_\phi H_\phi$, where $H_r H_r$, H_θ, and H_ϕ are its r-, θ-, and ϕ-components. With the fields in place, we can derive the expression for the total power radiated by the CNP when it is excited by the EHD, and this is given by the expression:

$$P_\mathrm{t} = \frac{\pi}{\omega k_0} \sum_{n=1}^{N_\mathrm{max}} \sum_{m=-n}^{n} 2\frac{n(n+1)}{2n+1} \frac{(n+|m|)!}{(n-|m|)!} \left[\frac{|\alpha_{nm}|^2}{\varepsilon_0} + \frac{|\beta_{nm}|^2}{\mu_0} \right], \tag{14.5}$$

where the coefficients $\alpha_{nm} = a_{nm}^{(4)} + A_{4,nm}$, $\beta_{nm} = b_{nm}^{(4)} + B_{4,nm}$ if the EHD is outside the CNP, and $\alpha_{nm} = A_{4,nm}$, $\beta_{nm} = B_{4,nm}$ if the EHD is inside the CNP. The symbol N_max is the truncation limit in a practical numerical implementation of the infinite summation in the exact solution and is chosen in a manner that ensures the convergence of this expansion.

The power radiated by the EHD, when situated alone in free space, is given by (14.5) with $\alpha_{nm} = a_{nm}^{(4)}$ and $\beta_{nm} = b_{nm}^{(4)}$, which reduces to the simple expression:

$$P_\mathrm{EHD} = \frac{\eta_0 \pi}{3} \left| \frac{p k_0}{2\pi} \right|^2. \tag{14.6}$$

In our investigations, the so-called normalized radiation resistance (NRR) is examined; this is the radiation resistance of the dipoles radiating in the presence of the CNP normalized by the radiation resistance of the dipoles radiating in free space, and in dB it reads:

$$\mathrm{NRR(dB)} = 10 \cdot \log_{10} \left(\frac{P_\mathrm{t}}{P_\mathrm{EHD}} \right). \tag{14.7}$$

In addition to the near-field distributions and the NRR values, the derived field solutions to the problem in Fig. 14.8 will also be used to study the power flow density inside and outside of the CNP. This will be achieved in terms of the Poynting vector (both magnitude and direction):

$$\mathbf{S} = \frac{1}{2}\mathrm{Re}\left\{\mathbf{E}\times\mathbf{H}^*\right\} = \mathbf{a_r}S_r + \mathbf{a}_\theta S_\theta + \mathbf{a}_\phi S_\phi, \tag{14.8}$$

where the r-, θ-, and ϕ-components of the Poynting vector are explicitly given by $S_r = 0.5\,\mathrm{Re}\left\{E_\theta H_\phi^* - E_\phi H_\theta^*\right\}$, $S_\theta = 0.5\,\mathrm{Re}\left\{E_\phi H_r^* - E_r H_\phi^*\right\}$, and $S_\phi = 0.5\,\mathrm{Re}\left\{E_r H_\theta^* - E_\theta H_r^*\right\}$, and the asterisk designates the complex conjugate.

The general field solutions obtained in [7] are straightforwardly specialized for the far-field observation points in order to derive an expression for the directivity of the configuration in Fig. 14.8. In particular, the directivity takes on the form,

$$D(\theta,\phi) = \frac{2\pi}{\eta_0}\frac{1}{P_\mathrm{t}}\left(\left|\mathbf{F}_{\mathrm{t},\theta}\right|^2 + \left|\mathbf{F}_{\mathrm{t},\phi}\right|^2\right), \tag{14.9}$$

where

$$\mathbf{F}_{\mathrm{t},\theta} = \sum_{n=1}^{N_{\max}}\sum_{m=-n}^{n} j^{n-1}e^{jm\phi} \times\left[\frac{1}{\omega\varepsilon_0}\alpha_{nm}\frac{d}{d\theta}P_n^{|m|}(\cos\theta) + \beta_{nm}\frac{1}{k_0}\frac{jm}{\sin\theta}P_n^{|m|}(\cos\theta)\right] \tag{14.10}$$

and

$$\mathbf{F}_{\mathrm{t},\phi} = \sum_{n=1}^{N_{\max}}\sum_{m=-n}^{n} j^{n-1}e^{jm\phi} \times\left[\frac{1}{\omega\varepsilon_0}\alpha_{nm}\frac{jm}{\sin\theta}P_n^{|m|}(\cos\theta) - \beta_{nm}\frac{1}{k_0}\frac{d}{d\theta}P_n^{|m|}(\cos\theta)\right] \tag{14.11}$$

are, respectively, the θ- and ϕ- components of the total radiation vector, $\mathbf{F}_\mathrm{t}(\theta,\phi)$, which is related to the total far-field $\mathbf{E}_\mathrm{t}$ through the relation $\mathbf{E}_\mathrm{t} \approx \mathbf{F}_\mathrm{t}(\theta,\phi)\exp(-jk_0 r)/r$. The function $P_n^{|m|}$ is the associated Legendre function of the first kind of degree n and order $|m|$.

14.3.3 *CNP Materials and Gain Model*

In contrast to the plane wave excitations, we focus our attention for the EHD excitations only on the active CNPs consisting, in all cases, of a SiO_2 nano-core covered by a plasmonic nano-shell. For the latter, three materials have been employed: silver (Ag), gold (Au), and copper (Cu). The corresponding CNPs will be referred to later as the Ag, Au, and Cu-based CNPs. Although results will be presented for all three cases, the emphasis will be put on the Ag-based CNP design. As with the plane wave case, the geometrical size of the CNP has the radius of the silica nano-core set to $r_1 = 24$ nm, and the outer radius of the plasmonic nano-shell set to $r_1 = 30$ nm. Thus, the plasmonic nano-shell is 6 nm thick in all of the cases.

Owing to the nano-scale dimensions of the CNP, accurate modelling of its optical properties requires one to take into account the size dependence of the materials used in making these structures. In the present case, the plasmonic nano-shell exhibits significant intrinsic size dependencies, which arise when the size of the material approaches and becomes less than the bulk mean free path length of the conduction electrons in the material (see, for example [11, 12], and the references therein). They are then incorporated into the Drude model as a size-dependent damping frequency. Empirically determined bulk values for the permittivity of silver (as well as gold) at optical wavelengths (λ) between 200 nm and 1800 nm were obtained in [46, 54] and successfully used in [11, 12].

The main focus here will be on the behavior of the CNP in the visible wavelength range spanning the interval from 450 nm to 650 nm. In this interval, Fig. 14.9 shows, for the 6 nm thick Ag, Au, and Cu nano-shells, the real part of the permittivity (normalized with the free-space permittivity) and the associated values of their loss tangents as functions of the excitation wavelength; here, we recall that the nano-shell permittivity is denoted by $\varepsilon_2 = \varepsilon_2' - j\varepsilon_2''$, with ε_2' being its real part, and ε_2'' being the negative of its imaginary part, and thus the loss tangent being defined by $\text{LT} = \varepsilon_2''/|\varepsilon_2'|$. It can be observed in Fig. 14.9 that the real part of the permittivity of the various plasmonic materials under consideration is negative in the depicted wavelength range, and moreover, that they all are

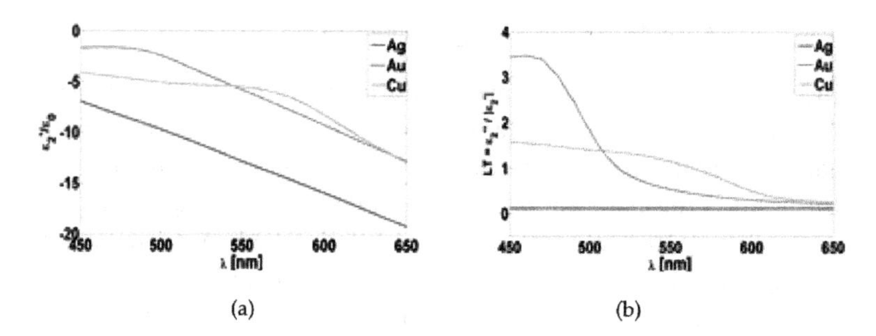

Figure 14.9 The real, ε_2', part of the 6 nm thick Au, Ag, and Cu nano-shells normalized to the free-space permittivity ε_0, and (b) the corresponding loss tangents $\text{LT} = \varepsilon_2''/|\varepsilon_2'|$.

lossy with Ag being the least lossy case. In contrast to the plasmonic nano-shells, there are no size-dependent effects at the considered wavelengths inside the dielectric silica nano-core.

In our investigations of EHD excitations of active CNPs, we have also considered the corresponding passive CNPs for reference purposes. For a passive CNP, the silica nano-core will be assumed lossless. As representative values of its permittivity and permeability, we take $\varepsilon_1 = 2.05\varepsilon_0$ and $\mu_1 = \mu_0$, respectively, implying $k_1 = \omega\sqrt{\varepsilon_1\mu_1} = k_0\sqrt{2.05} = k_0 n$, with $n = \sqrt{2.05}$ being the refractive index of silica. For an active CNP, we will consider a canonical gain model in which the gain is incorporated into the lossless silica nano-core. According to such a model, the permittivity of the silica nano-core reads $\varepsilon_1 = (n^2 - \kappa^2 - 2jn\kappa)$, where n is the refractive index (maintained at the silica nano-core value of $n = \sqrt{2.05}$), and the parameter κ determines the nature of the nano-core: the nano-core is lossless and passive for $\kappa = 0$, it is lossy and passive for $\kappa > 0$, and it is active for $\kappa < 0$. The parameters n and κ enter into the expression for the wave number according to the relation $k_1 = k_0(n - j\kappa)$.

14.3.4 *Resonance Effects of Active CNPs for Single EHD Excitation*

The present section illustrates interesting resonant properties of electrically small active CNPs excited by a single EHD. Throughout

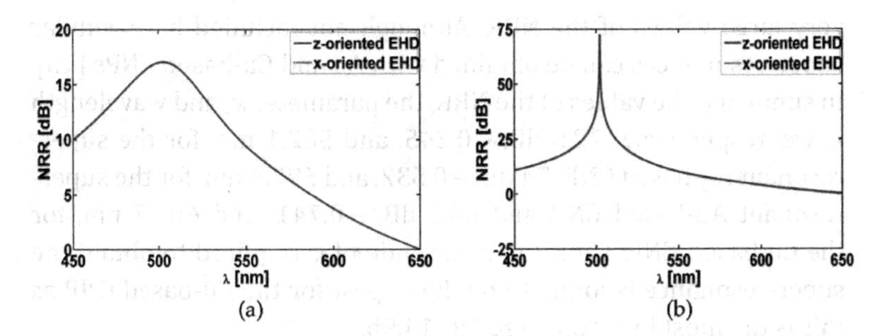

Figure 14.10 The NRR as a function of the wavelength, λ, for the spherical Ag-based passive (a) and super-resonant (b) CNP. The results are shown for both *z*- and *x*-oriented EHDs, and in all cases, the EHD is located in the silica nano-core at 12 nm along the positive *x*-axis.

the investigations, the EHD was taken to be either z or x-oriented, and is placed along the positive x-axis with the coordinates (r_s, $\theta_s = 90°$, $\phi_s = 0°$). Moreover, the magnitude of the dipole moment was set equal to $p = 5$nA-m. The present results represent a summary of [14–16]; for further details please refer to these works.

14.3.4.1 NRR results

Figure 14.10 shows the NRR as a function of the wavelength, λ, for the Ag-based CNP for (a) $\kappa = 0$ (corresponding to a lossless and passive CNP) and (b) $\kappa = -0.245$ (corresponding to an active CNP). In both of the cases, the results for the two EHD orientations are shown, and the EHD is located in region 1 at $r_s = 12$ nm.

For a given value of κ, identical NRRs result in the depicted wavelength range for the two EHD orientations. For the active CNP, $\kappa = -0.245$ was found to lead to the largest NRR value (around 72.5 dB at 502.1 nm) for both EHD orientations. This enhanced behavior corresponds to the super-resonant state found in the plane wave case, in which the NRR values are significantly increased and the intrinsic plasmonic losses are vastly overcome, relative to the case of the corresponding passive Ag-based CNP for which largest NRR is around 17 dB (at 502.7 nm), see Fig. 14.10a. Thus, as in the case of a plane wave excited CNP discussed previously, the gain inclusion helps overcoming the losses in the CNP when it is excited by an EHD. In addition, it thus leads to enhanced resonance phenomena with very large values of the NRR. Although not included here, similar super-resonances can be obtained with Au and Cu-based CNPs [15]. In summary, the values of the NRR, the parameter κ, and wavelength λ are, respectively: 72.5 dB, −0.245, and 502.1 nm, for the super-resonant Ag-based CNP; 74 dB, −0.532, and 597.4 nm, for the super-resonant Au-based CNP, and 68.5 dB, −0.741, and 601.7 nm, for the Cu-based CNP. Thus, the magnitude of κ required to obtain the super-resonance is found to be the largest for the Cu-based CNP as this is the most lossy CNP, see Fig. 14.9b.

14.3.4.2 Near-field distributions and directivity

The properties of the Ag-based CNP are next further illustrated via its near-field behavior. To this end, we depict the quantity $20\log_{10}|E_{t,\theta}|$, where $E_{t,\theta}$ is the θ-component of the total electric field normalized by 1 V/m, in the following plots. The plane of observation is the xz-plane, and the field will be shown in a circular region with a radius of 90 nm. Figure 14.11a and 14.11b show the electric field of the super-resonant Ag-based CNP ($\kappa = -0.245$ and $\lambda = 502.1$ nm) for the z and x-oriented EHDs, respectively. For comparison, Fig. 14.11c and 14.11d show the electric field of the corresponding lossless and passive Ag-based CNP configurations ($\kappa = 0$ and $\lambda = 502.7$ nm). For all results, the curves representing the spherical surfaces of the CNP are also shown. The excited modes in the Ag-based CNP reported in Fig. 14.11a and 14.11b are clearly seen to correspond, respectively, to those of a z-oriented EHD (located at the origin), and an x-oriented EHD (located at the origin). Moreover, the two modes are found to be very strongly excited and of comparable magnitude. This very strong excitation of the dipole mode inside the CNP is the reason for the super resonance effects illustrated previously in Fig. 14.11b, thus further supporting the fact that the inclusion of gain indeed helps overcoming the intrinsic plasmonic losses in the particle. In contrast hereto, the fields for the two EHD orientations in Fig. 14.11c and 14.11d are only weakly dipolar as the CNP is passive; this is consistent with the corresponding significantly lower values of the NRR reported in Fig. 14.10a. Although not included in here, similar near-field results are obtained for the Au-based and Cu-based CNPs.

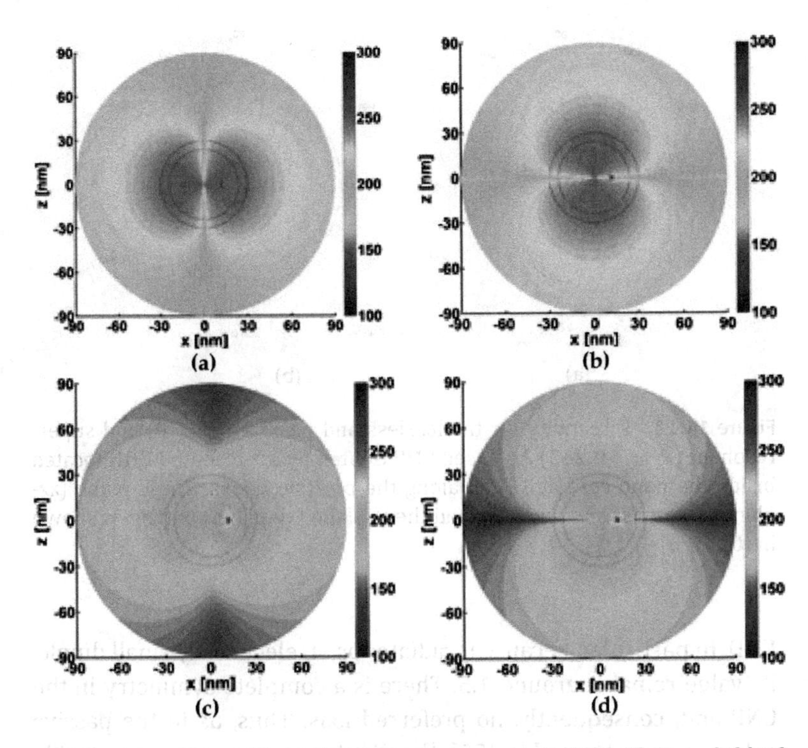

Figure 14.11 The magnitude of the θ-component of the electric field of the super-resonant spherical Ag-based CNP for (a) z-oriented and (b) x-oriented EHDs. The corresponding results for the passive Ag-based CNP are found in (c) and (d). The EHD is inside the nano-core at 12 nm along the positive x-axis. The results are shown in the xy-plane in a circular region of a radius of 90 nm. The curves representing the spherical surfaces of the CNP are likewise shown in all figures.

In addition to the above near-field results, we next show the directivity, as calculated from (14.9), for the Ag-based CNP. This is shown in Fig. 14.12 for lossless and passive ($\kappa = 0$) and super-resonant ($\kappa = -0.245$) Ag-based CNP in the case of a z-oriented EHD being located inside the nano-core at 12 nm along the positive x-axis. Specifically, the E-plane (xz-plane) pattern results are shown in Fig. 14.12a and the H-plane (xy-plane) pattern results are shown in Fig. 14.12b. Despite the large levels of the NRR reported in Fig. 14.10b, the directivity of the super-resonant state, which is clearly dipolar, is not enhanced relative to that of an isolated z-oriented

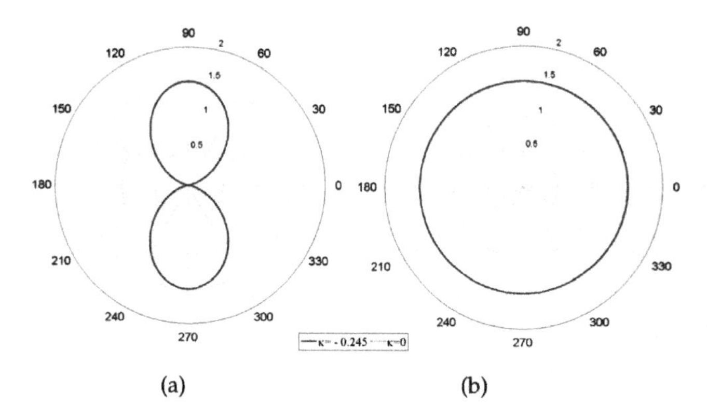

Figure 14.12 Directivity for the lossless and passive ($\kappa = 0$) and super-resonant ($\kappa = -0.245$) Ag-based CNP excited by a z-oriented EHD located inside the nano-core at 12 nm along the positive x-axis. The E-plane (xz-plane) pattern is shown in (a), and the H-plane (xy-plane) pattern is shown in (b).

EHD. In particular, because it radiates as an electrically small dipole, its value remains around 1.5. There is a complete symmetry in the CNP and, consequently, no preferred axis. Thus, as in the passive spherical case treated in [55], the dipolar resonance excited inside the active spherical CNPs does not modify the directivity pattern of the exciting Hertzian dipole; to this end, the higher-order modes need to be excited inside the CNPs. Apart from the flipping of the E and H-plane patters by 90°, the same results hold true for the case of x-oriented EHD excitation of the currently investigated CNPs.

14.3.4.3 Influence of EHD orientation

The super-resonances in Fig. 14.10b are not restricted for the EHD locations inside the nano-core of the CNP. Figure 14.13 shows the NRR as a function of the EHD distance r_s from the center of the super-resonant Ag-based CNP, as well as the corresponding Au and Cu-based CNPs, for a z-oriented (Fig. 14.13a) and x-oriented (Fig. 14.13b) EHDs—from these results, it is clear that large NRR values are also obtained for the EHD locations inside the nano-shell and outside the CNP.

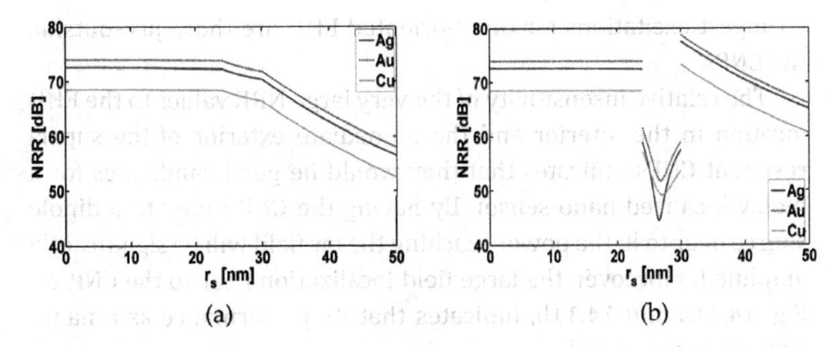

Figure 14.13 The NRR as a function of the EHD location along the positive x-axis for the super-resonant Ag, Au, and Cu-based CNPs. (a) z-oriented EHD and (b) x-oriented EHD.

Although the behaviors of the NRR for the two EHD orientations both qualitatively and quantitatively resemble one another for locations inside the nano-core (where an almost constant NRR is observed), this is not the case for other locations. In particular, two notable differences are observed. First, the NRR of the x-oriented EHD is well below that of the z-oriented EHD when the EHDs are inside the respective nano-shells. For the x-oriented EHD, some of its field gets trapped inside the nano-shell (i.e., the nano-shell acts as a waveguide excited by the EHD oriented orthogonal to its walls), and thus does not get radiated, resulting in reduced NRR values. The second notable difference between the two orientations occurs for the EHD locations outside the CNPs. For these locations, the NRR for the z-oriented EHD drops below its values obtained when the EHD is inside the CNPs, whereas the NRR for the x-oriented EHD attains its maximum as the x-oriented EHD moves just outside the CNP; this maximum is also larger than any of the NRR values obtained with the z-oriented EHD. The x-oriented dipole not only excites the resonant mode of the electrically small core–shell cavity, but it also directly drives the SPPs on the outer shell. Because the shell is thin, these surface waves extend into the core and in turn are amplified by the presence of the gain medium, and this explains the increased NRR values for the x-oriented EHD outside the CNP. In summary, the strongest excitation of the resonant dipole mode requires the z-oriented EHD to be inside the nano-cores, while the locations of the strongest excitations for the x-oriented EHD are those just outside the CNPs.

The relative insensitivity of the very large NRR values to the EHD location in the interior and the immediate exterior of the super-resonant CNPs indicates that they would be good candidates for a highly localized nano-sensor. By having the CNP tuned to a dipole source near to it, the power reaching the far field will be significantly amplified. Moreover, the large field localization near to the CNP, cf., Fig. 14.11a and 14.11b, indicates that its performance as a nano-antenna is very good.

Although the work in this section was devoted to active CNPs consisting of a 24 nm radius silica nano-core radius covered with a 6 nm thick silver nano-shell, the overall conclusions on their performance characteristics qualifying them as elements for localized nano-sensors and nano-antennas also hold true for a number of other active CNP configurations. As a brief illustration of this point, we note that two additional active Ag-based CNP configurations were studied: one in which the silica nano-core radius was set to 16 nm, with the plasmonic layer thickness being set to 4 nm, and the other in which the silica nano-core radius was set to 8 nm, with the plasmonic layer thickness being set to 2 nm. In both cases, the parameter κ was tuned to achieve the maximum NRR values, and it was found that they both lead to approximately the same peak NRR values as the currently examined 6 nm thick silver nano-shell. Nonetheless, as one might expect, the optimal value of the parameter κ leading to the super-resonant behavior was found to increase as the active CNP size was decreased, that is, more gain is needed to achieve the super-resonant state for smaller active CNP configurations.

14.3.5 *Jamming Effects of Active CNPs for Single and Multiple EHD Excitations*

Apart from the above-reported super-resonant properties of specific active CNPs, in which the power radiated by the EHD is significantly enhanced, there are additional interesting results when the EHD is located outside the CNPs. In [14], it was demonstrated that identical CNPs can lead to large enhancements as well as to large reductions of the NRR for altering locations of the z-oriented EHD outside the CNPs. When the radiated power is significantly decreased, the EHD is effectively cloaked to a far-field observer. This property might be of great potential use in biological fluorescent assays wherein the suppression of certain emission lines might be desired while simultaneously enhancing the emission of other, desired, lines.

This section reviews and further addresses the ability of active CNPs to effectively cloak not only one but also several EHDs (quantum emitters), for example, a set of fluorescing molecules. As they are traditionally treated in semi-classical models, the molecules and their emissions will be represented here by EHDs radiating at one of their resonance frequencies. Apart from a discussion of a single EHD excitation to introduce the solution approach and its prediction of outcomes associated with the basic physics, we devote our attention in this study to cases in which two or four EHDs are radiating simultaneously in the presence of a properly designed active spherical CNP. It is shown that the power radiated in the multiple-EHD configurations is decreased even further than is the case with a single EHD, that is, the signals radiated by the EHDs are effectively jammed (or cloaked) by the presence of the active CNP. The present results represent a summary of [14] and [56]; for further details please refer to these works.

14.3.5.1 Configuration

The configuration used to illustrate the transparency effects of active CNPs is shown in Fig. 14.14. The CNP in question has the same characteristics as the one described previously in Sections 14.3.1 and 14.3.3, that is, it consists of a silica nano-core (of 24 nm radius) covered concentrically by a 6 nm thick silver nano-shell. It is excited by one or more EHDs located exterior to it. The EHDs have, in principle, their own dipole moments (magnitude and direction), as well as locations outside the CNP. In here, three specific excitations of the CNP in Fig. 14.14 are investigated: Case 1, in which only EHD A excites the CNP; Case 2, in which EHD A and C excite the CNP; and Case 3, in which all four dipoles excite the CNP simultaneously. As before, a spherical coordinate system (r, θ, ϕ) and the associated rectangular coordinate system (x, y, z) are introduced such that the

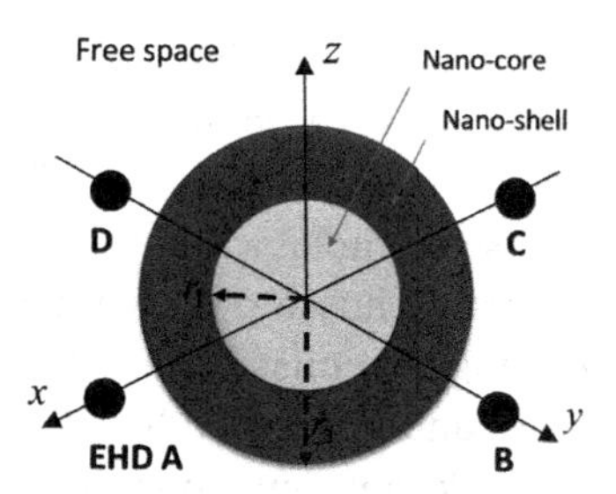

Figure 14.14 Single/multiple electric Hertzian dipoles (EHDs) illumination of a CNP. The EHDs are outside the CNP and are represented by the black-colored circles. Three excitation cases are considered: Case 1 (EHD A), Case 2 (EHD A+C), and Case 3 (EHD A + B + C + D). Please see the main text for further explanations.

origin of these coincide with the center of the CNP. The exciting EHDs are located symmetrically at positions in the *xy*-plane outside the CNP, that is, EHDs A and C are on opposite sides of the CNP along the *x*-axis, whereas EHDs B and D are on opposite sides along the *y*-axis. The coordinates of the observation point are (r, θ, ϕ). The locations of the EHDs are specified by $(r_S, \theta_S = \pi/2)$ and $\phi_S = 0$(EHD A), $\pi/2$(EHD B), π (EHD C) and $3\pi/4$(EHD D).

14.3.5.2 Far-field results

For the following discussion, the EHDs in any of the three excitation cases, Case 1, 2, or 3, were taken to be *z*-oriented, and the magnitudes of their dipole moments were assumed identical and equal to $p = 5$ nA-m. The EHDs are located at their respective places in the *xy*-plane at a radial distance of $r_S = 40$ nm, see Fig. 14.14 and the end of Section 14.3.5.1. The quantity NRR from (14.7) has been generalized to handle the multiple-EHD excitation and will be used currently to illustrate the far-field results.

As noted in the beginning of Section 14.3.5 above, the so-called super-resonant state, which leads to large radiated powers, was found for $\kappa = -0.245$ in the Case 1 excitation [14, 56] (this likewise being in line with the results reported in Section 14.3.4). The response of the super-resonant state is illustrated in Fig. 14.15 in which the NRR is shown as a function of the excitation λ for the super-resonant Ag-based CNP. The resonance, resulting in the

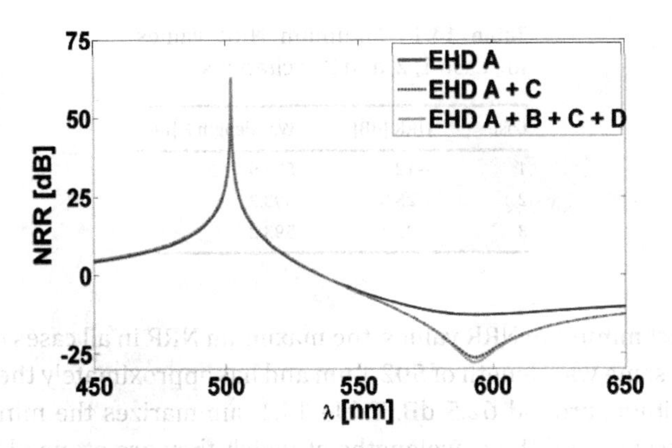

Figure 14.15 The NRR (dB) as a function of the free-space wavelength, λ, for the three excitation cases of the super-resonant Ag-based CNP.

distinctive, very large NRR value (62.5 dB) is found for Case 1 at 502.1 nm. As noted before, this is due to a strong excitation of the resonant dipolar mode inside the active CNP. What is even more interesting here for the Case 1 excitation is the appearance of the "dip" in the NRR (down to -12.7 dB) at 594.9 nm. At this dip, the power radiated by the EHD A is effectively 12.7 dB below the value when the EHD A is radiating alone in free space.[a]

As we move to Case 2, in which the EHDs A and C are radiating simultaneously in the presence of the CNP, the dip for the NRR is significantly lowered (down to -25.8 dB) relative to the Case 1 case. As can be inferred from Fig. 14.15, it now occurs at a slightly displaced wavelength equal to 593.2 nm. This additional lowering of the NRR clearly indicates that the active CNP jams the signal of the two EHDs, effectively cloaking them to a far-field observer. Adding an additional set of dipoles, EHDs B and D, and thus arriving at the Case 3 excitation, we only find a further lowering of the dip by an additional 1.3 dB relative to that of Case 2, while maintaining the wavelength at which it occurs. Contrary to the

[a]Two remarks are in order at the time being. First, the "dip" in the NRR is also observed for the super-resonant Au- and Cu-based CNPs for Case 1 excitation, i.e., for a single z-oriented EHD. Second, although not included here, we note that in the corresponding case of an x-oriented EHD, only the Au-based CNP exhibits a dip in the NRR for the investigated EHD locations. Contrary to the results for the z-oriented EHD (Case 1), the said dip occurs at wavelength lower than the one at which the super-resonance is attained for the case of an x-oriented EHD.

Table 14.1 Minimum NRR values for Case 1, 2, and 3 excitations

Case	NRR [dB]	Wavelength λ [nm]
1	−12.7	594.9
2	−25.8	593.2
3	−27.1	593.2

distinct minimum NRR values, the maximum NRR in all cases occurs at the same wavelength of 502.1 nm and has approximately the same magnitude, around 62.5 dB. Table 14.1 summarizes the minimum NRR values and the wavelengths at which they are attained in the three excitation cases. The dips in the NRR values found in Fig. 14.15 can be said to correspond to non-radiating, or transparent, or cloaked states, where the quantum emitters (EHD) are effectively cloaked or jammed to a far-field observer.

Although the active Ag-based CNP acts as a super-resonant CNP at the wavelength of 502.1 nm, where the peak in the NRR in Fig. 14.15 is observed, this is not so at the wavelengths found in Table 14.1 at which the respective dips in NRR are found. Nevertheless, it is interesting to observe how an active CNP, initially designed to be super-resonant at a certain wavelength, can in fact significantly suppress the EHD(s) radiation at another wavelength.

The fact that this suppression is indeed gain-assisted was confirmed by examining the present CNP configuration in the cases in which the silica nano-core is first lossless and then lossy (and passive), although the material parameters of the silver nano-shell remained unchanged. To this end, we expressed the silica nano-core permittivity as $\varepsilon_1 = \varepsilon_1' - j\varepsilon_1'' = 2.05\varepsilon_0(1 - j\tan\delta_e)$, where the loss tangent is $\tan\delta_e = \varepsilon_1''/(2.05\varepsilon_0)$, and examined the properties of the resulting Ag-based CNP for four cases: 1) $\tan\delta_e = 0$, 2) $\tan\delta_e = 0.01$, 3) $\tan\delta_e = 0.1$ and 4) $\tan\delta_e = 1$. For all excitations: Cases, 1, 2, and 3, of the CNP, the case of a lossless silica nano-core provides the lowest NRR values. For the Case 1 excitation, the NRR reduces to about 11 dB, which is comparable to the active CNP design. However, the two remaining passive cases result in a NRR value around -16 dB, which are still orders of magnitude above the minimum values obtained for the corresponding active Ag-based CNP.

14.3.5.3 Power-flow density distribution

To further illustrate the transparent properties of the super-resonant Ag-based CNP, we next show the power flow density, as determined from (14.8), for the Case 1 and 3 excitations. The power flow density is shown in the *xy*-plane in Fig. 14.16a and 14.16b, respectively. Specifically, the quantity $10\log_{10}|S/1(\mathrm{W/m^2})|$ is used to represent the normalized magnitude (color) of the power flow density, while the arrows depict its direction. The corresponding free-space results are shown in Fig. 14.16c and 14.16d. Although the results for the EHD(s) in free space show an expected outward propagating power flow density, those of the two excitation cases of the active Ag-based CNP are rather interesting. Clearly, effects due to the presence of the CNP are in evidence for the Case 1 excitation in Fig. 14.16(a), two source-like spots exist for the power flow: one on the negative *x*-axis and one at its center. The CNP is seen to act like an effective EHD, which is located at its center. The second effective source appears to be an image-like dipole whose presence is induced by the fields associated with the EHD and CNP. The power flow originating from it is seen to cancel the power flow generated by the effective dipole located at the center of the CNP. For the Case 3 excitation, only a single source-like spot is found, this being at the center of the CNP. However, from the outside, it is clearly seen that the entire four-EHD and Ag-based CNP system acts like an effective sink; the arrows outside the CNP all point inwards to the center of the CNP. This sink-like behavior confines the power from the EHDs to this region, that is, this power never reaches the far-field region. These power density results correspond nicely to their reduced NRR levels reported in Fig. 14.15. Similar results as those in Fig. 14.16 are obtained for the Case 2 excitation of the CNP.

As noted above, the dips in the NRR reported here for the active Ag-based CNPs can be interpreted as the active CNP cloaks of the EHDs.

This result is effectively the spherical version of the cloaking effects discussed in [57, 58] for coated cylinders. The non-radiating states observed here can also be connected directly the corresponding low-frequency antenna configuration [6] and to the transparency/cloaking effects introduced by Alù and Engheta (see, for instance [59–61]).

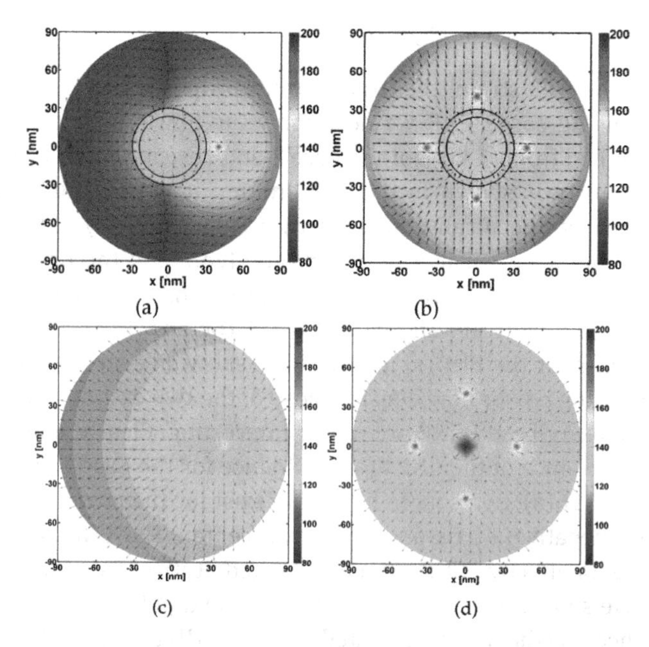

Figure 14.16 Magnitude (color) and direction (arrows) of the power flow density for Case 1(a) and 3(b) excitations of the Ag-based CNP corresponding to the respective dips in the NRR shown in Fig. 14.15. The corresponding results for the EHD(s) radiating in free space only are shown in (c) and (d). The results are shown in the *xy*-plane in a circular region of a radious of 90 nm. In all cases the respective EHDs are located at their respective places in the *xy*-plane at $r_s = 40$ nm. The curves representing the spherical surfaces of the CNP are likewise shown in (a) and (b).

The ability of the active CNP to significantly suppress the radiation of EHDs offers an interesting route toward the jamming of quantum emitters/nano-antennas for a variety of applications, for example, bio-sensors. This may be of particular interest in biological fluorescent arrays to suppress the emission of certain emission lines while enhancing that of other, desired, lines.

14.4 Conclusion

Properly designed passive and active spherical CNPs consisting of a silica nano-core alone or impregnated with gain medium, and covered concentrically with an ENG nano-shell and its IO alternative, offer interesting canonical nano-amplifier and nano-antenna configurations for optical applications. Both plane wave and EHD excitation examples were discussed. It was illustrated with power flow streamlines how these resonant electrically small scatterers physically behave as much larger objects, leading to smaller gain value requirements. Canonical and quantum dot gain media examples exemplified how insensitive these DPS-ENG configurations are to the actual gain medium choices. It was also shown how these core–shell configurations can be designed to amplify or to jam the signal radiated by one or more quantum emitters.

The nano-amplifier core–shell configurations offer the ability to enhance significantly the power radiated by a quantum emitter. These nano-antennas are highly localized and multi-layer designs offer multi-frequency applications. The jamming configurations illustrate how active CNPs also can prevent the power radiated by a quantum emitter from reaching a far-field observer. Thus, these active CNPs offer an interesting route, for instance, to enhancing the results of biological fluorescence assays where selectively enhancing or suppressing the emission of certain emission lines is highly desirable since these effects would allow one to have the ability to discriminate between different biological chemicals and entities simultaneously.

References

1. Engheta, N. and Ziolkowski, R.W. (2006) *Metamaterials: Physics and Engineering Explorations* (Hoboken, NJ: IEEE Press-Wiley Publishing).
2. Ziolkowski, R.W. (2006). Ultra-thin metamaterial-based laser cavities, *J. Opt. Soc. Am. B*, **23**, pp. 451–460.
3. Ziolkowski, R.W. and Kipple, A.D. (2003). Application of double negative metamaterials to increase the power radiated by electrically small antennas, *IEEE Trans. Antennas Propagat.*, **51**, pp. 2626–2640.
4. Ziolkowski, R.W. and Kipple, A.D. (2005). Reciprocity between the effects of resonant scattering and enhanced radiated power by electrically small antennas in the presence of nested metamaterial shells, *Phys. Rev. E.*, **72**, 036602.

5. Ziolkowski, R.W. and Erentok, A. (2006). Metamaterial-based efficient electrically small antennas, *IEEE Trans. Antennas Propagat.*, **54**, pp. 2113–2130.
6. Ziolkowski, R.W. and Erentok, A. (2007). A hybrid optimization method to analyze metamaterial-based electrically small antennas, *IEEE Trans. Antennas Propag.*, **55**, pp. 731–741.
7. Arslanagić, S., Ziolkowski, R.W. and Breinbjerg, O. (2007). Radiation properties of an electric Hertzian dipole located near-by concentric metamaterial spheres, *Radio Sci.*, **42**, doi:10.1029/2007RS003663.
8. Gordon, J.A. and Ziolkowski, R.W. (2008). Colors generated by tunable plasmon resonances and their potential applications to ambiently illuminated color displays, *Solid State Comm.*, **146** , pp. 228–238.
9. Campbell, S.D. and Ziolkowski, R.W. (2012). Simultaneous excitation of electric and magnetic dipole modes in a resonant core–shell particle at infrared frequencies to achieve minimal backscattering, *IEEE J. Sel. Topics Quantum Electron.*, doi: 10.1109/JSTQE.2012.2227248.
10. Ziolkowski, R.W. and Erentok, A. (2007). At and beyond the Chu limit: Passive and active broad bandwidth metamaterial-based efficient electrically small antennas, *IET Microwaves Antennas Propagat.*, **1**, pp. 116–128.
11. Gordon, J.A. and Ziolkowski, R.W. (2007). The design and simulated performance of a coated nano-particle laser, *Opt. Exp.*, **15**, pp. 2622–2653.
12. Gordon, J.A. and Ziolkowski, R.W. (2007). Investigating functionalized active coated nano-particles for use in nano-sensing applications, *Opt. Exp.*, **15**, pp. 12562—12582.
13. Gordon, J.A. and Ziolkowski, R.W. (2008). CNP optical metamaterials, *Opt. Express*, **16**, pp. 6692–6716.
14. Arslanagić, S. and Ziolkowski, R.W. (2010). Active coated nano-particle excited by an arbitrarily located electric Hertzian dipole—resonance and transparency effects, *J. Opt. A.*, **12**, 024014.
15. Arslanagić, S. and Ziolkowski, R.W. (2011). Active coated nano-particles: Impact of plasmonic material choice, *Appl. Phys. A*, **103**, pp. 795–798.
16. Arslanagić, S. and Ziolkowski, R.W. (2012) Directive properties of active coated nano-particles, *Adv. Electromagn.*, **1**, pp. 57–64.
17. Campbell, S.D. and Ziolkowski, R.W. (2012). Impact of strong localization of the incident power density on the nano-amplifier characteristics of active coated nano-particles, *Opt. Commun.*, **285**, pp. 3341–3352.

18. Campbell, S.D. and Ziolkowski, R.W. (2012). The performance of active coated nanoparticles based on quantum-dot gain media, *Adv. OptoElectron.*, **2012**, 368786.
19. Ziolkowski, R.W., Jin, P. and Lin, C.-C. (2011). Metamaterial-inspired engineering of antennas, *Proc. IEEE*, **99**, pp. 1720–1731.
20. Geng, J. and Ziolkowski, R.W. (2011). Numerical study of active open cylindrical coated nano-particle antennas, *IEEE Photon.*, **3**, pp. 1093–1110.
21. Geng, J., Ziolkowski, R.W., Jin, R. and Liang, X. (2012). Detailed performance characteristics of vertically polarized, cylindrical, active coated nano-particle antennas, *Rad. Sci.*, **47**, RS2013, doi:10.1029/2011RS004898.
22. Ziolkowski, R. W., Arslanagić, S. and Geng, J. (2012). Where high-frequency engineering advances optics: Active nano-particles as nanoantennas, in *Optical Antennas*, Eds. A. Alu and M. Agio (New York: Cambridge University Press) Chap. 4.
23. Zayats, A.V., Smolyaninov, I.I. and Maradudin, A.A., (2005). Nano-optics of surface plasmon polaritons, *Phys. Rep.*, **408,** pp. 131–314.
24. Bergman, D.J. and Stockman, M.I. (2003). Surface plasmon amplification by stimulated emission of radiation: Quantum generation of coherent surface plasmons in nanosystems, *Phys. Rev. Lett.*, **90**, 027402.
25. Klar, T Noginov, M.A., Zhu, G., Drachev, V.P., and Shalaev, V.M. (2007). Negative-index metamaterials: Going optical, *IEEE J. Sel. Topics Quantum Electron.*, **12,** pp. 1106-1115.
26. Noginov, M.A., Zhu, G., Drachev, V.P., and Shalaev, V.M. (2007). Surface plasmons and gain media, in *Nanophotonics with Surface Plasmons*, Eds. S. Kawata and V.M. Shalaev (Amsterdam: Elsevier), Chap. 5, pp. 141–169.
27. Noginov, M.A. (2008). Compensation of surface plasmon loss by gain in dielectric medium, *J. Nanophotonics*, **2**, 021855.
28. Sámson, Z.L., MacDonald, K.F. and Zheludev, N.I. (2009). Femtosecond active plasmonics: Ultrafast control of surface plasmon propagation, *J. Opt. A: Pure Appl. Opt.*, **11**, 114031.
29. MacDonald, K.F., Sámson, Z.L., Stockman, M.I. and Zheludev, N.I. (2009). Ultrafast active plasmonics, *Nat. Photonics*, **3**, pp. 55–58.
30. Li, D.B. and Ning, C.Z. (2009). Giant modal gain, amplified surface plasmon-polariton propagation, and slowing down of energy velocity in a metal–semiconductor–metal structure, *Phys. Rev. B*, **80**, pp. 153304.
31. Ramakrishna, S. and Pendry, J. (2003). Optical gain removes absorption and improves resolution in a near-field lens, *Phys. Rev. B*, **67,** 201101(R).

32. Sivan, Y., Xiao, S., Chettiar, U.K., Kildishev, A.V. and Shalaev, V.M. (2009). Frequency-domain simulations of a negative-index material with embedded gain, *Opt. Express*, **27**, 024060.

33. Dolgaleva, K., Boyd, R.W. and Milonni, P.W. (2009). The effects of local fields on laser gain for layered and Maxwell–Garnett composite materials, *J. Opt. A: Pure Appl.*, **11**, 024002.

34. Lagarkov, A.N., Kisel, V.N. and Sarychev, A.K. (2010). Loss and gain in metamaterials, *J. Opt. Soc. Am. B*, **27**, pp. 648–659.

35. Zheludev, N.I., Prosyirnin, S.L., Papasimakis, N. and Fedotov, V.A. (2008). Lasing spaser, *Nat. Photonics*, **2**, pp. 351–354.

36. Plum, E., Fedotov, V.A., Kuo, P., Tsai, D.P. and Zheludev, N.I. (2009). Towards the lasing spaser: Controlling metamaterial optical response with semiconductor quantum dot, *Opt. Express*, **17**, pp. 8548–8551.

37. Xiao, S., Drachev, V.P., Kildishev, A.V., Ni, X., Chettiar, U.K., Yuan, H.K. and Shalaev, V.M. (2010). Loss-free and active optical negative-index metamaterials, *Nature*, **466**, pp. 735–738.

38. Hill, M.T. (2010). Status and prospects for metallic and plasmonic nano-lasers, *J. Opt. Soc. Am. B*, **27**, B36–B44.

39. Noginov, M.A., Zhu, G., Bakker, R., Shalaev, V.M., Narimanov, E.E., Stout, S., Herz, E., Suteewong, T. and Wiesner, U. (2009). Demonstration of a spaser-based nanolaser, *Nature*, **460**, pp. 1110–1112.

40. Stockman, M.I. (2010). The spaser as a nanoscale quantum generator and ultrafast amplifier, *J. Opt. A*, **12**, 024004.

41. Bohren, C.F. (1982). How can a particle absorb more than the light incident on it? *Am. J. Phys.*, **51**, pp. 323–327.

42. Fischer, R. and Paul, H. (1983). Light absorption by a dipole, *Sov. Phys. Usp.*, **26**, pp. 923–926.

43. Wang, Z.B., Luk'yanchuk, B.S., Hong, M.H., Lin, Y., and Chong, T.C. (2004). Energy flow around a small particle investigated by classical Mie theory, *Phys. Rev. B*, **70**, 035418.

44. Bashevoy, M.V., Fedotov, V.A. and Zheludev, N.I. (2005). Optical whirlpool on an absorbing metallic nanoparticle, *Opt. Express*, **13**, pp. 8372–8379.

45. Bohren, C.F. and Huffman, D.R. (1983). *Absorption and Scattering of Light by Small Particles* (New York: John Wiley).

46. Johnson, P.B. and Christy, R.W. (1972). Optical constants of the noble metals, *Phys. Rev. B*, **6**, pp. 4370-4379.

47. Neogi, A., Morko, H., Kuroda, T. and Tackeuchi, A. (2005). Coupling of spontaneous emission from GaN-AIN quantum dots into silver surface plasmons, *Opt. Lett.*, **7**, pp. 93–95.
48. Song, J.H., Atay, T., Shi, S., Urabe, H. and Nurmikko, A.V. (2005). Large enhancement of fluorescence efficiency from CdSe/ZnS quantum dots induced by resonant coupling to spatially controlled surface plasmons, *Nano Lett.*, **5**, pp. 1557–1561.
49. Choy, W., Chen, X., He, S. and P. Chui, P. (2007). The Purcell effect of silver nanoshell on the fluorescence of nanoparticles, in *Optical Fiber Communication and Optoelectronics Conference*, Proceedings of the 2007 Asia, pp. 81–83.
50. Goodrich, G.P., Johnson, B.R. and Halas, N.J. (2007). Plasmonic enhancement of molecular fluorescence, *Nano Lett.*, **7**, pp. 496–501.
51. Jin, Y. and Gao, X. (2009). Plasmonic fluorescent quantum dots, *Nature Nanotech.*, **4**, pp. 571–576.
52. Noginov, M.A., Li, H., Barnakov, Yu. A., Dryden, D., Nataraj, G., Zhu, G. and Bonner, C.E., Mayy, M., Jacob, Z. and Narimanov, E.E. (2010). Controlling spontaneous emission with metamaterials, *Opt. Lett.*, **35**, pp. 1863–1865.
53. Pustovit, V.N. and Shahbazyan, T.V. (2012). Fluorescence quenching near small metal nanoparticles, *J. Chem. Phys.*, **136**, 204701.
54. Ashcroft, N.W. and Mermin, N.D. (1976) *Solid State Physics* (New York: Holt, Rinehart, and Winston).
55. Alú, A. and Engheta, N. (2008). Enhanced directivity from subwavelength infrared/optical nano-antennas loaded with plasmonic materials or metamaterials, *IEEE Trans. Antennas Propagat.*, **11**, pp. 3027–3039.
56. Arslanagić, S. and Ziolkowski, R. W. (2013). Jamming of quantum emitters by active coated nano-particles, *IEEE J. Sel. Topics Quantum Electron.*, **19**(3), 4800506.
57. Milton, G.W. and Nicorovici, N.-A. (2006). On the cloaking effects associated with anomalous localized resonance, *Proc. R. Soc.*, **462**, pp. 3027–3059.
58. Milton, G.W., Nicorovici, N.-A., McPhedran, R. C., Cherednichenko, K. and Jacob, Z. (2008). Solutions in folded geometries, and associated cloaking due to anomalous resonance, *N. J. Phys.*, **10**, 115021.
59. Alú, A. and Engheta, N. (2005). Achieving transparency with plasmonic and metamaterial coatings, *Phys. Rev. E*, **72**, 016623.

60. Alú, A. and Engheta, N. (2007). Plasmonic materials in transparency and cloaking problems: Mechanism, robustness, and physical insights, *Opt. Exp.*, **15**, pp. 3318–3332.

61. Alú, A. and Engheta, N. (2008). Plasmonic and metamaterial cloaking: Physical mechanisms and potentials, *J. Opt. A: Pure Appl. Opt.*, **10**, doi: 10.1088/1464-4258/10/9/093002.

Chapter 15

Plasmonic Nanostructures for Nanoscale Energy Delivery and Biosensing: Design Fabrication and Characterization

Remo Proietti Zaccaria,[a] Alessandro Alabastri,[a] Andrea Toma,[a] Gobind Das,[a] Andrea Giugni,[a] Salvatore Tuccio,[a] Simone Panaro,[a] Manohar Chirumamilla,[a] Anisha Gopalakrishnan,[a] Anwer Saeed,[a] Hongbo Li,[a] Roman Krahne,[a] and Enzo Di Fabrizio[b,c]

[a]*Nanostructures Department, Istituto Italiano di Tecnologia, via Morego 30, 16163 Genova, Italy*
[b]*KingAbdullah University of Science and Technology (KAUST), Thuwal 23955, Saudi Arabia*
[c]*BIONEM Laboratory, University of Magna Graecia, Campus S. Venuta, Germaneto, viale Europa, I88100 Catanzaro, Italy*
enzo.difabrizio@kaust.edu.sa

In this chapter, we present a set of applications that exploit surface plasmon (SP) generation, propagation, and concentration. We discuss how plasmonic properties of metallic nanostructures can be utilized in different fields, including nanoscale energy delivery, hot electrons generation, hybrid metal–semiconductor systems, and biosensing.

Singular and Chiral Nanoplasmonics
Edited by Svetlana V. Boriskina and Nikolay I. Zheludev

ISBN 978-981-4613-17-0 (Hardcover), 978-981-4613-18-7 (eBook)
www.panstanford.com

15.1 Introduction

The possibility to couple electromagnetic waves to surface electronic oscillations has been known for a long time. Nevertheless, only latest fabrication techniques give us the opportunity to translate theoretical predictions into practical applications. Moreover plasmonics, as the merger between photonics and electronics, offers new tools for investigating optical and electrical properties of matter at the same time. At this purpose, in Section 15.2, we present a device that couples and squeezes light by means of propagating SPs compression in a conical metallic nanostructure. The resulting instrument, set in contact with samples surfaces, can analyze their optical and electrical response on the nanometric scale in a stable configuration without the need of vacuum conditions.

Section 15.3 refers to the near-field coupling between the emitted light by colloidal core–shell nanorods to metallic nanowires. In this case, we make use of the relatively long range propagation of SPs to collect light at one end of a metallic nanowire, where nanorods are excited, and to detect the signal re-emitted at the other end, thanks to a low background noise. This study gives insights on the light-to-plasmon coupling in which randomly emitters, and not a focused laser beam, are considered.

In Section 15.4, the attention is brought to specific applications such as surface-enhanced Raman scattering (SERS). In this context, we exploit the high electric fields that localized plasmons (LPs) can promote in extremely confined nanovolumes. Since Raman scattering depends on the fourth power of the electric field, it is clear that electric field increasing may widely enhance the collected signal. Moreover, hot spots are generated only where the features of the structures get sharper; this provides great excitation selectivity, which is fundamental for biosensing applications.

Given hot spots importance, finally in Section 15.5, different antenna configurations are studied with the aim of understanding how coupled metallic structures behave and how the near-field response can be molded to obtain resonances in a specific range of frequencies. Increasing the length of antennas together with the aspect ratio, indeed, also THz regime can be reached and this can lead to various applications such as chemical identification and innovative imaging techniques.

15.2 A Plasmonic Device for Multidisciplinary Investigation

We address the problem of efficient electromagnetic field energy delivery at nanoscale and we propose a plasmonic device that makes use of photo-excited surface plasmon polaritons (SPPs) to harvest and subsequently deliver electromagnetic (e.m.) energy, either as a subdiffraction-limited bright photonic source or as a quasi DC ideal current source, offering nanometric spatial resolution in both cases. We report on the design, optimization, and characterization of such nanoplasmonic device.

Light waves coupling to free electron oscillations at the metal surface, that is, SPPs, overcome the fundamental propagation limit given by the electron's mean free path in the metal, providing a precious method to guide and localize e.m. [1, 2]. Tapered metallic structures allow the energy concentration at the nanoscale with minimal losses field [3–5]. The proposed device makes use of a grating coupler tailored upon a convergent geometry to exploit the e.m. field to SPPs coupling and a guiding cone structure to concentrate the e.m. field, hence acting as a subdiffraction-limited optical focusing lens or as a sensing electrode. When posed in contact with a planar semiconductor sample, it realizes a Schottky photodiode [6]. The point contact geometry at the nanoscale strongly departs on the 1D barrier potential model, leading to specific current to voltage characteristic that reflects the cylindrical symmetry of the electric problem [7–9]. The performance of such plasmonic concentrator is qualitatively described in terms of radiative and nonradiative losses for the excited SPPs. Radiated energy and excited energetic electrons can be observed in our originally developed optoelectronic setup. Validating the expectations, the photoemission and the photocurrent depend on light polarization orientation and result in linear proportion with the incident power, showing a conversion efficiency improvement for optimal grating coupling condition. We performed specific numerical simulations to obtain a realistic description of the device. We studied the dependence of the coupled energy in function of the grating parameters, that is, grooves periodicity, width, and depth. Obtained numerical results indicate an overall power coupling efficiency of about 5% to the downstream of the SPPs.

15.2.1 *Historical Perspective*

The theme of direct photon energy conversion into electrical energy by means of an efficient physical process draws back to the discovery of the photovoltaic effect in metal oxide crystals (solar cells of this type have a long history, dating to 1883, when Charles Fritts coated selenium with a thin layer of gold to make one of the world's first solar cells). Conventional technology, based on crystalline silicon properties, requires wafer thickness of hundreds microns; however, the energy band gap threshold and the minority diffusion lengths issue in the semiconductor pose severe limitations either to the intrinsic efficiency of the process or to the lower photon energy value [10, 11]. Practically, the costs of silicon materials and processing itself raise considerably the price production for large-scale application, determining the still relative tiny amount of energy created by solar energy extraction technology [12, 13].

Typical applications have focused on obtainable current through the optimization and engineering of semiconductor material and geometry; an appealing adopted technique uses surface texture or multilayer structures to increase the free effective path length in the cell [11, 14], but more often, budget reasons make acceptable a low efficiency value technology as low as 10%. Recent solutions, based on metamorphic materials [15] or thin cell [16], have disclosed new efficiency levels, reaching the theoretical combined-processes conversion efficiency limit for silicon-based cells up to 35–40% of solar power. In these cases, material-synthesis considerations and engineered structural properties are strongly dictated by opposing requirements for optical absorption thickness and carrier collection length.

Researchers have known for years that the efficiency of solar cells can be significantly improved by harvesting energetic electrons produced in the direct photo absorption process before they "cool down" and lose most of their energy. No matter what is the adopted technique, the common feature is the presence of nonequilibrium current carrier concentrations, an energy barrier that acts as the rectifying element in the circuit and a smart way to collect the charges before recombination takes place.

Relaxing the hypothesis of the semiconductor presence, new fundamental ideas are investigated today [17–19], and in this sense, a plasmonic approach is certainly the most promising. In fact, it naturally solves the challenging paradigm of an optical thick but physically very thin absorber. In this sense, it takes all the advantages of the dual nature of surface plasmons (SPs): electromagnetic waves confined laterally to the plane surface of a metal–dielectric interface well below the diffraction limit, a collective free electron excitation in a noble metal layer, aspects that are revolutionizing the opto-electronic devices design.

A further step ahead can be done moving from a planar 2D geometry to a full 3D structure, that it is known outperforming in terms of guiding, manipulating, and concentrating energy on the nanometre length scale. For such structures, it is easily possible to recognize new pathways to extract energy from the coupled e.m. field taking the advantage of new effective damping processes. For exemplum, the intrinsic limitations imposed by the linear momentum conservation for smooth Schottky contacts can be reduced if the diode is formed by a nanometric size rough interface that leads to the relaxation of momentum conservation, thus to much higher quantum efficiency for the plasmon to Schottky current conversion [20].

Recently, an adiabatic concentrator has been demonstrated to produce such enhancement of the internal photoemission process in a MS Schottky device, feature especially valuable for IR radiation excitation or for high band gap semiconductors investigation such as GaN and SiC, supposed to substitute the silicon in most demanding technology applications in the next future. The confinement of this photo voltage conversion, close to the point contact, and the efficient energy extraction from the coupled SPPs allowed the authors to propose the process as the basis of a novel scanning probe [21].

A plasmonic approach to the IPE process can open a new scenario: provided light coupling via Kretschmann configuration

[22] or structuring the metallic thin film's surface with a proper grating [23–25], propagating SPPs can flow with respect to the optical wave with own properties, with a propagation length up to tens of micron, depending only on the metal properties thus offering the potentiality to guide light toward the active photo conversion area, under external dark field illumination condition. Furthermore, controlling the surface roughness and structure geometry, it is also possible to stimulate local specific SPP damping channels [26–30], and it is possible to elect the radiative losses or the generation of hot electrons as the primary decay channel at the apex. It can happen stimulating the propagation of the TM or the HE mode, provided a device with the proper geometry.

Here, we want to describe a specific photoexcitable plasmonic device, characterizing it in terms of design and realization parameters for visible, 670 nm, radiation line. Then, we present the far-field emission measurements, followed by the current to voltage conduction characteristic recorded in the dark and under illumination condition when it was shined with a laser light directly on the coupler.

15.2.2 *Description of the Device*

The device is defined by a micrometric metallic cone with nanometric apex supported by an AFM tip provided by a grating coupler. The grating acts as unidirectional SPPs launcher allowing the conversion between light and SPs. Coupled SPPs (the energy reservoir) move down to the tapered guiding structure of the cone, allowing the concentration of energy at the nanometric scale of its apex where SPPs damp out.

The coupling surface used to harvest incident light into propagating SPPs was directly realized milling with focused ion beam (FIB) technique sub-wavelength grooves on the front facing part of a gold-coated pyramidal AFM tip (μ-masch snc38, n-type silicon tip, height $= 20\ \mu$m, full tip cone angle 40°). The sharp micrometric cone (height: 2500 nm, base radius: 300 nm) was realized with FIB-induced deposition technique, as a Pt–C cone structure with an apex radius of about 10 nm. Subsequently, the whole structure was coated with a metallic layer of controlled thicknesses by a plasma evaporation process of gold metal (~ 25 nm) providing a smooth coupling between tip and cone geometry.

15.2.3 *Design*

The challenge of efficient launcher from a freely propagating light has attracted increasing attention over the past years [24, 31, 32], either to realize an efficient unidirectional coupling/decoupling SPP [33] device or to achieve bright illumination in confined volume [25]. The coupling performance is usually described in terms of the launching efficiency of the SPP into the desired direction. It has been demonstrated that a global approach to the electromagnetic field scattered by a collection of groove can maximize the total SPP launching efficiency, up to a value exceeding 50% if the depths, widths, and locations of each groove are treated independently [33]. In our specific case, the efficiency for an equally spaced grating has been estimated via numerical approach. The results were in general agreement with those obtainable within an analytical model of the grating [24].

The grating-coupled excitation of SPPs undergoes the constraints of energy and momentum conservation laws. Thus, assumed a linear model that certainly holds here for the low-level energy density involved, the first condition fixes the frequency, while the second, known as the frequency-dependent SP wave-vector dispersion [34], determines the amount of momentum that must be provided to couple light and SP modes.

According to the scattering geometry of Fig. 15.1, we calculated the grating period a from

$$\frac{2\pi}{a} = \Delta K_z = K_z^{\mathrm{SPP}} - K_z^{\mathrm{Ph}} = \frac{\omega}{c}\sqrt{\frac{\varepsilon_\mathrm{d}\varepsilon_\mathrm{m}}{\varepsilon_\mathrm{d}+\varepsilon_\mathrm{m}}} - \frac{2\pi}{\lambda}\sin\theta_\mathrm{in} \qquad (15.1)$$

being ε_m and ε_d the frequency-dependent permittivity of the metal and the dielectric material [35], λ the free space wavelength, θ_in the incident angle in respect to the surface normal, z the SPP propagation direction. The calculation gives a grating period a ~1500 nm, at an incidence of 36° from the normal axis to the surface and for the wavelength of 670 nm. Efficient coupling also requires the accomplishment of optimal generation strength

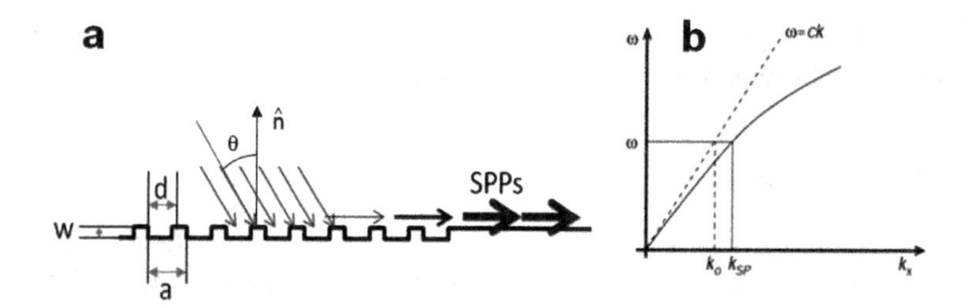

Figure 15.1 (a) Scattering geometry in which the four parameters object of the simulation are indicated. (b) The dispersion curve for an SP mode shows the momentum mismatch that must be overcome in order to couple light to SP modes.

that macroscopically depends on the duty cycle and depth of the grating grooves. The initial point for our 2D numerical optimization has been obtained following a literature analytical model that allows to calculate the scattering coefficients between an incident plane wave and the SPP modes, launched at the subwavelength metallic slit aperture [24]. Within the model, it is possible to determine the optimal generation strength in terms of the width to wavelength ratio, and the wavelength-dependent strength. Formally, the strength coefficients are expressed as the squared modulus of overlap integrals, between the scattered light and the SPP normal modes of the surface, practically the cited analytical expression stands on the observation that the SPPs generation results from a two-stage mechanism: a purely geometric diffraction problem followed by the launching of a SPP-bounded mode on a flat interface. Thus, to estimate the optimal grating parameters, we conducted iterative 2D simulations to maximize the coupled power, considering six grooves equally spaced. We swept routinely, one each time, all the three groove parameters: period, duty cycle, and depth for a fixed wavelength and geometry configuration, obtaining the results of Fig. 15.2. From Fig. 15.2a, we find a groove step of about 1550 nm accordingly with the result of (1. 1), while from Fig. 15.2b, we get an optimal width to wavelength ratio about 0.3, confirmed by the analytical methods of Ref. [24]. The groove depth appears the less critical parameter. All the simulations have been conducted considering a Gaussian beam defined with $E_0 = 1$ V/m on the axis. Fig. 15.2.3.1d shows the spatial intensity of the Poynting vector, the simulated launcher for the optimized grating parameters case. The

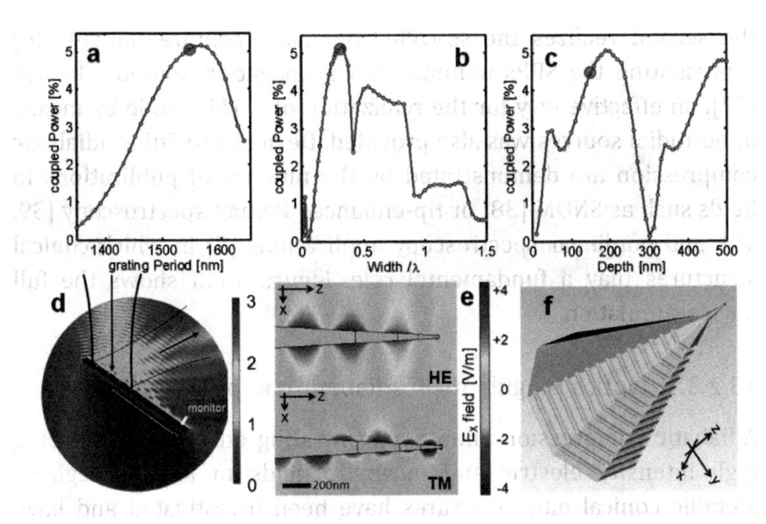

Figure 15.2 (a)–(c) Coupled power as a function of the groove period, the width to wavelength ratio, the groove depth for a wavelength of 760 nm, and an incidence angle of 36°, corresponding to the one adopted in this experiment. (d) Spatial intensity of the Poynting vector for the simulated grating of six grooves equally spaced; the parameters values are indicated as a red point in each of the upper insets (width = 220 nm, depth = 100 nm, groove period = 1550 nm). Arrows represent the main directions of the power flow. The position of monitor is indicated. (e) TM and HE x-component of the electric fields modes along the conical tip geometry, excited at 670 nm. In all cases, fields are normalized to the input amplitude, $E_0 = 1$ V/m. (f) Full device structure simulated, considering optimal value of the parameters.

monitor position extended along the metallic slab 5 μm apart the grating.

Figure 15.2e shows the component of the electric field on a symmetry plane of the cone at its terminus when an axially aligned plane wave or a radial source, respectively, impinges on the base of the cone. The simulated cone here reported has an apex radius of 10 nm (to get better evidence of the focusing phenomenon), a base $b = 300$ nm, and height $h = 2.5$ μm quite similar to the realized one. The mesh resolution considered for the present simulation was taken equal to 0.3 nm. The two pictures exemplify the fundamental difference between HE_1-like mode (dipole-like mode) and the TM_0 mode (radial mode) excited along a metallic conical structure: the first radiates energy progressively while reaching the apex, and the second realizes the so-called adiabatic compression [4, 36] representing the SPPs minimal loss propagation solution. In Ref. [37], an effective way for the realization of a TM_0 mode by means of no-radial sources was also provided. Devices providing adiabatic compression are demonstrated by the number of publications in fields such as SNOM [38] or tip-enhanced Raman spectroscopy [39, 40], and nonlinear spectroscopy applications [5] in which conical structures play a fundamental role. Figure 15.2f shows the full device simulation.

15.2.3.1 Optical singularity in adiabatic compression regime

Adiabatic compression allows concentrating energy and inducing high intensity electric and magnetic fields in narrow regions. Metallic conical nanostructures have been investigated and have been proved supporting this regime [3, 4, 41]. The localized near field obtained at the apex of such structures is extremely useful for sensing applications: devices exploiting adiabatic compression were fabricated and tested giving reliable results [39, 40]. Therefore, it is important to study the electromagnetic field behavior in such extreme conditions. In fact, the total electric field above the tip end is, in general, given by different contributes: the incident light, the scattered light by the whole structure, and the near field generated by SPPs" decay at the apex. In such conditions, the local fields may undergo optical singularities [42, 43]. In particular for a conical metallic nanostructure, it was found that destructive interference between incident and scattered field leads to a phase singularity, see Fig. 15.2a.4. The figure shows how a tilted incident wave (at 37° in this case) can induce adiabatic compression while a wave at 0° cannot. The tilting is measured as the angle between the cone's axis and the direction of propagation of the incident light starting from the base of the cone. In order to show the different behavior in the two cases, a parameter was introduced:

$$U = \frac{(|E\ (37^\circ)| - |E\ (0^\circ)|)}{(|E\ (37^\circ)| + |E\ (0^\circ)|)} \qquad (15.2)$$

where E is the electric field calculated at the edge of the cone, from the base to the apex for the two incident angles. From Fig. 15.3a,

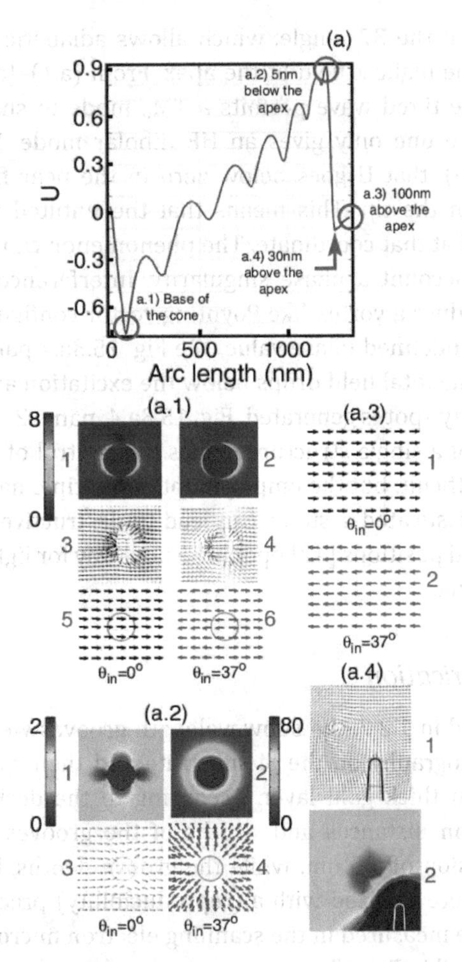

Figure 15.3 (Color online) (a) The U-profile along the cone edge is plotted. (a.1) Panels 1, 2: Norm of total (incident + scattered) electric field for the excitation wave at angle θ_{in} equal to 0° and 37°, respectively. Panels 3, 4: Vectorial total electric field representation for the 0° and 37° excitation waves, respectively. Panels 5, 6: Vectorial incident electric field representation for the 0° and 37° excitation waves, respectively. (a.2) Scalar and vectorial representation of the total electric field calculated for excitation waves at 0° and 37°. The TM0 and HE_1-like mode shapes are easily recognized. (a.3) Total vectorial field calculated 100 nm above the cone tip for the two excitation angles 0° and 37°. (a.4) Panel 1: Poynting vector around the cone tip. Panel 2: Norm of the total electric field. Both figures refer to a 37° excitation wave. Above the apex the total field drops below the excitation amplitude. The phase of the optical field is undefined and the Poynting vector goes to a vortex-like shape [41].

it is clear that the 37° angle, which allows adiabatic compression [37], gives the highest field at the apex. From (a.1)–(a.3), it can be seen how the tilted wave permits a TM_0 mode to shape on while the 0° degree one only gives an HE dipolar mode. Nevertheless, we see, in (a), that U goes below zero in the near field at about 100 nm from the tip. This means that the untilted wave gives a higher signal at that coordinate. The phenomenon can be explained taking into account a phase singularity. Interference in the near field can produce a vortex-like Poynting vector configuration which leads to an undefined phase value, see Fig. 15.3a.4 panel 1. In such conditions, the total field drops below the excitation amplitude and a low-intensity spot is generated, Fig. 15.3a.4, panel 2.

The use of a single structure makes the control of the phenomenon very difficult, but the employment of multiple nanostructures in more sophisticated systems can lead to destructive interference at pre-defined positions [42] opening a new way for light generation and processing.

15.2.4 *Fabrication*

As introduced in 1.2.2, the subwavelength grooves were fabricated with FIB lithography on the Si substrate and then sputtered with a 80–100 nm thick gold layer. According to the designed values, the separation distances and widths of the grooves were milled with a precision of ± 5 nm, while the groove depths, being harder to control, were realized with a (reproducibility) precision of ± 15 nm, as can be measured in the scanning electron microscope (SEM) image reported in Fig. 15.4.

Fabrication errors have an unavoidable impact on the experimental value of η. We estimated it by means of the mapped numerical results. For fixed groove spacing, equal to the nominal value, we found that a 10% overall error on all the groove widths or depths degrades the coupling efficiency but keeping it around $\sim 3\%$. Thus, we can realistically expect an experimental value of η of the order of the computed value of about 4%, Additionally, experimental misalignment could be also a critical point, as reported in literature [25], a 2–3° mismatch between the laser-grating alignment is expected to decouple the device.

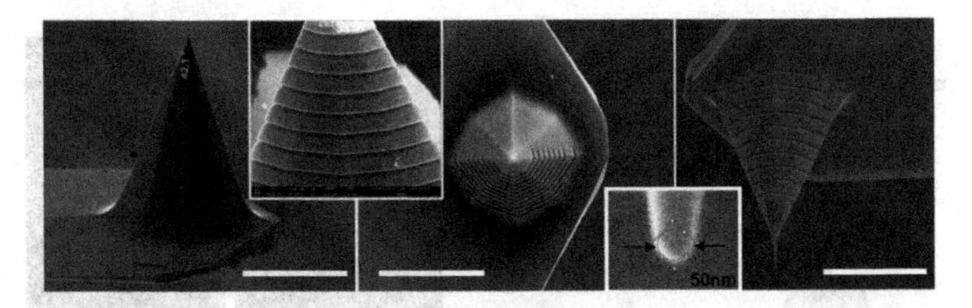

Figure 15.4 Scanning electron micrographs of the device illustrating the fabrication steps. The grating, used for nonlocal excitation of the tip apex, and the sharp cone were prepared by focused ion beam milling and induced deposition techniques, lead and followed by gold sputtering coating procedure. Unit bars indicate the 10 m length. Zoomed-in image from the apex of the coated cone demonstrate a 25 nm tip radius.

15.2.5 *Characterization*

To characterize the plasmonic properties of the device, we adopted an optical methodology. In particular, we analyzed the polarization state of the outcoming far field light from the apex of the cone with the optical setup sketched in Fig. 15.5a. It makes use of a linearly polarized laser source (670 nm—the one that match the grating for the SPP coupling) a half wave retarder to fix the polarization state of the incoming light in respect to the axis of the device. Two microscopy objectives were used for focalize the light on the grating and to collect it form the tip focal plane, a linear polarizer was used to analyze the radiated light. We identified the fingerprint of radial mode in the two lobes spot of Fig. 15.4e,f. Quite similar to the spherical emitting source case, these images show symmetric patterns in respect to the polarization analyzer axis, although the polarizer can be rotate arbitrarily around its axis. The far field optical propagation properties are inherited from the polarization state of the SPP mode in the tapered structure that propagate toward the dielectric bulk. This observation lets identify the propagated SPP with a TM_0 mode. This is illustrated in Fig. 1.4 d–f, where we reported the spots imaged on a CCD when the tip was observed through (or without) a polarization analyzer in two mutual normal configurations, namely V and H (U indicates the non polarized case). These results were obtained when excitation was polarized in the vertical (V) plane. For a horizontal (H) laser polarization it is still

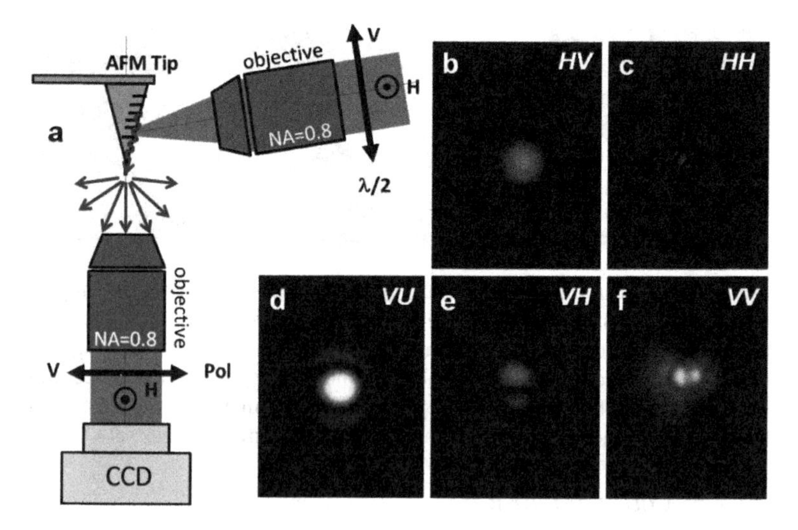

Figure 15.5 (a) Sketch of the optical set up. Two long working distance objectives (50×) in ~100° geometry supply the laser excitation to the grating and the collection of the scattered light to a CCD detector. (b, c) HV and HH geometries reveal the HE_1 mode. (d–f) VU, VV, and VH geometries identifying the TM_0 mode.

expected a SPP coupling through the grating which substantially maintain the polarization characteristic of the source propagating down to the structure as a HE_1 mode. This is confirmed by images of Fig. 15.5b,c that identify a substantially linearly polarized mode.

15.2.6 *Conclusions*

We projected and realized an efficient plasmonic device capable of collect and subsequently deliver energy at nanoscale, either as radiated photons or as emitted (through a Schottky barrier) energetic electrons. We provided the optical and electric response characterization methods leading to different practical applications.

It's worthwhile mention that this plasmonic approach to AFM spectroscopy introduces a quite new tool for chemio-physical semiconductive devices investigation.

Furthermore, the mapping of the photoelectric signal recorded with a spatial resolution close to the tip contact diameter could be obtained with unsurpassed S/N ratio, thanks to the charges confinement and to the large amount of hot electrons generated, both properties specific of point contact geometry with nanoscale dimensions.

15.3 Coupling Colloidal Nanocrystal Emission to Plasmons Propagating in Metallic Nanowire Structures

Plasmonics in metallic nanostructures is generating increasing interest for the development of optics that beat the diffraction limit of standard dielectric optical components. In this respect photonic circuits based on surface plasmon–polariton (SPP) propagation in metal nanowires have been explored and the exciton-plasmon coupling mechanisms have been investigated, using semiconductor [44–46] or metal [47] nanoparticles positioned in the vicinity of metal wires. Another approach consisted in photonic–plasmonic routing based on SnO_2 nanoribbons and Ag nanowires [48], and the electrical detection of the near-field coupling between a Ag and a Ge nanowire has been demonstrated [49]. In particular, quantum dot emission from colloidal semiconductor nanocrystals has been successfully coupled into Ag nanowires, and both experimental and theoretical investigations have shown that the emission properties of QDs can be significantly modified near metallic nanostructures [45], that is within the range of the evanescent SP mode tail. Here the spontaneous emission rate coupled into SPs is proportional to $(\lambda/d)^3$ [50], while the free-space (far-field) emission rate can be enhanced by at most a factor of four. Thus, for an optimally placed emitter, the spontaneous emission rate into SPs can far exceed the radiative, which results in highly efficient coupling to SPs and enhancement of the total decay rate compared to that of an uncoupled emitter. Radiative energy transfer from small core (donor) CdSe/ZnS QDs to large core ones was achieved via Ag nanowire arrays with several hundred nm in nanowire length [51].

In this section we will explore the near-field coupling of the light emitted from ensembles of nanorods to SPPs in line-shaped gold nanowires. To this aim we positioned "dot-in-a-rod" colloidal core–shell nanorods in the vicinity of one tip of Au nanowires with 150 nm width and several microns in length fabricated by electron beam lithography. The exciting laser beam was coupled off-axis into the confocal imaging optics, which allowed us to illuminate the rod-functionalized tip of the nanowire while recording the emission from the opposite end. Spatially resolved conventional confocal luminescence mapping was used to verify the presence of nanorods solely at one tip of the nanowires. The architecture of the nanorods was chosen such that their absorption was in the UV–blue spectral range and the band edge emission of the core material in the range between 600–650 nm. The nanorods were excited locally by laser light in resonance with the band gap of the shell, and the far field light emission at the opposite end of the nanowire was detected. By this approach we were able to spectrally decouple the laser excitation from the band in which the plasmon coupling occurred. The experimental data will be compared with the results obtained from finite elements (FEM) calculations on the propagation of SPPs in the metallic nanowires.

15.3.1 *Optical Properties of Rod-Shaped Colloidal Core–Shell Nanocrystals*

Colloidal CdSe/CdS core–shell nanorods with a dot-in-a-rod architecture [52] were used as light emitters, since this material shows very bright and stable emission and because the emission wavelength can be tuned via the size of the CdSe core [53]. The optical absorption and emission spectra of such nanorods is displayed in Fig. 15.6, where the onset in absorption at 500 nm is dominated by the CdS shell material, while the weaker absorption of the CdSe cores can be noted in the regime from 650–500 nm. The strong photoluminescence centered at 650 nm is slightly red-shifted with respect to the ground state absorption due to the Stokes shift. For a detailed review on the optical properties of nanorods see Chapter 2 in Ref. [54]. The core–shell architecture and the associated energy levels in the valence and conduction band are sketched in the inset of Fig. 15.6.

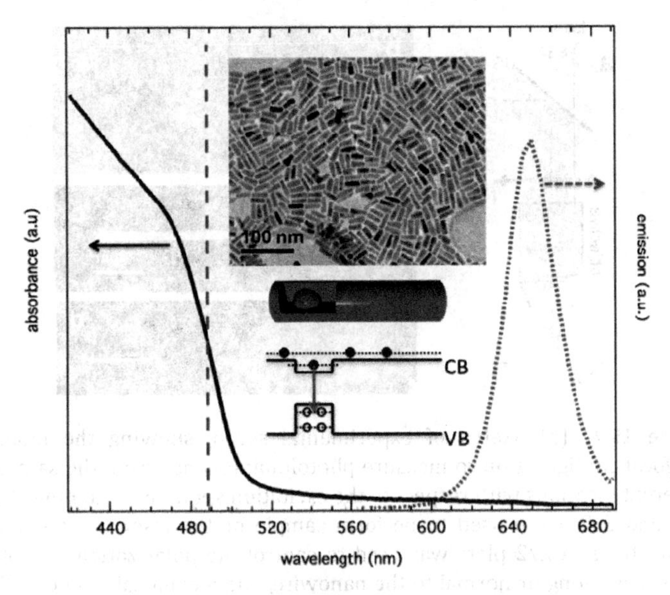

Figure 15.6 Optical absorbance (black solid line) and emission (red-dotted line) recorded from colloidal CdSe/CdS nanorods in solution with a spectrophotometer (Cary 500) The blue dashed lines indicate the wavelength of the linearly polarized laser used in the coupling experiments (ArKr laser lines 488 nm and 647 nm). The inset shows a transmission electron microscopy (TEM) image of the nanorods and sketches the dot-in-a-rod architecture.

15.3.2 *Device Fabrication and Experimental Setup*

The Au nanowires with length $L = 5$ μm and width $d = 150$ nm were fabricated with electron-beam lithography (EBL) and metal evaporation of Ti/Au with 5/60 nm thickness. Then micron size areas in the vicinity of the nanowire tips were patterned by EBL exposure of 300 nm thick PMMA resist. In the next step the nanorods were deposited from toluene solution via drop casting, and excess nanorods were removed by lift-off of the PMMA layer in acetone. Figure 15.7c shows a scanning electron microscopy (SEM) image of a typical device structure where the nanorods were positioned near one tip of the Au nanowire.

The experimental setup for studying the optical emission of the nanorod–nanowire system (displayed in Fig. 15.7a) is based on a

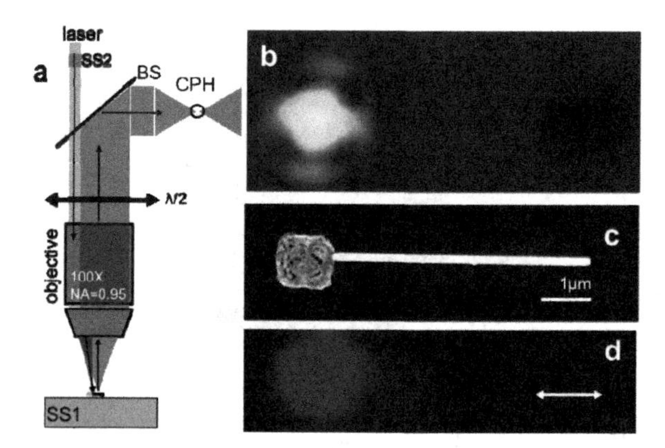

Figure 15.7 (a) Sketch of experimental setup, showing the modified confocal configuration to measure photoluminescence from the sample at different locations with respect to the excitation spot. Two scanning stages, SS1 and SS2, were used to perform sample or the laser point scanning acquisitions. A $\lambda/2$ plate was used to control the polarization axis of the laser light along or normal to the nanowire, and a confocal pin hole (CPH) was inserted to get spatial resolution better than 0.6 μm (b) Bright field optical image of the nanowire region (the size of the unsaturated laser spot is smaller than 0.7 μm). (c) SEM images of the nanowire with nanorods at one end (d) Conventional confocal photoluminescence map of the same region as displayed in (c) demonstrating that nanorods are solely located at the left end of the nanowire.

modified confocal microscope with two scanning stages with nm precision. One (SS1) was used in confocal imaging mode to map the region around the nanowires in 250 nm steps in both x- and y-directions of the sample surface plane, the other (SS2) allowed to scan or set the position of the laser spot in off-axis positions within the field of view of the $100\times$ objective lens. In particular, this configuration enabled laser excitation ($\lambda = 488$ nm) at the left end of the nanowire where the nanorods were deposited, and the recording of the emission spectra at the opposite nanowire end.

15.3.3 *Theoretical Modeling of Plasmon Propagation in Au Nanowire*

The computational simulations were performed by using the FEM method (Comsol Multiphysics) with perfectly matched layer boundary conditions. The cell volume size was shaped as a cylinder (height: 6 μm, diameter: 1 μm) coaxial with the nanowire. The nanowire (length: 5 μm, width: 150 nm, thickness: 60 nm) was excited via a Gaussian e.m field (beam waist: 1 μm, amplitude: 1 V/m) set on one wire end representing the rods emission. The norm of the electric field generated by SPPs was monitored at the opposite end at 10 nm distance from the nanowire edge. Sharp edges were smoothed with a 5 nm fillet in order to avoid unphysical peaks in the near field.

The refractive index of the surrounding medium was set to 1 and the complex dielectric constants of the Au were taken from Ref. [55].

15.3.4 *Experimental Results*

We excited the nanorod ensemble at one end of the nanowire with a focused laser at 488 nm, in resonance with shell absorption of the nanorods (see Fig. 15.6), stimulating their emission from the core levels at longer wavelength of 650 nm (full-width half-maximum FWHM = 31 nm). The nanorod emission coupled to SP modes in the metal, was transmitted through the nanowire and finally outcoupled to far-field radiation at the opposite end, as illustrated in Fig. 15.9a. The emission spectra recorded from one end of the nanowire with nanorods (position 1) and at the opposite end (position 2) are depicted in Fig. 15.9b. We clearly observed appreciable emission at the opposite end with a peak position that was slightly red-shifted with respect to the nanorod emission peak. This red-shift can be explained by convolution of the nanorod emission with the transmission dispersion of the nanowire, and indeed the transmitted P-polarized spectrum can be well reproduced by multiplication of the dispersion represented by the dashed black line with the emission spectrum depicted in red. Furthermore we observed that in P-polarization the intensity of the transmitted light was significantly stronger which points to better coupling of the exciton emission from the rods to the SPP modes with respect to S-polarization. Coupling will be dominated by near-field properties since the nanorods are located in only few tens of nanometres distance from the nanowire tip. The above described polarization dependence can be rationalized by the spatial

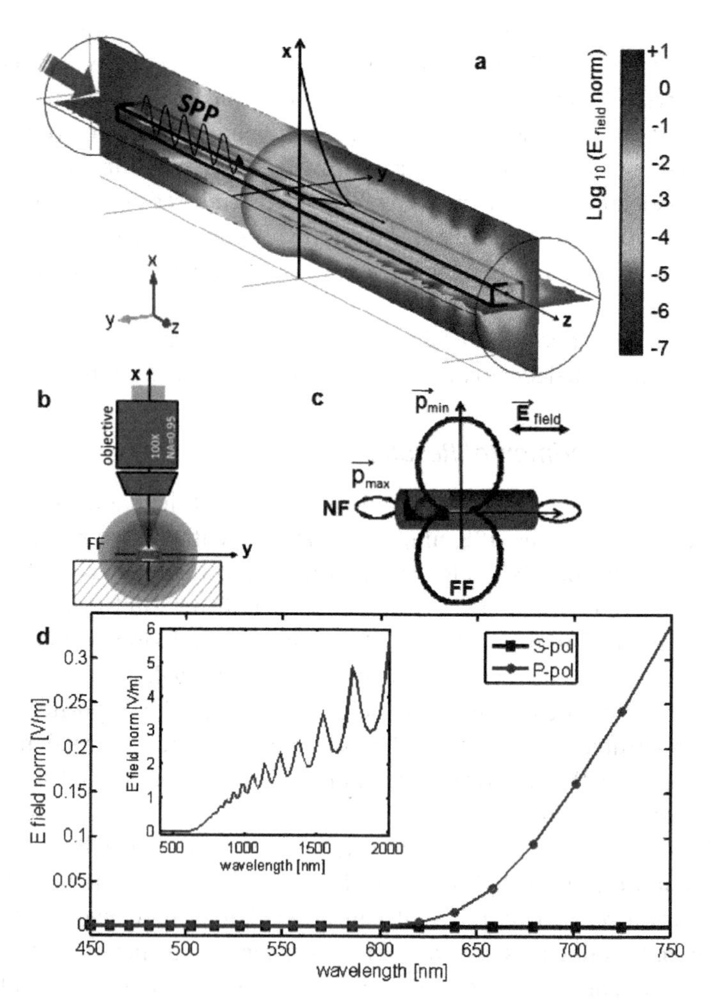

Figure 15.8 Aspects of the exciton–SPP–photon coupling process exploited via interaction of core–shell nanorods with a gold nanowire. Exciton relaxation to SPP is ruled by near field interaction, while SPP decoupling at the far end of the wire, is described via far field emission. (a) FEM simulation of a $L = 5$ μm long $d = 150$ nm wide wire when excited by a Gaussian beam at one end. (b) Directional emission pattern in respect to the axis of the nanowire at the emitting end. (c) Scheme illustrating the directional emission of a nanorod for near field (NF) and far-field (FF). (d) Electric field at the emitting far end as function of the polarization of the source respect to the wire axis.

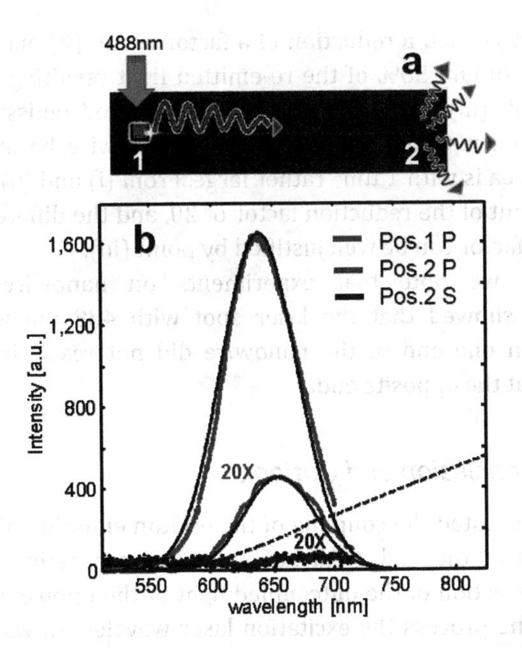

Figure 15.9 (a) Illustration of the plasmon excitation at position 1, propagation and outcoupling at position 2. (b) Experimental emission spectra from the nanorods at the left end of the nanowire (red markers), and recorded at the opposite end of the nanowire for P- (blue markers) and S- (black markers) polarization. The black dashed line shows the calculated wavelength dispersion for the transmitted light with P-polarization. The emission in P-polarization at the opposite nanowire end (blue) can be well reproduced by convoluting the nanowire transmission (black dashed) with the nanorod emission (red). P and S polarization refer to parallel and perpendicular directions with respect to the long axis of the nanowire, respectively.

near-field emission intensity distribution of the nanorods, as illustrated in Fig. 15.8 b, which results in effective coupling for nanorods oriented along the long axis of the nanowire. Furthermore the nanorod absorption and emission is known to be significantly enhanced in parallel polarization [52, 54, 56]. Comparing the emission intensities between positions 1 and 2 (i.e., of the light emitted by the rods and the light transmitted and collected at the other nanowire end) we obtain a reduction of a factor 80, which is reasonable considering the following rough estimation: (i) simulation yielded a reduction of a factor of 10; (ii) our optics can collect maximum 50% of the re-emitted light resulting in another factor of 2; (iii) only a fraction of the nanorod emission can be expected to couple to the plasmons in the nanowire, because the rod covered area is with 1 μm^2 rather large. From (i) and (ii) we obtain a lower limit of the reduction factor of 20, and the difference to the observed factor can be well justified by point (iii).

Finally we note that experiments on nanowires without nanorods showed that the laser spot with 488 nm wavelength focused on one end of the nanowire did not result in any light emission at the opposite end.

15.3.5 *Conclusion and Outlook*

We demonstrated the coupling of the exciton emission of nanorods positioned at the end a Au nanowire to propagating SP modes and the detection of the outcoupled light at the opposite nanowire end. For the process the excitation laser wavelength was tuned to the absorption of the nanorod material in the UV–blue spectral band, while the coupling occurred in the red triggered by the nanorod emission. This approach enabled the decoupling of the spectral bands of laser excitation and light transmission thanks to the core–shell architecture of the nanorods. The EBL defined fabrication allows for precise control over the position and shape of the Au nanowire. This will allow further investigations with more complex geometries that promise deeper insight into the coupling phenomena.

15.4 Plasmonics SERS Devices

Recently, there has been much attention for surface enhanced Raman scattering (SERS) because of its potential application in various fields [57–60]. The generation of SPs, due to the interaction of light and metal surface electrons, causes the focusing of hot spots, which depend on size, shape, interparticle gap, dielectric constant of the surrounding medium [61–65]. These parameters should be chosen carefully in order to achieve significant plasmon focusing, that is, enhanced Raman. Various SERS substrates were fabricated by means of different techniques such as chemical and lithographic technique [39, 40, 62, 66–68]. Since, chemical route doesn't guarantee the reproducible fabrication of SERS substrates so lithographic methods were opted.

Two different kinds of nanostructured plasmonic devices were fabricated (i) nanocuboids and (ii) nanostars. Various characterization techniques, such as optical absorption measurements to understand plasmon resonance, AFM and SEM measurements for morphological understanding, theoretical simulation on the fabricated structures in order to understand the electric field distribution and SERS measurements in order to demonstrate the sample as a biosensor, were performed over fabricated devices to optimize the geometrical parameters of the nanostructure and to fabricate a SERS device with remarkable enhancement factor.

15.4.1 *Nanocuboids*

Au nanocuboid structures of 4×4 array with varying edge size from 40 nm to 70 nm were fabricated by using electron beam lithography (EBL–Raith 150 two). The thickness was kept fixed as 25 nm for all structures and interparticle gap was around 20 nm. Polymethyl methacrylate (PMMA, Micro Chem Corp., 950 K) was spin-coated at 3000 rpm for 60 s onto the cleaned n-type c-Si (100) wafer and, thereafter, pre-baking was done at 180°C for 8 min on a hot plate. Nanocuboid features were patterned by using EBL machine operated at 30 kV accelerating voltage with 600 $\mu C/cm^2$ dosage and 20 pA current. Once the exposure was complete, the resist was developed in a solution of methyl isobutyl ketone (MIBK) and isopropanol (IPA) of 1:3 mixture ratios at 4°C for 40 s W. To improve adhesion of Au over Si, 3 nm Ti was evaporated before deposition of 22 nm Au film by electron beam evaporation (Kurt J Lesker system) with a low deposition rate of 0.3 Å/s. Au nanocuboid structures were obtained by performing ultrasonically assisted lift-off process. Finally, oxygen plasma treatment was performed to remove any organic contamination and residual resist.

SERS measurements were carried out by means of Renishaw InVia microscopy, using excitation laser 633 nm (power 0.55 mW

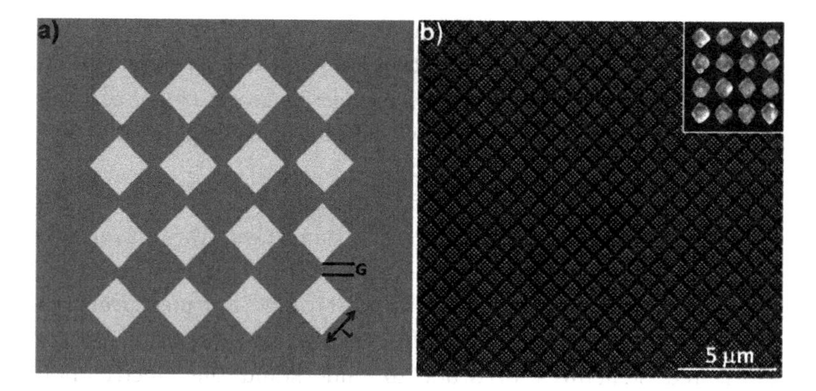

Figure 15.10 (a) Schematic representation of nanocuboid array SERS device. (b) Panoramic SEM image of 4 × 4 SERS substrate. In the inset, a zoomed SEM image of 4 × 4 nanocuboid device is also demonstrated.

and acquisition time 20 s) and 150× objective (NA 0.95). The spectral resolution of the instrument was about 1.1 cm^{-1}. R6G molecules (1 μM) were chemisorbed over the device as a Raman probe for the SERS enhancement investigations.

Au nanocuboid arrays with various edge sizes from 40 to 70 nm were fabricated by EBL in step of 5 nm. Normal-incidence SEM images were acquired on the fabricated samples. Figure 15.10a shows the schematic representation of the cuboid nanostructures with edge size, 'L' and interparticle separation 'D'. In particular, SEM image of 4 × 4 arrays with $L = 70$ nm and $D = 20$ nm is reported in Fig. 15.10b. Reference Raman measurements were carried out on Au markers, fabricated on the same substrate.

In order to investigate the electric field distribution for nanocuboid structures, theoretical calculations were performed. In particular, the gold nanocuboids were designed to resemble the fabricated structures and comparative simulations were performed using CST software, keeping all simulation parameters equal in both cases. Electric field distribution on both kinds of structures are shown in Fig. 15.11. It can be observed that the SPs were focused at the opposite corners of the nanostructures. In addition, it is also found that the electric field reaches up to about 26 V/m for sharp edge cuboid structure against about 10 V/m for truncated cuboid.

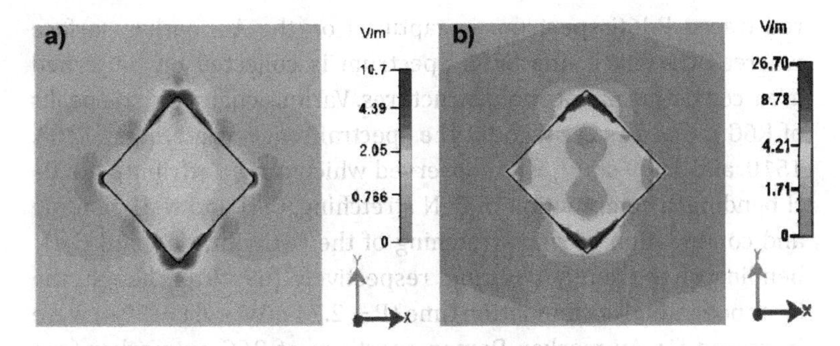

Figure 15.11 Theoretical simulation for nanocuboid structure with and without truncated corners. (a) Nanocuboid structure, resembling the fabricated structures, whereas (b) the ideal condition.

Raman measurements were firstly performed at different position of the patterned area without depositing any molecule, shown in Fig. 15.12a. Figure shows no Raman features. In order to understand the effect of Au nanocuboid size on SERS enhancement, R6G molecules were deposited on Au cuboid arrays and on Au marker, by means of chemisorption technique. SERS spectrum and the reference Raman spectrum of Rd6G were reported in Fig. 15.12b in the range of 1080–1750 cm^{-1}. This optical range was selected to avoid the c-Si Raman peak centered at 521 cm^{-1} (first order) and the broad band around 965 cm^{-1} (second order). In the figure,

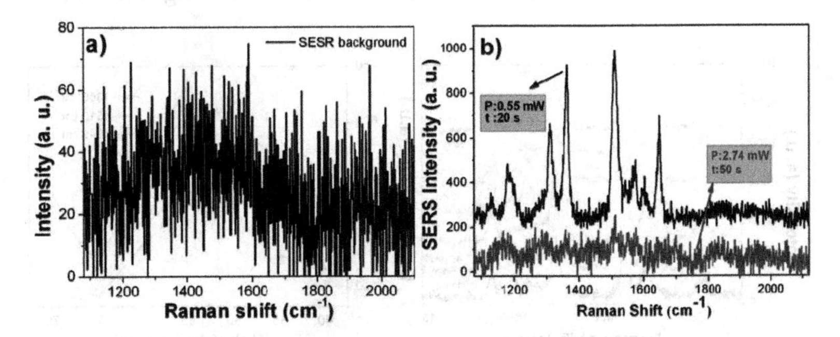

Figure 15.12 (a) Background measurements on patterned area without any molecule over the nanostructures. (b) SERS spectrum of R6G molecule on patterned (black) and unpatterned area (red). Measurements parameters are mentioned in the figure.

red traced Rd6G spectrum is captured on the Au-marker surface whereas the black line SERS spectrum is collected on patterned area, consisting nanocuboid structures. Various characteristic peaks of R6G molecules throughout the spectral range, centered at 1360, 1510, and 1649 cm^{-1}, were observed which can be attributed to C–H bending, a combination of C–N stretching, C–H and N–H bending and combination of ring stretching of the C–C vibration and C–H_x bending of the xanthenes ring, respectively [69, 70]. Though, the laser power and accumulation time ($P = 2.74$ mW and $t = 50$ s) were increased for Au marker, Raman spectrum of R6G on marker (red trace) was found to be very noisy without any distinguished spectral features.

SERS measurements were also performed on patterned nanocuboids with varying cuboid side length, shown in Fig. 15.13a. The variation of different peaks' intensity is clearly observed. Figure 15.13b shows the variation of SERS intensity for different peaks with cuboid size. SERS intensity is found to be increasing with cuboid size in the range from 40 to 55 nm and then decreasing with further growth of the nanostructure size. The optical excitation of LSPRs in metal nanoparticles is tunable throughout the visible and near-infrared region of the spectrum simply acting on the nanoparticle size. Therefore, we can assume that the plot observed in Fig. 15.13b corresponds to the resonant behavior of the cuboid nanostructure centered about $L = 55$ nm. In fact, the excitation of LSPR is one of the main mechanisms allowing for signal

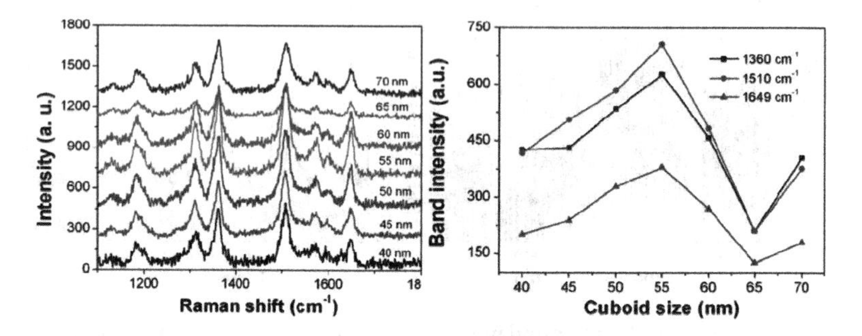

Figure 15.13 (a) SERS spectra of all SERS device with varying cuboid edge size from 40 to 70 nm. (b) SERS intensity for the bands 1360, 1510, and 1649 cm^{-1}.

enhancement in Raman measurements [65]. A similar behavior was also observed in the past for nanocylinder structures [61]. SERS enhancement factor was calculated by considering the laser beam focal radius of 1 μm that the molecules were deposited only onto metal surface, that the laser and that the molecules were closely packed, the SERS enhancement factor for the reference band centered at 1510 cm^{-1} can be estimated to 10^4 with respect to the Au marker [71].

In order to correlate the SERS and the electric field enhancement of the cuboid nanostructures, numerical simulations were performed on 4 × 4 array of 55 nm size Au cuboid. Convergence mesh analysis was applied in order to reach stationary results. Au and Si dielectric constant employed in the present simulation were $\varepsilon_{Au} = -9.79 + 1.97i$ and $\varepsilon_{Si} = 15.21$, respectively [72]. These values were taken for $\lambda = 633$ nm. Figure 15.14 shows the amplitude of the electric field distribution, |E|, on the *xy*-plane when a source was polarized along *x*-direction. In particular, a convergence analysis was adopted in order to assure the validity of

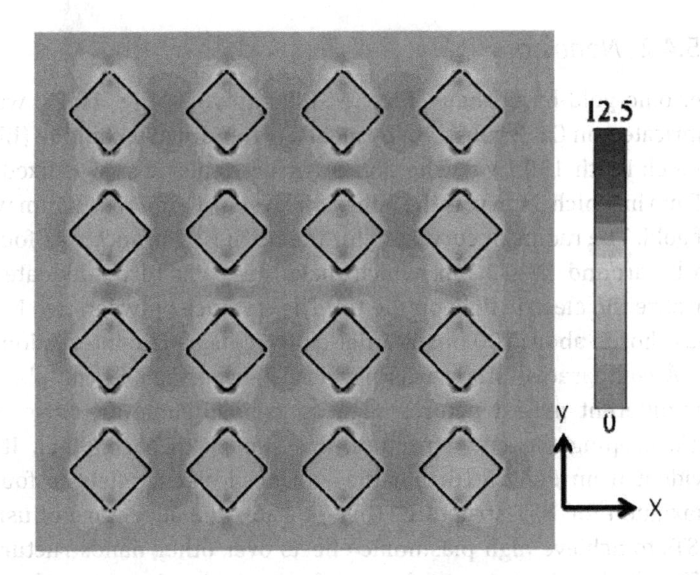

Figure 15.14 CST simulation for electric field distribution on 4 × 4 nanocuboid array with edge size 55 nm and gap 20 nm. The incident light polarization is along *x*-axis.

Figure 15.15 Periodic Au-NSTs SERS device. In the Inset single nanostar is shown.

the results. The distribution of the electric field presents a maximum field enhancement about 12.5 localized at the opposite corners of the nanocuboids.

15.4.2 *Nanostars*

Periodic gold-based nanostars (NSTs), shown in Fig. 15.15, were fabricated on CaF2 substrate by means of top-down technique (EBL, model: Raith 150 two). The nanostructures thickness was fixed to 25 nm in which 3 nm was the adhesion layer of Ti and rest 22 nm was of gold. The radius of curvature for the tip of NST branch was found to be around 10–12 nm, which make this structure adequate to localize the electric field on the tip. The distance between two NSTs was chosen about 100 nm in order to avoid near-field interaction.

A comparative study was performed by using CST simulation for different nanostructures such as, nanoantenna, nanodisc and NST, keeping the nanostructure size similar to each other. It is evident from Fig. 15.16 that the enhanced electric field is found maximum for NST structure. This indicates the advantage of using NSTs to achieve high plasmonic effects over other nanostructures. CST calculations were performed for excitation laser wavelength of 633 nm for all the nanostructures. The maximum electric field reached to around 6.5 V/m for NST structure from around 2.3 V/m

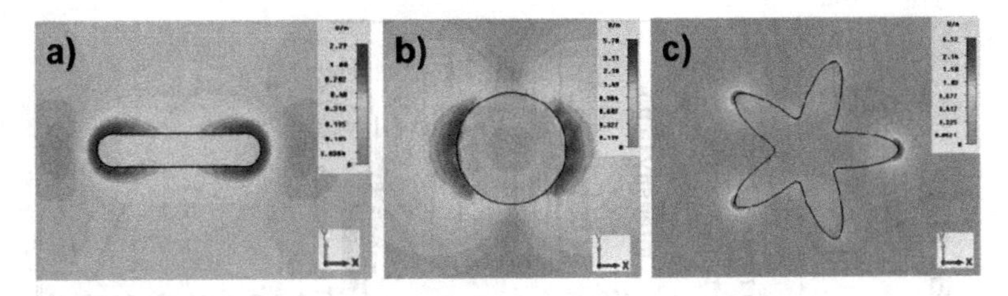

Figure 15.16 Comparative simulation of electric field distribution for different nanostructures. Reproduced from G. Das, M. Chirumamilla, A. Gopalakrishnan, A. Toma, S. Panaro, R. Proietti Zaccaria, F. De Angelis, E. Di Fabrizio. Plasmonic nanostars for SERS application, *Microelectron. Eng.*, **111**, pp. 247–250, Copyright (2013), with permission from Elsevier.

for nanoantenna plasmonic structure. Focusing of electric field is found very close to NST tip.

After confirmation from simulation results regarding the advantage of NST structure, optical transmission and SERS measurements were performed for the periodic NSTs. In the past, few research works were reported for gold NSTs, which were produced by chemical technique [73]. Results report the variation of plasmonic behaviors, depending on the polarization [68] and by changing the molecules [74]. Herein, a reproducible periodic gold NSTs based SERS device was reported. The optical transmission spectrum (inset of Fig. 15.17) shows its plasmon resonance, centered around 900 nm. SERS background measurement without any molecules was investigated, shown in the inset of same Fig. 15.17. It shows a featureless Raman spectrum, confirming the SERS nanostructures contamination free. Cresyl violet (CV) is deposited over nanostructures by means of chemisorption technique. SERS spectra of CV, deposited over NSTs device and on Au marker, are shown in Fig. 15.17. Various characteristic Raman bands of CV molecules can be found throughout the experimental range [75, 76]. In the figure, it can be found the Raman signal enhancement when deposited over patterned area, though the laser power and integration time are reduced with respect to the SERS measurement on marker. Taking into consideration all the experimental parameters and the number of molecules deposited close to NST tip, SERS enhancement factor is 2.80×10^4 with respect to the Au marker.

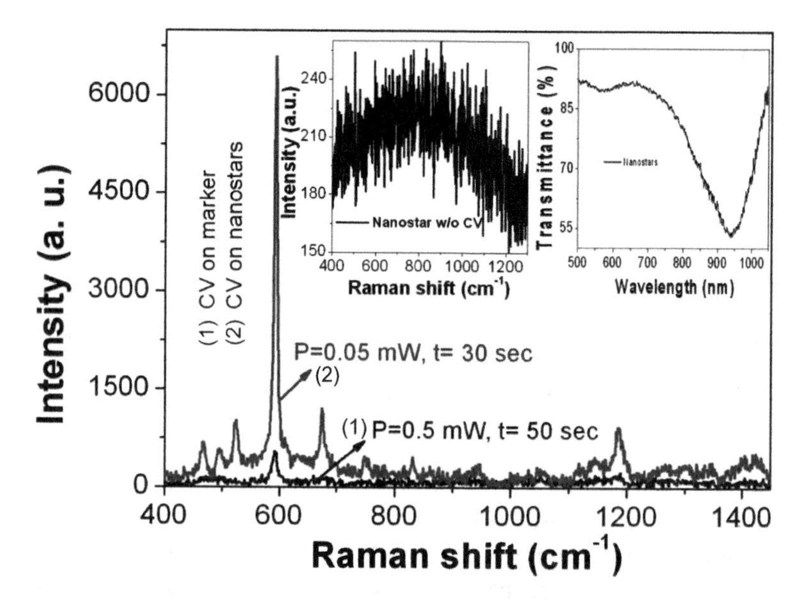

Figure 15.17 SERS spectra of cresyl violet (CV), deposited over Au-NSTs (2) and over Au-marker (1). All the experimental parameters are also indicated. In the inset, optical transmission spectrum of NST and SERS background measurement without any molecule are shown.

In conclusion, since NSTs having five petals with tip radius of curvature close to 10–12 nm, it can localize electric field at the tip. Various fabrication modifications are needed to modify the nanostructure (regarding the petal length and also the total area) in order to achieve a SERS device with higher enhancement factor to carry out single molecule detection and therefore, for further application in biomedical field.

15.5 Nanoantenna-Based Devices for Highly Efficient Hot Spot Generation

15.5.1 *Introduction*

Plasmonics is a branch of nanophotonics which deals with the energy transfer from free-propagating electromagnetic (EM) radiation to surface electrons in metallic nanostructures giving rise to collective charge oscillations called SPPs [77].

One of the nanostructures that more efficiently supports such modes is the plasmonic nanoantenna [78], a noble metal elongated structure able to work as optical concentrator. When the incident light is polarized along the nanoantenna main axis and the wavelength is matching the localized SP resonance (LSPR) an enhancement of the local electric field can be observed around the apexes of the structure [79]. Consequently energy carried by free propagating EM field can be concentrated in two spatial regions of sub-wavelength dimension, the so-called hot spots [65], breaking in this way the diffraction limit [80]. The combination of huge field enhancement and nanoscale localization makes this structure a good candidate for efficient light-matter energy transfer, opening the way to a wide variety of applications, for example, biosensing [66], single-molecule detection [81], enhanced photovoltaic conversion [19] and nanoantenna assisted IR spectroscopy [82].

In order to achieve a reduction in the hot spot dimension and a consequent increase in the local-field enhancement, a linearly coupled antenna dimer (Fig. 15.18a) has been extensively investigated for its peculiar configuration composed by two face-to-face aligned plasmonic antennas. By illuminating this system it is possible to generate, in resonance condition, a bright hot spot within the interparticle gap whose enhancement factor results amplified with respect to the single antenna case [80]. The enhancement observed in the nanocavity region indeed is related to the excitement of two aligned single antenna Localized SPs (LSPs) with prominent in-phase dipolar momenta (Fig. 15.18b). This phenomenon, coherently with the clear anisotropy of the system, is polarization-dependent. In fact incident light polarized along the short axis of the dimer can excite two LSPs at highest energy, according to a Fabry–Pérot-like model [83]. These two plasmonic charge oscillations induce only a residual electric field enhancement in the gap region (Fig. 15.18c), which is lower than the one achieved in the parallel configuration (Fig. 15.18b). The reason of that has to be ascribed to the mismatch between the dipolar momenta associated to long and short axes resonances.

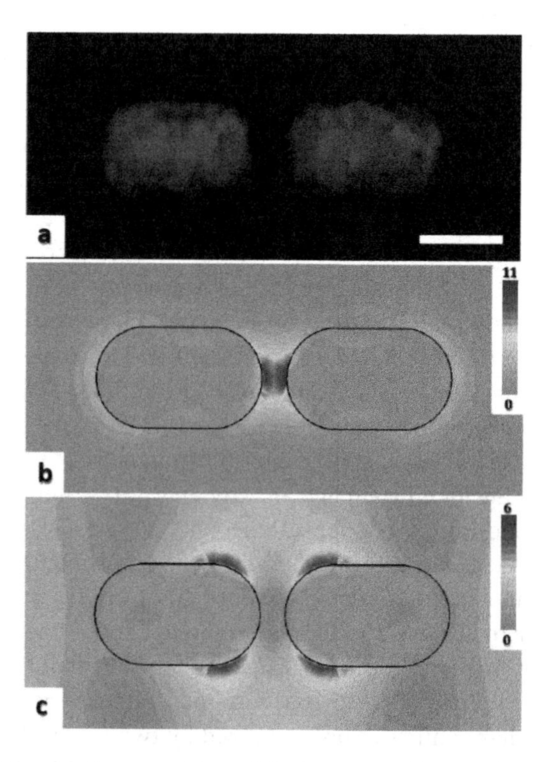

Figure 15.18 (a) Representative SEM image of Aligned Antenna Dimer (Single Antenna Arm Dimensions: 180 nm × 120 nm ×60 nm; Interparticle Gap: 20 nm) (Scalebar: 100 nm); (b,c) 2D field enhancement (ratio between the local and incident electric field) plots generated, at the resonance, on a plane parallel to the substrate that cuts the structure in the center, in condition of normal incident plane wave polarized respectively parallel to structure long and short axis.

In perspective of engineering a plasmon-based device for non-linear spectroscopy, for example, surface-enhanced Raman spectroscopy (SERS), the chance of having a parameter, able to modulate the hot spot intensity, ideally from a maximum value to zero, would be a great advantage in order to reduce the Raman signal-to-noise ratio by collecting, at zero-field condition, a precise background signal. On the other hand the necessity of further amplifying hot spot intensities requires decreasing the interparticle spacing towards the sub-5 nm range, that is, in the strong-coupling condition.

In order to drastically increase the hot spot's intensity, nanostructure spacing can be reduced by changing the nanocavity orientation, that is, forcing the interparticle coupling along the out-of-plane direction. 3D nanostructures have been fabricated, presenting thin separation dielectric layers evaporated with controlled and spatially uniform deposition thicknesses under 10 nm. The device proposed assumes the geometrical configuration of Stacked Optical Antennas which present gaps that can easily reach the sub-5 nm regime. Complementarily, by rotating one of the two antenna arms of 90° with respect to the center of the gap, it is possible to obtain a new plasmonic L-shaped antenna device supporting, at precise polarization direction, the combination of single antenna out-of-phase oscillating LSPs, inducing in its gap region a "zero-field spot".

In conclusion, within the context of non-linear spectroscopy for bioapplications, great interest is held in the low energy molecular vibrational/rotational investigation, avoiding photo-damaging in biological tissues [84]. The typical energies involved in these light-matter interaction processes lead to the necessity of moving towards the terahertz frequency domain. In the future perspective of terahertz-enhanced spectroscopy [85] arrays of high aspect ratio plasmonic antennas resonating in THz range are designed, fabricated and optically characterized.

15.5.2 *Stacked Optical Antenna*

The capacitive coupling between two plasmonic antennas starts to be significant and influences the near-field response of the system when interparticle gap is smaller than the typical dimensions of the dimer single arms [80]. Considering structures that resonate in the visible range, it immediately follows that a reasonable coupling value has to be less than 50 nm. Moreover, in order to observe high enhancements factors, energy has to be concentrated in really small hot spots and, since hot spot dimensions are defined by the gap parameter, structures have to be drawn up under a reciprocal distance of at least 10 nm. The big issue related to this spatial threshold consists in the fact that common planar

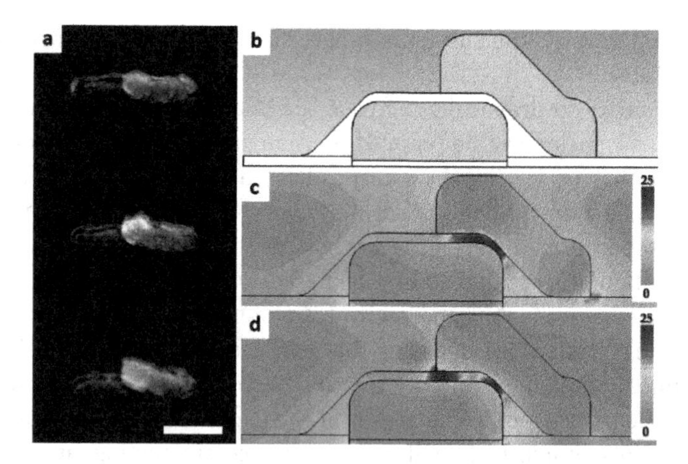

Figure 15.19 (a) Representative 30° tilt angle SEM image of stacked optical antennas (single antenna dimensions: 250 nm × 90 nm × 60 nm; interparticle gap: 9 nm; overlapping length: 50 nm) (scalebar: 200 nm); (b) SOA lateral cross section; (c,d) 2D field-enhancement (ratio between the local and incident electric field) plots generated, respectively at the longer and shorter resonance wavelengths, on a profile plane that cuts the structure in the center.

top-down fabrication techniques, typically lithographic approaches, present lacks of accuracy control in the sub–10 nm regime. A way to overcome the problem consists in expanding the gap along the vertical direction (i.e., out of plane), by a double patterning process interspersed with the deposition of a thin dielectric layer working as spacer. This combined process results in an overlapped configuration, that is, in the fabrication of stacked optical antennas (SOAs).

The design of SOAs has been carried out recurring to a simulation software that solves EM equations by means of a finite element (FE) code. A lateral cross section of the nanostructures is sketched in Fig. 15.19b which clearly elucidates the nanocavity configuration under study. The evolution of the gap profile suggests in principle this device could support both stationary and propagating electromagnetic modes and, besides, there could exist variability in the response of the system as function of light incidence angle. Simulations have been performed, at first, for normal incidence incoming plane wave and polarization parallel to the structure long

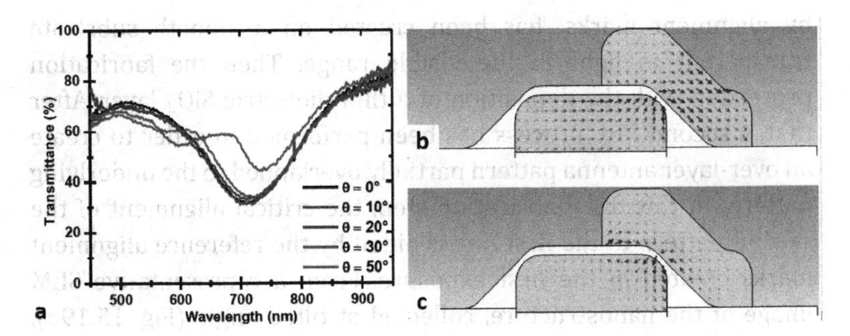

Figure 15.20 (a) SOAs far-field transmittances measured, for polarization aligned to structure long axis, at increasing tilt angles with respect to the normal to the substrate; (b,c) 2D current density plots generated, respectively at the longer (stationary plasmonic mode) and shorter (guided photonic mode) resonance wavelengths, on a profile plane that cuts the structure in its center.

axis for different wavelengths in the visible range. From far-field analysis the extinction efficiency spectrum shows an intense peak related to a mode that efficiently couples to light (Fig. 15.20a). A near-field analysis of that resonance (Fig. 15.19c) with the help of electric field plots shows the rising of a bright hot spot in the middle of the gap very similar to what occurs in aligned nanoantenna dimers. In fact, for incoming light at normal incidence angle, upper and lower antennas can be in practice considered as aligned gap-coupled nanostructures. From simulations at increasing tilt angles another resonance peak starts emerging in extinction spectrum (Fig. 15.20a) and electric field plots calculated at that resonance show evidence of a mode propagating along the gap (Fig. 15.19d). In practice the same device supports two completely different modes: the former presents a stationary plasmonic nature and the latter originates from a guided photonic behavior. In fact, for fixed polarization and different tilt angles light starts seeing the system not only as an aligned dimer, but also as a configuration similar to a waveguide, that is, a metal–insulator–metal (MIM) system, that can support guided modes [86].

Following the designed morphology, stacked antennas have been fabricated by means of electron beam lithography (EBL) in a two-fold sequential process. At first lower antenna pattern, enclosed by alignment marks, has been created on a smooth substrate transparent to light in the visible range. Then the fabrication proceeded with the deposition of a thin dielectric SiO_2 layer. After that, a second EBL process has been performed in order to create an over-layer antenna pattern partially overlapped to the underlying pattern. In this final fabrication step, the critical alignment of the second pattern to the first one is aided by the reference alignment marks created in the first exposure. From a representative SEM image of the nanostructure, collected at tilted angle (Fig. 15.19a), it is possible to appreciate the result of this complex and delicate fabrication process. Antennas result aligned and the overlapping length is estimated around 50 nm.

Optical characterization in far-field transmission configuration has been carried out on the device for different incident light tilt angles. Initially Stacked Antennas have been illuminated at normal incidence with polarization parallel to system long axis and then, for fixed polarization, sequential spectra measurements have been performed at increasing tilt angles up to 50° (Fig. 15.20a). Beyond a threshold angle between 20° and 30° light starts perceiving the structure not only as simple antenna dimer, but also as a waveguide and its coefficient of transmission [87] through the interstitial cavity increases as demonstrated by the comparison at higher energy of an intense extinction peak in the transmission spectrum. In conclusion this theoretical work and optical spectral analysis as function of light incidence angles show a significant evolution in transmission spectrum, suggesting SOAs as a good candidate for bright hot spot supporting metamaterial engineering [88].

15.5.3 *L-Shaped Antenna*

Complementarily to the fabrication of plasmonic nanodevices generating bright hot spots for the detection of molecules at low concentration [89], there is an effort towards the investigation of novel nanoantenna arrangements for a complete control over local field distribution. In perspective of engineering plasmonic nanodevices, a good candidate consists in a light polarization-sensitive nanostructure in which charge and consequently electric field distributions can be easily controlled. The interparticle nanocavity is a sub-wavelength region that, in resonance condition, experiments the presence of a very bright hot spot due to the superposition of two strong in-phase LSPs in both dimer antenna arms. By rotating the polarization from long to short axes direction the hot spot intensity decreases. Although a residual EM field remains active in the close proximity of the face-to-face apexes, related to the excitement of short axis LSPs oscillating in-phase. The key point in order to nullify the electric field in the gap region consists in forcing charge distributions in both antenna arms to contemporarily converge towards the gap in a complete out-of-phase plasmon mode. This effect is not possible, in normal incidence light condition, for aligned antenna dimers in dipolar approximation since all the charges in the system perceive, in every instant, the same external electric field. Therefore, in perspective of arising the out-of-phase mode in antenna dimer, creating a device more sensitive to light polarization, a solution consists in rotating one of the two dimer antenna arms of 90° with respect to gap center, inducing a symmetry breaking directly on the device geometry [90].

This configuration assumes the form of an L-shaped antenna dimer. EM simulations by means of a FE code software have been carried-out both in far- and near-field range to study the spectral response of this device to incoming EM plane wave as function of its polarization. In particular the study has been focused on two specific polarization directions: along the planar inversion-symmetry axis ($\theta = 45^\circ$ with respect to horizontal antenna long axis) and along its normal ($\theta = -45^\circ$ with respect to horizontal antenna long axis). In fact, for $\theta = -45^\circ$ charge distributions in both antenna arms are induced to oscillate in-phase similarly to what happens in aligned configuration but, for $\theta = 45^\circ$ charges in both arms are contemporarily pulled/pushed by external electric field towards/far from the gap region in an out-of-phase collective mode. Moreover, in this device configuration, both in-phase and out-of-phase LSPs modes present non-zero associated dipolar momenta and for this reason both resonances efficiently radiate in far field. From simulated far-field extinction efficiency spectra (defined as the ratio between the extinction and the geometrical cross sections) at $\theta = -45^\circ$ and $\theta = 45^\circ$ polarization angles, it can be observed how both spectra show a proper resonance peak, the second one

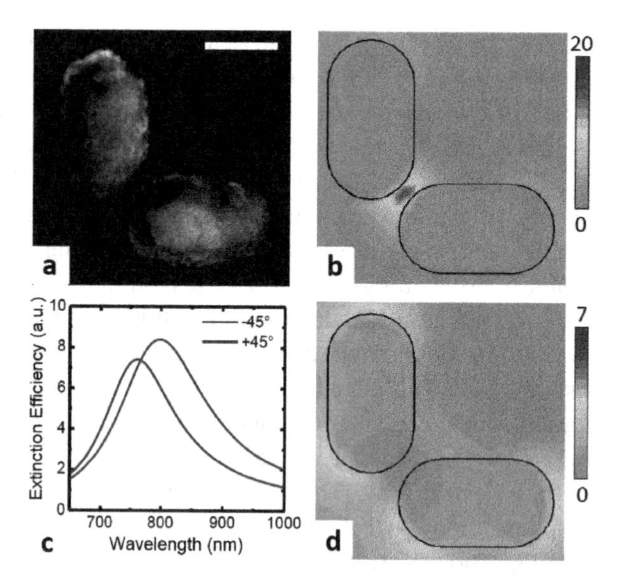

Figure 15.21 (a) Representative SEM image of L-shaped antenna dimer (single antenna arm dimensions: 180 nm × 120 nm × 60 nm; Interparticle gap: 20 nm) (scalebar: 100 nm); (b) electric field intensity 2D plots, in resonance condition, for normal-incidence light polarized at -45° valued at $\lambda = 810$ nm; (c) Theoretical extinction efficiency spectra of the L-shaped antennas; (d) electric field intensity 2D plots, in resonance condition, for normal-incidence light polarized at $+45^\circ$ valued at $\lambda = 760$ nm.

blue-shifted with respect to the first one (Fig. 15.21c). This should not surprise since a configuration in which free charges are forced to converge towards the same point is energetically inconvenient. Near-field analysis performed at the two fore mentioned collective resonances shows that the $\theta = -45^\circ$ polarization induces, in the gap region, the rising of a similar hot spot supported by aligned dimer (Fig. 15.21b). In the case of $\theta = 45^\circ$ the preceding hot spot switches off without presenting any residual EM field around the close antenna apexes (Fig. 15.21d).

L-shaped antennas have been fabricated on CaF_2 (100) substrates by means of EBL achieving an interparticle gap between 10 nm and 20 nm as it is possible to appreciate from SEM close-up image in Fig. 15.21a. Far-field optical transmission spectroscopy for $\theta = \pm 45^\circ$ polarization angles has been performed on L-shaped

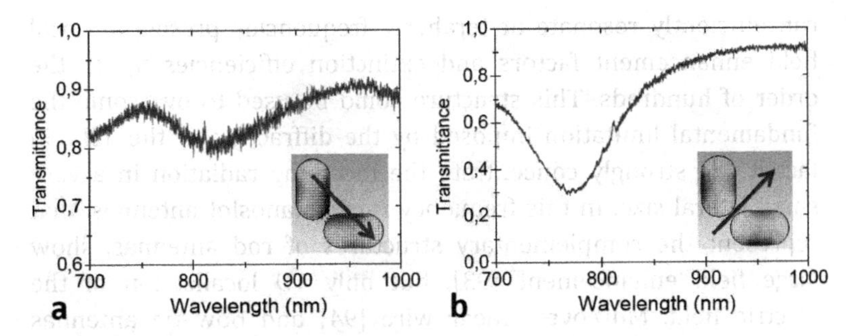

Figure 15.22 (a,b) measured transmittances of the L-shaped antennas for normal incidence light polarized respectively at $-45°$ and $+45°$ with respect to the horizontal antenna arm main axes (Insets: 2D plots of current density distributions, in resonance conditions, induced respectively by normal incidence light polarized respectively at $-45°$ and $45°$ with respect to horizontal antenna arm main axes).

antenna matrices confirming theoretical previsions by comparison of simulated (Fig. 15.21c) and measured spectra (Fig. 15.22a,b). As result of this work, from near-field analysis and correlated far-field measurements it has been demonstrated how L-shaped antenna device is a strong polarization-sensitive plasmonic system supporting in its interparticle gap both hot and zero-field spot respectively induced by in-phase and out-of-phase collective LSPs modes.

15.5.4 *Resonant Terahertz Dipole Nanoantenna*

A relatively unexplored and promising branch in nanophotonics is constituted by the design and the fabrication of nanostructured devices resonating in the terahertz frequency domain (tipically 0.1–10 THz) due to the fascinating applications, for example, in spectroscopy [91], chemical identification [92] and non-destructive imaging [84]. In fact, most dielectric materials, opaque to visible light, are transparent to THz radiation. Moreover its low photon energy enables the excitation of vibrational/rotational modes in several molecules, avoiding photoionization in biological tissues. Anyway, the spatial resolution of THz spectroscopy is limited by the large wavelength associated with this radiation. A metallic nanostructure in a shape of a half-wavelength dipole nanoantenna can efficiently resonate at terahertz frequencies presenting local field enhancement factors and extinction efficiencies up to the order of hundreds. This structure could be used to overcome the fundamental limitation imposed by the diffraction at the THz, in fact it can strongly concentrate the incoming radiation in a very small lateral size. In this frequency range, nanoslot antennas, that represent the complementary structures of rod antennas, show large field enhancement [93], but only 1D localization of the electric field. Moreover, linear wire [94] and bow–tie antennas [95] at THz frequencies have large dimensions and limited field enhancement. The resonant dipole nanoantennas combine, instead, much higher local enhancement factor in a nanoscale sized hot spot, with significant far field properties thanks to the extremely large extinction efficiencies.

By using high-resolution electron beam lithography technique a 2D array of aligned planar gold nanoantennas was fabricated on a high resistivity silicon substrate, as shown in Fig. 15.23a. Si has been chosen due to its high transparency and a constant value of refractive index in the THz region. With the use of electromagnetic simulations, the nanoantenna has been designed to be 40 μm long, thus presenting a resonant behavior at the frequencies close to 1.5 THz, the width and height being respectively 200 nm (presenting a very high aspect ratio of 200) and 60 nm [83]. The distance of the nanoantennas in the array is fixed to 20 μm in both directions [96] (Fig. 15.23b).

Standard zinc telluride source, generating the terahertz beam, was employed in a time domain spectroscopy setup to carry out the characterization of the sample. The nanoantenna array was illuminated by a collimated terahertz beam of 7 mm beam diameter.

THz pulses transmitted through the array with normal incidence have been measured for the polarizations parallel and perpendicular to the long axis of the nanoantennas. The nanoantenna covering factor (ratio of the area covered by the nanoantennas divided by the total illuminated area) is less than 0.5%, so that the array transmission in the case of polarization set along the short axis of the nanoantennas is practically the same to that of a reference silicon substrate without nanoantennas.

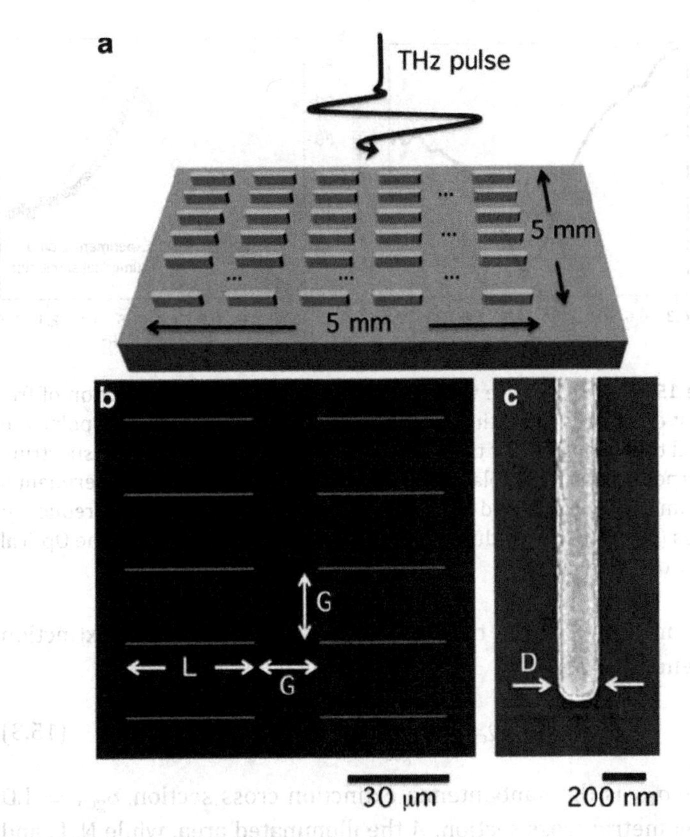

Figure 15.23 (a) Sketch of the fabricated terahertz nanoantenna arrays. (b) SEM image of a small portion of the array. (c) Magnification of the nanoantenna end. (L and D are, respectively, the length and the width of the nanoantenna and G is the array spacing). Reproduced from Ref. 96, with permission of Springer.

Hence, it is possible to obtain the resonance properties of the arrays by dividing the power spectrum of the transmitted pulse for long axis excitation by the one taken for short axis excitation. Finally, the quantity extracted is called "relative transmittance" (T_{rel}) and is reported in Fig. 15.24a. We can notice that, despite the low covering factor, the transmission is significantly reduced down to the 60%, together with a broad resonance behavior.

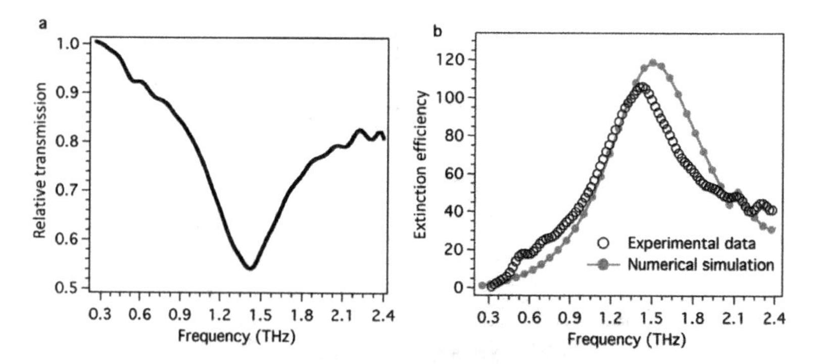

Figure 15.24 (a) Relative transmission of the sample as a function of frequency, defined as the ratio of the transmitted spectrum for light polarized parallel to the long axis of the nanoantennas over the transmitted spectrum for perpendicular light polarization. (b) Comparison between experimental and numerically calculated extinction efficiencies as a function of frequency. Figures (a) and (b) reproduced from Ref. 83, with permission of the Optical Society of America.

From T_{rel}, we can then estimate the nanoantenna extinction efficiency Q_{ext} as:

$$Q_{ext} = \frac{\sigma_{ext}}{\sigma_{geo}} = \frac{A(1\text{-}T_{rel})}{NLD} \tag{15.3}$$

where σ_{ext} is the nanoantenna extinction cross section, $\sigma_{geo} = LD$ the geometric cross section, A the illuminated area, while N, L, and D are the number, length and width of the illuminated nanoantennas, respectively.

In Fig. 15.24b experimental "Q_{ext}" (black trace) shows a clear peak at 1.4 THz, in correspondence to the extinction peak of the related transmittance spectrum, with a maximum efficiency of more than 100.

To better understand the resonance characteristics of terahertz nanoantennas array we performed numerical simulations. The nanoantenna design resembles the fabricated structure and all the simulations parameters are the same as the ones used in [83]. The calculated extinction efficiency, in good agreement with experimental values, is represented in Fig. 15.24b (green trace).

Simulations also provide near field properties of the nanoantenna. Figure 15.25a depicts the electric field norm around the

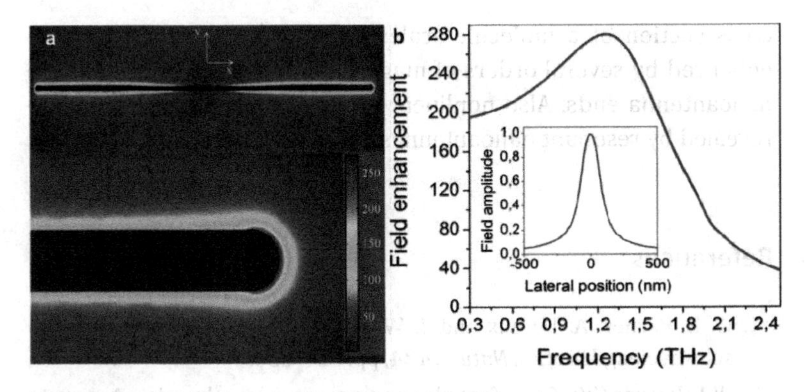

Figure 15.25 (a) Contour plot of the absolute value of the electric field around the nanoantenna under resonant condition. (b) Field enhancement factor 'F' at the nanoantenna end as a function of frequency. Inset: normalized electric field profile close to the nanoantenna end. Reproduced with permission from Ref. 97, Copyright © 2012 Wiley-VCH Verlag GmbH & Co. KGaA, Weinheim.

nanoantenna on a cut plane normal to the direction of the illuminating wave and positioned at the antenna half height [97]. The field is highly localized at the nanoantenna ends. The full width at half maximum of the field distribution 1 nm away from the nanoantenna end is 180 nm (inset in Fig. 15.25b, so that the incoming radiation is confined on a lateral size smaller than $\lambda/1000$. This result clearly demonstrates that the considered device efficiently acts as a nanoconcentrator. Field enhancement factor as a function of frequency, per definition F, is reported in Fig. 15.25b. A wide resonant behavior can be observed, with a maximum enhancement value of about 280 at around 1.3 THz. It's interesting to note the broadening of the near field peak toward low frequencies and the red shift when compared to the far field resonance. These phenomena, observed at optical wavelengths, has been explained with the plasmon damping [98]. Summing up, the nanostructure proposed here can be used to concentrate the THz radiation on the nanoscale and the high field enhancement could be successfully employed to improve the sensitivity for applications in the THz few-molecules spectroscopy, promising unprecedented spatial resolutions and extending the use of nanoantennas for the sensing in a new region of frequency. In fact, the effective absorption cross section of a molecule scales with $|F|^2$ and would thus be enhanced by several orders of magnitude in close proximity of the nanoantenna ends. Also nonlinear interaction at the THz could be revealed by resonant nanoantennas.

References

1. W. L. Barnes, A. Dereux and T. W. Ebbesen (2003) Surface plasmon subwavelength optics, *Nature*, **424**, pp. 824–830.
2. W. L. Barnes (2006) Surface plasmon-polariton length scales: A route to sub-wavelength optics, *J. Opt. A: Pure Appl. Opt.*, **8**, pp. S87–S93.
3. A. J. Babadjanyan, N. L. Margaryan and K. V. Nerkararyan (2000) Superfocusing of surface polaritons in the conical structure, *J. Appl. Phys.*, **87**, pp. 3785–3788.
4. M. Stockman (2004) Nanofocusing of optical energy in tapered plasmonic waveguides, *Phys. Rev. Lett.*, **93**, p. 137404.
5. S. Schmidt, B. Piglosiewicz, D. Sadiq, J. Shirdel, J. S. Lee, P. Vasa, N. Park, D.-S. Kim and C. Lienau (2012) Adiabatic nanofocusing on ultrasmooth single-crystalline gold tapers creates a 10-nm-sized light source with few-cycle time resolution, *ACS Nano*, **6**, pp. 6040–6048.
6. K. K. N. Simon M. Sze (2006) *Physics of Semiconductor Devices*, 3rd Ed., John Wiley, USA.
7. C. Donolato (1995) Electrostatic problem of a point-charge in the presence of a semiinfinite semiconductor, *J. Appl. Phys.*, **78**, pp. 684–690.
8. C. Donolato (2004) Approximate analytical solution to the space charge problem in nanosized Schottky diodes, *J. Appl. Phys.*, **95**, pp. 2184–2186.
9. G. D. J. Smit, S. Rogge and T. M. Klapwijk (2002) Scaling of nano-Schottky-diodes, *Appl. Phys. Lett.*, **81**, pp. 3852–3854.
10. M. A. Green (2000) Photovoltaics: Technology overview, *Energ. Policy*, **28**, pp. 989–998.
11. L. El Chaar, L. A. Lamont and N. El Zein (2011) Review of photovoltaic technologies, *Renewable & Sustainable Energy Reviews*, **15**, pp. 2165–2175.
12. L. L. Kazmerski (2006) Solar photovoltaics R&D at the tipping point: A 2005 technology overview, J. Electron Spectrosc. Relat. Phenom., **150**, pp. 105–135.

13. H. A. Atwater and A. Polman (2010) Plasmonics for improved photovoltaic devices, *Nat. Mater.*, **9**, pp. 205–213.
14. H. W. Deckman, C. B. Roxlo and E. Yablonovitch (1983) Maximum statistical increase of optical absorption in textured semiconductor films, *Opt. Lett.*, **8**, pp. 491–493.
15. M. W. G. Wanlass (2011) High-efficiency, monolithic, multi-bandgap, tandem photovoltaic energy converters (Alliance for Sustainable Energy, LLC, Golden, CO, USA).
16. P. Andrew and W. L. Barnes (2004) Energy transfer across a metal film mediated by surface plasmon polaritons, *Science*, **306**, pp. 1002–1005.
17. C. Robert Kostecki (2010) Surface plasmon-enhanced photovoltaic device (The Regents of The University Of California, USA).
18. M. A. Green and S. Pillai (2012) Harnessing plasmonics for solar cells, *Nat. Photon.*, **6**, pp. 130–132.
19. A. Polman and H. A. Atwater (2012) Photonic design principles for ultrahigh-efficiency photovoltaics, *Nat. Mater.*, **11**, pp. 174–177.
20. B. D. I. Goykhman, J. Khurgin, J. Shappir and Uriel Levy (2012) Waveguide based compact silicon Schottky photodetector with enhanced responsivity in the telecom spectral band, *Opt. Express*, **20**, p. 28594.
21. A. Giugni, B. Torre, A. Toma, M. Francardi, M. Malerba, A. Alabastri, R. Proietti Zaccaria, M. I. Stockman, E. Di Fabrizio (2013) Hot-electron nanoscopy using adiabatic compression of surface plasmons, *Nat. Nanotechnol.*, **8**, pp. 845–852.
22. Z. D. Genchev, N. M. Nedelchev, E. Mateev and H. Y. Stoyanov (2008) Analytical approach to the prism coupling problem in the Kretschmann configuration, *Plasmonics*, **3**, pp. 21–26.
23. S. R. J. Brueck, V. Diadiuk, T. Jones and W. Lenth (1985) Enhanced quantum efficiency internal photoemission detectors by grating coupling to surface plasma waves, *Appl. Phys. Lett.*, **46**, p. 915.
24. P. Lalanne, J. P. Hugonin and J. C. Rodier (2005) Theory of surface plasmon generation at nanoslit apertures, *Phys. Rev. Lett.*, **95**, p. 263902.
25. C. Ropers, C. C. Neacsu, T. Elsaesser, M. Albrecht, M. B. Raschke and C. Lienau (2007) Grating-coupling of surface plasmons onto metallic tips:? A nanoconfined light source, *Nano Lett.*, **7**, pp. 2784–2788.
26. K. Hasegawa, J. U. Nockel and M. Deutsch (2004) Surface plasmon polariton propagation around bends at a metal-dielectric interface, *Appl. Phys. Lett.*, **84**, pp. 1835–1837.

27. K. Hasegawa, J. U. Nöckel and M. Deutsch (2007) Curvature-induced radiation of surface plasmon polaritons propagating around bends, *Phys. Rev. A*, **75**, p. 063816.
28. A. Wiener, A. I. Fernandez-Dominguez, A. P. Horsfield, J. B. Pendry and S. A. Maier (2012) Nonlocal effects in the nanofocusing performance of plasmonic tips, *Nano Lett.*, **12**, pp. 3308–3314.
29. S. Link and M. A. El-Sayed (1999) Spectral properties and relaxation dynamics of surface plasmon electronic oscillations in gold and silver nanodots and nanorods, *Journal of Physical Chemistry B*, **103**, pp. 8410–8426.
30. A. Melikyan and H. Minassian (2004) On surface plasmon damping in metallic nanoparticles, *Applied Physics B-Lasers and Optics*, **78**, pp. 453–455.
31. A. V. Zayats and I. I. Smolyaninov (2003) Near-field photonics: Surface plasmon polaritons and localized surface plasmons, *Journal of Optics A: Pure and Applied Optics*, **5**, p. S16.
32. H. Raether (1988) *Surface Plasmons on Smooth and Rough Surfaces and on Gratings*, Springer Tracts in Modern Physics, Vol. 111, Springer Berlin Heidelberg.
33. A. Baron, E. Devaux, J.-C. Rodier, J.-P. Hugonin, E. Rousseau, C. Genet, T. W. Ebbesen and P. Lalanne (2011) Compact antenna for efficient and unidirectional launching and decoupling of surface plasmons, *Nano Lett.*, **11**, pp. 4207–4212.
34. R. H. Ritchie, E. T. Arakawa, J. J. Cowan and R. N. Hamm (1968) Surface-plasmon resonance effect in grating diffraction, *Phys. Rev. Lett.*, **21**, pp. 1530–1533.
35. E. D. Palik (1985) *Handbook of Optical Constants of Solids*, Part II, Elsevier Science & Technology.
36. D. K. Gramotnev, M. W. Vogel and M. I. Stockman (2008) Optimized nonadiabatic nanofocusing of plasmons by tapered metal rods, *J. Appl. Phys.*, **104**, p. 034311.
37. R. P. Zaccaria, F. De Angelis, A. Toma, L. Razzari, A. Alabastri, G. Das, C. Liberale and E. Di Fabrizio (2012) Surface plasmon polariton compression through radially and linearly polarized source, *Optics Letters*, **37**, pp. 545–547.
38. F. De Angelis, M. Patrini, G. Das, I. Maksymov, M. Galli, L. Businaro, L. C. Andreani and E. Di Fabrizio (2008) A hybrid plasmonic–photonic nanodevice for label-free detection of a few molecules, *Nano Lett.*, **8**, pp. 2321–2327.

39. F. De Angelis, G. Das, P. Candeloro, M. Patrini, M. Galli, A. Bek, M. Lazzarino, I. Maksymov, C. Liberale, L. C. Andreani and E. Di Fabrizio (2010) Nanoscale chemical mapping using three-dimensional adiabatic compression of surface plasmon polaritons, *Nature Nanotechnology*, **5**, pp. 67–72.
40. F. De Angelis, F. Gentile, F. Mecarini, G. Das, M. Moretti, P. Candeloro, M. L. Coluccio, G. Cojoc, A. Accardo, C. Liberale, R. P. Zaccaria, G. Perozziello, L. Tirinato, A. Toma, G. Cuda, R. Cingolani and E. Di Fabrizio (2011) Breaking the diffusion limit with super-hydrophobic delivery of molecules to plasmonic nanofocusing SERS structures, *Nature Photonics*, **5**, pp. 683–688.
41. R. Proietti Zaccaria, A. Alabastri, F. De Angelis, G. Das, C. Liberale, A. Toma, A. Giugni, L. Razzari, M. Malerba, H. B. Sun and E. Di Fabrizio (2012) Fully analytical description of adiabatic compression in dissipative polaritonic structures, *Phys. Rev. B*, **86**, p. 035410.
42. S. V. Boriskina and B. M. Reinhard (2012) Molding the flow of light on the nanoscale: From vortex nanogears to phase-operated plasmonic machinery, *Nanoscale*, **4**, pp. 76–90.
43. B. S. Luk'yanchuk and V. Ternovsky (2006) Light scattering by a thin wire with a surface-plasmon resonance: Bifurcations of the Poynting vector field, *Physical Review B*, **73**, p. 235432.
44. Y. Fedutik, V. V. Temnov, O. Schoeps, U. Woggon and M. V. Artemyev (2007) Exciton-plasmon-photon conversion in plasmonic nanostructures, *Phys. Rev. Lett.*, **99**, p. 136802.
45. A. V. Akimov, A. Mukherjee, C. L. Yu, D. E. Chang, A. S. Zibrov, P. R. Hemmer, H. Park and M. D. Lukin (2007) Generation of single optical plasmons in metallic nanowires coupled to quantum dots, *Nature*, **450**, pp. 402–406.
46. H. Wei, D. Ratchford, X. Li, H. Xu and C.-K. Shih (2009) Propagating surface plasmon induced photon emission from quantum dots, *Nano Lett.*, **9**, pp. 4168–4171.
47. M. W. Knight, N. K. Grady, R. Bardhan, F. Hao, P. Nordlander and N. J. Halas (2007) Nanoparticle-mediated coupling of light into a nanowire, *Nano Lett.* **7**, pp. 2346–2350.
48. R. X. Yan, P. Pausauskie, J. X. Huang and P. D. Yang (2009) Direct photonic-plasmonic coupling and routing in single nanowires, *Proceedings of the National Academy of Sciences of the United States of America*, **106**, pp. 21045–21050.
49. A. L. Falk, F. H. L. Koppens, C. L. Yu, K. Kang, N. d. L. Snapp, A. V. Akimov, M.-H. Jo, M. D. Lukin and H. Park (2009) Near-field electrical detection

of optical plasmons and single-plasmon sources, *Nature Physics*, **5**, pp. 475–479.

50. D. E. Chang, A. S. Sorensen, P. R. Hemmer and M. D. Lukin (2006) Quantum optics with surface plasmons, *Phys.Rev. Lett.*, **97**, p. 053002.
51. Z.-K. Zhou, M. Li, Z.-J. Yang, X.-N. Peng, X.-R. Su, Z.-S. Zhang, J.-B. Li, N.-C. Kim, X.-F. Yu, L. Zhou, Z.-H. Hao and Q.-Q. Wang (2010) Plasmon-mediated radiative energy transfer across a silver nanowire array via resonant transmission and subwavelength imaging, *ACS Nano*, **4**, pp. 5003–5010.
52. L. Carbone, C. Nobile, M. De Giorgi, F. D. Sala, G. Morello, P. Pompa, M. Hytch, E. Snoeck, A. Fiore, I. R. Franchini, M. Nadasan, A. F. Silvestre, L. Chiodo, S. Kudera, R. Cingolani, R. Krahne and L. Manna (2007) Synthesis and micrometer-scale assembly of colloidal CdSe/CdS nanorods prepared by a seeded growth approach, *Nano Lett.*, **7**, pp. 2942–2950.
53. R. Krahne, M. Zavelani-Rossi, M. G. Lupo, L. Manna and G. Lanzani (2011) Amplified spontaneous emission from core and shell transitions in CdSe/CdS nanorods fabricated by seeded growth, *Appl. Phys. Lett.*, **98**, p. 063105.
54. R. Krahne, G. Morello, A. Figuerola, C. George, S. Deka and L. Manna (2011) Physical properties of elongated inorganic nanoparticles, *Physics Reports*, **501**, pp. 75–221.
55. A. D. Rakic, A. B. Djuricic, J. M. Elazar and M. L. Majewski (1998) Optical properties of metallic films for vertical-cavity optoelectronic devices, *Appl. Opt.*, **37**, pp. 5271–5283.
56. F. Pisanello, L. Martiradonna, G. Lemenager, P. Spinicelli, A. Fiore, L. Manna, J. P. Hermier, R. Cingolani, E. Giacobino, M. De Vittorio and A. Bramati (2010) Room temperature–dipole-like single photon source with a colloidal dot-in-rod, *Appl. Phys. Lett.*, **96**, art. no. 033101.
57. P. G. Etchegoin and E. C. Le Ru (2008) A perspective on single molecule SERS: Current status and future challenges, *Physical Chemistry Chemical Physics*, **10**, pp. 6079–6089.
58. M. J. A. Canada, A. R. Medina, J. Frank and B. Lendl (2002) Bead injection for surface enhanced Raman spectroscopy: Automated on-line monitoring of substrate generation and application in quantitative analysis, *Analyst*, **127**, pp. 1365–1369.
59. H. Ko, S. Singamaneni and V. V. Tsukruk (2008) Nanostructured surfaces and assemblies as SERS media, *Small*, **4**, pp. 1576–1599.

60. B. Sagmuller, B. Schwarze, G. Brehm and S. Schneider (2001) Application of SERS spectroscopy to the identification of (3,4-methylenedioxy) amphetamine in forensic samples utilizing matrix stabilized silver halides, *Analyst*, **126**, pp. 2066–2071.

61. A. Gopinath, S. V. Boriskina, W. R. Premasiri, L. Ziegler, B. M. Reinhard and L. Dal Negro (2009) Plasmonic nanogalaxies: Multiscale aperiodic arrays for surface-enhanced Raman sensing, *Nano Lett.*, **9**, pp. 3922–3929.

62. L. Gunnarsson, E. J. Bjerneld, H. Xu, S. Petronis, B. Kasemo and M. Kall (2001) Interparticle coupling effects in nanofabricated substrates for surface-enhanced Raman scattering, *Appl. Phys. Lett.*, **78**, pp. 802–804.

63. M. Rycenga, M. H. Kim, P. H. C. Camargo, C. Cobley, Z.-Y. Li and Y. Xia (2009) Surface-enhanced Raman scattering: Comparison of three different molecules on single-crystal nanocubes and nanospheres of silver, *Journal of Physical Chemistry A*, **113**, pp. 3932–3939.

64. H. Chen, X. Kou, Z. Yang, W. Ni and J. Wang (2008) Shape- and size-dependent refractive index sensitivity of gold nanoparticles, *Langmuir*, **24**, pp. 5233–5237.

65. B. Fazio, C. D'Andrea, F. Bonaccorso, A. Irrera, G. Calogero, C. Vasi, P. G. Gucciardi, M. Allegrini, A. Toma, D. Chiappe, C. Martella and F. B. de Mongeot (2011) Re-radiation enhancement in polarized surface-enhanced resonant Raman scattering of randomly oriented molecules on self-organized gold nanowires, *ACS Nano*, **5**, pp. 5945–5956.

66. G. Das, F. Mecarini, F. Gentile, F. De Angelis, M. H. G. Kumar, P. Candeloro, C. Liberale, G. Cuda and E. Di Fabrizio (2009) Nano-patterned SERS substrate: Application for protein analysis vs. temperature, *Biosensors and Bioelectronics*, **24**, pp. 1693–1699.

67. A. Toma, G. Das, M. Chirumamilla, A. Saeed, R. Proietti Zaccaria, L. Razzari, M. Leoncini, C. Liberale, F. De Angelis and E. Di Fabrizio (2012) Fabrication and characterization of a nanoantenna-based Raman device for ultrasensitive spectroscopic applications, *Microelectronic Engineering*, **98**, pp. 424–427.

68. C. L. Nehl, H. W. Liao and J. H. Hafner (2006) Optical properties of star-shaped gold nanoparticles, *Nano Lett.*, **6**, pp. 683–688.

69. M. L. Coluccio, G. Das, F. Mecarini, F. Gentile, A. Pujia, L. Bava, R. Tallerico, P. Candeloro, C. Liberale, F. De Angelis and E. Di Fabrizio (2009) Silver-based surface enhanced Raman scattering (SERS) substrate fabrication using nanolithography and site selective electroless deposition, *Microelectronic Engineering*, **86**, pp. 1085–1088.

70. F. Gentile, G. Das, M. L. Coluccio, F. Mecarini, A. Accardo, L. Tirinato, R. Tallerico, G. Cojoc, C. Liberale, P. Candeloro, P. Decuzzi, F. De Angelis and E. Di Fabrizio (2010) Ultra low concentrated molecular detection using super hydrophobic surface based biophotonic devices, *Microelectronic Engineering*, **87**, pp. 798–801.

71. M. Chirumamilla, G. Das, A. Toma, A. Gopalakrishnan, R. P. Zaccaria, C. Liberale, F. De Angelis and E. Di Fabrizio (2012) Optimization and characterization of Au cuboid nanostructures as a SERS device for sensing applications, *Microelectronic Engineering*, **97**, pp. 189–192.

72. A. D. Rakic, A. B. Djurisic, J. M. Elazar and M. L. Majewski (1998) Optical properties of metallic films for vertical-cavity optoelectronic devices, *Applied Optics*, **37**, pp. 5271–5283.

73. L. Rodriguez-Lorenzo, R. de la Rica, R. A. Alvarez-Puebla, L. M. Liz-Marzan and M. M. Stevens (2012) Plasmonic nanosensors with inverse sensitivity by means of enzyme-guided crystal growth, *Nature Materials*, **11**, pp. 604–607.

74. E. N. Esenturk and A. R. H. Walker (2009) Surface-enhanced Raman scattering spectroscopy via gold nanostars, *Journal of Raman Spectroscopy*, **40**, pp. 86–91.

75. G. Das, N. Patra, A. Gopalakrishnan, R. P. Zaccaria, A. Toma, S. Thorat, E. Di Fabrizio, A. Diaspro and M. Salerno (2012) Fabrication of large-area ordered and reproducible nanostructures for SERS biosensor application, *Analyst*, **137**, pp. 1785–1792.

76. E. Vogel, A. Gbureck and W. Kiefer (2000) Vibrational spectroscopic studies on the dyes cresyl violet and coumarin 152, *Journal of Molecular Structure*, **550**, pp. 177–190.

77. A. Demming, M. Brongersma and D. S. Kim (2012) Plasmonics in optoelectronic devices, *Nanotechnology*, **23**, p. 440201.

78. L. Novotny (2007) Effective wavelength scaling for optical antennas, *Phys. Rev. Lett.*, **98**, p. 266802.

79. R. Blanchard, S. V. Boriskina, P. Genevet, M. A. Kats, J. P. Tetienne, N. F. Yu, M. O. Scully, L. Dal Negro and F. Capasso (2011) Multi-wavelength mid-infrared plasmonic antennas with single nanoscale focal point, *Optics express*, **19**, pp. 22113–22124.

80. P. Biagioni, J. S. Huang and B. Hecht (2012) Nanoantennas for visible and infrared radiation, *Rep. Prog. Phys.*, **75**, p. 024402.

81. J. J. Xu, L. Zhang, H. Gong, J. Homola and Q. M. Yu (2011) Tailoring plasmonic nanostructures for optimal SERS sensing of small molecules and large microorganisms, *Small*, **7**, pp. 371–376.

82. F. Neubrech, D. Weber, J. Katzmann, C. Huck, A. Toma, E. Di Fabrizio, A. Pucci and T. Hartling (2012) Infrared optical properties of nanoantenna dimers with photochemically narrowed gaps in the 5 nm regime, *ACS Nano*, **6**, pp. 7326–7332.
83. L. Razzari, A. Toma, M. Shalaby, M. Clerici, R. P. Zaccaria, C. Liberale, S. Marras, I. A. I. Al-Naib, G. Das, F. De Angelis, M. Peccianti, A. Falqui, T. Ozaki, R. Morandotti and E. Di Fabrizio (2011) Extremely large extinction efficiency and field enhancement in terahertz resonant dipole nanoantennas, *Optics Express*, **19**, pp. 26088–26094.
84. J. F. Federici, B. Schulkin, F. Huang, D. Gary, R. Barat, F. Oliveira and D. Zimdars (2005) THz imaging and sensing for security applications—explosives, weapons and drugs, *Semiconductor Science and Technology*, **20**, pp. S266–S280.
85. A. Berrier, M. C. Schaafsma, G. Nonglaton, J. Bergquist and J. G. Rivas (2012) Selective detection of bacterial layers with terahertz plasmonic antennas, *Biomedical optics express*, **3**, pp. 2937–2949.
86. J. Park, H. Kim, I. M. Lee, S. Kim, J. Jung and B. Lee (2008) Resonant tunneling of surface plasmon polariton in the plasmonic nano-cavity, *Optics express*, **16**, pp. 16903–16915.
87. J. S. Huang, T. Feichtner, P. Biagioni and B. Hecht (2009) Impedance matching and emission properties of nanoantennas in an optical nanocircuit, *Nano Lett.*, **9**, pp. 1897–1902.
88. D. W. Pohl, S. G. Rodrigo and L. Novotny (2011) Stacked optical antennas, *Appl. Phys. Lett.*, **98**, p. 023111.
89. J. N. Anker, W. P. Hall, O. Lyandres, N. C. Shah, J. Zhao and R. P. Van Duyne (2008) Biosensing with plasmonic nanosensors, *Nature Materials*, **7**, pp. 442–453.
90. B. Luk'yanchuk, N. I. Zheludev, S. A. Maier, N. J. Halas, P. Nordlander, H. Giessen and C. T. Chong (2010) The Fano resonance in plasmonic nanostructures and metamaterials, *Nature Materials*, **9**, pp. 707–715.
91. S. L. Dexheimer (2008) *Terahertz Spectroscopy: Principles and Applications*, CRC Press, New York.
92. D. Mittelman (ed.) (2010) *Sensing with THz Radiation*, Springer-Verlag, Germany.
93. M. A. Seo, H. R. Park, S. M. Koo, D. J. Park, J. H. Kang, O. K. Suwal, S. S. Choi, P. C. M. Planken, G. S. Park, N. K. Park, Q. H. Park and D. S. Kim (2009) Terahertz field enhancement by a metallic nano slit operating beyond the skin-depth limit, *Nat. Photon.*, **3**, pp. 152–156.

94. F. Blanchard, A. Doi, T. Tanaka, H. Hirori, H. Tanaka, Y. Kadoya and K. Tanaka (2011) Real-time terahertz near-field microscope, *Opt. Express*, **19**, pp. 8277–8284.
95. A. Berrier, R. Ulbricht, M. Bonn and J. G. Rivas (2010) Ultrafast active control of localized surface plasmon resonances in silicon bowtie antennas, *Opt. Express*, **18**, pp. 23226–23235.
96. L. Razzari, A. Toma, M. Clerici, M. Shalaby, G. Das, C. Liberale, M. Chirumamilla, R. Zaccaria, F. Angelis, M. Peccianti, R. Morandotti and E. Fabrizio (2012) Terahertz dipole nanoantenna arrays: Resonance characteristics, *Plasmonics*, pp. 1–6.
97. A. Gopalakrishnan, M. Malerba, S. Tuccio, S. Panaro, E. Miele, M. Chirumamilla, S. Santoriello, C. Dorigoni, A. Giugni, R. Proietti Zaccaria, C. Liberale, F. De Angelis, L. Razzari, R. Krahne, A. Toma, G. Das and E. Di Fabrizio (2012) Nanoplasmonic structures for biophotonic applications: SERS overview, *Annalen der Physik*, doi: 10.1002/andp.201200145.
98. B. M. Ross and L. P. Lee (2009) Comparison of near- and far-field measures for plasmon resonance of metallic nanoparticles, *Opt. Lett.*, **34**, pp. 896–898.

Index

absorption 2, 3, 5, 9, 14, 15, 17, 25, 30, 31, 49, 178, 403, 412, 415, 447, 448, 466, 472
  differential 4, 12, 28, 70
  ground state 466
  heat 29
  negative 415
  optical 466, 495
  two-photon 100
absorption spectrum 31, 42
active CNPs *see* active coated nano-particles
active coated nano-particles (active CNPs) 411, 413, 416–422, 424, 427, 431–433, 438, 439, 441, 442, 444, 445
adiabatic compression 460, 462, 497
adiabatic taper 377
analysis
  convergence 477
  convergence mesh 477
  eigenmode 195
  generalized Bloch 116–118
  optical spectral 486
  pair correlation 348
  rigorous coupled wave 245
  spatial Delaunay triangulation 350
  wavelet-based multifractal 347
angular momentum 96, 234, 240, 256, 312, 314, 321, 326, 329, 385, 386, 407
antennas 84, 122, 215, 216, 263, 447, 452, 493
  bow–tie 490
  broad-band 187
  bull eye 84
  chiral bull eye's 84
  circular 83
  nanoslot 490
apertures 241
  angular 356
  beam 316
  circular 242, 252
  metallic 239
  quasiperiodic 239 255
applications 46–48, 50–52, 69, 70, 167, 169, 170, 178–180, 233, 260, 281, 336, 369, 370, 390, 451–454, 480, 481, 498, 499
  biomedical 130
  biosensing 6
  jamming 427
  metamaterial 101
  multi-frequency 445
  sensing 12, 460, 500
  spectroscopy 460
approximation 194
  effective medium 186
  electrostatic 399
  long wave 30
  paraxial 61, 70
  quasistatic 29, 143
  rotating wave 8, 9

arrays 93, 95, 99–102, 217, 222, 336, 339–343, 347, 348, 353, 355, 360, 363, 473, 474, 477, 490, 491
achiral plasmonic nanohole 83
aperiodic plasmon 368
aperiodic spiral plasmonic 352
biological fluorescent 444
homogeneous particle 340
metal-dielectric 342
microneedle 324, 325
nanocuboid 477
planar 101, 371
quasi-periodic Penrose 360
square 188, 189, 200, 215
terahertz nanoantennas 492
azimuthal direction 224, 243, 324
azimuthal symmetry 351, 379

backward scattering (BS) 288, 291, 293, 298, 303, 305
band
broad 475
conduction 466
continuous energy 336
infrared 47
molecular 4, 7, 12, 42, 47
negative dispersion 197
negative group velocity 196
spectral 472
surface plasmon resonance absorption 39
visible optical 7
bandgap 117, 119, 187, 208, 211, 466
band structure 195–197, 199, 200, 206, 208, 209
Bessel functions 88, 171, 189, 194, 223, 224, 353, 377
Bragg scattering 199, 200, 351
broadband wave plate 185, 186, 199, 201, 219
BS *see* backward scattering

CD *see* circular dichroism
CD mechanisms 3, 6, 7, 29
CD signals 3–7, 9, 10, 12, 16–22, 27, 30, 31, 33, 34, 36, 37, 39, 41, 42, 44, 45
CD spectra 3, 4, 10–12, 21, 27, 28, 31, 34, 35, 39, 40, 44, 47
chiral channel 188, 203–208, 210–214
chiral effects 3, 97, 99, 119, 130
chirality 2, 39–42, 57, 58, 60, 61, 71–74, 76–78, 127, 152–157, 168, 169, 186–188, 218, 219, 221, 222, 239–241, 245, 330–332
artificial 2
geometrical 70
intrinsic 6
structural 59, 222
chirality parameter 134, 135, 153, 154, 157, 158, 164, 165
chiral ladders 205, 206
chiral medium 2, 5, 12, 14–16, 22, 26, 58, 77, 78, 80, 81, 133, 134, 136, 143, 145, 146, 171, 173
chiral meta-atoms 131–133
chiral metamaterials 131, 132, 134–136, 144, 147, 186, 217, 219, 235
planar 133, 134, 313
three-dimensional 101
chiral microspheres 147, 173
chiral molecules 6, 7, 9–11, 13, 26, 39–41, 51, 52, 128, 131, 135–137, 141–151, 153–167, 171–173, 175, 179, 222, 223
chiral nanoantenna 136, 158, 165–167
chiral nanoneedles 313, 325–330
chiral nanoparticles 147, 152, 165, 167, 168, 170

chiral nanostructures 1–4, 6, 8, 10, 12, 14, 16, 18, 20, 22, 24, 40–42, 46–48, 86, 311
chiral objects 2, 18, 30, 60, 69, 74, 77, 75, 85, 128, 183, 221, 222, 233
chiral parameter 5, 16, 22, 23, 129
chiral particles 132, 151, 158, 159, 162, 163, 179
chiral properties 15, 22, 130, 131, 144, 245
chiral systems 61, 68, 69, 74, 75, 77, 81, 186
chiral waveguide 204
circular dichroism (CD) 1–4, 6, 7, 9, 11–13, 16, 19, 22, 25, 26, 30, 42, 44, 46, 50–55, 128, 129, 221, 222
circular polarization 99, 108, 116, 223, 312, 324
circulating powerflows 264, 265, 267, 269–272, 277
CNP *see* coated nanoparticle
coated nanoparticle (CNP) 410, 411, 414, 416–420, 423, 424, 427–431, 433–444, 446
confocal pin hole (CPH) 468
Coulomb interactions 7, 8, 41
coupling 101, 102, 112, 113, 116, 204, 228, 230, 231, 242, 264, 272, 288, 290, 386, 427, 469, 472
  capacitive 483
  chiral stack-plasmon 39
  cross-polarization 100
  dipolar 40
  interparticle 483
  light-matter 338
  light-to-plasmon 452
  optical 272
  orbital 41
CPH *see* confocal pin hole
cylindrical structures 300, 301, 303
cylindrical symmetry 228, 230, 298, 301, 307, 453
cylindrical wire 226, 376–379, 381

DANS *see* deterministic aperiodic nano structures
decay rates 144, 155, 157, 162, 163
  nonradiative 151, 424
  radiative 150, 152, 164
  spontaneous 144, 156
defects 20, 50, 384
deterministic aperiodic nano structures (DANS) 337
DFT *see* discrete Fourier transform
dielectric constant 17, 26, 30, 35, 314, 322, 469, 472
dielectric functions 15, 23, 30, 31
dielectric material 267, 457
dielectric particles 178, 288
dielectric PCs *see* dielectric photonic crystals
dielectric permittivity 66, 289, 298, 299, 301
dielectric photonic crystals (dielectric PCs) 186, 187, 203, 214
dielectric plate 210, 213
dielectric slab 211, 212
dielectric tensor 274
diffraction 93, 215, 218, 294, 338, 357, 370, 490
dipolar approximation 101, 112, 487
dipole radiation 140, 275, 276
dipole source 62, 141, 274–277, 383, 438
direct laser writing (DLW) 100, 124
discrete Fourier transform (DFT) 346
dispersion 200, 273, 274, 282, 469

frequency-dependent optical rotatory 38
frequency-dependent SP wave-vector 457
optical rotary 129
photonic band 195
wavelength 471
DLS *see* dynamic light scattering
DLW *see* direct laser writing
dynamic light scattering (DLS) 35

EBL *see* electron beam lithography
EELS *see* electron energy loss spectroscopy
EFS *see* equi-frequency surfaces
electron beam lithography (EBL) 48, 100, 362, 395, 466, 467, 472, 474, 478, 485, 488
electron energy loss spectroscopy (EELS) 376, 380–384, 386, 388–390
emission 43, 143, 228–230, 233, 277, 280, 424, 439, 444, 445, 466, 467, 469, 471
angle-dependent 233
nanorod 469, 471, 472
quantum dot 465
enantiomers 2, 42, 69, 73, 83, 86, 135, 147, 155, 157, 158, 167–170, 181, 221
equi-frequency surfaces (EFS) 198, 201
excitation 101, 105, 107, 111, 143, 152, 226, 227, 229, 233, 234, 236, 241, 242, 250, 410, 416, 437–444
background 233
chiral plasmon 32
endfire 229
free electron 455
grating-coupled 457
polarized 103, 113
excitation wavelength 43, 233, 275, 276, 431
excitation waves 461
exciton 7, 8, 11, 16, 41, 42, 51
resonances 1–4, 6, 8, 10, 12, 14, 16, 18, 20, 22, 24, 26, 28, 30, 40–46

fabrication 2, 20, 47, 48, 50, 101, 325, 329, 336, 425, 452, 462, 484, 486, 489, 499, 500
Fabry–Pérot resonances 229
Fano resonances 285–287, 289–293, 295, 297–308, 501
far-field radiation 300, 363, 364, 469
FDTD *see* finite-difference-time-domain
FDTD simulations 211, 214
FEM *see* finite element method
FEM simulations 230, 470
FHD *see* Fourier–Hankel decomposition
FIB *see* focused ion beam
field distributions 248, 254, 267, 270, 271, 493
field intensity 251, 261, 364
field localization 269, 336, 411
finite-difference-time-domain (FDTD) 207, 214, 383
finite element method (FEM) 227, 229, 397, 404, 466
FMM *see* Fourier modal method
focused ion beam (FIB) 48, 86, 134, 456, 463
forward scattering (FS) 288, 291–293, 303–305
Fourier–Hankel decomposition (FHD) 353–355, 364, 365
Fourier modal method (FMM) 245, 258
Fourier spectrum 339, 340, 346, 351, 352

Fourier transform 357
discrete 346
fractional 359
standard 357
frequencies 98, 99, 196, 212–214, 262, 267, 270, 273, 289, 290, 295, 297, 300–302, 314, 316, 490, 492, 493
angular 266, 377
circular 191
collision 289
cut-off 200
discrete spatial 360
dominant spatial 346
optical 100, 115, 122, 217, 222, 390, 391, 410, 412, 413
plasma 265, 267, 288, 302, 397
resonant 290, 306, 423
FS *see* forward scattering

gain 410, 411, 418, 427, 432, 434, 438, 447, 448
effective 411
optical 412, 447
gain medium 318, 409–413, 415, 416, 420, 423, 424, 437, 444, 447
gain model 416, 431
canonical 415, 427, 432
gain region 411, 413, 419, 421, 426
gammadion 59, 72, 74, 75, 82, 239, 241, 245, 247, 249–251, 255
gap 130, 158, 163–167, 196, 198, 200, 214, 264, 339, 345, 380, 477, 482–485, 487
interparticle 472, 473, 481, 483, 484, 488, 489
Gaussian beam 198, 228, 317, 324, 325, 332, 458, 470
Green's function 62, 177, 382, 383, 386

handedness 21, 22, 57, 60, 63–65, 78, 84, 86, 196, 200, 204–206, 228, 240, 241, 246, 247, 250–252, 255, 256
helical pattern 227, 230, 231, 233, 243
helical structures 19, 32, 187, 218, 240
helical symmetry 186, 197, 199, 206
helices 18, 20–22, 37, 38, 70, 72, 74, 75, 78, 133, 134, 183, 187–190, 196, 197, 200, 204, 217, 222
double 37, 49
left-handed 21, 22, 37, 38
right-handed 21, 22, 37, 38
symmetric 58
hole 84, 85, 134, 329, 402
confocal pin 468
hot spots 1, 35, 36, 47, 223, 452, 472, 481
hyperbolic metamaterials 264, 272–275, 277

impedance 110, 136, 160, 378, 428
incident angles 26, 198, 229, 457, 460
incident beam 8, 9, 15, 18, 229, 234, 247, 250
incident field 17, 29, 133, 242–244, 249, 253, 255, 411, 415, 417–421, 425, 426
incident light 7, 30, 82–84, 226, 229, 233, 234, 239, 241–244, 247, 249, 250, 300, 397, 404, 460, 481
incident plane wave 359, 414, 416, 425, 458
incident power flow 418–420, 426
incident waves 196, 202, 262, 403, 405

Intensity distribution 89, 90, 246–248
intensity patterns 246, 247, 250, 358
interactions 4, 6, 7, 9, 11, 13, 15, 41, 42, 46, 131, 135, 411, 413, 417, 418, 470, 472
  coherent 358
  complex wave matter 98
  dipolar particle-stack 39
  electronic 35
  exciton–exciton 6, 41
  exciton–plasmon 33
  inelastic 336
  laser–matter 320
  light-chiral objects 234
  near-field 342
  nonlinear 494
  plasmon–exciton 26
  plasmon–plasmon 3
interference 28, 154, 215, 225, 229, 230, 287, 289, 295, 297, 299, 315, 462
  coherent 227
  destructive 261, 290, 460, 462
  dipole-octupole 300
  dipole-quadrupole 302

Jones matrix 61–65, 67, 68, 70–75, 77, 82, 84–86
  enantiomorphic 75
  planar chiral 77

Laguerre–Gaussian modes 312, 314–317
lattice constants 101, 134, 189, 195, 196, 199, 200, 207, 351
lattices 45, 338
  crystal 384
  periodic 338
  Pinwheel 340
  rectangular 101
layers 23, 24, 26, 113, 116–118, 177, 186, 423
  adhesion 396, 478
  chiral 136
  dielectric 484
  dielectric coating 230
  metallic 456
LCP *see* left-circular polarization
LCP incidences 246, 248, 250
LEDs *see* light-emitting diodes
left-circular polarization (LCP) 2, 5, 9, 15, 59, 65, 101, 107, 195, 202, 241–243, 246, 247, 249, 250, 253, 255
light absorption 49, 130, 448
light beams 5, 47, 62, 64, 240, 331
light emission 181, 337, 342, 384, 389, 472
light-emitting diodes (LEDs) 6
light scattering 18, 177, 263, 264, 282, 285–287, 295, 297, 303, 304, 309, 336, 361, 368, 497
linear combinations 23, 136, 189, 224, 367
localized plasmons (LPs) 381, 452
localized surface plasmons (LSPs) 3, 260, 262, 378, 412, 481, 496
LPs *see* localized plasmons
LSPs *see* localized surface plasmons

magnetic dipole moments 10, 133, 138, 142, 148, 149, 151, 153–155, 159, 160
magnetic fields 23, 24, 105, 106, 131, 132, 136–139, 141, 148–150, 160, 161, 171–173, 189, 190, 195, 196, 266, 268, 272, 274, 399
magnetic plasmon 384, 385, 387, 388
magnetic response 376, 380, 381, 388, 391

materials 2, 6, 137, 139, 260, 265, 266, 275, 277, 278, 281, 302, 307, 370, 410, 424, 431
active 424
artificial 97–99, 131, 375
dielectric 489
homochiral 2
inorganic 45
lossless 266
magnetic 214
metamorphic 454
nanostructured 376
nonchiral 152
nondissipative 288, 299
semiconductor 454
mechanisms 1, 2, 7, 32, 33, 40–42, 46, 48, 84, 186, 200, 259, 260, 268, 287, 416, 450, 476
angular momentum multiplexing 278
chirality-transfer 41
energy storage 269
exciton-plasmon coupling 465
metal–dielectric interfaces 259, 265, 269, 271, 455
metal films 86, 340, 371, 495
metallic helix array 188, 189, 191, 193, 195, 197, 199, 201, 214, 218, 219
metallic nanoparticles 258, 286, 335, 337, 352, 355, 371, 394, 496, 502
metallic nanowires 96, 189, 192, 194, 223, 226, 232, 233, 236, 388, 452, 466, 497
metal nanoparticles 54, 55, 165, 235, 236, 262, 263, 279, 361, 476
metamaterials 95, 98, 99, 120–122, 124, 125, 131, 133, 186, 187, 218, 260, 273–278, 282, 283, 388–390, 405, 407, 448, 449
gold helix 187
helicoidal 187
metallic photonic 412
multilayered 273, 274
optical 97, 98, 125, 178, 412, 446
planar 330
metamaterial slab 274, 275, 277
metasurfaces 98, 100, 102, 105, 108–111, 113–116, 118, 120
adjacent 117
cascaded 114, 117
lossless 111
reciprocal 111
ultrathin 109, 120
microscope 89, 168
confocal laser-scanning 322, 326
homemade 85
infrared 397
modified confocal 468
scanning 167
scanning electron 326, 462
model 8, 14, 18, 21, 22, 28, 31, 34, 97, 100, 113, 115, 120, 329, 478, 481
linear 457
semiclassical 4 439
transmission-line 111, 114, 177
modes 20, 21, 24, 195, 196, 198, 224–233, 259, 260, 317, 318, 361, 364, 381, 384, 456, 459, 481, 485
azimuthal 351
coexcited fundamental 225
counter-propagating chiral 204
dark 289, 384
dipolar 422, 462
free-space 379
Gaussian laser 361
hybridized 230
lasing 372
longitudinal 196
out-of-phase 487

plasmonic-guided 242
whispering gallery 156, 295, 351
MST *see* multiple scattering theory
multiple scattering theory (MST) 187, 197

nanoantennas 170, 264, 447, 478, 481, 489–494, 500, 501
nanocuboids 473, 476, 478
nanoneedle 313, 324–328
fabricated 328, 330
twisted 313, 327
nanoparticles 49, 50, 98, 100, 131, 153–155, 157, 158, 163, 168, 169, 260, 280, 282, 293, 295, 309, 340, 341
active core–shell 409
chiral spherical 152, 156, 158, 165
coated 410
nonchiral 153
nanorods 19, 32, 304, 452, 465–472, 496
nanostructures 7, 12, 13, 25–28, 33, 36, 47, 260, 286, 306, 382, 388, 473–475, 478–481, 483, 484, 486
achiral 223
anisotropic 47
cuboid 474, 476, 477
gap-coupled 485
hybrid 9
metallic conical 460
photonic-plasmonic 338
spherical 307
twisted 330
vortex-pinning 264
nanowire 224–227, 229–234, 282, 298, 301–303, 305, 306, 309, 377, 390, 466, 468–472, 497
branch 234
cylindrical metal 223
grooved metal 234
substrate-supported 230
normalized radiation resistance (NRR) 429, 433–437, 439–444
NRR *see* normalized radiation resistance
NRR values 430, 433, 437, 442

OAM *see* orbital angular momentum
OAM peaks 354, 355, 366, 367
OAM spectra 335, 343, 355, 367
OAM states 353, 361
OAM values 353, 355
optical beams 88, 96, 169, 234, 256, 335, 352, 353, 367, 368
optical chirality 13, 15, 16, 57, 58, 60, 62, 66–74, 76–80, 82, 84, 86, 88, 90, 92, 94, 257
optical energy 259, 264, 265, 268, 270, 390, 394, 494
optical fields 240, 241, 304 356, 360, 361, 461
optical modes 336, 355
optical motors 403
optical power 261, 264, 360, 411, 414, 418
optical powerflow 259, 260, 262, 264, 268, 269, 275–277
circulating 269, 277
global 260
instantaneous 269
local 260
time-averaged 267, 268, 270–272
optical singularity 240, 393, 395, 397, 399, 401, 403, 460
optical torque 359, 393, 403–405
optical trapping 278, 332, 367, 368, 394, 406
optical tweezers 311, 312, 321, 408

optical vortex (OV) 259, 262–264, 277, 278, 281, 283, 285, 286, 303, 304, 307, 309, 311–317, 320–323, 325–327, 329–333, 393, 394, 405–407
orbital angular momentum (OAM) 86–88, 234, 239, 240, 243, 244, 256, 311, 312, 330–332, 335, 336, 342, 354–356, 360–362, 364–368, 372–374, 384, 385, 388–390
oscillators 289–291, 318, 404
OV *see* optical vortex
OV helicity 313, 326, 329, 330
OV laser ablation 313, 321, 322, 324, 326, 329, 330
OV pulses 323, 325–327

particles 17–19, 21, 35, 37, 42, 132, 133, 152, 153, 156, 157, 159–167, 262, 313, 314, 342, 343, 347, 348, 415, 418–420
  conjugated 35
  finite-size 347
  gold 37, 38
  molecular stack-coated 39
  nonchiral 151, 155, 156, 164
  silicon 28
patterns 53, 211, 243, 247, 248, 250–255, 338–340, 343, 345, 347, 349, 361, 372, 394, 436, 486
  directivity 436
  element 347
  far-field 353, 364
  metallic 241
  topological 241
PCs *see* photonic crystals
perfectly matched layers (PMLs) 253
permeability 102, 134, 137, 144–146, 153–155, 158, 165–167, 191, 375, 427, 432
permittivity 134, 137, 145, 146, 153–155, 165–167, 181, 182, 191, 224, 225, 233, 266, 377, 380, 424, 427, 431, 432
  electric 63
  free-space 102, 431, 432
  frequency-dependent 457
  negative 375, 376
phase singularities 260, 261, 278, 311, 312, 314, 315, 321, 325, 332, 384, 394, 405, 460, 462
phase velocity 128, 129, 377, 381
photonic crystals (PCs) 185, 186, 216, 369
planar chirality 53, 59, 61, 73, 74, 80, 83, 92, 235
  optical 83, 90
plane wave 61, 63, 66, 103, 217, 221, 262, 269, 356, 295, 380, 409–411, 413, 415, 417, 419, 421, 423, 425, 433, 445
plasmon frequencies 4, 12, 37
plasmonic effects 115, 260, 264, 265, 278
plasmonic materials 267, 278, 281, 291, 427, 431, 449, 450
plasmonic metasurfaces 97, 99, 100, 120
plasmonic modes 242, 260, 383, 390
plasmonic nanocrystals 28, 29, 31
plasmonic nanolenses 263, 264
plasmonic nanoparticles 101, 125, 130, 131, 295, 308, 389
  dissipative 287
  nondissipative 291, 298
plasmonic nanostructures 1, 48, 54, 264, 280, 286, 295, 297–299, 301, 307, 313, 330, 497, 500, 501
plasmonic resonances 10–12, 16, 18, 152, 388, 393, 395, 403, 405

plasmonic structures 47, 86, 230, 243, 287, 291, 298, 300, 307
plasmonic waveguides 224, 234
plasmon resonances 11, 31, 37, 179, 270, 292, 404, 413, 473, 479, 501
PMLs *see* perfectly matched layers
polarization 117, 119, 201, 202, 205, 208, 211, 234, 240, 249–251, 325, 326, 403, 405, 470, 471, 485, 487, 488
  elliptical 187, 209
  inclined 229
  left-circular 241
  left-handed 117
  linear 84, 202, 203
  oblique 226
  right-circular 241
  right-handed 117
  x-linear 207, 208
polarized light 20, 58, 128, 129, 222, 223, 225, 241–243, 288, 314, 325
polarizers 85, 463
  near-infrared 397
  symmetric 320
powerflow 261–263, 265, 275–277, 418
  reactive 268, 269
  time-averaged 268, 275
  density 430, 443, 444
  saddle points 262
Poynting vector 268, 295, 298, 303, 304, 306, 314, 393, 394, 399, 405, 414, 430, 458, 459, 461

QDs *see* quantum dots
quantum dots (QDs) 6, 41, 42, 45, 228, 274, 377, 388, 411, 421, 423–425, 465, 497
  bio-conjugated 45
  core/shell 423
quantum emitters 274, 411, 427, 439, 442, 445, 449
quasicrystals 338, 342, 343, 370–372

radiation 131, 135, 136, 143–145, 147, 149–155, 157–159, 161–163, 165, 166, 168, 171, 173, 178, 180, 442, 444
radiation power 150
radiation rates 142, 144, 146
radiation resistance 429
  normalized 429
radiative decay rate 151, 154–157, 161, 162, 164–168, 424
radiative heat transfer 274, 278
rational approximations 343, 353–355, 366
RCP *see* right-circular polarization
RCP incidence 246, 248, 250
RCWA *see* rigorous coupled wave analysis
reciprocal space 337–342, 346, 360
reciprocity 60, 61, 63, 65–67, 69, 71, 73, 74, 77, 78, 80, 82, 90, 91, 102, 110, 445
reciprocity theorem 61–63, 65, 75
reflection coefficients 104–108
refractive index 137, 139, 228, 230, 231, 299, 314, 337, 397, 432, 469, 490
regime 460, 466, 483, 484, 501
  dipole interaction 19
  hyperbolic 274–278
  infrared 388
  mid-infrared 100
  monochromatic 66
  near-infrared 83, 376
  off-resonance interaction 12
  paraxial 64
  subwavelength 376
  visible 298, 375, 377, 380

resonance condition 135, 481, 487–489
resonance frequency 380, 416, 418, 423, 439
resonances 27, 28, 163, 164, 288, 299, 375, 378, 380, 411, 413, 420, 422, 423, 466, 469, 485, 487
  chiral-plasmon 143, 152, 154, 158
  dipolar 436
  higher-order 287
responses 108, 243, 246–248, 252–254, 375, 416–418, 440, 484
  electrical 452
  far-field 409
  local 256
  magnetic plasmon 376
  nonreciprocal 101
  paramagnetic 380
  reciprocal 108
  spectral 487
  time-dependent 248
right-circular polarization (RCP) 2, 5, 9, 15, 58, 59, 65, 101, 195, 202, 241, 243, 246, 247, 249, 250, 253, 255
rigorous coupled wave analysis (RCWA) 245
rotational symmetry 338–343, 345–347, 349, 351

scale 21, 303, 305, 347, 350, 375, 376
  logarithmic 304
  macroscopic 393
  millimeter 83
  nanometer 83, 222, 388
  nanometric 452, 456
  subwavelength 376, 381
scanning electron microscope (SEM) 89, 326, 362, 462, 467
scattered light 240, 251, 300, 363, 364, 458, 460, 464
scattered radiation 355, 356, 358, 361, 363–366
scattering 3–5, 25, 27, 28, 30, 49, 50, 147, 178, 182, 282, 285, 288, 292, 295, 415–417, 425
  far-field 286, 287, 307
scattering amplitudes 287–289, 291, 292, 298
scattering efficiencies 291, 293, 415, 425
scattering forces 313
scattering losses 415
scattering probability 382, 384, 386, 387
scattering resonances 292, 373
SEF *see* surface-enhanced fluorescence
SEM *see* scanning electron microscope
SERS *see* surface-enhanced Raman scattering
SERS device 473, 476, 480, 500
SERS intensity 476
SERS measurements 473, 476, 479
shell 5, 14, 416, 424, 437, 466
  chiral spherical 147
  dye-doped silica 421
silver nanoparticles 3, 52–54
silver nano-shell 438, 439, 442
simulations 12, 21, 206, 209, 212, 214, 227, 252, 398, 401, 458, 459, 472, 477, 484, 485, 492
  full-wave 97, 112, 114, 117, 118
  theoretical 473, 475
singular optics 86, 278, 281, 286, 295, 309, 336, 342, 356, 367, 368, 393, 394, 406
SLMs *see* spatial light modulators
SOAs *see* stacked optical antennas
SP *see* surface plasmon
spatial light modulators (SLMs) 361

spectroscopy 91, 130, 260, 489, 500
  electron energy loss 376, 381, 383, 384, 389
  few-molecules 493
  non-linear 482, 483
  surface-enhanced Raman 130, 482
  terahertz-enhanced 483
  tip-enhanced Raman 460
spherical particles 147–150, 155, 159, 161, 162, 166, 301, 310, 406
  chiral dielectric 156
  left-handed 156
  nonchiral dielectric 163
spiral arrays 341, 343, 346–349, 352
spirals 74, 75, 78, 84, 89, 129, 131–133, 135, 222, 341, 344–346, 348, 350, 351, 354–357, 360, 362–367
spiral structures 344, 346, 349, 350
spontaneous emission 136, 141, 146–148, 150–152, 155–157, 159, 162, 164–166, 168, 179–181, 282, 449
spontaneous radiation 136, 137, 142, 145–148, 159, 179
SPP coupling 270, 271, 463, 464
SPP frequencies 259, 267, 268
SPP modes 223, 225, 226, 228, 229, 232, 233, 265, 266, 268–271, 277, 458, 463, 469
SPPs *see* surface plasmon polaritons
SPP waves 259, 260, 264, 265, 267–269, 272, 278
SPs *see* surface plasmons
stacked optical antennas (SOAs) 483–486, 501
stacks 111, 114–118, 120, 362
  coupled 120
  finite 115, 117
  infinite 116
  multilayer 276
  thin metal 362
substrate 82, 222, 327, 328, 362, 404, 406, 474, 482, 485, 488
  fused silica 362, 363, 395
  high resistivity silicon 490
  patterned 396
  planar 342
  quartz 362
  silicon 490
surface-enhanced fluorescence (SEF) 130
surface-enhanced Raman scattering (SERS) 130, 180, 279, 280, 452, 472, 477, 482, 499–501
surface plasmon (SP) 50–52, 94, 96, 221–228, 232–237, 239, 240, 248–250, 256–260, 262, 298, 376–378, 406, 407, 447, 449, 494–498
surface plasmon polaritons (SPPs) 223–225, 228, 230, 233, 234, 236, 240–244, 258–260, 322, 326, 453, 454, 457, 458, 460, 465, 466, 469, 470, 495–497
surface plasmons (SPs) 60, 83, 84, 86, 94, 222, 236, 237, 249, 250, 257, 389, 412, 447, 449, 455, 456, 465, 495, 496
surface waves 223, 376, 378, 437

topological optics 285–288, 290, 292, 294–296, 298, 300, 302, 304, 306–308
topological transitions 273, 274, 283
transition 42, 116, 142, 151, 154–156, 260, 342, 347, 358
  excitonic 45
  light-induced 42

molecular 41
nonradiative 151, 162
optical 129
transmission coefficients 112, 113, 117
transmission-line approach 109, 115, 116
transmission spectra 203, 206, 214, 397
transmittances 115, 116
far-field 485
residual 210

Vogel spiral arrays 341, 345, 353
Vogel spirals 335, 340, 342–344, 346–353, 355, 356, 358, 360, 361
aperiodic 343, 344, 351, 367, 368, 371
arbitrary 258, 372
disordered 347
Vogel's spiral patterns 251–253, 255
vortex beam 384, 385, 395
vortices 240, 260, 262, 263, 269, 277, 278, 285–287, 295–297, 302–305, 307, 312, 326, 333, 337, 393, 394, 413, 419, 420

waves 137, 138, 143, 178, 197, 204, 210, 218, 260, 265, 267, 269, 283, 308, 353, 376, 377
diffracted 342
evanescent 60
harmonic 192
left-polarized 136
radiated 242
reflected 138, 149, 160, 205, 211
refracted 198
right-polarized 155
scattered 27
transmitted 199, 202, 203, 209
wave vectors 8, 63, 64, 67, 199, 201, 224, 226, 266, 273, 305, 412

9789814613170